Kathrin Dörfler, Jan Knippers, Achim Menges, Stefana Parascho, Helmut Pottmann, Thomas Wortmann (Eds.)

Advances in Architectural Geometry 2023

Also of interest

AutomationML
The Industrial Cookbook
Edited by Rainer Drath, 2021
IISBN 978-3-11-074592-4, e-ISBN (PDF) 978-3-11-074597-9

Soft Computing and Optimization Techniques for Sustainable Agriculture
Debesh Mishra , Suchismita Satapathy, Prasenjit Chatterjee, 2022
ISBN 978-3-11-074495-8, e-ISBN (PDF) 978-3-11-074536-8
in Smart Computing Applications
ISSN 2700-6239

Blockchain 3.0 for Sustainable Development
Edited by Deepak Khazanchi , Ajay Kumar Vyas , Kamal Kant Hiran, Sanjeevikumar Padmanaban, 2021
ISBN 978-3-11-070245-3, e-ISBN (PDF) 978-3-11-070250-7
in De Gruyter Frontiers in Computational Intelligence
ISSN 2512-8868

Artificial Intelligence of Things in Smart Environments
Applications in Transportation and Logistics
Edited by Mariyam Ouaissa , Zakaria Boulouard , Mariya Ouaissa, Yassine Maleh, 2022
ISBN 978-3-11-075533-6, e-ISBN (PDF) 978-3-11-076047-7
in De Gruyter Series on the Internet of Things
ISSN 2626-5451

Systems Reliability Engineering
Modeling and Performance Improvement
Edited by Amit Kumar, Mangey Ram, 2021
ISBN 978-3-11-060454-2, e-ISBN (PDF) 978-3-11-061737-5
in De Gruyter Series on the Applications of Mathematics in Engineering and Information Sciences
ISSN 2626-5427

Advances in Architectural Geometry 2023

Edited by
Kathrin Dörfler, Jan Knippers, Achim Menges, Stefana Paraschо, Helmut Pottmann and Thomas Wortmann

DE GRUYTER

Editors
Prof. Dr. Kathrin Dörfler
Technical University of Munich
School of Engineering and Design
Department of Architecture
Arcisstr. 21
80333 München
Germany
doerfler@tum.de

Prof. Dr.-Ing. Jan Knippers
University of Stuttgart
Institute of Building Structures
and Structural Design
Keplerstr. 11
70174 Stuttgart
Germany
info@itke.uni-stuttgart.de

Prof. Achim Menges
University of Stuttgart
Institute of Computational Design
and Construction
Keplerstr. 11
70174 Stuttgart
Germany
mail@icd.uni-stuttgart.de

Prof. Dr. Stefana Parascho
École Polytechnique Fédérale de Lausanne
ENAC IA CRCL
GC G1 567 (Bâtiment GC)
Station 18
1015 Lausanne
Switzerland
stefana.parascho@epfl.ch

Prof. Dr. Helmut Pottmann
Vienna University of Technology
Applied Geometry
Wiedner Hauptstraße 8–10/104
1040 Vienna
Austria
pottmann@geometrie.tuwien.ac.at

Prof. Dr. Thomas Wortmann
University of Stuttgart
Institute of Computational Design
and Construction
Keplerstr. 11
70174 Stuttgart
Germany
thomas.wortmann@icd.uni-stuttgart.de

ISBN 978-3-11-116011-5
e-ISBN (PDF) 978-3-11-116268-3
e-ISBN (EPUB) 978-3-11-116347-5
DOI https://doi.org/10.1515/9783111162683

Bibliographic information published by the Deutsche Nationalbibliothek
The Deutsche Nationalbibliothek lists this publication in the Deutsche Nationalbibliografie; detailed bibliographic data are available on the Internet at http://dnb.dnb.de.

Cover image: Cluster of Excellence IntCDC, University of Stuttgart
Printing and binding: CPI books GmbH, Leck
www.degruyter.com

Preface

Geometry is the center of the architectural design process and omnipresent from the initial stages of form finding to novel fabrication techniques, construction and monitoring during the use of the building. It does not only influence the modeling of environments and the processing of sensor data, but also enables new, efficient design, analysis and fabrication of complex forms through advanced geometric calculations. Furthermore, algorithmic processes play an increasingly crucial role in the integration of disciplinary input and open up new horizons for architecture. As the architectural context constantly poses new challenges, the research field of architectural geometry is in ongoing evolution and of steady interest.

This book contains a collection of 34 technical papers presented at the "Advances in Architectural Geometry Conference 2023", organized by the Cluster of Excellence IntCDC (Integrative Computational Design and Construction for Architecture) and held in Stuttgart from October 4 to October 7, 2023. The AAG is a biennial event. It was held first in 2008 and 2010 in Vienna, 2012 in Paris, 2014 in London, 2016 in Zurich, 2018 in Gothenburg and 2020 in Paris.

The conference series is renowned for the presentation of both theoretical and practical work related to new geometric developments. Over the years, the symposium has become a reference point for research and practice, as well as a platform for sharing cutting-edge advancements and fostering collaboration between academia and industry. It also aims to connect researchers from architectural and engineering practices, academia, and industry. It provides an interdisciplinary link between architecture, applied geometry, computational design, mathematics, and manufacturing and attracts architects, engineers, computer scientists, mathematicians, software and algorithm developers, and contractors.

The papers in this book reflect the wide range of expertise and interests of the authors. The main topics include architecture, geometry, engineering, and the application of artificial intelligence in construction. All papers offer unique insights and perspectives and contribute to the advancement of architectural geometry as a discipline. This book, as a valuable resource, presents a broad and diverse collection of contributions that highlight innovative approaches, novel applications, and emerging trends in the field of architectural geometry. It includes a broad spectrum of topics, from parametric design strategies to advanced computational methods, tracing the role of geometry in shaping our built environment.

We would like to thank all the authors whose contributions have made this publication possible. Their commitment and expertise have ensured the high quality and relevance of this book. Furthermore, we would like to thank the organizing committee, reviewers and sponsors of the AAG 2023 for their invaluable support. We are honored

https://doi.org/10.1515/9783111162683-202

to take you on this intellectual journey through the pages of "Advances in Architectural Geometry 2023". May they provide you with insights into cutting-edge research and innovation in architectural geometry and inspire new ideas, foster interdisciplinary collaboration, and contribute to the advancement of the architectural design process.

The Editors

Contents

Andrei Nejur

Structurally Aware Fabrication for Large-Scale, Curved Architectural Skins

Abstract: This paper presents a new construction technique for large complex curved surfaces built from very low thickness sheet materials, reinforced through the structural activation of the dedicated assembly parts. The proposed technique works in several steps. 1) A large architectural curved surface is evaluated using FEA for local buckling and bending under self-weight and the assumption that it will be built from thin sheet material and without an additional structure or reinforcements. 2) The graphical representation of the principal bending moment stress field in the surface is simplified and used as the basis for the generation of a discretization pattern. 3) The discretization pattern is used to generate assembly elements (i. e., flaps) normal to the base surface. 4) The flaps are connected through a system of fins that engage them into a cellular structural system aligned both with the discretization and the stress patterns of the surface. Figure 1 presents a full-scale built prototype demonstrating the workflow.

Fig. 1: General view of the research demonstrator pavilion showcasing the proposed technique. (Image credit Caroline St-Hilaire)

https://doi.org/10.1515/9783111162683-001

1 Introduction

For large architectural surfaces, discretization and approximation of some kind is usually the only viable way of construction. Regardless of complexity and chosen material, to be constructed, architectural scale surfaces must be divided into pieces or discretized. The segmentation is determined by the limitations of the fabrication process, dimensions and properties of the chosen materials, transport and assembly possibilities (Eigensatz et al. 2010).

1.1 Discretization as a necessity

Discretization is an important method to manage the costs and time required for the construction. It acts as an approximation tool that unlocks the use of cheap planar sheet materials for the construction of complex curved surfaces. However, even if partitioning and approximating architectural curvature and complexity is an effective way of rendering it constructible, it is not a "panacea." In fact, there is a fine line between "just enough" discretization and approximation to make a form feasible for construction and "too much discretization" that wipes off the benefits of manageable dimensions and planarity through unreasonably expensive part multiplication and impossibly long assembly times. Additionally, not all discretizations are built equal. For example, double curvature approximations like PQ meshes (Pottmann 2013) or ruled surface rationalizations (Flöry and Pottmann 2010) are more fabrication friendly due to their planarity and node valence for the former and the simple unfolding on planar sheets for the latter.

1.2 Disadvantages of discretization

Beyond being a necessary step, discretization brings a series of shortcomings to the architectural form. From a structural point of view, the most important is material discontinuity. The discontinuity of the building material induces fault lines in the surface where the pieces are joined together. This can be mitigated using specialized techniques like multilayered surfaces (Fornes 2016; Stanojevic and Takahashi 2019), discrete piece overlaps (Schleicher et al. 2015; La Magna, Schleicher, and Knippers 2016), finger joints (Magna et al. 2013), folds (Demin 2015), and several others. However, with any of those, a comparable material coherence with a single-piece surface is difficult to achieve.

Discretizing a curved architectural surface for fabrication and construction from sheet-based materials requires some sort of approximation. This translates in turn to a deviation from the initial surface that can be simulated in the digital version of the form in the process of design. This deviation is often negligible at the scale of the full surface and/or is often incorporated in the design intent (Pottmann et al. 2015). The processes that aim to minimize this deviation from the curved base form can reduce the size of discrete parts, thus increasing their number and potentially increasing the costs through assembly complexity. Alternatively parts can be deformed (i. e., by bending or rolling) (Rossi and Nicholas 2018) to make them approximate more closely the subdomain of base surface they represent. They can be deformed in the process of part fabrication (Cai et al. 2012), or they can be a deformed during assembly (Fornes 2016). The latter process combines two steps into one, thus reducing construction time and costs. Deformation during assembly requires additional information to guide the deformation process. Cutting and engraving can encode in some cases this extra information but in others, reference elements (i. e., assembly guides) are required that increase construction complexity and costs.

For flexible and deformable materials that are subjected to structural loads, the simulated and assumed deviation is often accompanied by a deformation due to discretization-induced material discontinuity (Lienhard and Knippers 2013). During conception, this deformation is much harder to predict because it requires high-fidelity (computationally expensive) simulation especially for the joints. During construction, assembly conditions and quality of craftmanship can amplify these uncontrolled deformations leading to undesired structural and aesthetic defects.

Other less desired effects of discretization are the added weight due to the extra connection components, and a limitation of the aesthetic freedom because the architectural image must include, or in other cases take steps to hide, the seams between the discrete parts.

1.3 Self-supporting or supported surfaces

All the above-stated advantages and shortcomings of discretization for large curved architectural surfaces apply in equal measure to self-supported surfaces (i. e., able to carry the self and applied loads) and to surfaces that require a support structure (i. e., a structural network resolving the major load collection and transport to the supports).

Self supported architectural surfaces, beyond inherent material properties (i. e., rigidity or flexibility), rely on folds (Friedman 2016), active bending (Lienhard and Knippers 2015), curvature (Martín-Pastor and García-Alvarado 2019), or multiple layers (Nicholas et al. 2016) to add structural rigidity to the surface. The notable exception here is inflatables that leverage positive internal pressure to gain rigidity (Ayres, Vestartas, and Ramsgaard Thomsen 2018). These methods that add pseudo-thickness to the surface can program surface rigidity through an activation of a specific structural form. The effect of those techniques is still limited by the surface material properties, is local, and implies a certain aesthetic. This aesthetic usually means a constrained proportion between the surface dimensions and the employed curvature radii or fold amplitudes. If the proposed design cannot accommodate such formal constraints, or the structural demands are greater than the surface material limits, an additional structure is still required.

Supported curved architectural surfaces, even if for the moment we ignore the complexity incurred by an additional structural network (Schling 2018), require a surface-to-structure connection strategy (Schmieder and Mehrtens 2013) that, depending on the rigidity of the chosen material, could imply additional weight, added assembly complexity, and/or increased fabrication costs.

2 Problem statement

Even with all the shortcomings enumerated in the previous section, discretization is the most cost-effective and sometimes the only available solution to construct large-scale architectural curved surfaces from sheet materials. Discretized, self-supported, thin sheet surfaces relying on form for rigidity are well studied and the same can be said about supported surfaces. Contrarily, surfaces that integrate structural reinforcement with the discretization pattern and use the assembly material as support are not covered in the surveyed literature. We can therefore conclude that a gap currently exists in the construction/design techniques used today for large scale structural architectural surfaces built from thin sheet materials. This gap limits the materialization possibilities of self-supported architectural surfaces that do not rely on high curvature and/or local folding for structural reinforcement.

3 Proposed contribution

This paper proposes a new method of construction for architectural-scale curved surfaces from sheet metal, through a unified process of discretization and structural optimization. The proposed technique aims to stand in between methods of construction for self-supported surfaces and methods that rely on a separate structural support for the architectural surface. The technique works by directly deriving the discretization pattern from the structural analysis and then using this pattern together with the material dedicated to discrete piece assembly to reinforce the base surface. Engaging structurally the extra material necessary for joining the discrete parts can improve a) Precision: By using the engaged joints as assembly guides; b) The modulation of the structural properties of the construction independent of the shape of the architectural surface; c) The possibilities to construct smooth and low curvature surfaces without the use of an additional structural system; d) Assembly times, costs and material use compared to architectural surfaces supported by an additional structure.

4 Methodology

The proposed method is the result of research undertaken at University of Montreal's School of Architecture in 2021 (Nejur and Balaban 2022) and was partly developed in a masters research studio in an iterative process between physical prototypes and digital tool development. The method is tested on the construction of a large scale (4.5 m high) thin sheet metal prototype named A(fin)ne. Even though the process presented here is linked to the final prototype, the resulting technique has a much broader application in terms of scale and form. The research and subsequently the method introduced with this paper are presented in several steps: surface pre-processing (Sec. 4.1), structural evaluations (Sec. 4.2), discretization for assembly and structural optimization (Sec. 4.3), assembly geometry generation (Sec. 4.4), fabrication data (Sec. 4.5), and implementation and testing (Sec. 4.6). They are detailed below highlighting the innovations brought to the process.

4.1 Surface pre-processing

The proposed workflow requires the produced architectural surface to be presented as a mesh for discretization but also for structural analysis (i. e., FEA). The chosen discretization technique is a stripification process that decomposes the base mesh into one mesh-face wide ribbons. To facilitate the decomposition and the structural analysis a mesh with constant triangles is required. The triangle meshes with quasi-equal face sizes are appropriate for stripification and FEA as their topology and geometry do not excessively influence the two processes. For our prototype, the base surface was first constructed as a loft then meshed and relaxed as quasi-minimal surface with Kangaroo and it was finally remeshed with an average edge length of 16 cm. This ensured that the average discrete mesh ribbon would end up with an approximate width of 12 cm. Figure 2 shows the first steps in preparing the surface.

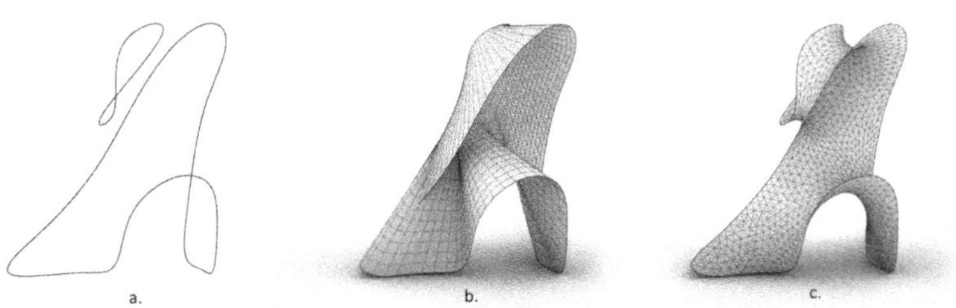

Fig. 2: The form-finding process for the base surface. a) Input curves, b) Meshed loft between curves, c) Relaxed mesh with the soap bubble goal in Kangaroo for Grasshopper. The relaxed mesh is also remeshed with Grasshoppers tri-remesh with a 16 cm average edge.

4.2 Structural evaluation

Next the structural behaviour of the found mesh was tested using the Grasshopper extension Karamba (Preisinger and Heimrath 2014). The mesh was evaluated as an aluminium shell with a constant thickness of 1 mm, rigid supports with no degrees of freedom, and subjected to self-weight. See Fig. 3a for the testing setup. The model exhibited a large deformation mostly caused by the weight of the cantilevering part. The model deformed around an axis defined by a linear area of close to zero Gaussian curvature in the midsection of the surface. Figure 3b shows the simulated deformation of the base mesh based on the self-weight loading scenario as well as the clustering of high bending moment values in the same midsection area. To rigidify the surface, instead of adding curvature in the problem area, we decided to selectively increase the virtual thickness of the metal membrane and thus "program" rigidity into the structure of the pavilion only where it was required.

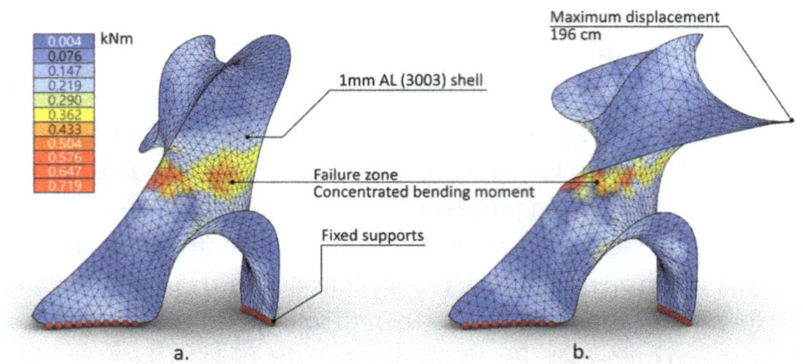

Fig. 3: Karamba simulation of the base mesh under self-weight. The red spheres represent the supports with restricted translations and rotations. a) Un-deformed model, b) Deformed model showing close to 200 cm of displacement.

4.3 Discretization and structural optimization

To increase membrane thickness without the added increase in weight we decided to use the extra material required to assemble the discrete parts of the model (i. e., the assembly tabs). If this extra material can be structurally engaged, it can act as a thicker surface without the added weight. To do this we decided to connect neighbouring tabs with strips of laser-cut metal sheet thus creating a series of "fins" running perpendicular to the surface. Figure 4 shows an early investigative prototype.

Two important issues were identified early in the development process: structural anisotropy and depth. As the rigidity requirement was limited to certain areas and to a certain direction it was clear that not all areas needed the same level of reinforcement

Fig. 4: Early connection prototype. Showing a) Surface pieces and assembly/reinforcement guides (fins), b), c) Fins acting as curvature guides for the assembly.

or "skin depth" and that connections between the tabs had to preferentially follow a certain direction on the surface. As the assembly pattern depends on the discretization of the base mesh into segments, a process for directional-bias stripe segmentation was implemented for the base surface as shown in Fig. 5. Using Karamba, the principal bending moment field was extracted for the curved surface as a network of curves on the base mesh Fig. 5a. After a reduction in complexity the resulting curves shown in Fig. 5b were used to instruct the stripification process (Fig. 5c) for the base mesh using a workflow developed on top of the Ivy add-on for Grasshopper (Nejur and Steinfeld 2016; 2017).

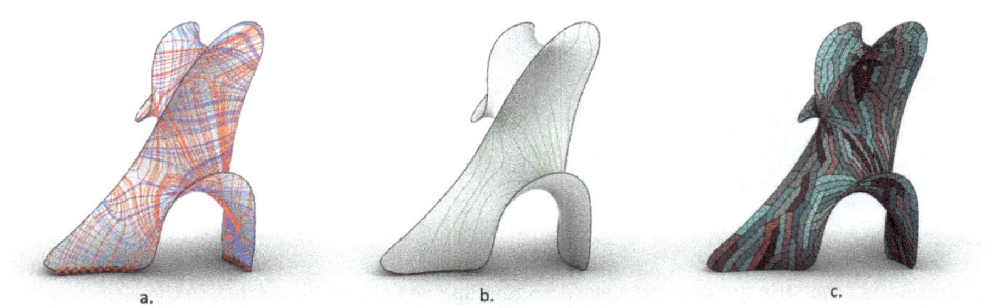

Fig. 5: From FEA simulation to segmentation. a) The graphical representation of the principal bending moments for the aluminum shell under self-weight, b) The simplified bending moment field to be used by the stripification algorithm. c) The resulting segmentation with Ivy Agents.

Using the Ivy API, a series of MeshGraph agents (Nejur 2016) is created on the base mesh. The first agent spawns on a random mesh face and walks on the mesh at each step integrating a mesh face into its subgraph. The agent walks on the dual graph (MeshGraph) of the mesh. In this graph each face of the mesh is a node in the graph and each mesh edge is an edge in the graph. Figure 6 shows the walk of a mesh agent on the base mesh. The MeshGraph representation of the base mesh is not shown to make the figure more legible. In the agent's subgraph the nodes are named $n_0 \dots n_{12}$ and the edges $e_0 \dots e_{11}$. The decision on what mesh face to include in the subgraph at each step is made based on the angle between the next graph edge direction and the tangent direction of the closest simplified bending moment surface curve. In Fig. 6 the direction curves are shown in green as are the tangent vectors $v_0 \dots v_{11}$ corresponding to each graph edge. The decision mechanism for the next step is shown in the same figure. To walk from n_{11} to the two possible n_{12} variants (red and blue) the algorithm selects the one with the smallest dot product between e_{11} and v_{11} among the red and the blue choices. This ensures that the agent loosely follows the locally closest guide curve and thus the geometry of the principal bending moment network computed by Karamba. The dot product calculation for each edge of the MeshGraph is precomputed for the whole mesh and it is stored as weight (a value between 0 and 1) for each MeshGraph edge. In the end the walk of the agent is simply following the "least weight" choice at every step.

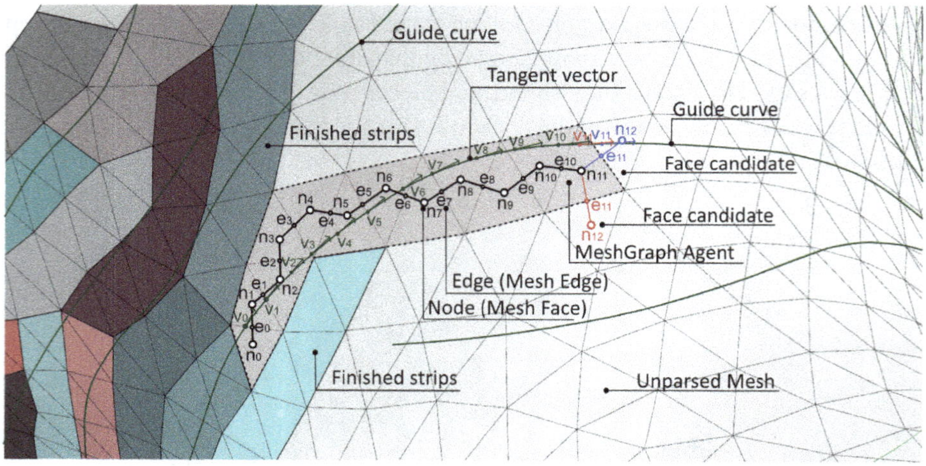

Fig. 6: Detail of the strip growing process. An agent spawns on the base mesh in the vicinity of a recent stripe and walks on the mesh using the local direction (tangent vector) of the nearest guide curve as a reference.

Each agent expires after a preset number of steps (15), thus limiting the maximum number of triangles in a strip. After each agent expiry, a new one is spawned on a node in the immediate vicinity of recently walked nodes (mesh faces). The process continues until all the graph nodes (base mesh faces) are parsed. The shortcoming of this approach is that started strips often run out of mesh faces to parse and result into stubs (strips with fewer than 5 triangles). This happens because the directional field steers the MeshGraph agent into already parsed faces. To overcome this, another algorithm runs in parallel with the stripification process and seeks to unify short strips end-to-end into larger strips with a maximum length of 25 triangles. Simultaneously strips over 22 triangles long are split approximately in half thus greatly reducing both the stubs and the long strips of faces.

4.4 Secondary structure (the fins)

With the discrete parts of the mesh created, the margins of the strips are used to create a set of fins locally perpendicular to the base mesh and connected to two discrete mesh strips at a time. The creation of the fins happens in several steps. Figure 7 illustrates the process.

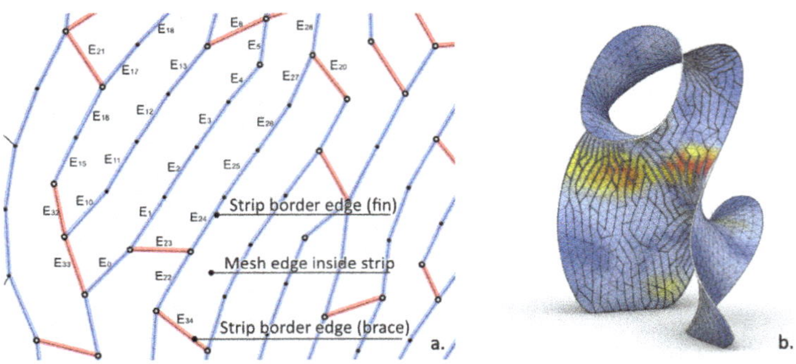

Fig. 7: Secondary structure creation. a) The strip border graph is walked to create the fins and the braces. b) The width of the secondary structure (fins and braces) is controlled by the bending moment values calculated through the FEA analysis.

Step 1 A graph of interconnected strip border edges is created. The new graph contains all the naked edges of the mesh strips with the nodes of the graph being the mesh vertices shared by those edges. The edges of the new graph retain the weight values stored in the corresponding MeshGraph edges used previously to produce the strips.

Step 2 The new graph is "walked" with a preference for the highest weight edges (shown in blue in Fig. 7a) and a constraint of walking on edges with no more than 30 degrees between their walk-oriented directions. This is to avoid large fin deformations during fabrication due to bending. In Fig. 7a mesh edges E0, E1, E2, E3, E4 are part of the same walk that will become a continuous fin. E4 and E5 are split due to the angle deviation. The length of the walks is limited to 10 consecutive edges to facilitate fabrication and assembly. The process produces a series of unconnected edges that will remain singletons (braces) between the fins. They are shown in red in Fig. 7a.

Step 3 The resulting polylines are extruded in the normal direction of each contained mesh vertex. The amount of the extrusion at each vertex is based on the vertex flexion value averaged from the faces touching the vertex. The averaged values are remapped into usable millimeter values (25–50 mm) for the extrusion process. Figure 7b shows the correlation between fin height and bending moment.

With the secondary structure in place and fused to the base mesh a new structural simulation with Karamba reveals that the maximum deformation has been reduced from 1960 mm to just 2 mm.

4.5 Fabrication data

For fabrication, the discreet parts of the prototype (i. e., strips and fins) were unfolded flat and tagged using Ivy. To enable manual assembly the strips were clustered in 10 sectors each not exceeding 2 m × 2 m and max 20 kg. The clustering was done using Ivy's k-means algorithm implementation to produce 10 compact groups of mesh faces. The strip clusters are created through the allocation of each strip to the cluster where the largest number of its faces reside. Figure 10 shows the result of the clustering process.

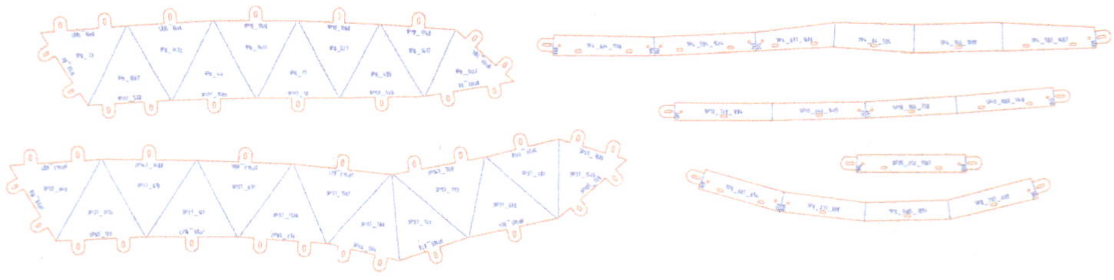

Fig. 8: The 2D fabrication data produced by the script. Left: several typical strips with connection tabs and assembly tags to be cut from 1 mm aluminium sheet. Right: Typical fins and a brace with longitudinal tabs for continuity and oblong holes for the connections to the strip tabs. The fins are cut from 1.5 mm aluminium sheet.

The unfolded strips are tagged using the standard Ivy workflow (Nejur 2016). Each mesh strip receives text tags in all face centers and next to the border edges. Each tag encodes the sector, strip, and face number. The center tag numbers refer to the current face and strip while the border tags identify the connecting ones. Each strip naked edge is fitted with a connection tab with strip end edges receiving two tabs for added stiffness in the higher strain areas and a reduced assembly complexity. To accommodate the nut and bolt manual assembly strategy the tabs are fitted with oblong holes. To accommodate the curvature radius of the bent 1 mm thick tabs each unfolded stripe contour is offset inwards with 1.5 mm.

The perpendicular fins are constructed as individual pieces from the extruded polylines presented in Sec. 4.4. The fins use the strip tabs to connect to the base surface and are intercalated between each pair of strip tabs. To connect to the tabs, the fin geometry includes oblong holes corresponding to each tab hole. To match fins with the relevant strips and tabs each fin segment is tagged with the number of the adjacent mesh faces also present on the strips. Fins are fitted with end tabs to ensure the structural continuity of the fin network.

4.6 Implementation and testing

The workflow was tuned and tested during a research studio at University of Montreal School of Architecture. Because the ambition was to build the large-scale demonstrator by hand, the physical prototype to digital development approach was useful in tuning the scale of the final build towards something buildable by non-specialists. The final full-scale prototype called A(fin)ne was built in September 2021 over the course of 5 days by 12 students and two tutors with hand tools only. The final prototype stood 4.5 m tall, had 202 laser-cut stripes, 465 fins and braces, approximately 3200 bolts and a surface area of 16.7 m^2. The prototype remained on display for 3 months inside the faculty building.

Fig. 9: The assembly process lasted 5 days. First the cut strips were pre-processed a) and assembled into sectors b) that were brought together in the full prototype c) and d). (Image credit Caroline St-Hilaire)

5 Discussion, limitations, and future work

The proposed method while very effective at reinforcing the 1 mm aluminum shell for the proposed scope, also induces fabrication and assembly problems and its use revealed a series of limitations during construction. For instance, it was difficult to manage the large-scale assembly of the sectors, due to their limited rigidity and their weight-induced deformations coupled with the interlocking, finger-like geometry shown in Fig. 10. Suspending the individual sectors and assembling them in mid-air was the only way to overcome the hurdle. The programed structural anisotropy did not account enough for lateral efforts. If pushed firmly the built prototype would sway and vibrate and had to be attached to a walkway for safety. To address this, future iterations of the method will include more complex bracing. Other considered avenues of investigation are multilayer structures. Through the inclusion of additional partial layers of strips on the back (fin) side of the surface, a more controlled aesthetic for both sides of the surface and an improved rigidity will be possible.

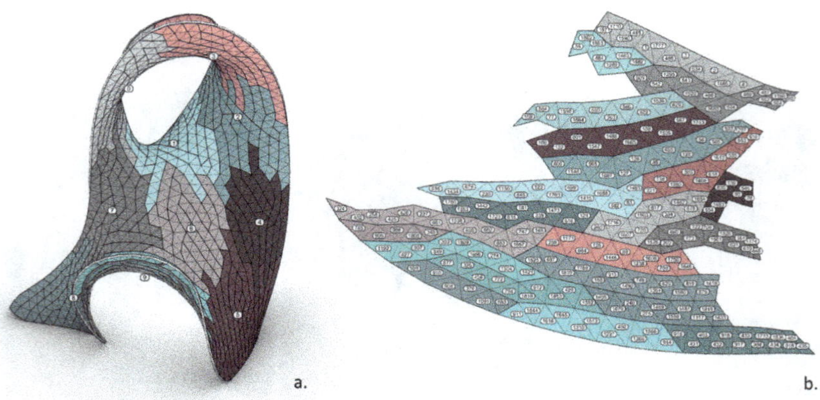

Fig. 10: The geometry of the sectors was designed to hide the seams and avoid folding due to faulty assembly. However, the sector fingers were a lot harder to match than anticipated during the actual construction process. a) A general view of the sector distribution. b) Detail view of sector 3 showing mesh face numbers used for assembly refence.

6 Conclusion

In this paper we have shown, a new method for building self-supported, large, curved, architectural surfaces from ultra-low thickness metal sheet using the segmentation of the surface for fabrication and engaging the dedicated assembly material for structural rigidity. Beyond augmented structural stiffness, the method we proposed improves assembly times and assembly precision especially for low-tech, low budget construction scenarios. Our method significantly reduces weight compared to surfaces supported by a dedicated structure and unlocks an interesting aesthetic for large curved architectural surfaces. The method we proposed successfully addresses an existing gap in the existing research of large, curved, self-supported architectural surfaces, built from sheet metal, that cannot rely on curvature or folds for rigidity.

The initial base surface (without the fins) weighed an estimated 55 kg including the bolts and the connection tabs and displayed a simulated 1960 mm deformation. The proposed thick skin including the fins, bolts and the connection tabs weighed an estimated 80 kg and displayed a deformation of only 2 mm similar to a supported surface but with a fraction of the weight and associated costs.

Acknowledgement

The author of the paper acknowledges as co-authors of the built pavilion, the team of students and tutors of the ARC 6801H 2021 research studio. They are as follows. Students: Alphonso Francis, Belanger Nicolas, Berube Camille, Faltas Youstina, Gamachio Eva, Langelier Marc-Antoine, Larouche-Wilson Kevin, Mansy Mariam, Mireault Pierre-Alexandre, Morissette Olivier, Nahra Ellie, Poulin Delphie, St-Hillaire Caroline; Tutors: Thomas Balaban and Andrei Nejur. The author also acknowledges the support offered for the building of the pavilion by the School of Architecture at University of Montreal, Aluquebec, and Industries B37.

References

Ayres, P., P. Vestartas, and M. Ramsgaard Thomsen. 2018. Enlisting clustering and graph-traversal methods for cutting pattern and net topology design in pneumatic hybrids. In *Humanizing Digital Reality: Design Modelling Symposium Paris 2017*, 285–94. Springer.

Cai, Z.-Y., Y.-W. Lan, M.-Z. Li, Z.-Q. Hu, and M. Wang. 2012. Continuous sheet metal forming for doubly curved surface parts. *International Journal of Precision Engineering and Manufacturing* 13: 1997–2003.

Demin, D. 2015. The rose pavilion. In *Proceedings of IASS Annual Symposia*, 2015:1–9. International Association for Shell and Spatial Structures (IASS).

Eigensatz, M., M. Deuss, Al. Schiftner, M. Kilian, N. J. Mitra, H. Pottmann, and M. Pauly. 2010. Case studies in cost-optimized paneling of architectural freeform surfaces. *Advances in Architectural Geometry* 2010: 49–72.

Flöry, S., and H. Pottmann. 2010. Ruled surfaces for rationalization and design in architecture.

Fornes, M. 2016. The art of the prototypical. *Architectural Design* 86 (2): 60–67.

Friedman, T. 2016. Foldfinding – a novel approach to the design and fabrication of folded structures. In *Proceedings of IASS Annual Symposia*, 2016:1–10. International Association for Shell and Spatial Structures (IASS).

La Magna, R., M. Gabler, S. Reichert, T. Schwinn, F. Waimer, A. Menges, and J. Knippers. 2013. From nature to fabrication: Biomimetic design principles for the production of complex spatial structures. *International Journal of Space Structures* 28 (1): 27–39.

La Magna, R., S. Schleicher, and J. Knippers. 2016. Bending-active plates. *Advances in Architectural Geometry* 2016: 170–87.

Lienhard, J., and J. Knippers. 2013. Considerations on the scaling of bending-active structures. *International Journal of Space Structures* 28 (3-4): 137–48.

Lienhard, J., and J. Knippers. 2015. Bending-active structures. *Bautechnik* 92 (6): 394–402. DOI: 10.1002/bate.201500007

Martín-Pastor, A., and R. García-Alvarado. 2019. Developable wooden surfaces for lightweight architecture: bio-dune pavilion. *Digital Wood Design: Innovative Techniques of Representation in Architectural Design,* 1481–1500.

Nejur, A. 2016. Ivy for Grasshopper Manual: Version 0.802. *Digital Design Research Repository Andrei Nejur.*

Nejur, A., and T. Balaban. 2022. The A(Fin)Ne pavilion. In *Co-Creating the Future: Inclusion in and through Design. Proceedings of the 40th Conference on Education and Research in Computer Aided Architectural Design in Europe* (ECAADe 2022), 2:507–16.

Nejur, A., and K. Steinfeld. 2016. Ivy bringing a weighted-mesh representation to bear on generative architectural design applications. In *ACADIA* 2016, 140–51. Ann Arbor MI: CUMINCAD.

Nejur, A., and K. Steinfeld. 2017. Ivy: Progress in developing practical applications for a weighted-mesh representation for use in generative architectural design. In *ACADIA* 2017, 446–55. Boston, MA: CUMINCAD.

Nicholas, P., D. Stasiuk, E. Nørgaard, C. Hutchinson, and M. Ramsgaard Thomsen. 2016. An integrated modelling and toolpathing approach for a frameless stressed skin structure, fabricated using robotic incremental sheet forming. *Robotic Fabrication in Architecture, Art and Design* 2016, 62–77.

Pottmann, H. 2013. Architectural geometry and fabrication-aware design. *Nexus Network Journal,* 15 (2): 195–208.

Pottmann, H., M. Eigensatz, A. Vaxman, and J. Wallner. 2015. Architectural geometry. *Computers & Graphics* 47: 145–64.

Preisinger, C., and M. Heimrath. 2014. Karamba — a toolkit for parametric structural design. *Structural Engineering International* 24 (2): 217–21.

Rossi, G., and P. Nicholas. 2018. Modelling a complex fabrication system: New design tools for Doubly curved metal surfaces fabricated using the english wheel. In *Proceedings of ECAADe 2018: Computing for a Better Tomorrow,* 811–20.

Schleicher, S., A. Rastetter, R. La Magna, A. Schönbrunner, N. Haberbosch, and J. Knippers. 2015. Form-finding and design potentials of bending-active plate structures. In *Modelling Behaviour,* 53–63. Springer.

Schling, E. 2018. Design and construction of curved support structures with repetitive parameters. In *Advances in Architectural Geometry* 2018.

Schmieder, M., and P. Mehrtens. 2013. Cladding freeform surfaces with curved metal panels — a complete digital production chain. *In Advances in Architectural Geometry* 2012, Eds. L. Hesselgren, S. Sharma, J. Wallner, N. Baldassini, P. Bompas, and J. Raynaud, 237–42. Vienna: Springer Vienna.

Stanojevic, D., and K. Takahashi. 2019. Strip-based double-layered lightweight timber structure. *Proceedings of IASS Annual Symposia* 2019 (11): 1–8.

Lorenzo Santelli

A Generative Approach towards the Design of a Spherical Structural Envelope

Abstract: This article aims to describe the geometrical challenges that the structural design team encountered while approaching the structural modelling of the superstructure of the Apple Store at Marina Bay Sands in Singapore. Finite element analysis modelling followed a generative and parametric approach throughout the schematic, detailed and construction design stages. This enabled the structural design team to manage the geometrical complexity of the superstructure while fulfilling the architectural ambition. This study also focuses on how the structural design team made use of intermediary software (API) to streamline the modelling and analysis process. Several models were generated with the intention of studying the global structural behaviour and evaluating deflections, movements, buckling modes and the level of stress in the structural elements. In parallel, local detailed models were generated to evaluate stress concentration in the conical and spherical glass panels. Finally, this essay also retraces all the finite element models created to evaluate the alternative structural configurations that were explored during the design phase and helped define the path that led to the ultimate structural scheme.

Keywords: structural glass, finite element modelling, parametric design, spherical structural typologies

1 Introduction

This study focuses on the modelling challenges encountered by the structural glass engineers during the design phase of the Apple Store in Marina Bay Sands in Singapore. The iterative analysis process that led to develop the 30-metre diameter dome superstructure, which sits on a concrete podium surrounded by water, proved to be demanding, especially in terms of time and resources. This paper aims to explore how the analysis process was streamlined to maximise efficiency and cope with a challenging schedule throughout a generative approach.

The reader will find a review of selected spherical glass structures as well as advances on structural glass geometry typologies. The paper then describes the challenges of applying glass as main lateral force resisting structural system and the interactive methodology applied which culminated in an integrated analysis and parametric design workflow that empowered the spherical structural glass system design. With the methodology in place the author explores advantages or otherwise of the diverse geometrical morphologies tested, finally describing the structural glass approach taken for manufacturing and construction. The paper provides the results achieved and a

https://doi.org/10.1515/9783111162683-002

Fig. 1: Apple Store retail unit in Marina Bay Sands, Singapore. (Photo by Finbar Fallon)

reflection upon the workflow that enabled the daring design and technology advances which together enabled achieving even more challenging frontiers on structural glass geometries.

2 Relevant precedents

2.1 Spherical glass structures

The Biosphere Environment Museum in Montreal, also known as Montreal Biosphere, designed by Buckminster Fuller and built as part of Expo 67 [1], established a fundamental precedent in dome structures. Although there are some precedents in geodesic domes, such as the Zeiss Planetarium in Jena [2], it remains an important example to be inspired by. It was originally covered by acrylic cells. The cladding system was lost during the 1976 fire. The building has a diameter of 76 m and height of 62 m. The structural system is based on a geodesic polyhedron tessellation.

Fig. 2: Montreal Biosphere structure in 1968. (Archives de la Ville de Montréal, VM94-EX136-779)

Fig. 3: Biosfera Porto Antico, Genoa, Liguria. (Stock photo 1157076005)

Renzo Piano Building Workshop's Biosfera [3] established an important precedent in terms of glass spheres surrounded by water. The glass sphere has a diameter of 20 m and a height of 15 m and was completed in 2001 [4]. The structure is characterised by steel tubes organized in parallels and meridians, following the principal cuvature spherical lines, while bracing is ensured by steel diagonal rods. The glass is spherical and fixed to the steel superstructure by mechanical fixings.

The Montreal Biosphere, the first selected precedent, although represents a historic advance in spherical structures given its shear scale and geodesic geometry tessellation that has driven the structural members distribution, prior to the fire, had acrylic panels that were significantly smaller in scale, planar and used as cladding, i. e. with no structural performance. The second selected precedent, Biosfera, represents a similar location context in relation to the presented project. The glass panels, albeit spherical and following the spherical principal stress lines, have no structural load bearing and are supported by a large structure in relation to the case studied in this paper.

2.2 Advances in structural glass geometry

Recently, glass manufacturing processes and bending techniques rapidly evolved leading to the introduction of extra-large curved panels. The improvements across the industry, together with the increased desire for transparency, led to use of glass as structural element. Structural glass engineering had to follow at pace.

Apple Zorlu created a decisive precedent in terms of structural glass design [5]. The four glass walls are 10-metre wide and 3-metre tall. The walls are formed of 3 plies of laminated glass with ionoplast interlayers, while the roof is made of a single carbon fibre reinforced plastic panel. The glass walls work structurally to ensure vertical and horizontal stability of the whole lantern.

Another important reference in terms of structural glass design is given by the Steve Jobs Theatre in Cupertino [6], where the glass works structurally and is responsible for

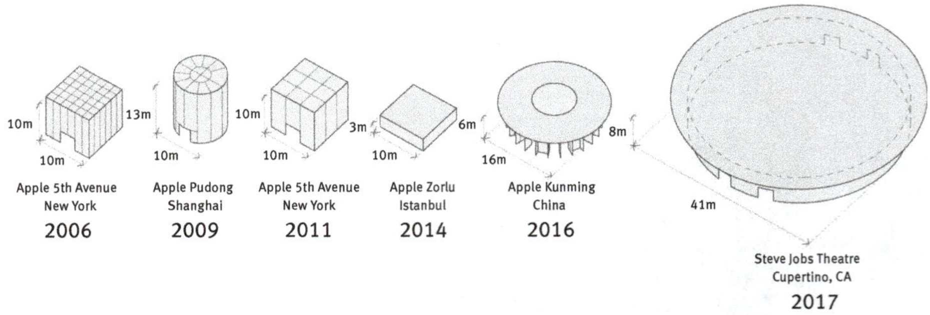

Fig. 4: Evolution of structural glass design.

Fig. 5: Left: Apple Zorlu, Istanbul (Photo by Hufton+Crow); Right: Steve Jobs Theatre, Cupertino (Photo by James O'Callaghan).

the vertical and lateral stability of the whole envelope in a highly seismic zone. The glass panels are arrayed in a 41-metre diameter circle.

3 Glass as main lateral force resisting system

Foster and Partners' architectural and engineering teams developed the original concept alongside EOC Engineers. Working together allowed the team the opportunity to approach the design holistically.

The whole design team embraced the idea of pushing the glass design to its limit by considering the possibility of adopting unconventional structural solutions. This led to the use of the glass as the main lateral force resisting system being considered, without relying on more standard bracing strategies.

Glass is a brittle material. A structural glass failure could potentially lead to a catastrophic scenario if adequate countermeasures are not introduced. This is even more relevant when the glass is responsible for the lateral stability of the entire building.

Therefore, a detailed understanding of the structural behaviour of the dome superstructure was necessary in the early design stages. Several potential accidental scenarios had to be evaluated in parallel with the standard structural design in order to transform a given architectural concept into a valid structural scheme. This modus operandi led to the necessity of implementing the way we worked on each given architectural proposal by speeding up the whole analysis process.

Consequently, a systematic generative approach to the analysis process was finally considered to streamline our structural studies, hoping this would help facilitate working to an extremely demanding schedule. On the other hand, this choice was benefited by the proposed spherical architectural arrangement, possibly an ideal form to deploy this methodology.

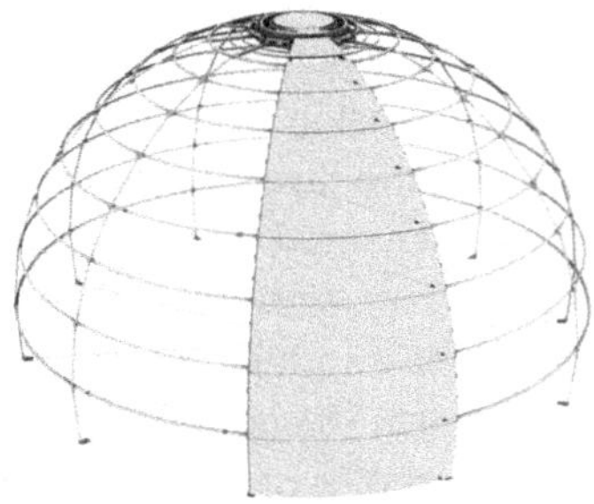

Fig. 6: Horizontal layout, system to resist horizontal loads.

4 Structural analysis and design: an iterative journey

Firstly, the dome superstructure consisted of a pure sphere, a relatively complex form from a construction point of view. Secondly, further complexity was added by the adopted panelisation scheme driven by the two following criteria:

- Maximization of transparency,
- Current glass manufacturing limitations.

Several architectural configurations emerged from the combination of these two criteria. Unavoidably, this approach did not necessarily lead to the most optimal structural arrangements. Thus, further detailed studies were required to transform a potential concept into a valid structural proposal.

Extensive studies were carried out in parallel on several schemes. Finite element analysis was performed with the intent of understanding the global structural behaviour under the imposed loading scenarios and estimating movements, natural frequencies and the level of stress in the structural elements. Special attention was then given to the buckling modes since the boost for transparency led to critically slender structural arrangements with a considerable number of elements prone to buckle. Each glass panel, which works as a tension and compression bracing element had to be checked for local buckling too.

Modelling had to capture the complexity of the details surrounding the glass and replicate the correct load path, which varied according to the analysed loading scenario. Different geometrical configurations had a strong impact on the required resources. Steel slender elements were conveniently approached as beam elements (1D), while the glass was generally approximated as shell elements (2D). Sometimes, brick elements (3D) were instead required to add granularity, but this was carefully balanced with the expected computational weight generated by the brick elements. Moreover, the supports in direct contact with the glass required special attention. Small structural elements called bearing blocks were used at the interface between glass panels; intermediate elements are necessary because glass should not come into direct contact with other glass elements. The bearing blocks have no tension capacity, hence, non-linear material analysis had to be performed, adding an extra layer of complexity.

Shear transfer between the glass and the steelwork was ensured through use of structural silicone, which could be modelled in different ways with appropriately associated computational weights. The first approach was to use springs (1D) leading to the creation of several elements spaced at intervals of between 50 and 100 mm. The second approach was to use brick elements (3D). Again, the choice of approach was based on computational efficiency. However, this led to geometrically complex 3D models, where a lot of extra smaelements had to be managed within a relatively large model in a short amount of time, increasing the risk of losing control over the model.

Before starting the analysis process, EOC Engineers' design team carefully evaluated various potential tools and software. Rhinoceros 5 (Grasshopper) and Strand7 R2 were selected as the main two software to create the global model and perform the required finite element analysis. The computational design workflow choice was based on experience and potential: Rhinoceros and Grasshopper gave us the possibility of generating and handling complex 3D models, while Strand7 R2 provided the level of detail we believe was required to tackle this project from schematic design to detailed design. Other tools were integrated, but with a minor impact on the analysis process.

The main challenge behind this strategy was the missing link between the two software. Drawing directly within the finite element modelling (FEM) environment was not feasible considering the complexity of the structural arrangement. On the other hand, importing models from the CAD environment into the FEM environment proved not to be a viable due to the tight schedule. It was essential to generate models within the CAD environment while preserving the freedom to manipulate them within the FEM environment. The Application Programming Interface (API) gave us the chance to establish a solid link between the CAD and the FEM environments. This synergy gave us the ability to bypass the FEM interactive environment and perform functions which were not directly available within the FEM software. This approach enabled us to streamline the analysis process, seamlessly generate options and variations and cope with the complexity of the structural glass design within a tight schedule.

5 Preliminary design options

At schematic design stage several architectural and structural schemes were analysed and assessed. This section aims to describe the challenges encountered during the finite element modelling phase of some of the options.

5.1 Geodesic layout

The geodesic arrangement tended to replicate a more traditional grid-shell structure with rather small panels. Glass manufacturing limitations did not allow for larger panels leading to a critical compromise in terms of transparency. Considering the proposed size of the building, an option with no steel superstructure was not a practicable one. A framing system had to be introduced and the reduced size of the glass panels led to a low glass to steel ratio.

From a modelling perspective the frame was approximated as beam elements, while the glass as shells. The glass provided lateral stability to the whole dome, and this was assured by a stiff interface between the glass and the steelwork. All local glass supports had to be included within the finite element model with adequate stiffnesses in order to reflect different materials, such as nylon or hard rubber. Each panel had up to twelve setting blocks. Special attention was also given to shear transfer and thermal expansion.

In the end this scheme was not pursued due to the lack of transparency and the difficulties to minimize the steelwork frame.

(a) (b) (c)

Fig. 7: Preliminary design options: (a) geodesic layout, (b) slanted layout, (c) vertical layout.

5.2 Slanted layout

In parallel to the development of the geodesic scheme, an alternative design with slanted glass panels took shape. The glass arrangement was characterised by long rhomboidal panels arrayed around the dome. This proposal took advantage of the glass manufacturing capabilities of the time leading to a higher degree oftransparency.

The glass panels were designed to be responsible for the lateral stability of the dome superstructure. A steel frame was added to restrain the glass and help transfer shear between the panels. Modelling proved to be challenging due to the complex panelisation and system to resist the lateral loads. Side setting blocks between the glass panels assured shear transfer at different levels and to the concrete podium.

Even though assuring a higher degree of transparency, this scheme proved to be too challenging for several factors including glass fabrication, installation and replacement. The structural system had to temporarily braced in case a glass panel had to be replaced. Thus, alternative options were investigated.

5.3 Vertical layout

A simpler version of the slanted scheme emerged where all panels were oriented vertically. This led to a simplification of the structural and geometrical arrangements. A slender steel frame had to follow the vertical glass joints to provide additional robustness to the system. This proposal proved to be simpler from a modelling perspective since the complexity linked to the glass local supports largely decreased.

However, the system to resist the horizontal loads did not benefit from the vertical orientation of the glass panels. While the structural scheme had some potential, the steel superstructure had to be increased to address buckling and frequency issues.

Ultimately, this scheme did not encounter the necessary support to be brought into the following stage and alternative options were investigated.

6 Final Design

6.1 Horizontal layout

The 30-metre diameter sphere was equally divided into 10 sectors and 9 gradually decreasing horizontal bands and completed by a 3.1 m diameter oculus manufactured as a single glass panel with a spherical curvature. This architectural arrangement led to the creation of an almost uninterrupted 3.2-metre tall transparent band at eye level. This was possibly the most optimal solution in terms of transparency as well as the glass manufacturing limitations of the time. The concrete podium was cladded with two additional fully fritted glass bands.

All glass panels between band 1 and band 8 were designated to be responsible for the lateral stability of the whole dome. These were supported on a slender steel frame organised in segmented mullions and circular transoms. Transom to mullion connections provided full moment continuity, while the interfaces between the concrete and the steelwork superstructure worked as hinges. Each glass panel was pre-bonded with structural silicone in the factory to two stainless steel conical plates along the top and the bottom edges. The glass was supported on bearing blocks for in-plane loads. The stainless steel plates were then connected to the steel superstructure.

All the structural elements mentioned above had to be included within the global model leading to a substantial drawing effort. The use of generative and parametric tools helped manage the geometrical complexity necessary to develop the full structural glass concept. All structural glass supports were generated and transformed into structural elements such as beam, shell and brick elements within the Grasshopper environment. The analysis process was also recalled within the Rhino-Grasshopper environment by using the Strand7 API. Adjustments based on the initial results were again performed in the Rhino-Grasshopper environment without the need to modify the model within the FEA environment. This approach helped optimise the whole analysis process.

Additional models were analysed in parallel to the main one in order to evaluate several potential accidental scenarios where some structural glass panels were alternatively removed and the ultimate behaviour of the dome superstructure examined.

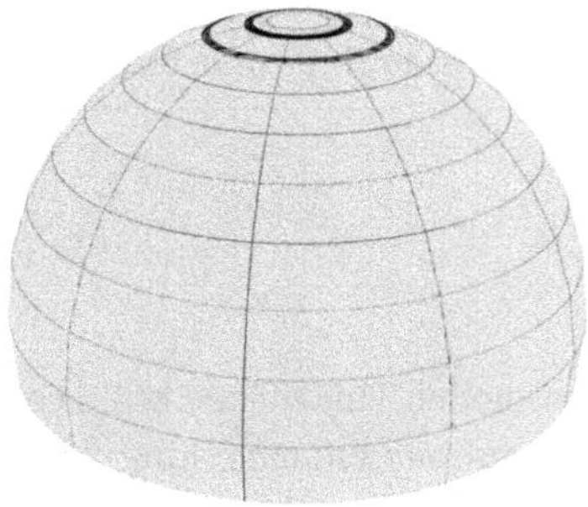

Fig. 8: Horizontal Layout.

6.2 The glass

All glass panels, except for the oculus, were conical in shape (singly curved with varying radii). This form represented the most optimal synthesis between the two leading criteria of enhanced transparency and glass manufacturing limitations.

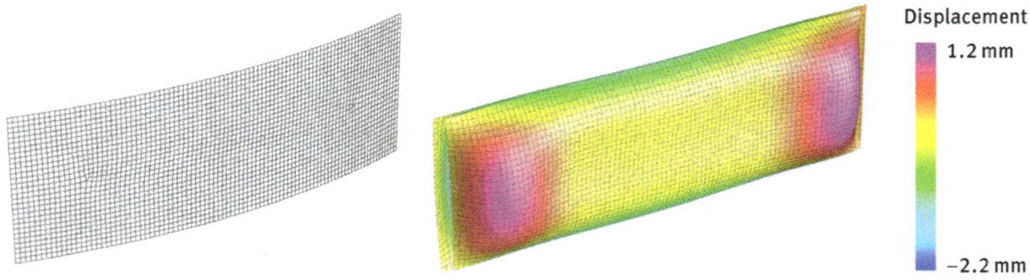

Fig. 9: Horizontal layout. Deflection under climatic loads, detail of local FEA model.

The conical shape was achieved through the cold bending during lamination manufacturing procedure (sometimes also referred as warm bending), where the individual glass plies were forced into shape on a mould prior to lamination with the ionoplast interlayer. After the lamination process, the glass mantained its conical shape thanks to the shear strength of the interlayer.

Modelling of the glass panels was carried out globally and locally. A local model of each conical glass panel was generated to simulate the bending process and estimate the expected spring-back when the panel would be released from its mould. Due to the conical nature of the geometry, spring-back varied across each panel.

Detailed models of the conical panels were also used to evaluate the level of stress in the insulated glass units (IGU) caused by climatic loads, such as variation in pressure and temperature.

The *oculus*, placed at the top of the dome, was the only spherical panel. It was manufactured differently from all the other panels due to its doubly curve form. It was slumped and then laminated with ionoplast interlayer.

6.3 Modelling approach

The structural modelling began with a simplified stick model. This helped build confidence in the global scheme and created the basis for the subsequent modelling strategy. The latter was characterised by some calibrated steps aimed at reducing the gap between the digital and the physical environment by adding complexity without losing control.

Modelling progressed in a few steps:

- The glass was incorporated into the global model as plate elements with simplified connections between the glass and the steelwork.
- A conical door was added within a sector at 'ground' level (band 1). This generated an asymmetrical weakness that required local strengthening and two additional mullions next to the opening.
- Further detail was added to the glass to steel connections, including silicone patches, toggles and stainless steel clamps that helped restrain the glass panels

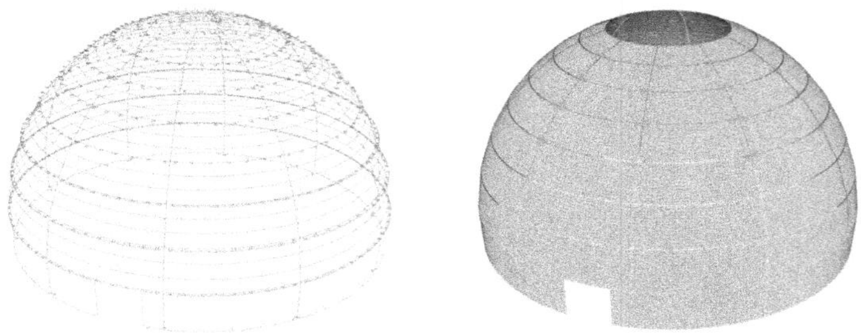

Fig. 10: Left, horizontal layout. Detail of FEA model, with beam elements, right, horizontal layout. Detail of FEA model, with beam and shell elements.

in and out of plane. At this stage all additional items were modelled as beam elements.
- All glass supports were later modelled as plate elements. Local 3D brick models of the glass to steel connections were generated in order to evaluate the level of stress concentration in the glass panes at each connection.
- Construction stages and accidental scenarios that simulated potential patterns of glass breakages were investigated in parallel.

The chosen generative approach helped retain control over a global model which was eventually characterised by a high degree of complexity. Global and local models were developed within the CAD environment and analysed in the FEA environment. Results were then retrieved in the CAD environment using the application programming interface. This process also helped calibrate the model during the various design iterations and construction stages. Cross-sections, materials properties, units and loads were all defined and directly controlled through components in Grasshopper, visualisation of all the results was also managed in this environment.

6.4 An alternative structural approach

An option based on local mechanical fixings for shear transfer between the glass and the steelwork was explored as a potential alternative to the use of structural silicone. The glass had to be manufactured with holes in order to ensure shear transfer to the steel superstructure throughout stainless steel bolts.

Modelling proved to be more challenging than any other option due to level of detail required by each mechanical fixing point, where the interface between the glass and the steel had to be modelled thoroughly. Several mechanical fixings were placed along the top and bottom edges of each glass panel at a distance of 50 cm. Each bolt had to be surrounded by either grout or nylon to resist in-plane loads and capped with machined stainless steel washers for out of plane loads.

In the end, due to the complexity of the joinery system and the high level of stress concentrations in the glass around the mechanical fixings, this option was discarded.

7 Results

While the use of the API undoubtedly helped optimise the modelling and analysis process, a significant amount of time had to be spent to adequately manage the link between the CAD and the FEA environments.

Several levels of control had to be introduced to thoroughly check the whole digital workflow. This helped expand our experience and give the senior members of the team the right level of confidence to endorse the proposed structural design.

We can affirm that, even though some challenges had to be faced, the team unquestionably benefited from the approach described in this paper.

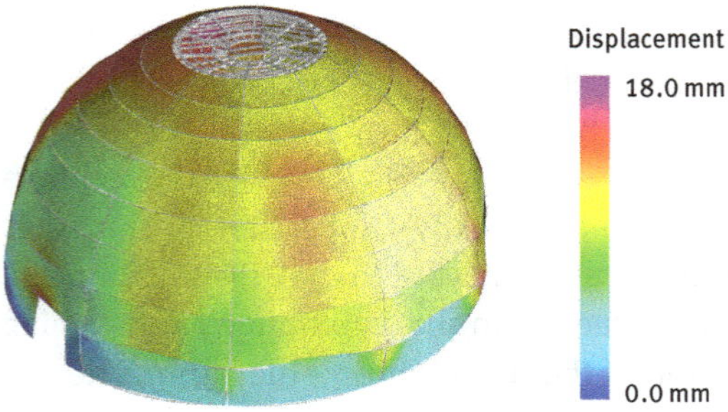

Fig. 11: Horizontal layout: Deflection under wind load.

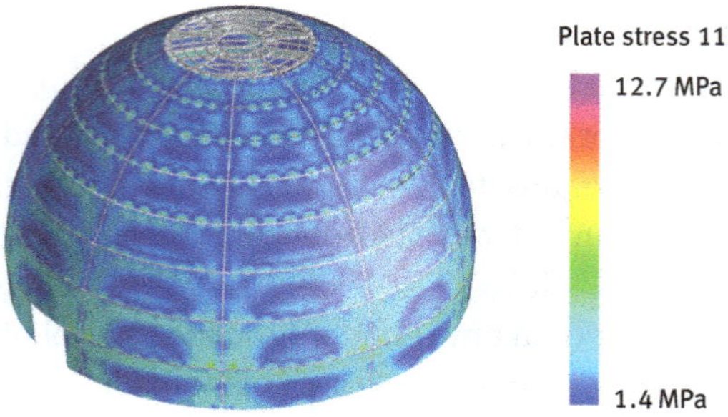

Fig. 12: Horizontal layout: Principal stress 11 in the glass elements, ULS envelope.

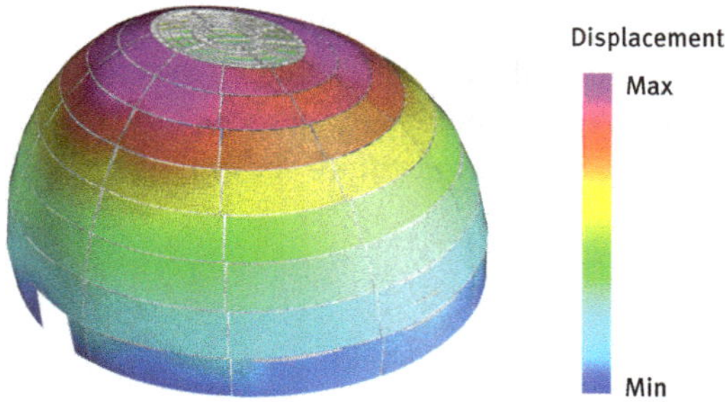

Fig. 13: Horizontal layout: Natural frequency analysis, first mode 2.1 Hz.

8 Manuscript Reflection

It was the first time that we adopted the use of the API for a such high-profile project. Even though it was considered a success, this methodology was questioned several times internally due to the tough challenges we encountered, especially at the very beginning of implementation.

It should be noted that the spherical architectural arrangement facilitated the application of the API strategy. On the other hand, the complexity linked to the glass structural design posed arduous challenges.

We can observe that some design stages were more appropriate than others to exercise this methodology. Schematic and detailed design proved to be the most efficient ones. During these two phases the number of options to be evaluated was narrowed down to a small group, while an adequate level of detail was required to give the whole team the right degree of confidence to proceed.

There are no doubts that this approach will be applied again in the future, but not all projects are suitable for this. The experience gained through the analysis and design process of the dome superstructure will guide us to decide when to adopt this strategy for any future applications.

The designers of Apple Marina Bay Sands pushed the boundaries of the industry in favour of lightness and transparency [8]. The latest advances in the glass manufacturing industry accompanied by the integrated analysis method outlined in this article helped develop a structural system that responded to the given challenge.

The project leaves us with the next frontier on the horizon. At the time of writing (May 2023), no manufacturing technology has advanced to produce doubly curved spherical glass panes at the same scale of the conical ones utilized. However, supposing the manufacturing technology was possible, what would be the structural as well as optical challenges that may result from the development? How would these benchmark against systems that provide more conventional load bearing systems and materials? These questions suggest further investigations.

The desire for lightness and transparency is far from over, and, as designers, we now face even bigger challenges, such as sustainability, logistics, limited resources etc. Based on the experience gained on this project, a further integrated structural design process could benefit the development of the design of any future challenging project.

9 Conclusions

In conclusion we have explained the challenges and introduced a method to integrate and interactively design and analyze spherical glass structures utilizing the glass itself as main lateral force resisting system. We have tested the method through the iteration of several options and variations and could therefore evaluate the benefits or otherwise of these that culminated in the manufacturing and construction of the chosen structural typology. Finally, the manuscript provides a reflection upon the methodology and provides a reflection for further investigations.

Acknowledgements

I would like to express my gratitude to Prof. Miriam Dall'Igna for her contribution and expertise, without which this article would not have been possible. I would also like to acknowledge the following parts for the successful collaboration that helped bring this project to life: Apple, Foster + Partners, SH NG Consultant PTE Ltd, Gartner GmbH, Sedak GmbH & Co. KG. Finally, I would like to thank Alexandros Cannas and Chay Graham for their contribution to the project and Alistair Frost for his help reviewing this manuscript.

References

[1] Massey, J., 2016 Buckminster Fuller's cybernetic pastoral: the United States Pavilion at Expo 67, *The Journal of Architecture,* 21(5), 795-815, DOI: 10.1080/13602365.2016.1207433

[2] Gáspár, O., 2022. The optimization process leading to the tessellation of the first geodesic dome structure, the first Planetarium of Jena. *International Journal of Space Structures,* 37(1), 49–64.

[3] Rossi, M., 2012. The Fuller's heritage: organic models and artificial microcosm in XX century domes. In Domes in the world (pp. 1-11). Nardini Editore.

[4] Fondazione. (2022). Recupero del Porto Antico di Genova. https://www.fondazionerenzopiano.org/it/project/recupero-del-porto-antico-di-genova/

[5] www.eocengineers.com. (n.d.). Apple Zorlu – Eckersley O'Callaghan. https://www.eocengineers.com/projects/apple-zorlu-86/

[6] www.eocengineers.com. (n.d.). Steve Jobs Theater – Eckersley O'Callaghan. [online] Available at: https://www.eocengineers.com/projects/steve-jobs-theater-293/ [Accessed 30 May 2023].

[7] Coult, G., A. Cannas, S. Gregson, L. Santelli. (2022). Apple Marina Bay Sands: utmost transparency. *Glass Structures & Engineering* 7, 1–18. DOI: 10.1007/s40940-022-00196-8.

Lawson Spencer, Matthew T. Reiter, Leslie Lok, Sasa Zivkovic

The Finite Element Method (FEM) of the Unlog Tower

Abstract: The Unlog Tower uses finite element analysis (FEA) to simulate the structural, material, and geometric limitations necessary to transform locally available, Emerald Ash Borer (EAB) infested White Ash timbers into bending-active trusses through elastic kinematics. Like other Elastic kinematic structures, the Unlog Tower relies on elastic deformation through inherent material properties to create structurally stable and lightweight bending-active components that can easily be deployed and disassembled. Through robotic kerfing techniques, halved logs are stretched along threaded rods and connected into larger panels with a custom slip washer splice connection. To inform the fabrication of these panels, a finite element method (FEM) was developed to simulate the elastic deformation of individual panels as they were stretched, and a second FEM was developed to test the whole structure against gravity and wind loads. This research uses EAB-infested white ash, to simulate and assemble roundwood timbers into bending-active structural components for a fully permitted outdoor installation.

Keywords: bending active structures, digital timber, deployable structures, finite element analysis, elastic kinematics

1 Introduction

Situated on the Arts Quad at Cornell University in Ithaca, NY, the Unlog Tower (Fig. 1) uses emerald ash borer (EAB) infested White Ash logs in response to the ongoing EAB epidemic (Herms and McCullough 2014). The EAB threatens to eradicate most of the 8.7 billion Ash trees in North America and as of 2018, the epidemic has spread to 35 US states and several Canadian provinces (Flower, Knight, and Gonzalez-Meler 2012; USDA Forest Service 2022). In New York State where ash trees constitute approximately 8 % of the overall tree population, the EAB was first discovered in 2009 (USDA Forest Service 2022). Consequently, many ash trees that were traditionally used for cabinetry, hardwood floors, and baseball bats are now discarded as waste or used as firewood, thereby releasing their carbon store back into the atmosphere. Despite being a locally available timber resource and maintaining their structural properties up to two years after infestation (Persad et al. 2013), ash trees (infected or not) are rarely used in the construction industry. The Unlog Tower expands upon the Unlog method (Lok et al. 2023) to transform EAB infested ash wood logs into a triangular and lightweight timber tower using robotic fabrication, mixed reality (MR) workflows, and augmented reality (AR) instruction.

https://doi.org/10.1515/9783111162683-003

Fig. 1: Unlog Tower. (Photo by Cynthia Kuo)

The Unlog Tower uses EAB infested ash wood logs that would otherwise be discarded as waste, with diameters above 12 inches and lengths between 8, 10, and 12 feet. Such logs fall within standard parameters preferred by regional loggers. By utilizing EAB-infested ash wood as a structurally viable, materially efficient, leaf-spring wall component, the Unlog Tower, challenges present EAB-related logging practices, supply chains, fabrication methodologies, and architectural tectonics as an advocate for the use of EAB infested ash logs in timber construction. The research for the Unlog Tower develops and deploys a Finite Element Method (FEM) to simulate the deformation of individual ash wood boards as they are stretched into leaf-spring panels. Then a second FEM is used to simulate the gravity and wind loads affecting the installation. Both literally and figuratively, the Unlog Tower investigates how far this discarded material resource can be stretched through robotic kerfing and elastic kinematics.

2 Background

In woodworking, kerfing is a relief cut technique that reduces the stiffness of a material, allowing the object to be more pliable through a geometric change (Mansoori et al. 2019). Though, kerfing is often applied to sheet materials to generate surface curvature (Menges 2011), others have kerfed round wood timbers with a track bandsaw, which were then steam bent (Kuo 2022). However, steam bending alters the mechanical properties of wood and thus also weakens the material (Svilans et al. 2019). Controlled elastic deformation through pliable materials is a bending-active method to achieve geometric curvature, while maintaining the structural viability of the material as long as the elastic deformation is reversible (Lienhard et al. 2010). This structural principal is evident in the Multihalle Mannheim built in 1974 and the ReciPlyDome, in which the timber grid shells were able to elastically deform according to the horizontal translation of the supports until they were pinned in place (Lienhard et al. 2013; Brancart et al. 2017). Within this context, researchers have used finite element analysis (FEA) and other computational physics solvers to simulate the bending of individual plywood panels, which were used for form-finding and fabrication (Schleicher et al. 2015).

Bending-active structures with elastic kinematic joints rely on elastic deformation through material pliability to create structurally stable and lightweight bending-active joints that can be easily stretched and bent (Lienhard et al. 2010; Schleicher et al. 2011). This principle is observed in the Unlog method (Lok, Zivkovic, and Spencer 2023), whereby, roundwood ash logs were robotically cut in half and then kerfed lengthwise to construct a leaf-spring joint between boards for each panel (Fig. 2). The robotic method and leaf-spring structural logic exhibited in Unlog and now, the Unlog Tower is comparable to that of the Torus Research Pavilion by CODA, in which individual plywood boards were face coupled with rigid connections allowing the boards to elastically deform when stretched (Tornabell, Soriano, and Sastre 2014). The Torus assembly was flat-packed and quickly unraveled on site to exhibit and validate the inherent bending-active structural principles of the plywood members. Both Torus and Unlog investigate the fabrication feasibility of bending-active, leaf-spring timber structures through their respective elastic kinematic assemblies. Instead of relying on a sheet material, the Unlog method unravels the volumetric roundwood into an operable leaf-spring component as opposed to assembling individual sheets that are then unraveled. By doing so the Unlog method attempts to reduce the amount of waste material produced through squaring of a log for dimensional lumber and/or reducing the amount of adhesive required for plywood. In this way, the robotically kerfed roundwood components produced through the Unlog method can easily be recycled at the end of use.

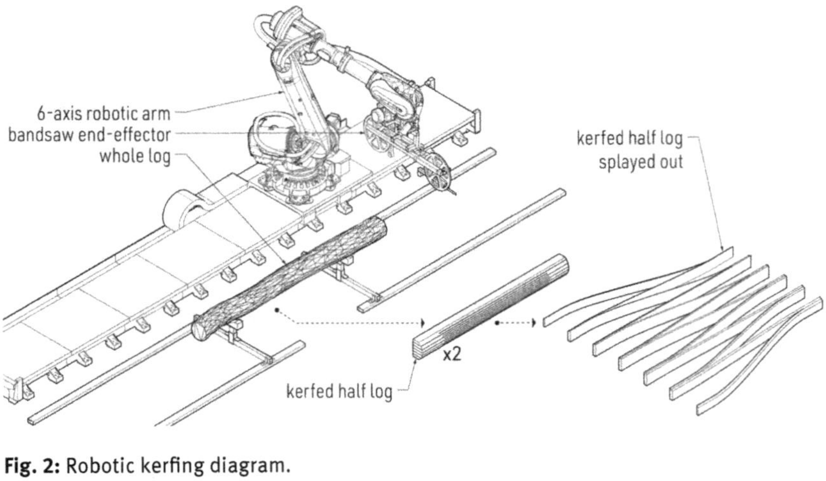

Fig. 2: Robotic kerfing diagram.

3 Methods

FEA is not only used to analyze the effects of gravity and wind loads on the pavilion, but also simulate the displacements of individual half logs (panels) as they stretch along two threaded rods into panels. This simulation is used to inform the robotic fabrication of each individual half log. In response to the ongoing EAB epidemic, the Unlog Tower uses EAB infested ash logs which maintain their structural properties for two years after infection (Persad et al. 2013). The research methods will present the following: the structural principals and code limitations that are used to inform the overall design (Sec. 3.1), the design of the foundation connection (Sec. 3.2), the difference between geometric curvature and elastic bending (Sec. 3.3), the design of the splice connection (Sec. 3.4 and its relation to the robotic fabrication and coordination (Sec. 3.5), with mention of the design of the tube steel frames (Sec. 3.6). Finally, the methods will conclude with illustrations of the parameters defining the analytical model for FEA (Sec. 3.7). The results of these methods will illustrate the displacements due to wind and gravity loads as well as additional coordination models developed from the FEM. The work presented demonstrates a FEM method to simulate and assemble EAB-infested roundwood timbers into a fully permitted outdoor tower installation through elastic kinematics.

3.1 Code Limitations and Design

The driving geometry from which the Unlog Tower is developed from the structural principals of a triangular column, a triangular prism. Based upon the Euler buckling formula below, the triangular prism has a critical buckling load (P_{cr}) of 1.2 times that of cylindrical prism and almost 1.2 times that of the rectangular prism (Sandaker, Eggen, and Cruvellier 2011).

$$P_{\mathrm{cr}} = \pi^2 EI/l^2$$

This observation assumes that between the three prismatic columns, the elasticity of the material (E), the length of the column (l), and the cross-sectional area (A) for the moment of inertia (I) are all constant (Sandaker et al. 2011). For the design of the Unlog Tower, the base of the triangular prism was to be accessible with a 60″ wheel chair turning diameter in accordance with the standards set by the American Disability Act (ADA) (DOJ and DOT 2014, Sec. 304.2.1). Additionally, the entry was to have a minimum clear opening of 32″ with no more than a 4″ projection into either side of the opening between 34″–80″ from the ground (DOJ and DOT 2014, Sec. 404.2.3). These codes informed the design of the entry openings at the base of the installation (Fig. 3).

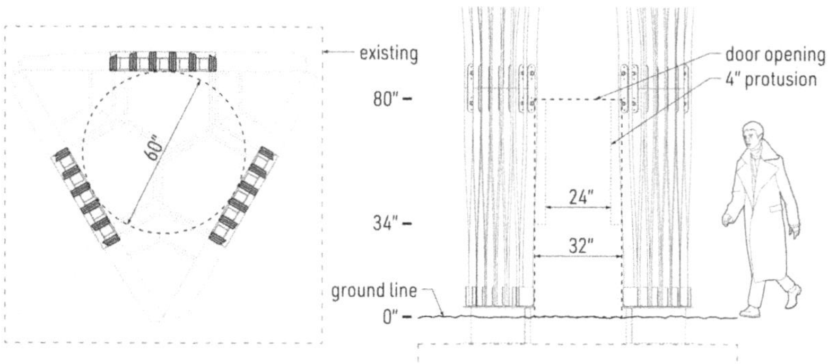

Fig. 3: Unlog Tower plan with ADA design parameters.

3.2 Foundation Connection

The openings at the base led to the design of the foundation design. This particular location on the Cornell Arts Quad had a buried concrete foundation from a previous art installation (Chong 2015). The foundation connection to the timber consisted of three steel I-beams (W14 × 30) that were secured to the concrete pad with eight 0.5″ diameter threaded rods per I-beam. Each threaded rod went 6″ deep into the foundation and was secured with an epoxy adhesive. Sitting atop the I-beam was a steel angle (L5 × 5 × 0.25″), welded on one side of the I-beam and bolt fastened on the other. Atop the steel angle, seven 0.25″ steel plates were welded to the angle with two 0.875″ holes in each. The six 1.5″ wide wood boards were bolted to these plates with 5 steel angles (L7 × 4 × 0.375″). This detail fixed translation in the local x, y, and z axis and resisted rotation around the local x and y axis, but not the z axis (Fig. 4). This relation was used to define the supports in the analytical model for FEA.

3.3 Geometric Curvature vs. Elastic Bending

Based upon geometric studies conducted in Rhino3D and Grasshopper (Robert McNeel & Associates 2022; Rutten 2022), it was known that if boards were spaced evenly along a threaded rod, the total height of the panel would decrease with respect to the spread distance. Additionally, if the bounding shape was a trapezoid as opposed to a rectangle, the boards towards the end would be longer than the boards in the center to have a parallel top and bottom for the threaded rods to pass through (Fig. 5). In Fig. 5, the change in board length per each panel is represented through a green to yellow to red gradient with Δl (delta length) as the total change in length between boards per panel. The change in length was defined to ensure the top and bottom edges of the panel were parallel for the threaded rods to pass through.

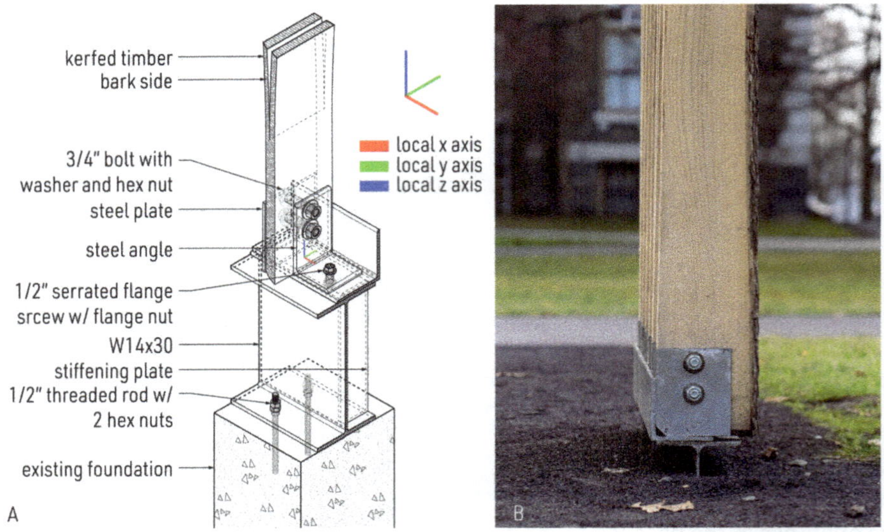

Fig. 4: Foundation Connection: (A) axonometric of foundation detail with support definition; (B) Photo of Ground Connection. (Photo by Cynthia Kuo)

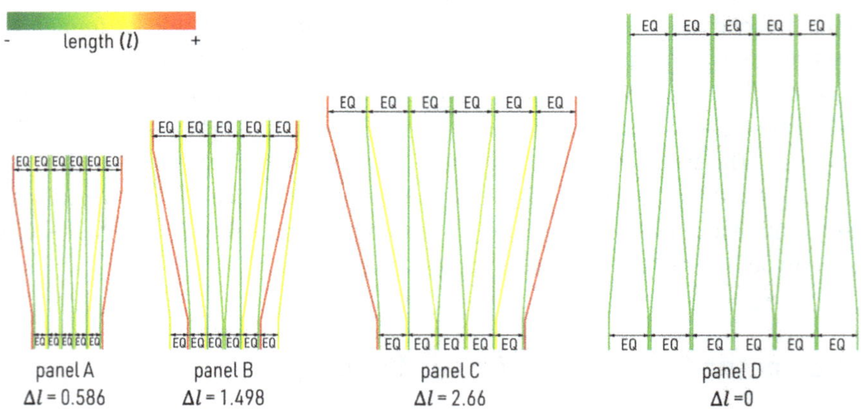

Fig. 5: Change in board length for trapezoidal panels (A-C) and rectangular panels panel D.

The curvature per board was initially assumed through the geometric construction of a Bezier curve which respected the board length relative to the stretching of the before-mentioned panels. However, through prototyping it was revealed that the actual board curvature from elastic deformation in the physical prototype was different than the assumed curvature based upon the Bezier curve construction. This issue affects the ability of the threaded rods to pass through the kerfed panel without bending. In order to simulate the actual bending and determine the length of each individual board more accurately, a FEM was developed to measure the elastic deformation of each individual panel as it was being stretched (Fig. 6). In this FEM, two fixed points were placed at the center-most board; all other boards could transition in the local x and y axis. Four-point loads were applied to the outer most boards on the top and bottom of the kerfed log.

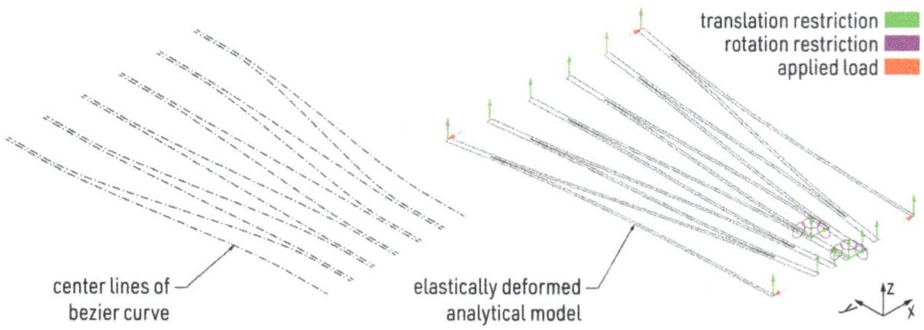

Fig. 6: Comparison between Bezier stretched model and elastic deformation model.

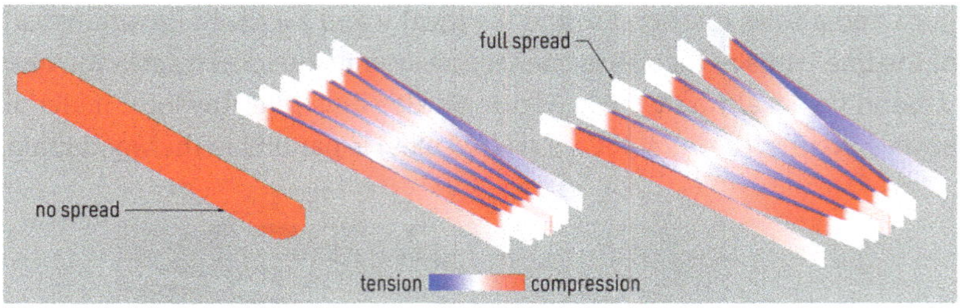

Fig. 7: Individual trapezoid panel being spread through FEA.

Curvature through geometry idealizes a particular material condition, whereas curvature designed through the simulated elastic properties of the material enables the geometry to adapt to a probable or likely condition, which is critical when integrating other connected assemblies. Both Fig. 6 and 7 describe this relationship, which was simulated in Karamba3d, a parametric FEA solver for Rhino3d and Grasshopper (Preisinger and Heimrath 2014). In this FEM, individual boards were defined as shells (meshes) whose orientation followed that of the wood grain. According EN 1912 (2012), White Ash, Fraxinus americana, falls under the Eurocode strength class D35; the material properties in the FEA were based upon such. A single point load was applied to the top and bottom edges of each board, which when increaded, stretched the kerfed log panel. The model was analyzed with the first order theories of small displacements (Rubin and Schneider 2002). The first-order theory assumes that the deflection of the shell is in-line with the mesh face orientation for orthotropic materials and that the cross-sections of the shells remain planar through deformation. This FEM effectively demonstrated how the robotically kerfed half logs would stretch, given the applied forces. This simulation was used to help stagger the finger joint location within each half log so that the top and bottom edges of each panel would be parallel when stretched.

3.4 Splice connection

The splice connection joins individual 1.5″-thick boards together between panels (Fig. 8). As panels are stretched along a 0.875″ diameter threaded rod with preplaced hex nuts, custom, temporary slip washers are placed between the board and the hex nut to temporarily pin the location of the boards. Each board has a 2″ diameter hole that the threaded rod with preplaced hex nuts would pass through. This allowed for quick assembly and disassembly of individual panels. Once both top and bottom threaded rods have passed through, panels can be joined together. When panels are joined together, the temporary slip washer is replaced with a custom, plasma-cut steel slip washer with a thickness of 0.25″. The custom steel slip washer is secured in place with bolts and hex nuts. The finger joints between individual boards allow translation only in the local *x* axis and allows rotation around the local *y* and *z* axis. In the analytical model for FEA, the line joints at the splice connection assume no local translation and only rotation around the local *y* and *z* axis. Figure 8b also illustrates the local material orientation between boards, in which the *x* axis (red axis) is parallel to the wood grain.

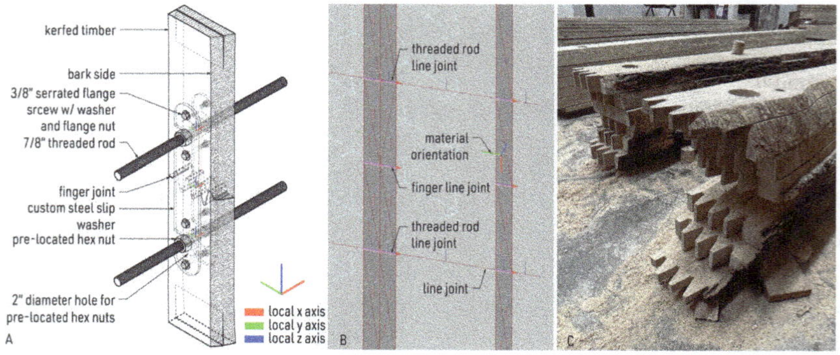

Fig. 8: Splice Connection: (A) detail of splice connection; (B) splice connection in analytical model; (C) Photo of staggered finger joints within a robotically kerfed half log.

3.5 Robotic and Fabrication Coordination

The finger joint location is staggered within the log and different between panels due to the elastic deformation and equal spacing between boards in the three trapezoidal panels. The robotic coordination between individual panels had to ensure (a) that the bark edge of each panel faced outward from the installation, (b) that the cross-section of each board met the minimum structural requirements, and (c) that finger joint location for both ends of each half log was correctly positioned. To accomplish these three tasks, coordination drawings were made to ensure each half log was correctly positioned in the robot cell. Then the finger joints were cut with an oscillating saw and the holes drilled with a hole saw (Fig. 8c).

3.6 Reciprocal steel frame

Because of the elasticity and diameter of the threaded rods, the tower required interior bracing to keep the sides from bending when exposed to wind loads. Six tube steel frames were designed and fabricated with 11-gauge tube steel to brace the interior of the tower. The tube steel frames were placed in pairs of two between the second and third panel, the third and fourth panel, and at the top of the fourth panel. Each frame was composed of 9 steel tubes with 3 different lengths and a cross section of 3″ by 1″. The interior most connection between 3 tubes was a welded reciprocal frame and the outer three most connections were welded with internal, plasma cut steel plates for added bracing. These 6 frames latched onto the threaded rods that only allowed rotation around the threaded rod axis and the vertical axis of the tube steel cross section.

3.7 Analytical Model

Based upon the detailing and design mentioned in the previous sections, an FEM was constructed that parametrically that related all the translation and rotation restrictions of each joint and support to the joint and support definitions in the analytical model (Preisinger and Heimrath 2014) (Fig. 9). Additionally, a gravity load was applied along with a 100 mph (miles per hour) wind load (Fw) according to the appropriate force to area ratio listed in the ASCE standards 2022. All the material properties for the threaded rods, the tube steel, and the kerfed timber half logs were accounted for, except for the weight of custom steel slip washers in the splice connection. Except for the kerfed timber half logs, all other materials were isotropic and therefore, did not require an added material orientation. The grain orientation for the kerfed logs was accounted for through the mesh face orientation of each board and defined as shell elements in the analytical model. The threaded rods and tube steel were defined as beam elements and the bracing plates between the tube steel members were defined as shells. Finally, a steel plate was placed at the three corner connections between the top two most panels (Fig. 10a). This had the same joint definition as the splice connection (Fig. 10b).

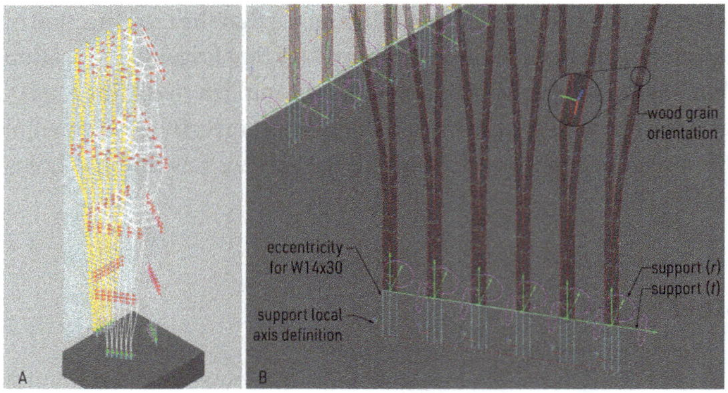

Fig. 9: FEA Model Parameters and Assumptions.

Fig. 10: Left: connection between panels C and D; Right: splice connection. (Photos by Cynthia Kuo)

4 Results

Based upon the reaction forces and reaction moments from the gravity and wind loads at the foundation detail, a force multiplier was created and applied to all loads and calculations in the FEM. The wind loads for were studied from three different directions, perpendicular to a flat face (Fig. 11a), parallel to the same flat face (Fig. 11b), and bisecting a corner condition (Fig. 11c) using the first order theory of small deflections (Rubin and Schneider 2002). The face perpendicular to the flat face proved to have the greatest amount of displacement (d) and was studied for continued displacements, with the following multipliers: $d(0)$, $d(1)$, $d(5)$, and $d(10)$ (Fig. 12). Here the utilization model showed principal compression and tension stress within the structure and was used to evaluate conditions where possible splitting of the wood could take place due to the vertical shear between adjacent boards. The utilization is calculated by taking the total average of the ratio between tension or compressive strengths and the material's comparative stress per mesh face in the case of shell elements and per any point along a curve in the case of beam elements.

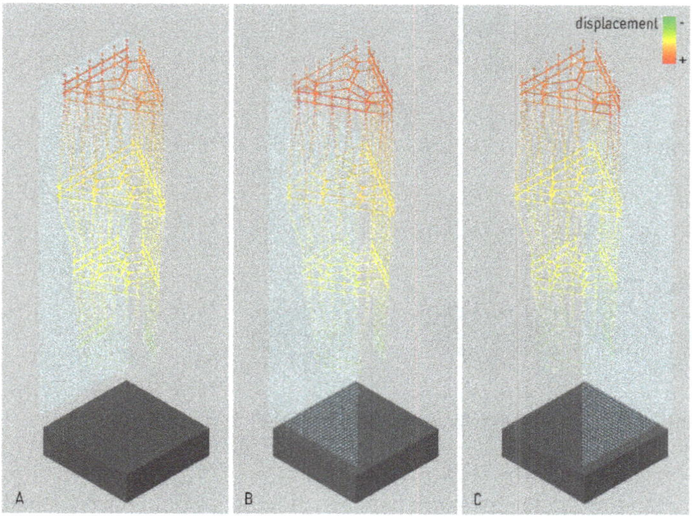

Fig. 11: Tower Displacements per wind load direction. (A) load perpendicular to flat face; (B) load parallel to flat face; (C) load bisecting a corner.

As the design went through several iterations, it became imperative to automate the workflow between design and simulation to quickly communicate the reaction forces at the foundation along with the first and second principal normal forces (Fig. 13a, b), the first and second principal moments (Fig. 13c, d), and the shear forces in the local *x* and *y* directions for the kerfed timber members (Fig. 13e, f). The first principal normal forces were used to check the bearing capacity of each board per mesh face and required reaction force to oppose uplift, while the secondary principal normal forces were used to check for any potential deformation as consequence of the lateral loads. In both cases the compression stresses (red) are negative, and the tension stresses (blue) are positive. Per board, the four highest values for both tension and compression stresses were automatically sorted into a combined list of the extremes for all boards in the model and graphically displayed. The extremes in the combined list were displayed through the red to blue gradient with the highest values being either absolute red or absolute blue in the largest font size. A similar process was repeated for all the moment and shear forces which were displayed through a green to yellow to red gradient. The moment forces in both the first and second directions were checked to see if any board exhibited significant bending, particularly at the unbraced lengths between the bottom two panels. The shear forces, particularly those parallel to the grain, were checked to ensure that there would be no splitting between board members. This simulation greatly assisted in evaluating the complex geometry for locations of overstress and enabled an iterative design process. Finally, if the board lengths in the digital model did not match those fabricated, as would have been the case if they were constructed

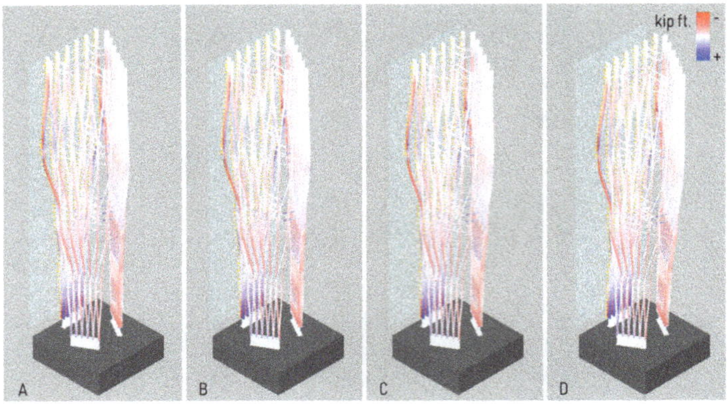

Fig. 12: Tower displacements *d* per wind and gravity loads: (A) $d \cdot 0$; (B) $d \cdot 1$; (C) $d \cdot 5$; (D) $d \cdot 10$.

through Bezier curves, then the change in length between members would have created added shear in the splice connection and potentially splitting as a consequence.

5 Conclusion

The FEM developed for the Unlog Tower was used both graphically and empirically to quickly coordinate between architectural design iterations and structural engineering. Ultimately, the collaboration between the two led to the submission and acceptance of a permitted building set for this installation with an engineer's stamp of approval. Each of the four different panels were assembled into one long prefab panel; this process was repeated twice more until all three sides of the tower were prefabricated. The installation was assembled on its side and then lifted into place with a boom forklift. During the lifting process, the topmost tube steel frame broke in transportation to the foundation pad, due to initially inadequate rigging. The finite element model was re-run to see whether or not the broken tube steel needed to be replaced. The simulation proved viable without the added bracing, and the tower was installed with one less reciprocal steel frame. The tower remained in service for seven months and performed as anticipated through winter conditions.

This research demonstrated how to stretch EAB infested timbers into a structurally viable wall assembly through bending-active elastic kinematics, which were simulated through the development of a custom FEM. The methods illustrated how the design of the structure was informed by code, which lead to the design of the foundation detail. This was followed by assumptions of Bezier curves compared to bending-active bending boards simulated through FEA, which lead to the design of the splice connection. To successfully assemble the trapezoidal panels according to the detail a custom robot coordination workflow was developed, which was followed by the design and detail of

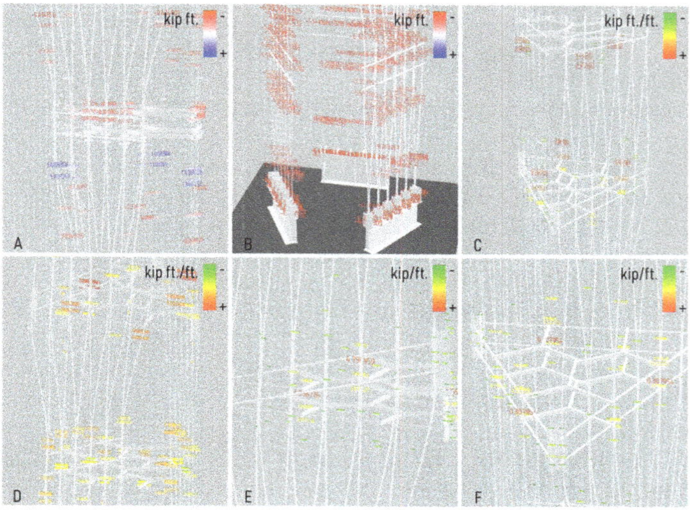

Fig. 13: Graphic display of forces for structural coordination: (A) first principal normal forces; (B) second principal normal forces; (C) first principal moment forces; (D) second principal normal forces; (E) shear force parallel to wood grain; (F) shear force perpendicular to wood grain.

several tube steel framed to brace the interior of the pavilion. Finally, all the design of all the joints and supports were defined in analytical model for the FEA. The FEM developed through this paper was used graphically and empirically coordinate between architectural design and structure. From the development of the FEM to the building approval process, to the successful install and performance (in service), the Unlog Tower demonstrates the viability of EAB infested ash wood as a novel bending-active leaf spring wall panel through FEA.

References

ASCEE. 2022. *Minimum Design Loads and Associated Criteria for Buildings and Other Structures*. Vol. 7–22. ASCE Standard. American Society of Civil Engineers.

Brancart, S., O. Larsen, L. De Laet, and N. Temmerman. 2017. Bending-active reciprocal structures based on equilateral polyhedral geometries. In *Proceedings of the IASS Annual Symposium 2017*, eds. A. Bögle and M. Grohmann.

Chong, Jaeho. 2015. A Needle Woman: Galaxy Was a Memory, Earth Is a Souvenir. *Space*, 114–18.

Department of Justice (DOJ), and Department of Transportation (DOT). 2014. U.S. Access Board - ADA Accessibility Standards (Enhanced Single File Version). https://www.access-board.gov/ada/.

European Committee for Standardization. 2012. EN1912: Structural Timber – Strength Classes – Assignment of Visual Grades and Species.

Flower, C. E., K. S. Knight, and M. A. Gonzalez-Meler. 2012. Impacts of the Emerald Ash Borer (Agrilus Planipennis Fairmaire) Induced Ash (Fraxinus Spp.) Mortality on Forest Carbon Cycling and Successional Dynamics in the Eastern United States. *Biological Invasions 15* (4): 931–44.

Herms, D. A., and D. G. McCullough. 2014. Emerald Ash Borer Invasion of North America: History, Biology, Ecology, Impacts, and Management. *Annual Review of Entomology* 59 (1): 13–30. DOI: 10.1146/annurev-ento-011613-162051

Kuo, E.-K. 2022. The eco bend steam-lam tree log. Now design spotlight. 2022. https://drivenxdesign.com/d100/project.asp?ID=16434

Lienhard, J., H. Alpermann, C. Gengnagel, and J. Knippers. 2013. Active Bending: A Review on Structures Where Bending Is Used as a Self-Formation Process. *International Journal of Space Structures* 28 (3–4): 187–96. DOI: 10.1260/0266-3511.28.3-4.187

Lienhard, J., S. Schleicher, J. Knippers, S. Poppinga, and T. Speck. 2010. Form-Finding of Nature Inspires Kinematics for Pliable Structures. In *Proceedings of the International Symposium of the International Association of Shell and Spatial Structures (IASS)*, 2545–54. Spatial Structures – Permanent and Temporary. Shanghai, China.

Lok, L., S. Zivkovic, and L. Spencer. 2023. UNLOG: A Deployable, Lightweight, and Bending-Active Timber Construction Method. *Technology|Architecture + Design* 7 (1): 95–108. https://doi.org/10.1080/24751448.2023.2176146.

Mansoori, M., N. Kalantar, T. Creasy, and Z. Rybkowski. 2019. Adaptive Wooden Architecture. Designing a Wood Composite with Shape-Memory Behavior. In *Digital Wood Design: Innovative Techniques of Representation in Architectural Design*, ed. F. Bianconi and M. Filippucci, 703–17.

Menges, A. 2011. Integrative Design Computation: Integrating Material Behaviour and Robotic Manufacturing Processes in Computational Design for Performative Wood Constructions. In *Proceedings of the 31st Annual Conference of the Association for Computer Aided Design in Architecture (ACADIA)*, 72–81. ACADIA 11: Integration through Computation. Banff, Alberta: ACADIA.

Persad, A. B, J. Siefer, R. Montan, S. Kirby, O. J. Rocha, M. E. Redding, C. M. Ranger, and A. W. Jones. 2013. Effects of Emerald Ash Borer Infestation on the Structure and Material Properties of Ash Trees. *Arboriculture and Urban Forestry* 39 (1): 11–16.

Preisinger, C., and M. Heimrath. 2014. Karamba—A Toolkit for Parametric Structural Design. *Structural Engineering International* 24 (May).

Robert McNeel & Associates. 2022. *Rhino3d.* Windows. Robert McNeel & Associates.

Rubin, H., and K. J. Schneider. 2002. *Baustatik: Theorie I. und II. Ordnung.* Werner-Ingenieur-Texte. Werner.

Rutten, D. 2022. Grasshopper. Windows. Robert McNeel & Associates. www.grasshopper3d.com/.

Sandaker, B. N., A. P. Eggen, and M. Cruvellier. 2011. *The Structural Basis of Architecture.* 2nd ed. New York: Routledge.

Schleicher, S., J. Lienhard, S. Poppinga, T. Masselter, T. Speck, and J. Knippers. 2011. Adaptive Façade Shading Systems Inspired by Natural Elastic Kinematics. In *Proceedings of the International Adaptive Architecture Conference IAAC*, 2545–54. London.

Schleicher, S., A. Rastetter, R. La Magna, A. Schönbrunner, N. Haberbosch, and J. Knippers. 2015. Form-Finding and Design Potentials of Bending-Active Plate Structures. In *Modelling Behavior*, ed. M. Ramsgaard Thomsen, M. Tamke, C. Gengnagel, B. Faircloth, and F. Scheurer, 53–63. Springer International Publishing. DOI: 10.1007/978-3-319-24208-8_5

Svilans, T., M. Tamke, M. Ramsgaard Thomsen, J. Runberger, K. Strehlke, and M. Antemann. 2019. New Workflows for Digital Timber. In *Digital Wood Design: Innovative Techniques of Representation in Architectural Design*, ed. F. Bianconi and M. Filippucci, 93–134. Cham: Springer International Publishing. DOI: 10.1007/978-3-030-03676-8_3

Tornabell, P., E. Nadal Soriano, and R. Sastre Sastre. 2014. Pliable Structures with Rigid Couplings for Parallel Leaf-Springs: A Pliable Timber Torus Pavilion. *WIT Transactions on the Built Environment* 136: 117–28.

USDA Forest Service. 2022. Emerald Ash Borer. Emerald Ash Borer Information Network. 2022. http://www.emeraldashborer.info/.

Dylan Wood, Laura Kiesewetter, Axel Körner, Kenryo Takahashi, Jan Knippers, Achim Menges

HYGROSHELL – In Situ Self-shaping of Curved Timber Shells

Abstract: Curved, surface-active, shell structures are known for material efficiency and slenderness but typically require complex manufacturing and formwork in combination with intricate on-site construction processes. The presented research proposes an alternative approach: a self-shaping building system for deploying lightweight, curved surface structures made from timber. The system uses the inherent hygromorphic properties of wood which naturally shrinks through drying. This anisotropic shape change is embedded into large-scale bilayer sheets – produced, machined, and shingle clad in a flat state with their later curved shape and connection detailing physically programmed within the material build-ups. When placed on-site, these sheets actuate through air drying to a final curved and interlocked geometry. Geometrically the structure is integratively designed from variable single curved surfaces using key material parameters (end grain angle and moisture content change) within a material stock, in relation to both the self-shaping and the final structural configuration. Each surface is modeled in the curved state using a board specific algorithmic calculation of curvature potential in parallel to a flat fabrication model. Emphasis is placed on investment in early-stage planning and intelligent material arrangement as a method to produce useful curvature. As a result, the curved shell shapes and interlocks without formwork or external mechanical force, with little onsite work. The outcome is a lightweight, long-span roof structure built from single curved wood surfaces with a thin cross-laminated build up. The project demonstrates a tangible new method of low impact, light touch self-construction and an ecologically effective use of material and geometry.

1 Introduction

1.1 Surface-active structures

The high ecological impact of the building industry urgently calls for the reduction of the carbon emissions in construction processes (United Nations Environment Programme 2022). In architecture and engineering this challenge requires innovation in integrating material and manufacturing with structure and geometry as well as development of new ways of working with bio-based, natural materials. Curved, surface-active structures, in which the geometry is used to reduce bending in favor of membrane action, offer elegant solutions for the construction of lightweight building structures (Engel 2013; Adriaenssens 2014). Historic and recent advances in engineering and design have

https://doi.org/10.1515/9783111162683-004

opened a range of applications for these types of geometries as long span, lightweight, material efficient roof structures. While designable and efficient in their final form, it is difficult to physically construct curved surface-active geometry. Most surfaces are made by first producing a laborious and material consuming formwork and support structure, or through complex three dimensional, high-precision manufacturing of the elemental parts (Sasaki 2014). Alternatively, many attempts have been made to design deployable structures that are pulled or lifted into shape, but most are made from linear elements that lack surface continuity and require secondary systems for enclosure (Lienhard et al. 2013; Burkhardt and Otto 1978; Rihaczek et al. 2022). Fundamentally, structures that transform face an inherent contradiction between weakening the material or surface for compliance and the required overall stability (Hoberman 2004). Similar conflicting constraints limit many "smart" stimuli responsive shape-changing materials to small scale applications, due to low actuation force, overall stiffness and availability (Tibbits and Cheung 2012).

1.2 Hygromorphic self-shaping

Many regenerative, natural, fibrous composite materials have the ability to change their shape passively based on external stimuli such as moisture. Wood is one of such materials, which exhibits hygromorphic self-shaping behavior, naturally shrinking and swelling anisotropically in response to changes in moisture (Hoadley 2000). Favorably for self-shaping, wood is also a viscoelastic material, its stiffness is lower when wet and increases during drying. When deployed in a bilayer cross-ply board configuration, the shrinking in one "active" layer of boards can be used to generate curved surfaces from flat sheets (Wood et al. 2018). Self-shaping wood has been implemented for predictably shaping bilayers through kiln drying for manufacturing large scale curved cross laminated timber (CLT) and for self-shaping flat pack furniture in ambient conditions (Grönquist et al. 2019; Aldinger et al. 2020; Bechert et al. 2021; Wood et al. 2020; Wood 2021; ICD - University of Stuttgart 2022).

1.3 Research objectives and implementation project

The current research implements self-shaping wood bilayers for on-site shaping of curved surface-active structures. Through this we develop a first of its kind approach to on-site construction, in which flat-pack components, including structure, cladding and connections, can be produced in thin wood layups – physically programmed to deform and interlock into to high curvature surface geometries. To achieve this, the research addresses two objectives. (1) How can specific material information be gathered and included in the computational design and fabrication process to generate structural curved geometries informed and enabled by the hygromorphic and mechanical qualities of wood? (2) How does the self-shaping method reveal a new type of building system for lightweight form active, deployable structures made from regenerative materials and without formwork or supporting structure? The research is investigated through the integrative design, development, and implementation of a research pavilion, the HygroShell (Fig. 1). The geometry of the HygroShell is comprised of three prefabricated, flat-packed components (packages), each containing two self-shaping bilayer surfaces (wings), connected along a shared lower edge (Fig. 2A–C). Through air drying the wings of each package transform from flat to curved. Collectively, the single curved wings interconnect – forming a roof supported at three points (Fig. 2D–F). The system is tested by prefabricating the packages and deploying them in a mock site to evaluate the functional and architectural potential at full scale.

Fig. 1: Interior view of the lightweight curved wood shell, constructed from intersected single curved surfaces.

2 Material driven computational design and fabrication

2.1 Programming self-shaping wood bilayers

Using a natural material such as wood in this context relies heavily on understanding the specific properties of the material and their relationship as both actuator and structure. A bilayer wood structure is used as the self-shaping mechanism and as the primary cross-ply buildup of the larger structure (Wood 2021). The bilayer consists of a thicker 'active' layer which serves as the hygromorphic actuator and a thinner 'restrictive' layer (Fig. 2A). As the moisture content (MC) of the active layer changes the mechanics of the bilayer cause the surface to curve. Curvature in the bilayer can be designed through controlling key parameters: the thickness ratio of the layers, orientation of the wood's fibrous structure in the end grain, described by the angle of the radial and tangential axis in relation to the board (R/T angle), and the change in MC. Material thickness can be defined precisely during the sawing process. The R/T angle in each board is defined by a combination of the sawing pattern, the concentric circular morphology of the log, and natural variation in the growth rings (Fig. 3A–B). When harvested, wood contains a high MC as a result of its living functions and starts to dry out as it equalizes with the relative humidity of the surrounding environment. Thus, the MC can be controlled by defining the point in the natural drying process at which the bilayers are constructed in relation to the desired service conditions. Methods for designing and predicting curvature in bilayers work well using generalized mechanical characteristics from literature and measurement, mainly material stiffness, and the coefficient of expansion but do not account more specifically for variation in the MC

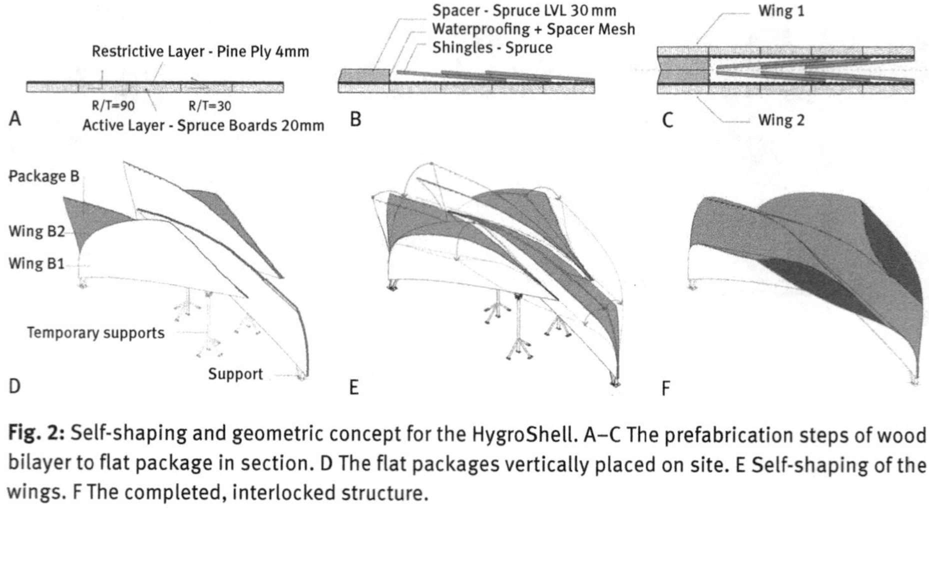

Fig. 2: Self-shaping and geometric concept for the HygroShell. A–C The prefabrication steps of wood bilayer to flat package in section. D The flat packages vertically placed on site. E Self-shaping of the wings. F The completed, interlocked structure.

and R/T angle per board in larger layouts (Rueggeberg and Burgert 2015; Grönquist et al. 2019). Given the natural variation in these parameters, the geometric usefulness of the models is directly limited by the coarseness of the available material input data.

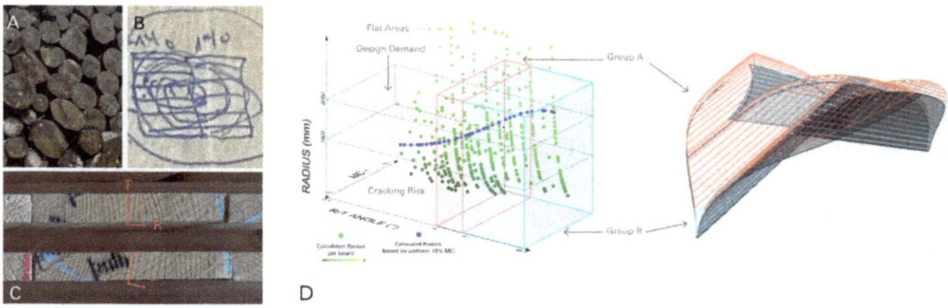

Fig. 3: The data driven approach to utilizing material and curvature. A-C Practical gathering of data in consultation with the existing resource stream. D The graph shows the calculated curvature potentials of each board at delivery (green), the ideal curvatures without variation in MC (dark blue), the global design demand (green box) derived from the ideal 3D design geometry. The sorting groups (red and blue) and optimized distribution of the groups in relation to the 3D design.

2.2 A practical data-driven approach relating material to curvature design and global geometry

As the basis for curvature design, a known analytical model for predicting curvature in wood bilayers is used (Rueggeberg and Burgert 2015). Rather than targeting one specific bilayer configuration and generalizing the natural material properties, a data driven approach was used to simultaneously understand the variation in stock material and its relation to variable curvature in the surface design. To evaluate the feasible surface curvatures, a design space was established using literature values for locally available materials and an estimated range of variation in the R/T and MC parameters. This design space was used to study architectural and structural aspects of the global design configurations, testing different wood species and build ups in relation to geometry. Working with a local sawmill, the species and distribution of R/T angles was refined in relation to a standard sawing pattern that was economical, materially efficient, and fast (Fig. 3C). At this stage, the thickness ratio, specifically, the thickness of the active layer, was fixed and the design space refined. The logs were selected, and the boards cut in 16 minutes, just in time, a week before production (empirical research shows speed is a critical parameter in working with partners from the timber industry).

2.3 Optimizing real word material and geometry

With the thickness ratio selected, the curvature potentials associated with each board depended on the combination of their R/T angle and MC. The R/T angles were not adjustable after cutting, while the MC slowly reduces and can be modulated by the storage conditions. Upon delivery, board specific data was gathered using manual measurements of the end grains and a moisture meter. Due to a high variation in the MC, the boards were stored in a climate chamber at 85 % relative humidity to equalize towards 18 % MC. The MC and R/T angle data was fed into the computational design model to calculate the curvature potential for each board and then compared with the curvature demand from the design geometry (Fig. 3D). Given the variation in MC and the physical space required for sorting, an initial grouping was required in place of board specific arrangement. Boards with low MC and or low curvature potential were sorted out for use in the flat connection areas. Similarly, boards with high MC and extreme tangential orientations were removed due to risk of cracking and used for transport packaging. The remaining boards were sorted into two groups – each representing two target curvatures from the design. A multi-objective optimization algorithm, RBFMOpt, was used to calculate the distribution of the two curvatures in relation to the design geometry while respecting the quantity and range of boards in each group (Wortmann 2017). This step determines the number boards from each group in each wing, and the boards for each wing are chosen at random from the groups to ensure an even distribution of variation. A final smoothing sequence assignment within the wing was made just before fabrication based on a final curvature prediction of each board at the time of the 2D layout (Figure 4A-B). Key to this process was that the global 3D design can be digitally updated and verified throughout the process, ensuring that the specific arrangement was within an acceptable range for the design and structural performance.

3 Development of prefabricated self-shaping wood building components

3.1 Prefabrication

The base elements of the building system are the bilayer wood plates (ca. 10 m × 3 m) of each wing, The wings are manufactured flat by laminating the 'restrictive' layer of 4 mm 3 ply pine veneer plywood (ZEG Zentraleinkauf Holz + Kunststoff eG) with a layer of 'active', 20 mm spruce wood boards with an high MC (Sägewerk Kolb GmbH) using a PUR adhesive (HBS 709, Henkel AG) (Fig. 6A). With a width of 140 mm the boards offer a fine resolution to grade curvature in the active layer. After lamination the bilayer surface is immediately machined on the worktable in the flat position using a 7-axis

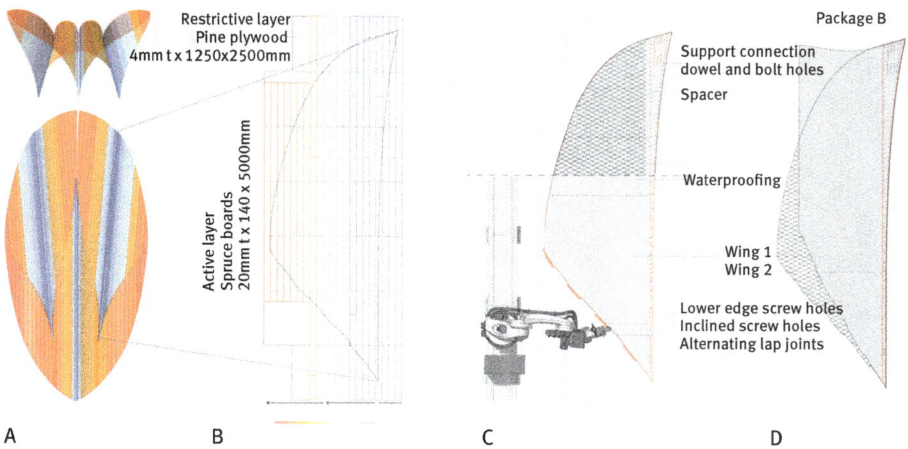

Fig. 4: Prefabrication steps. A, B Gradient in curvature on group distribution (blue/red) and sequence smoothing. B Bilayer lamination. C Machining of edge details, adding of waterproofing and wood shingles. D Completed flat package.

milling setup. The connections along the curved wing to wing intersections use an elongated finger joint with sheer blocks in each finger to compensate for tolerances in the lateral direction. The geometry of the curved connection is calculated from the 3D, curved geometry and translated to the flat geometry, considering the change in angle during shrinkage. Predrilled, angled screw holes are placed along the connection edges and shared lower edges. A waterproofing layer including a spacer mesh (Dörken Delta Trela-plus) is added to the bilayer followed by spruce wood shingles (Schindelzentrum Allgäu). The triangular tipped shingles, common for traditional buildings in South Germany, are applied along the ruling lines (zero curvature), allowing the self-shaping while the angled edges allow for directional water drainage. Sets of two wings are then connected along the lower edge with a softwood LVL spacer forming a package. The packages were wrapped to prevent further drying and, thus, shaping, nested in a single truck and transported to the simulated building site.

3.2 In Situ shaping through air drying

On site the packages were positioned vertically with one end on the ground support and one resting on an elevated temporary support. Once all three packages were in place, the wrapping was removed from the active layer and the shaping began (Fig. 5A). Iterative 3D laser scanning (Leica BLK360 G1) and inertial measurement unit (IMU) sensors embedded in the tip of each wing were used to monitor the curvature of the structure over time. Over 96 hours the components self-shaped in the unconditioned but enclosed environment of ca. 40–65 % RH, decreasing in MC ca. 8–10 % through air drying (Fig. 5).

Fig. 5: In situ self-shaping of the bilayer timber building components. Starting flat in a vertical configuration with high MC in the active layer, during shaping, and after full shaping and interlocking.

As the intended curvatures for each wing approached, the longitudinal connections were nudged into alignment and shear blocks and screws were installed. In parallel, two smaller edge components were self-shaped on the ground, lifted into place and attached to the structure connecting the open edges. Lastly, a 4 mm plywood "locking" layer was screw press laminated to the interior surface locking the curvature and stabilizing the structure from further deformation.

4 Geometric synthesis of self-shaping, structure, and architecture

While the self-shaping system can produce high curvature and is valuable for on-site assembly it is constrained geometrically to single curved surfaces and structurally to thin cross sections. The challenge structurally involved designing a geometric arrangement that functions structurally during shaping and, more critically, performs effectively in its final form. To address shaping stability, the packages were oriented vertically so as the self-shaping is assisted, rather than hindered by the self-weight as the wings cantilever out. Additionally, the partially symmetric packages balance along their shared axis reducing load on the supports during shaping. This stabilizes the larger thin sheets during transformation and favorably loads the asymmetric cross section of the bilayer with the outer restrictive layer in tension along the longitudinal fiber direction. The shaped structural geometry was designed through calculated intersection of the single curved surfaces of each package (Fig. 6). Where the surfaces intersect, one curved plate provides out-of-plane support to the adjacent plate. This intersection, combined with the resulting curved seam, effectively restricts flexural deformation of the surface and ensures sound structural behavior. The angle, pitch and spacing of the intersections was modeled parametrically using Rhinoceros and Grasshopper (Robert McNeel & Associates) and the structure studied extensively using

Finite Element Modeling (SOFiSTiK) throughout the design and manufacturing process ensuring its relation to the material and producible curvatures. Architecturally, the geometric system was designed to cover a wide spanning area with the least amount of material. Using a balanced tripod design on hinged supports maximizes the coverage area and reduces the need for bounding supports or walls. This feature was further enabled by the gradient curvature allowing the surfaces to flatten and act as columns as they approach the ground (increasing head height) while smoothly transitioning to curved overhead (increasing the span within the limited sheet width) (Fig. 7).

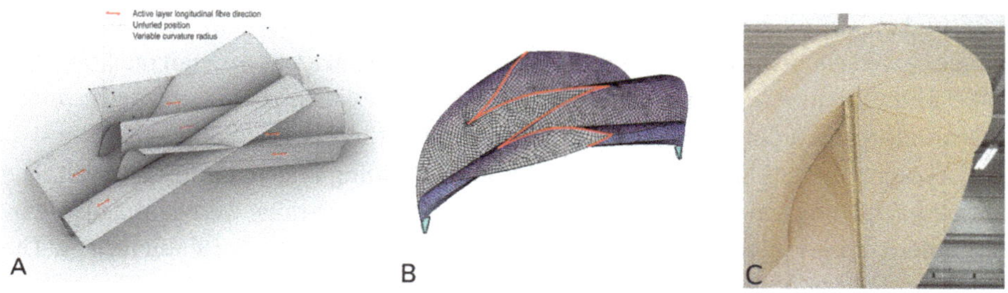

Fig. 6: Geometry of the surface-active structure. A Generation of connections from single curved surfaces of the unfurling packages. B FE model with curved connection seams between wings. C Hidden curved connection from interior.

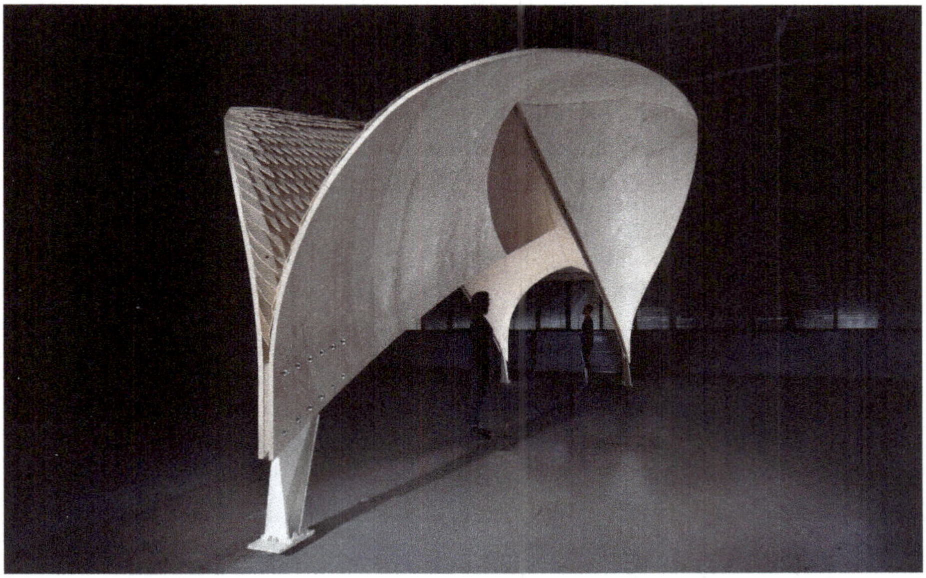

Fig. 7: The completed HygroShell.

5 Results and discussion

Through the presented integrative, material driven design approach a single curved roof structure spanning 9.5 m longitudinally, 2.8 m between the transverse supports and 4.5 m at the widest cross section is achieved using a 28 mm laminate made from medium grade soft wood. Wood is used as the primary material, including structure (2.58 cbm) and the cladding (0.98 cbm) with the majority sourced directly from a local sawmill. Covering an area of 40 m^2 the building system including cladding weights 39 kg/m^2 of floor area (28 kg/m^2 structural) and 18 kg/m^2 of shell surface (13 kg/m^2 structural), required only three temporary point supports for assembly. Variable curvature in the single curved elements was achieved transitioning from flat areas through radii between 1400 and 2100 mm. The designed curvature gradient was possible, through careful distribution and ordering of the natural variation in the R/T grain angles.

As a prefabrication method, the self-shaping building system performed well. The shaping itself happened relatively fast and geometric deviations were manageable by alignment of curved connection between wings and designed lateral tolerances. The primary shaping was achieved autonomously through air drying. Adding functional layers to the bilayers for weather protection and cladding in the flat state replaces on-site construction labor and assembly steps that are difficult on curved geometries. This approach has the advantage of finished modular prefabrication with the reduced complexity and lower transport volumes of panel-based prefabrication. Compared to even the lightest curved formwork systems, the unfurling self-shaping package concept requires no formwork, no heavy boundary conditions on site and no mechanical or pneumatic actuators, which are time and material intensive (Popescu et al. 2021). In a true exterior environment, actuation conditions would vary more and be harder to predict, but the vertical orientations of the components could protect from the most direct weather. It is further planned to study the long-term durability and form stability of the structure in outdoor conditions and further development will focus on including additional on-site steps, such as the locking layer application, as part of the self-shaping packages.

6 Conclusions and future potentials

The research investigates specific design challenges associated with prefabricated self-shaping building systems which are overcome through integrated exploration of material, geometry, and structure. The resulting form is enabled by in-depth material understanding and by utilizing curvature, material anisotropy, and hygromorphic actuation, showcasing a path towards a new class of lightweight, sustainably produced, surface-active structures. Considering the negative ecological and social impacts of on-site construction, as well as costs and complexity of on-site assembly, future forms of self-shaping system have a high potential impact through their simple embedded logics and on-site autonomy. The approach of programing geometry though material demonstrates how investment in the integration of material intrinsic properties of a low-tech and low-resolution material system can lead to an incredible reduction in complexity throughout the construction process. By learning to work with the, forces, forms, and intricacies of natural materials we chart a path towards intriguing new architectures derived computationally from regenerative and sustainable resources.

Acknowledgements

The research was partially supported by the German Research Foundation under Germany's Excellence Strategy – EXC 2120/1 – 390831618 and the BBSR Forschungsinitiative Zukunft Bau - Self-shaping cylindrical wood components for lightweight sustainable construction - 10.08.18.7-22.07. The project was developed within the Integrative Technologies and Architectural Design Research (ITECH) MSc. Program – Active Wood Morphologies design studio with students: Andre Aymonod, Wai Man Chau, Min Deng, Fabian Eidner, Maxime Fouillat, Hussamaldeen Gomaa, Yara Karazi, Arindam Katoch, Oliver Moldow, Ioannis Moutevelis, Xi Peng, Yuxin Qiu, Alexander Reiner, Sarvenaz Sardari, Edgar Schefer, Selin Sevim, Ali Shokri, Sai Praneeth Singu, Xin Sun, Ivana Trifunovic, Alina Turean, Aaron Wagner, Chia-Yen Wu, Weiqi Xie, Shuangying Xu, Esra Yaman, and Pengfei Zhang, with generous support from Henkel AG, Kolb Sagewerk, Scantronic, Brookhuis Technologies, BEC GmbH, and colleagues, Markus Rüggeberg, Philippe Grönquist, Simon Bechert, Michael Schneider, Michael Preisack, Sergej Klassen, Hendrik Köhler, Dennis Bartl, Katja Rinderspacher, Jakob Wagner, and Gabriel Kerekes.

References

Adriaenssens, S., P. Block, D. Veenendaal, and C. Williams, eds. 2014. *Shell Structures for Architecture: Form Finding and Optimization*. London: Routledge.

Aldinger, L., S. Bechert, D. Wood, J. Knippers, and A. Menges. 2020. Design and Structural Modelling of Surface-Active Timber Structures Made from Curved CLT – Urbach Tower, Remstal Gartenschau 2019. In *Impact: Design with All Senses*. Vol. 30, ed. Ch. Gengnagel, O. Baverel, J. Burry, M. Ramsgaard Thomsen, and S. Weinzierl, 419–32. Cham: Springer International Publishing. DOI: 10.1007/978-3-030-29829-6_61

Bechert, S., D. Sonntag, L. Aldinger, and J. Knippers. 2021. Integrative Structural Design and Engineering Methods for Segmented Timber Shells – BUGA Wood Pavilion. *Structures* 34: 4814-33. DOI: 10.1016/j.istruc.2021.10.032.

Engel, H. 2013. *Tragsysteme: Structure Systems*. 5. ed. Ostfildern: Hatje Cantz.

Grönquist, P., D. Wood, M. M. Hassani, F. K. Wittel, A. Menges, and M. Rüggeberg. 2019. Analysis of Hygroscopic Self-Shaping Wood at Large Scale for Curved Mass Timber Structures. *Science Advances* 5 (9). DOI: 10.1126/sciadv.aax1311.

Hoadley, R. B. 2000. *Understanding Wood: A Craftsman's Guide to Wood Technology* [Rev. ed.]. Newtown, Conn. Taunton Press.

Hoberman, Ch. 2004. Transformation in Architecture and Design. In *Transportable Environments 3: Proceedings of the 3rd International Conference on Portable Architecture and Design,* April 28-30, 2004. Toronto, Ont. Ryerson University.

ICD – University of Stuttgart. 2022. *HygroShape: Self-Shaping Wood Furniture.* https://www.icd.uni-stuttgart.de/projects/hygroshape/.

Popescu, M., M. Rippmann, A. Liew, L. Reiter, R. J. Flatt, T. van Mele, and P. Block. 2021. Structural Design, Digital Fabrication and Construction of the Cable-Net and Knitted Formwork of the Knit-Candela Concrete Shell. *Structures* 31:1287–99. DOI: 10.1016/j.istruc.2020.02.0130

Rueggeberg, M., and I. Burgert. 2015. Bio-Inspired Wooden Actuators for Large Scale Applications. *PloS one* 10 (3): e0120718. DOI: 10.1371/journal.pone.0120718

Sasaki, M. 2014. Strcutural Design of of Free-Curved RC Shells: An Overview of Built Works. In *Shell Structures for Architecture: Form Finding and Optimization,* Ed. S. Adriaenssens. London: Routledge.

Tibbits, S., and K. Cheung. 2012. Programmable Materials for Architectural Assembly and Automation. *Assembly Automation* 32 (3): 216–25. https://doi.org/10.1108/01445151211244348.

United Nations Environment Programme. 2022. 2022 Global Status Report for Buildings and Construction. Towards a Zero-Emissions, Efficient and Resilient Buildings and Construction Sector.

Wood, D. 2021. Material Programming for Fabrication – Integrative Computational Design for Self-Shaping Curved Wood Building Components in Architecture. Dr.-Ing., Institute for Computational Design and Construction, University of Stuttgart. http://dx.doi.org/10.18419/opus-11968.

Wood, D., P. Grönquist, S. Bechert, L. Aldinger, D. Riggenbach, K. Lehmann, M. Rrüggeberg, I. Burgert, J. Knippers, and A. Menges. 2020. From Machine Control to Material Programming: Self-Shaping Wood Manufacturing of a High Performance Curved CLT Structure – Urbach Tower. In *FABRICATE 2020: Making Resilient Architecture,* Eds. J. Burry, J. Sabin, B. Sheil, and M. Skavara, 50-57: UCL Press. DOI: 10.2307/j.ctv13xpsvw.11

Wood, D., C. Vailati, A. Menges, and M. Rüggeberg. 2018. Hygroscopically Actuated Wood Elements for Weather Responsive and Self-Forming Building Parts – Facilitating Upscaling and Complex Shape Changes. *Construction and Building Materials* 165:782–91. https://doi.org/10.1016/j.conbuildmat.2017.12.134.

Wortmann, T. 2017. Opossum – Introducing and Evaluating a Model-Based Optimization Tool for Grasshopper. In *Protocols, Flows, and Glitches – Proceedings of the 22nd CAADRIA Conference,* Eds. P. Janssen, P. Loh, A. Raonic, and M. A. Schnabel, 283–92: CAADRIA.

Mohammad Mansouri, Aly Abdelmagid, Zlata Tošić, Marta Orszt, Ahmed Elshafei

Corresponding Principal and Asymptotic Patches for Negatively-Curved Gridshell Designs

Abstract: A design model for negatively-curved gridshells whose primary and secondary structures are corresponding principal and asymptotic patches is presented. This correspondence is realized on minimal surfaces with their adjoint transform and on constant negative Gaussian curvature surfaces with their Bäcklund transforms. Based on the design model, we introduce a gridshell prototype utilizing the correspondence.

1 Introduction

It is known that principal patches have the advantage of giving rise to developable beams (ribs) as well as being the smooth continuous analogue of circular planar quadrilateral meshes cf. (Liu et al. 2006), (Abdelmagid et al. 2022) and (Bobenko and Tsarev 2007). Asymptotic patches, on the other hand, have the advantage of generating ribs that unroll (albeit with some deformation) to almost straight bands (Jiang et al. 2020). Thus having a viable model to associate them is beneficial for design purposes, more precisely that the two are diagonal to one another (in domains). From another side, it is also known that structural bracing in the sense of two intersecting grids defining a primary and a secondary structure is essential in gridshell design. In the article, we present a design model consisting of three degrees of design freedom through which the designer can generate a wide variety of morphological solutions, all of which maintain the correspondence between principal-asymptotic patches. Finally, we present a gridshell prototype whose primary and secondary structures are from the corresponding principal and asymptotic patches.

2 Geometry

Let us recall some facts from differential geometry cf. (Eisenhart 1909) and (Gray et al. 2006). In all that follows, we denote by $X(u, v)$ a parameterization patch on a smooth negatively-curved surface S in three-dimensional space $\mathbb{R}^3$ equipped with the Euclidean scalar product $\langle , \rangle$. We denote by $X_u, X_v, X_{uu}, \dots$ the partial derivatives and

https://doi.org/10.1515/9783111162683-005

by N the normal to S. The fundamental coefficients are given by

$$\begin{cases} F = \langle X_u, X_v \rangle \quad , \quad E = \langle X_u, X_u \rangle \quad , \quad G = \langle X_v, X_v \rangle \\ f = \langle X_{uv}, N \rangle \quad , \quad e = \langle X_{uu}, N \rangle \quad , \quad g = \langle X_{vv}, N \rangle \,. \end{cases} \tag{1}$$

2.1 Principal and asymptotic patches

We recall that a curve in a surface S is said to be principal if it is always tangent to a direction of principal curvature of S and is asymptotic if it has a vanishing normal curvature everywhere. Observe that the normals N along a principal curve form a developable strip, as seen in Fig. 1a. However, the normals along an asymptotic curve form a strip that unrolls (with deformation) to an almost straight band, as seen in Fig. 1b. Furthermore, a patch $X(u, v)$ on S is called principal (or orthogonal conjugate (OC)) if its coordinate lines are principal curves while a patch $Y(x, y)$ on S is asymptotic (or AS) if its coordinate lines are asymptotic curves. OC-patches are advantageous in giving rise to planar panels and orthogonal nodes.

$$\text{OC-patch:} \quad F_X = 0, f_X = 0 \tag{2}$$

$$\text{AS-patch:} \quad e_Y = 0, g_Y = 0. \tag{3}$$

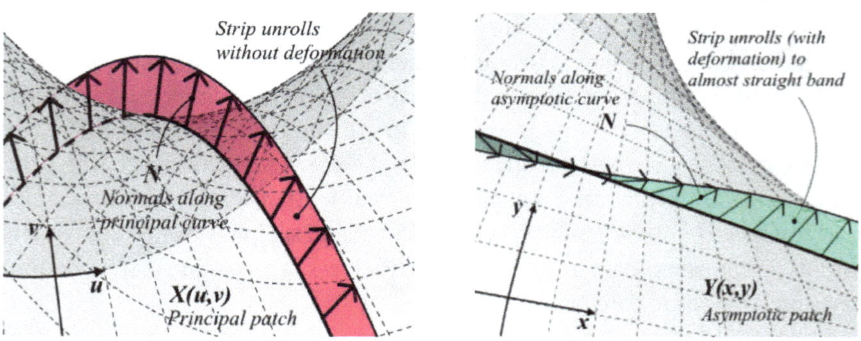

(a) (b)

Fig. 1: Behavior of normal strips along principal and asymptotic curves: (a) normals along principal, (b) normals along asymptotic.

2.2 Corresponding principal and asymptotic patches

We define the notion of correspondence between patches $X(u, v)$ and $Y(x, y)$ on the negatively-curved surface S to be given by (the linear change of coordinates) reparameterizations of the form

$$X(x+y, x-y) = Y(x,y) \quad \longleftrightarrow \quad X(u,v) = Y\left(\frac{u+v}{2}, \frac{u-v}{2}\right). \tag{4}$$

This construction establishes that the coordinate lines of the two patches are diagonal to one another in the parameter space, hence intersecting on S at exact points (corresponding to square lattices in the domains), as seen in Fig. 2. Observe that the reparameterization (4) creates a correspondence in the above sense between the following patches (described by their fundamental coefficients)

(C1) Y Iso-speed $(E_Y = G_Y)$ $\longleftrightarrow$ X Orthogonal $(F_X = 0)$

(C2) Y Conformal $(F_Y = 0, E_Y = G_Y)$ $\longleftrightarrow$ X Conformal $(F_X = 0, E_X = G_X)$

(C3) Y Tcheb-1 $(E_Y = G_Y = c)$ $\longleftrightarrow$ X Tcheb-2 $(F_X = 0, E_X + G_X = c)$

(C4) Y Asymptotic $(e_Y = g_Y = 0)$ $\longleftrightarrow$ X Iso-conjugate $(f_X = 0, e_X = -g_X)$

where c is a constant. There follows that combining any of the correspondences $C1, C2, C3$ with $C4$ yields a correspondence between principal and asymptotic networks on S, a property that in general is not true. More precisely, we have

$$\begin{array}{rcl} Y(x,y) & \longleftrightarrow & X(u,v) \\ \text{Iso-speed Asymptotic} & \longleftrightarrow & \text{Orthogonal Iso-conjugate.} \end{array}$$

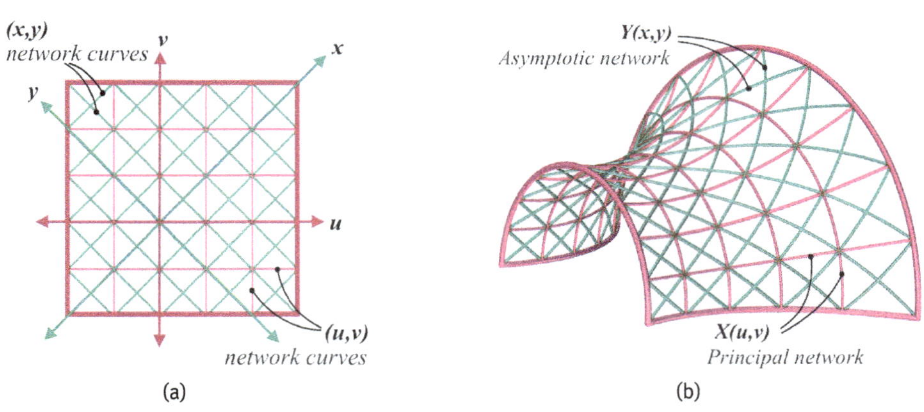

Fig. 2: Correspondence of principal and asymptotic networks: (a) intersecting in domain, (b) intersecting in surface.

2.3 Surfaces admitting principal-asymptotic correspondence

After defining the conditions for the correspondence, the following two realizations illustrate the combining of one the correspondences $C1, C2, C3$ with the correspondence $C4$.

Minimal Surfaces (MS) (C2 + C4)

The first type is the minimal surface (MS). The idea is to construct conformal iso-conjugate MS patches X starting from a conformal parameterization N of the sphere. We observe that when X is a conformal MS patch, its normal N is also conformal and it satisfies $e_X = -g_X$. Hence, the Codazzi equations reduce to the Cauchy-Riemann equations

$$\begin{cases} e_v = f_u \\ e_u = -f_v \end{cases} \quad \text{equivalently} \quad f_{uu} + f_{vv} = 0. \tag{5}$$

A general solution is given by two functions $\phi(u + iv)$ and $\psi(u - iv)$ such that

$$\begin{cases} f = \phi + \psi \\ e = -i(\phi - \psi) + c \end{cases}$$

where c is a constant and the quantity e is real when ϕ and ψ are complex conjugates. Thus, to each ϕ, ψ corresponds a conformal iso-conjugate MS patch X satisfying the differential system

$$X_u = -\frac{eN_u + fN_v}{\widetilde{\Lambda}}, \quad X_v = -\frac{fN_u - eN_v}{\widetilde{\Lambda}} \tag{6}$$

where $\widetilde{\Lambda}$ is the conformal factor of N. Precisely, if we take f equals 0 and e equals 1, then a solution $X(u, v)$ to Eq. (6) is conformal iso-conjugate, as seen in Fig. 3a and its corresponding reparameterization using (4) yields a conformal asymptotic patch $Y(x, y)$, as seen in Fig. 3b.

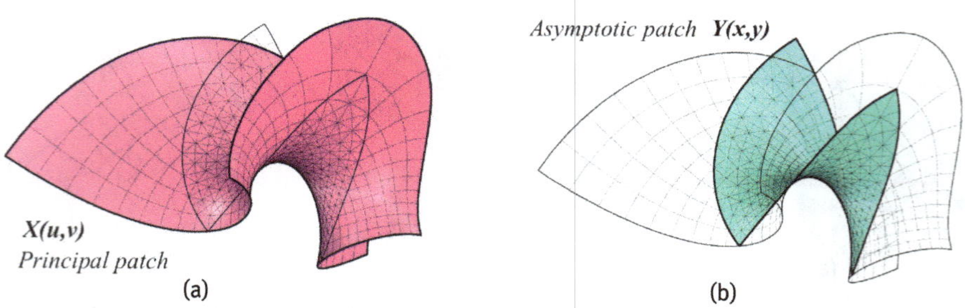

Fig. 3: Principal-Asymptotic correspondence realized on MS: (a) principal patch on MS, (b) asymptotic patch on MS.

Constant negative Gaussian curvature (CGC) surfaces (C3 + C4)

The second type of principal-asymptotic correspondence is in the context of surfaces of negative CGC. Recall that a surface S with negative CGC equals to $-1/\rho^2$ for $\rho > 0$

admits a Tcheb-2 iso-conjugate patch $X(u, v)$ characterized by

$$\begin{cases} F_X = 0\,, & E_X + G_X = (\rho\cos\theta)^2 + (\rho\sin\theta)^2 = \rho^2 \\ f_X = 0\,, & e_X = -g_X = \pm\rho\sin\theta\cos\theta \end{cases}$$

with angle function $\theta(u, v)$ satisfying the Sine-Gordon-2 equation

$$\theta_{uu} - \theta_{vv} = \sin\theta\cos\theta. \tag{7}$$

The function θ can be understood geometrically as half the angle ω between the asymptotic directions, and the Sine-Gordon equation (7) arises directly from using the expressions of the fundamental coefficients in the Gauss-Codazzi equations. Therefore, the corresponding reparameterization using Eq. (2) thus yields a Tcheb-1 asymptotic patch $Y(x, y)$ characterized by

$$\begin{cases} F_Y = \rho^2\cos\omega \;, & E_Y = G_Y = \rho^2 \\ f_Y = \mp\rho\sin\omega \;, & e_Y = g_Y = 0 \end{cases}$$

with angle function $\omega(x, y)$ satisfying the Sine-Gordon-1 equation

$$\omega_{xy} = \sin\omega. \tag{8}$$

We focus, in particular, on the Pseudo-Sphere (PS), Dini surface, and Kuen surface as seen in Fig. 4. The explicit parameterizations $X(u, v)$ of the surface in question are obtained by Bäcklund transformations explained in the next section.

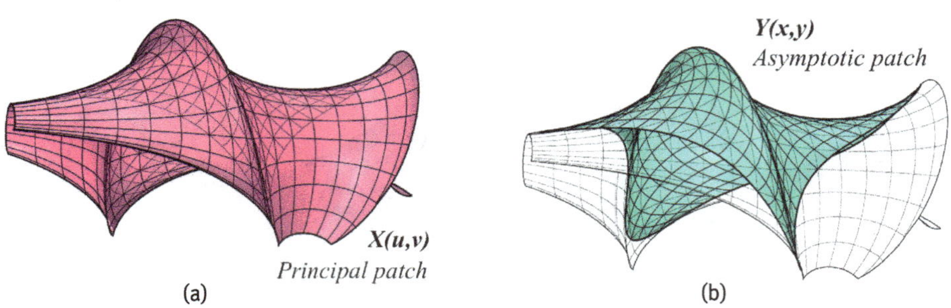

Fig. 4: Principal-Asymptotic correspondence realized on negative CGC: (a) principal patch on CGC, (b) asymptotic patch on CGC.

2.4 Transformations preserving the correspondence

Next, we present surface transformations that preserve the principal-asymptotic correspondence. For that, we need the transformation to preserve the orthogonality and iso-conjugacy of $X(u, v)$ (or equivalently the iso-speed and asymptotics of $Y(x, y)$), in the types of surfaces specified above. In this context, we present transformations only for surface type B (MS) namely, the adjoint transformation, and for surface type C (Negative CGC) namely, the Bäcklund transformation.

Adjoint transformation

To understand this transformation, we consider the Eqs. (6). We saw above that we chose f to be 0 and e to be 1, reducing the system to Eqs. (9) and obtaining a solution $X(u, v)$ which is a conformal iso-conjugate MS patch. Now, if we instead take f to be 1 and e to be 0, reducing the system to Eqs. (10) and thus obtaining a solution $X^*(u, v)$ which is a conformal asymptotic MS patch.

$$X_u = -\frac{N_u}{\widetilde{\Lambda}}, \quad X_v = \frac{N_v}{\widetilde{\Lambda}} \tag{9}$$

$$X_u^* = -\frac{N_v}{\widetilde{\Lambda}}, \quad X_v^* = -\frac{N_u}{\widetilde{\Lambda}}. \tag{10}$$

It can be seen that X and X^* are related by with Cauchy-Riemann equations

$$X_u = X_v^*, \quad X_v = -X_u^*. \tag{11}$$

thus are called adjoints of one another, as seen in Fig. 5. Finally, the conformal iso-conjugate patch $Y^*(x, y)$ is adjoint to the conformal asymptotic patch $Y(x, y)$.

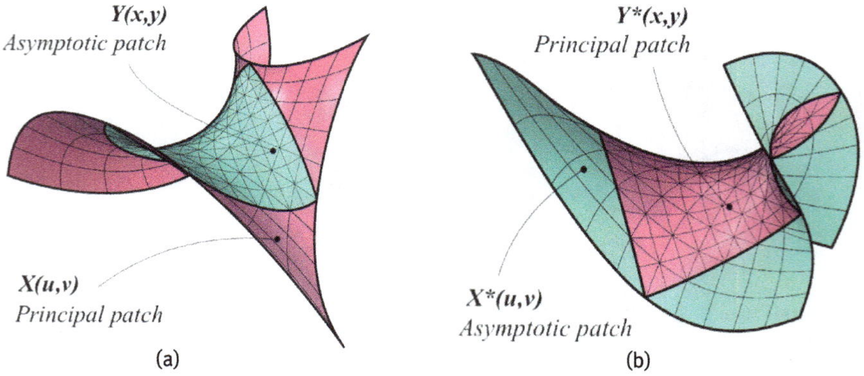

Fig. 5: MS surface and its adjoint transform MS: (a) X principal and Y asymptotic, (b) X^* asymptotic and Y^* principal.

Bäcklund transformation

The Bäcklund transform of a surface S with negative CGC $-1/\rho^2$ for $\rho > 0$ is a surface S^* of the same negative CGC defined as the locus of points drawn by moving a line L over the points of S. This line L is tangent to both surfaces and is of constant length $\rho \cos \sigma$, for some real constant σ. Moreover, the normals to S and S^* make a constant angle together of $(\pi/2 - \sigma)$. In particular, when σ is 0, the surface S^* is said to be a Bianchi transform, as seen in Fig. 6. Now, if S of negative CGC equals $-1/\rho^2$ defined by a Tcheb-2 iso-conjugate patch $X(u, v)$ of angle function $\theta(u, v)$, its Bäcklund transformation is given by

$$X^*(u, v) = X(u, v) + \left(\frac{\cos \sigma \cos \theta^*}{\cos \theta}\right) X_u + \left(\frac{\cos \sigma \sin \theta^*}{\sin \theta}\right) X_v \tag{12}$$

where the angle function $\theta^*(u, v)$ satisfying the Bäcklund-Darboux system

$$\begin{cases} \theta^*_u + \theta_v = \sec \sigma (\sin \theta^* \cos \theta + \sin \sigma \cos \theta^* \sin \theta) \\ \theta^*_v + \theta_u = -\sec \sigma (\cos \theta^* \sin \theta + \sin \sigma \sin \theta^* \cos \theta)\,. \end{cases} \tag{13}$$

In this context, we give the explicit Tcheb-2 iso-conjugate patches $X(u, v)$ for the PS, Dini, and Kuen surfaces as Bäcklund (and Bianchi) transformations. More precisely, the PS is obtained as a Bianchi transformation of the vertical line (which can be seen as a degenerate surface), and the Dini surface as a Bäcklund transformation of the vertical line. Finally, the Kuen surface is given as a Bianchi transformation of the PS, note that all these surfaces have negative CGC $-1/\rho^2$.

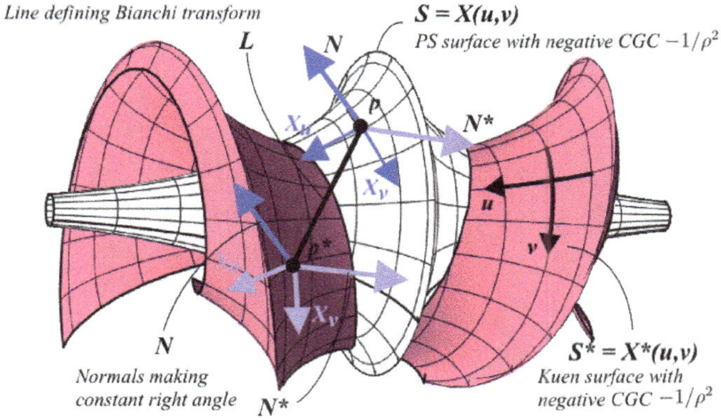

Fig. 6: Bäcklund (in fact Bianchi) transform of PS to Kuen surface.

3 Design Model

Now we explain how the geometric constructions described above are reinterpreted as three Degrees of Design Freedom (DF) constituting the design model

$$\begin{cases} \textbf{DF-1:} \text{ Surface type} \\ \textbf{DF-2:} \text{ Subtypes} \\ \textbf{DF-3:} \text{ Variations and transformations.} \end{cases}$$

DF-1 is the choice of one of two surface types explained above (A) Minimal surface or B) Negative CGC). Once a surface type is chosen, DF-2 generates subtypes through different ways of constructing the surface type in question. Finally, DF-3 provides variations and transformations by manipulating the different parameters in their equations. Note that DF-3 preserves the surface type and the correspondence. In the following part, we show the generation of two subtypes for surface type A (MS) and one subtype for surface type B (Negative CGC).

Surface type A (MS)

Subtype A01

Conformal N obtained by composing the holomorphic function $f(z) = (az + b)/(cz + d)$ where a, b, c, d are reals such that $ad \neq bc$, with the inverse of the stereographic projection, the variations of the resulting subtype (Fig. 7) is obtained by manipulating the parameters a, b, c, d in:

$$\frac{1}{6(ad - bc)} \begin{pmatrix} 3b^2u - 3d^2u + (a^2 - c^2)u(u^2 - 3v^2) + 3ab(u^2 - v^2) + 3cd(v^2 - u^2) \\ v(3b^2 + 3d^2 + 6abu + 6cdu + (a^2 + c^2)(3u^2 - v^2)) \\ -6bdu - 2acu(u^2 - 3v^2) - 3bc(u^2 - v^2) - 3ad(u^2 - v^2) \end{pmatrix}$$

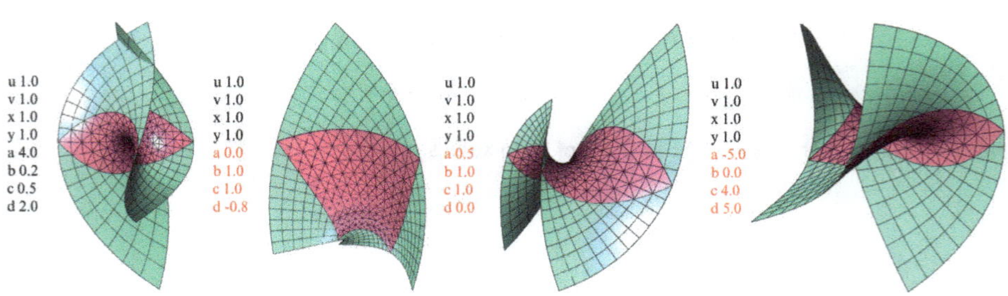

Fig. 7: Variations of Subtype A01.

The transformation (preserving the surface type and the OC-AS correspondence) is provided by the expression for the adjoint MS patches X^*, Y^* explained above.

Subtype A02

Conformal N obtained by composing the holomorphic function $f(z) = \alpha z^2$ where $\alpha = a + ib$, with the inverse of the stereographic projection, the variations of the subtype (Fig. 8) are obtained by manipulating the parameters a, b in:

$$\frac{1}{16(a^2+b^2)}\begin{pmatrix} (a^2+b^2)\left(4buv(v^2-u^2)+a(u^4-6u^2v^2+v^4)\right)-4b\arctan\left(\frac{v}{u}\right)-2a\log(u^2+v^2) \\ (a^2+b^2)\left(4auv(u^2-v^2)+b(u^4-6u^2v^2+v^4)\right)+4a\arctan\left(\frac{v}{u}\right)-2b\log(u^2+v^2) \\ 4(v^2-u^2)(a^2+b^2) \end{pmatrix}$$

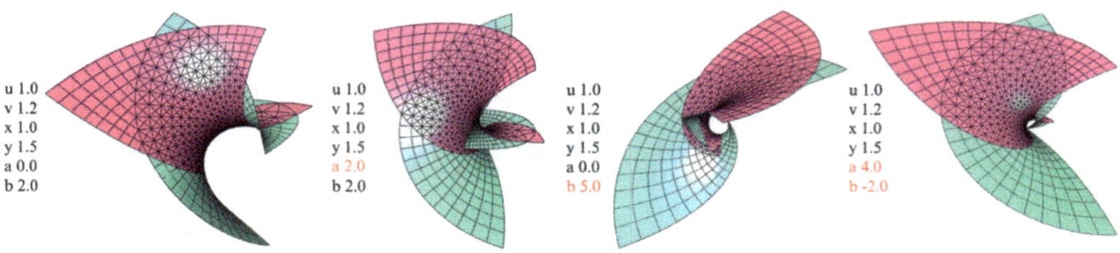

Fig. 8: Variations of subtype A02.

Surface type B (Negative CGC)

In this surface type, the subtypes are essentially the PS, Dini, and Kuen surfaces obtained as Bäcklund transforms. Notice that solving the Bäcklund-Darboux Eq. (13) gives us a constant of integration w, which creates a family of solutions seen as variations in the subtype (Fig. 9). Next, varying the angle σ (between 0 and $\pi/2$) results in different Bäcklund transforms, while the parameter ρ controls the general negative CGC of the surfaces.

Subtype B01

This subtype is obtained as a Bäcklund transform of the vertical line, and its variations are obtained by manipulating ρ, σ, w. In particular, when $\sigma = 0$ we obtain the PS.

$$\rho\left(\frac{\cos\sigma\cos v}{\cosh(u\sec\sigma - v\tan\sigma + w)}, \frac{\cos\sigma\sin v}{\cosh(u\sec\sigma - v\tan\sigma + w)}, u - \cos\sigma\tanh(u\sec\sigma - v\tan\sigma + w)\right)$$

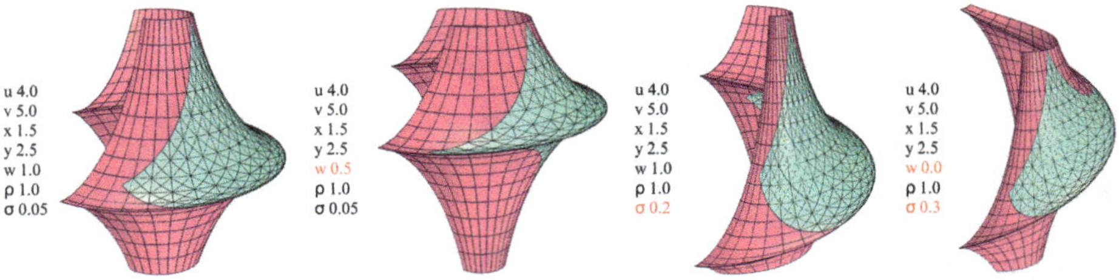

Fig. 9: Variations of Subtype B01.

4 Design application (Gridshell design)

Next, we present a gridshell prototype based on a variant of subtype A01 (surface type MS), where DF-1, DF-2, and DF-3 are used to generate the form and the structural/architectural elements utilizing the principal-asymptotic correspondence.

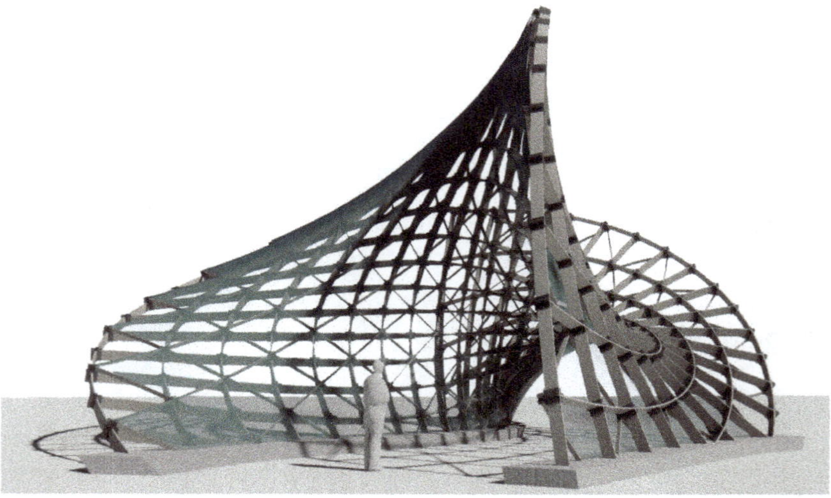

Fig. 10: Design prototype for a gridshell.

Elements assembly

In Fig. 11, we see a zoom-in on a detail from the proposed gridshell prototype that shows the assembly of the stacked ribs from the two patches (which generate the primary and secondary structures), the connector, and the quad panels.

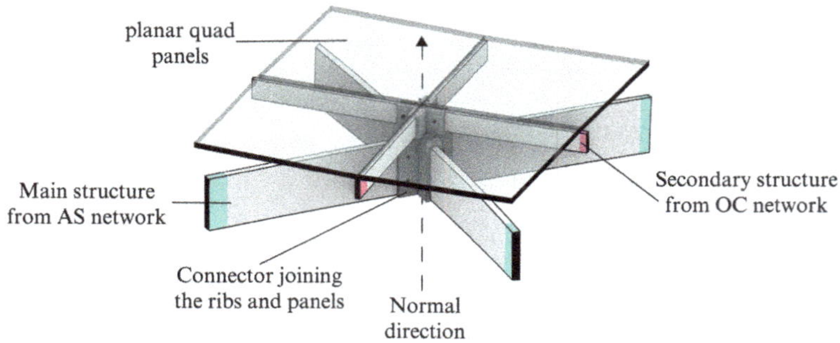

Fig. 11: Assembly of the three elements.

Element 01 (Ribs)

As shown in Fig. 12, both sets of ribs are generated from strips arising from the normals along their respective network curves. The ribs of the primary structure are from the asymptotic network, which unroll (albeit with deformation) to almost straight bands, cf. (Jiang et al., 2020). While the ribs of the secondary structure are from the principal network which are naturally developable. The two directions of each set of ribs are joined through a slit assembly to maintain the continuity of the ribs.

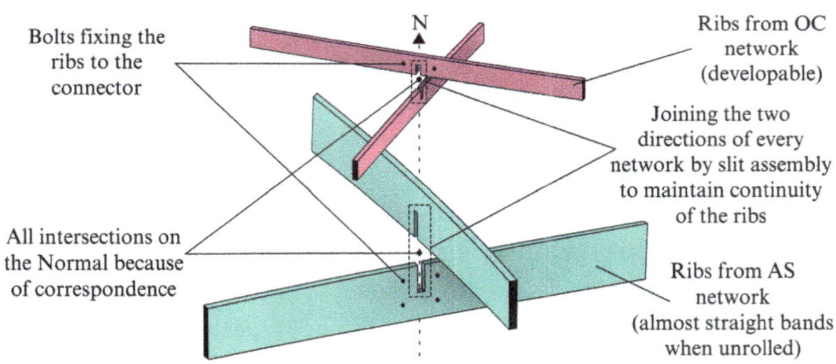

Fig. 12: Main and secondary structures.

Element 02 (Connector)

Since the strips forming the ribs in question are generated by the normals, their intersections will always coincide with the normal direction. Hence, simplifying the design of the connector joining all the elements of the structure, as seen in Fig. 13.

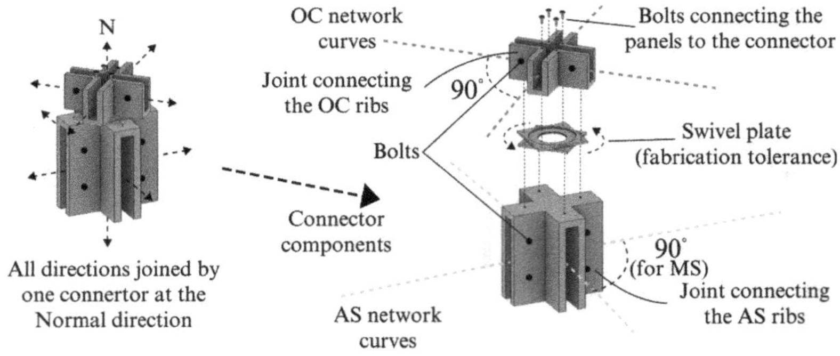

Fig. 13: Connector.

Element 03 (Quad panels)

Recall that OC-patches are the continuous smooth analogues of circular quad meshes. Thus, we can use planar quad panels (up to architectural tolerance) in our construction. These panels provide cladding and secondary bracing of the structure. The panels are bolted to the OC-joint at the normals, as seen in Fig. 11.

5 Conclusion

In this work, we presented a geometric setting for the correspondence between principal and asymptotic patches on surfaces of constant negative Gaussian curvature. That is since, on one hand, principal and asymptotic patches provide advantages for architecture fabrication. On the other hand, intersecting grids are structurally beneficial for grid shells, especially when these grids are principal and asymptotic networks. It is important to point out that the design model presented here provides an accessible way to realizing principal-asymptotic correspondence on two surface types and their variants, which is not a generically true condition. Moreover, the model created parameter spaces of rationalized morphological explorations that respect the geometric conditions of the correspondence. Finally, we presented a prototype of the morphological options.

Acknowledgements

The research of the fifth author Elshafei, A. was partially financed by Portuguese Funds through FCT (Fundação para a Ciência e a Tecnologia) within the Projects UIDB/00013/2020 and UIDP/00013/2020. The work of the fourth author Orszt, M. is financed by national funds through FCT - Fundação para a Ciência e a Tecnologia, I.P., under the Strategic Project with the references UIDB/04008/2020 and UIDP/04008/2020.

The research of third author Tošić, Z. is a part of the priority program SPP 2187: Adaptive Modular Construction with Flow Production Methods – Precision High-Speed Construction of the Future in the subproject Formwork-free Flow Production of Adaptive Supporting Structures from Variable Frame Elements – Adaptive Concrete Diamond Construction (ACDC) funded by the German Research Foundation (DFG).

References

Abdelmagid, A., Elshafei, A., Mansouri, M., and Hussein, A. (2022). A design model for a (grid)shell based on a triply orthogonal system of surfaces. In *Towards Radical Regeneration: DMS Berlin 2022*. Springer.

Bobenko, A. and Tsarev, S. (2007). Curvature line parameterization from circle patterns. https://arxiv.org/abs/0706.3221

Eisenhart, L. (1909). *A Treatise on Differential Geometry of Curves and Surfaces*. Ginn and Company, Boston.

Gray, A., Abbena, E., and Salamon, S. (2006). *Modern differential geometry of curves and surfaces with Mathematica*. 3rd Edition. Chapman & Hall/CRC.

Jiang, C., Wang, C., Schling, E., and Pottmann, H. (2020). Computational design and optimization of quad meshes based on diagonal meshes. *In Advances in Architectural Geometry 2020*. Springer.

Liu, Y., Pottmann, H., Wallner, J., Yang, Y., and Wang, W. (2006). Geometric modeling with conical meshes and developable surfaces. *ACM Tr., Proc. SIGGRAPH*, 25 (3): 681–89.

Moritz Niebler, Sylvain Usai, Martin Antemann, Fabian Scheurer, Evy Slabbinck

Bent-on-Site Flat-Pack Delivery of a Timber Shell

Abstract: Wisdome Stockholm is an extension to the Stockholm's Technical Museum due to open in December 2023 featuring an innovative timber roof. The free-form gridshell overarches 24 by 47 meters, consists of five crossing layers of doubly curved beams and is completely prefabricated from laminated veneer lumber (LVL). Beam segments were bent and twisted on-site from 2,190 flat, CNC-cut LVL-lamellae and locked in shape by 3,500 wooden dowels to form a stiff structure. Several challenges arose during the 1.5 year of planning, most of them induced by early design decisions such as material choice: despite being less common than glued-laminated (glulam) beams for doubly curved timber gridshells, the use of LVL was mandatory from the start. Furthermore, time and budget constraints did not allow for lamination and machining of doubly curved beams. The beams of the gridshell are composed of five layers of individually CNC-cut LVL-lamellae, thin enough to twist and bend. Only the lowest beam layer was pre-laminated to create a stiff base while the remaining lamellae were shipped to site flat-packed and bundled for assembly. The entire roof was parametrically planned in 3D. Lamellae were digitally unrolled, mapping all detailed connections from the twisted center surface to the flat lamella, and all were nested for 5-axis fabrication. The constructive system and digital process robustness was tested and validated using demonstrators as early as one month into the project, putting an early emphasis on fabrication and assembly. This paper discusses the workflow and problem-solving strategies implemented during the execution stage.

Keywords: timber structure, gridshell, laminated veneer lumber, free-form, digital planning, design-for-manufacture-and-assembly, digital fabrication

1 Introduction

The Wisdome Stockholm exhibition space, designed by Elding Oscarson Architects, permanently sits in the courtyard of the Tekniska Museet and houses a wooden dome dedicated to visualization technology. The insulated roof features a 24 × 47 × 18 meters freeform timber gridshell (see Fig. 1 left and right) built from CNC milled Laminated Veneer Lumber (LVL) lamellae. As the roof's edge is standing on 4 to 9-meter-high columns, loads normally transferred tangentially from the roof surface must be brought down vertically through reinforced "stilts".

On behalf of the timber contractor Blumer-Lehmann AG along with the structural engineers SJB Kempter Fitze and Création Holz, Design-to-Production was responsible

https://doi.org/10.1515/9783111162683-006

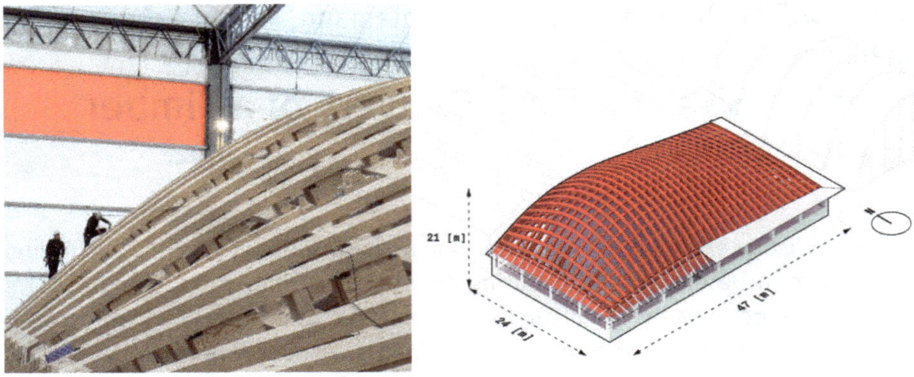

Fig. 1: Left: Wisdome Stockholm during assembly; Right: Axonometry of the roof.

for timber consulting, project management support and digital planning. The team already had extensive shared experience with freeform timber projects (Nine Bridges Golf Club in Yeoju, Venlo Casino in Venlo, Swatch Headquarter in Biel) using CNC milled glulam beam segments. However, the sponsoring of material by Stora Enso meant that the project was to pioneer the use of LVL panels in complex timber structures. Making a doubly curved blank from LVL panels and milling it afterwards to the final shape would not have been cost, material and time efficient. Instead, beams would be milled from a standard flat panel and mechanical-laminated.

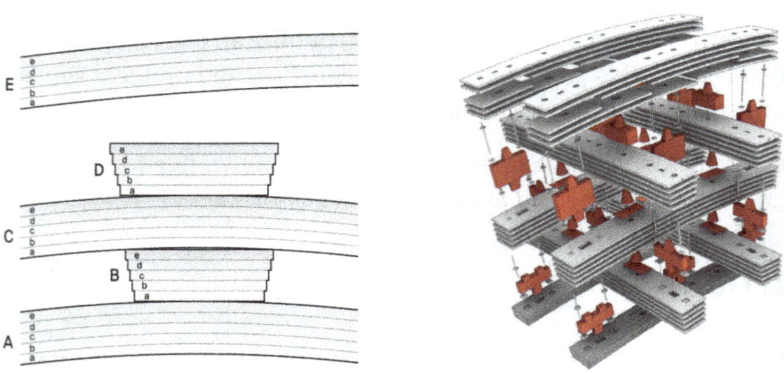

Fig. 2: Left: Beam layers composing the roof; Right: detailed roof build up.

The gridshell is composed of five beam layers running in two alternating directions (see Fig. 2, layers A to E). The beams are made of five stacked LVL lamellae, stacked and bent-on-site. Each lamella is 31 mm thick. Lamellae are held in shape and position by screws and wooden dowels, effectively turning two or three "parallel" beams into a Vierendeel truss. In order to be produced from a flat material, lamellae were digitally unrolled, including their complex detailing, and nested onto raw standard panel. Only the lowest beam layer is glued and pre-assembled off-site in order to precisely define the beam grid's shape and act as a stay in place falsework for all layers above. This paper discusses the digital process and methodology driving the modeling and production of fabrication data, the use of demonstrators to validate assumptions on structural, fabrication, and assembly concepts, the reconstruction of precise reference geometry as well as the parametric segmentation, nesting, detailing, and unrolling of all the individual lamellae.

2 Timber gridshells: state-of-the-art

There has been an increase in the design and construction of innovative and advanced wooden gridshell constructions over the past decade (Chilton and Tang 2017) (Lara-Bocanegra et al. 2018). Computational form-finding process or physical hanging-model method allows to determine the shape of elastic timber gridshells (Tomei et al. 2022). Most built gridshells use slender and thin elements, e. g. Multihalle Mannheim, Weald and Downland Open Air Museum and PEMADE gridshell (Tomei et al. 2022). By reducing the width of the elements, the geometric torsion is reduced (Slabbinck et al. 2017). Elastic timber gridshells often constitute the continuous façade of a building, starting at ground level and covering all wall and roof geometry, to be able to handle horizontal forces keeping the gridshell bent. Contrary to this, the Savill Garden gridshell is solely a roof on columns with a ring beam transferring the forces to the inclined steel columns, as in the Flimwell Woodland Enterprise Centre, while timber columns are used to transfer the loads to the ground in the tangential direction of the roof (Chilton and Tang 2017; Harris and Roynon 2008).

Following advances in digital fabrication, gridshell design has shifted away from bending elements on-site as a method of assembly. The evolutions of CAD, CAE, CAM and CNC technologies enables the production of precisely shaped curved beams from glulam blanks. This expands the design solution space while freeing the designer from the inherent difficulties associated with the design and construction of elastic gridshells (Lara-Bocanegra 2022). Alternative timber gridshell systems were developed using glulam beams, among others: Zollinger system (Toskana Thermal Springs), use of lap joints (Swatch Headquarters), hierarchical beam directions of beams (French Pavilion Expo 2015) and the use of timber dowels (Centre Pompidou Metz) (Chilton and Tang 2017).

In an effort to overcome the limited design freedom, construction constraints and complexity of the design, research has developed further by bending continuous elements in timber gridshells in a novel way. The use of single curvature locally in the elements was enabled in the bent-plate gridshell like the 2010 ICD/ITKE research pavilion (Quinn 2018). Ribbed and cross-ribbed shells, use false-work to construct the designed geometry and dowel or nail the elements together (Gliniorz et al. 2002). Or by using the hygroscopic qualities of wood to form surface-active plates like the Urbach Tower (Bechert et al. 2021).

In timber construction, the use of a single material for joint and structural members speaking the same language is key. The use of dowels to mechanically laminate timber has been a focus point of research in the last decades. Mechanical lamination of timber was already documented in the 19th century by Amand-Rose Emy, where he illustrates the bending of timber elements in a jig and held in place by the help of bolts and clamps (Emy 1837). More recent research shows ribbed shells with doweled planks, keyed beams where shear keys are used between timber layers to improve the overall beam stiffness and efficiency, and dowel laminated timber (DLT) (Miller and Bulleit 2011; Gliniorz et al. 2002; Sotayo et al. 2020; O'Ceallaigh et al. 2022).

3 Digital process and methodology

Off-site prefabrication has been around for centuries within the timber construction industry, allowing builders to shift complexity from the construction site to a controlled workshop environment. Nowadays, so called free-form project forces builders to shift part of the complexity once again from the workshop onto the digital drafting board (Scheurer et al. 2013). These projects make full use of digital tools both during planning (CAD) and fabrication (CAM). However, to fully bear the fruits of the digital revolution on the production side, the entire process chain, from procurement to fabrication and assembly, needs to be coordinated by "pulling" from the back-end (Scheurer and Stehling 2020). The objectives for digital modelling are defined by the final step: a level of detail that contains every screw-hole to be drilled by digital machinery, and a level of accuracy that matches that of digital fabrication. Multi-scalar parametric models, a stack of models with different resolutions as defined by Scheurer and Stehling (2020), containing minimal but accurate information is a mean toward both these ends. Each model embeds the learnings and results of a design stage and is used as input to parametrically detail the model of the next stage (see Fig. 3) (Scheurer and Stehling 2020).

Dependencies and limitations must be identified upfront to make the process viable and a high degree of systematization and clear interfaces between trades and building parts are required to achieve quality and consistency.

Entering the Wisdome project pre-tender allowed for an early consideration of logistics, material constraints, assembly feasibility and fabrication limitations. These dependencies were woven into the process during the revision of the Reference Geometry[1]

1 The Reference Geometry defines basic geometric properties of the project, including Reference surfaces and axes (in production quality), Main components (beams and panels) as continuous axes and surfaces and undetailed 3D volumes and Nodes positions.

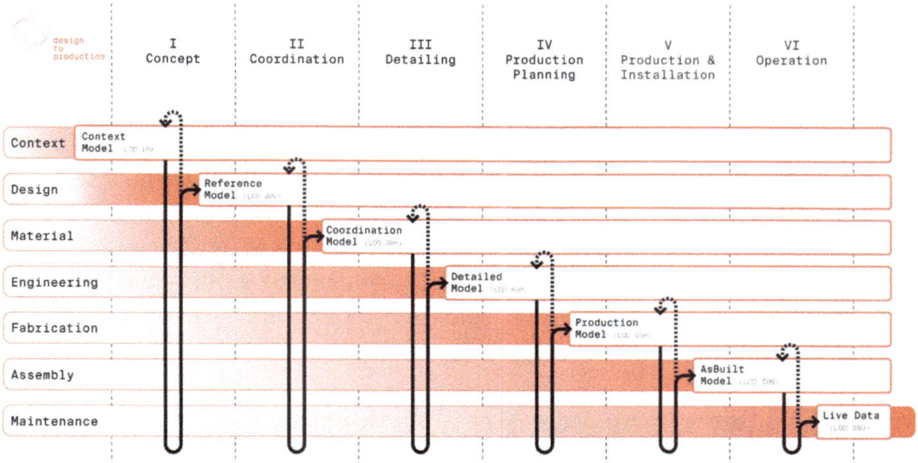

Fig. 3: Agile Design-to-Production process.

optimization. Additionally, lessons learned from three early built demonstrators involving all stakeholders and subcontractors proved an invaluable basis for decision-making.

A building element catalogue describes all existing element types in the building. It can be used to derive a detail matrix, describing the interfaces and dependencies between these element types. Such a building catalogue was created from an early stage and maintained during the entire project, allowing for a clear understanding of the dependencies between elements. The detail matrix was paramount to reach systematization of the Joint Catalogue, leading to the control and reduction of the amount of detail types, complexity, cost and time (see Fig. 4).

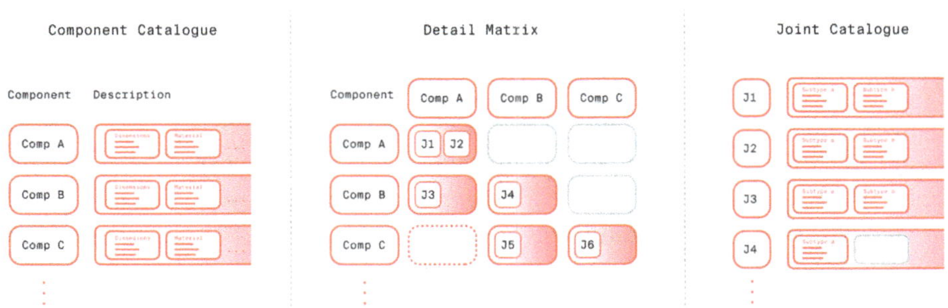

Fig. 4: Schematic catalogues concept overview.

4 Demonstrators

The construction system imagined for the roof was beyond the team's experience. Additionally, the forces required to bend the lamellae (which was to be performed repeatedly by the montage team on-site), the internal stresses and the behavior of the beams and dowels under bending and torsion were unknown. The beams with the highest torsion (8.4°/m) and the one with the smallest radius of curvature (10 m) were isolated and fully detailed for production in order to test bending and recoil. The lamellae were unrolled, CNC milled, positioned in a formwork jig made of milled-out beam profiles and fixed with LVL oblong dowels measuring 300 × 160 × 60 mm.

Fig. 5: Left: Torsion demonstrator; Right: Four crossing segments.

The demonstrators proved that bending lamellae was physically possible. Nonetheless, recoil out of the formwork exceeded 1 m for the bent demonstrator (see Fig. 5 left). Even if the spring back would have remained in check, the number of dowels and the manual labor required to insert them all could not be justified. In parallel, four crossing segments were also detailed and assembled to test and validate the aesthetic of the stacked lamellae and dowels (see Fig. 5 right).

The construction system was subsequently reworked and further tested with a second demonstrator. The beams on the lower layer were glued and shaped off-site and serve as stay-in-place falsework for the rest of the timber layers. This means that less dowels were required and recoil is limited because dowels were fastened to a stiff base. Additionally, the dowel shape was changed to allow for an easier assembly (see chapter 3D Detailing). This demonstrator consisted of 5 beam crossings in a high torsion area. It involved the entire team of subcontractors and served as a dry run for the actual building production and assembly. Effectively, most special details present in the building had to be solved for the mockup in a 2-week period. This demonstrator satisfied tolerance and precision expectation and the feedback from the montage team was highly positive, it remained so during the entire project assembly (see Fig. 6).

Fig. 6: Second demonstrator.

5 Definition and optimization of reference geometry

The reference surface is the basic geometric definition for all subsequent parametric modeling of beams and their connection details. The 3D-model provided by the architect as input for Wisdome, while transporting design ideas and concepts did not meet the requirements for execution and needed to be revised and rebuilt. An optimization of the reference geometry was carried out as part of the pre-project in close collaboration with the architect, structural timber engineer and the timber contractor. The entire stakeholder involvement allowed to cover and inform the optimization process with all aspects from architectural design intent, site boundary condition, material, and fabrication constraints.

The optimization problem revolved around the architectural design intent by extending the strict orthogonal grid basis into the roof construction. A key requirement was to establish a clean and matching interface between the base grid and the member axes of the roof structure, addressing both architectural and engineering considerations. This involved ensuring that the offset of the surface symmetry axes, and the beam axes grid align with the axes system of the columns. Additionally, material constraints related to maximum feasible bending and twisting need to be taken into account.

To address these challenges, an optimization workflow comprising multiple iterations was employed to determine the most suitable and producible surface and beam grid geometry. Given the manifold boundary conditions and the need for coordination with various parties, a manual approach was chosen to optimize and inform the geometry generation and formulation. Each iteration involved surface and beam grid optimization. The initial design surface was rebuilt from the architectural MESH input. And the beam grid on the surface was based on a strict orthogonal planar grid projected to the surface. Throughout the process, the resulting normal curvature and geodesic torsion were constantly monitored to ensure they remained within the range of the material properties.

The asymmetrical position of columns between North and South (long edge of the building, see Fig. 1 right) meant that the optimal on one side did not reflect on the other. Minor adjustments on both the surface, beam grid and column position were therefore necessary. Resulting in a slight deformed planar curve network.

Ultimately, five reference surfaces were created, one for each beam layer. The three lower surfaces (layer B-D) are strict offsets of the main surface, the last one (layer E) is only partially following the main surface on the center and deviates as it gets close to the border to define the outer roof and meet the horizontal eave lines (see Fig. 7, left). To reduce the number of individual dowels while reducing torsion and bending in the beams and maintaining geometric surface continuity, the offset between A and E was maintained as much as possible (see Fig. 7, right).

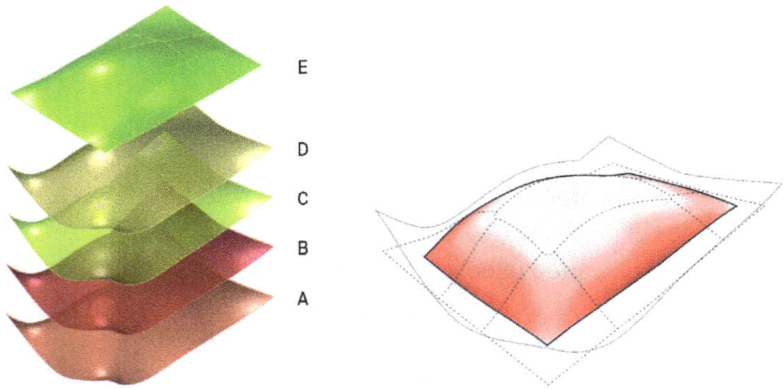

Fig. 7: Left: Master surface overview; Right: Master surface E offset deviation compared to A.

6 Segmentation

The beam length ranges from 25 to 53 m long while the maximum length of the standard panel from which beams are produced is 13.5 m. This meant that beams had to be produced in smaller segmented elements. In addition to raw material requirements, the segmentation strategy of a timber gridshell must comply with transport regulation and sizing, structural requirements, logistic, and engagement directions.

In an effort to minimize kinks in a lamella stack after on-site bending, the segmentation position is staggered within all the lamellae of the same beam. These segmentation joints are located between shear and cross dowel and span four consecutive dowels. Within one segment, these staggered joints can span between 2.9 and 4.2 m.

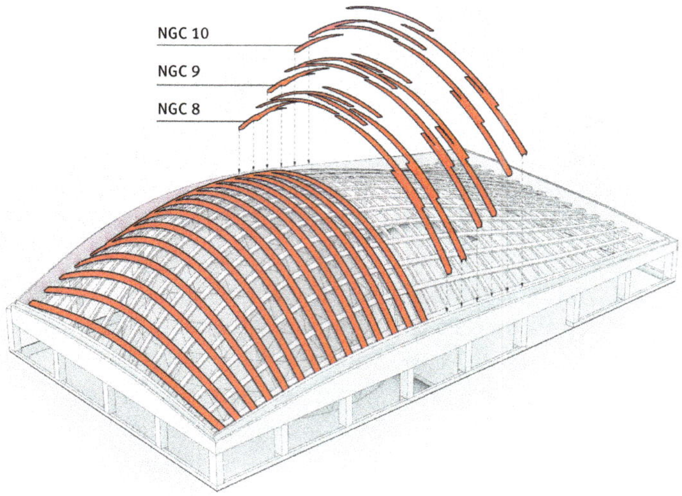

Fig. 8: Layer C assembly sequence.

To identify segmentation positions, a script ran from each opposite end of the beam, collecting cross and shear joints normal directions until the cumulative angle deviation exceeded 5° or 15° depending on the area of the roof. These threshold values relate to the slope of the dowel faces: once the angle value exceeds the threshold, it means that the lamella can't be inserted on the dowel, a segmentation joint is created, and the process is iterated further. This ensures that all joints on one segment share a common assembly direction. It is worth noting that this process simulates the translation of a rigid curved segment (all joints inserted at the same time) and ignores the bending motion (joints inserted one after the other), giving a comfortable margin for the team on-site.

If the angle value is still in range after 13 m, segmentation is triggered in order to fit in the 13,5-meter-long raw standard panels. The optimal solution was then adapted by the assembly team, favoring length homogeneity (see Fig. 8).

7 Nesting of 2651 components

Once segmented, all lamellae received a reasonable oversize (to compensate for the fact that most details were still undefined at this stage) and were nested into raw LVL panels and translated into a bill of quantity (BoQ). Available raw panel length ranged from 8 to 13 m with 200 mm increments, such variety of sizes increased the utilization rate of the raw material. This early BoQ was motivated by material order, production and shipping lead times. In a glulam free-form project, the 3D part is used as a basis for both the blank generation and the machining data. In Wisdome, the 3D geometry will only be "revealed" once bent into shape on-site. Therefore, all lamellae were unrolled into a "flat twin" model containing the parts as they are produced and shipped.

Within each beam layer, nesting panels are grouped into sectors according to the assembly sequence. The use of two different sheet materials LVL-S and LVL-X as well as two different visual qualities prompted further subdivisions within one sector. This resulted in 96 nesting batches for 2651 lamellae with a total of ca. 12,000 m^2 of sheet material (more than 10 times the footprint of the building). Taking the assembly sequence into account during the nesting allowed for lean production, keeping logistics and part sorting to a minimum on the workshop floor and simplifying the flat-packing of the lamellae into segments for a swift on-site handling and assembly. The resulting nesting has been kept untouched from the early undetailed stage until the final production export and proved to be a crucial part for the success of the project (see Fig. 9). Ultimately, every panel was exported individually as a Cadwork file containing named and detailed volumes for each nested lamellae and every relevant cutting operation.

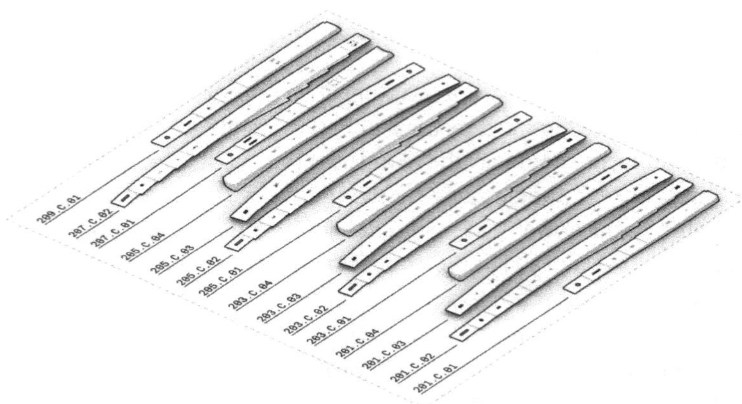

Fig. 9: Flat-packed beam segments.

8 3D Detailing

Joints are created at every crossing between two elements (cross joints) or within one element itself (continuous joint). Each joint is typed according to the elements it connects and subtyped according to structural requirements. Every joint defines their own coordinate system, used to parametrically build the details. Dependencies between neighboring joints are important to consider during detail development as moiré effect can appear when two details are instantiated on different coordinate systems close to each other. The joint between the column, edge panel and beam is a good example (see

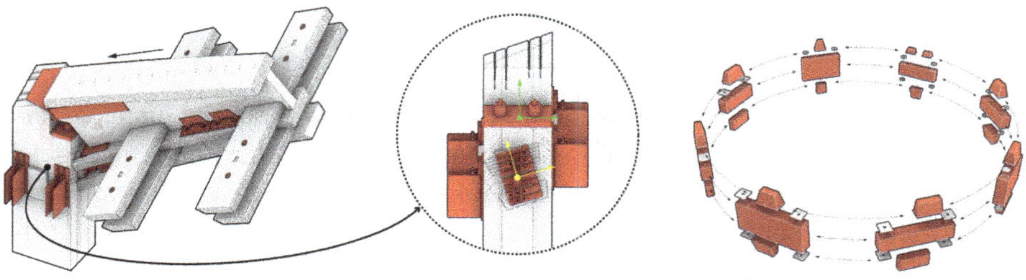

Fig. 10: Left: Column, edge panel, beam detail; Right: Shear-Dowel Configuration..

Fig. 10 left). Vertical tensioning rode (instantiated on the green plane) meets screws from the beam foot detail (instantiated on the yellow plane) and screws from the steel socket on the side of the column.

All details within layer B-E are instantiated in 3D and subsequently unrolled onto the lamella. The Flat twin model is synchronized after every change on the 3D models. Layer A did not require unrolling of details as the raw blanks were glued in shape and details were 5-axis milled out of the blank.

The entire gridshell contains 1820 cross dowels and 1756 shear dowels. Cross dowels are instantiated at every crossing and are made of truncated conical tenons, they are holding the lamellae into position. Shear dowel span between parallel horizontal beams (AC, BD, CE) and have 4 sided pyramidal tenons on the top and bottom face to shape the lamellae into form. Even though layers B, C, D are strict offsets from the reference A surface, local curvature and torsion configuration meant that the dowel height would have to be unique for every joint. For each subtype, the smallest existing dowel height configuration was identified and used as the baseline height. The remaining height difference was compensated by steel washers partially sunk in the lamella. This reduced the amount of different dowel types from 2859 to 9, simplifying dowel production, logistics, assembly time and reducing cost significantly (see Fig. 10 right). Working with repeating geometries also allowed the planning team/process to make full use of modeling “blocks”: the coordinate system of each block is parametrically defined within one connection and the block geometry is simply oriented to its final position, rather than parametrically encoding the base, plug, washer, bolts, drilling, cutters, etc. Additionally, these repeating dowels are made symmetrical so it would be impossible to assemble them “flipped”. For the individual dowels, positioning drillings facing toward the building edges are introduced to inform the montage team of the dowel orientation.

The E surface is following its own rules and is not an offset of the reference A surface around the edges (see Sec. 5), therefore, the cross dowels between layers DE and shear dowels between layers CE had to be unique anyway.

9 Flat part detailing

Once instantiated in 3D, all detail operations are unrolled with the flat lamellae. Two different unrolling strategies were developed consecutively for the unrolling of detail operations. For the demonstrator, operations were intersected with the top and bottom face of the lamella and pulled (normal projection) onto the middle surface of the lamella before being unrolled. After unrolling, the top intersection curves were moved up and the bottom intersection curves were moved down by half the lamella thickness. Cutters were then rebuilt in 3D by interpolating these curves. This process allowed to account for the cutter deformation induced by curvature and torsion during the unrolling, but it was discarded by the fabricator after the mockup as it significantly increased the complexity of milling operation for a marginal gain in precision. The alternative solution was to individually unroll the detail planes onto the flat lamella to map cutting operations from 3D plane to flat plane: the cutters for the dowel tenons can then be milled as a straight contour instead of a twisted one.

Unrolling detail planes instead of single operations allows to apply the same unrolling routine to any drillings, saw-cuts, pockets and block instances. This implies that every operation must use one of these detail planes as its local coordinate system. The unrolling of detail planes follows a three step process. They are serialized into a set of unrollable geometries with attributes, unrolled and deserialized back into a plane (see Fig. 11).

To serialize a plane in reference to the lamella mid surface, three points are projected normally to the lamella: the plane origin, origin + plane unit X and origin + plane unit Y. Distances from the point and its projected counterpart, as well as additional information regarding the operation type, where embedded as a dataset on a Text-Dot object. Text dots store an operation type string and a concatenation of height distance, Global Unique Identifier (GUID) and additional operation specific information. The GUID is common for all texts dots defining the same plane (origin, X and Y) and is later used to retrieve and group them together as dozens of planes get unrolled simultaneously. After unrolling, dots are sorted by GUIDs and deserialized to a local unrolled plane. To check deviations between the 3D and flat detail plane the angle between the lamella normal at origin and plane normal are compared. Keeping the initial detail plane as close as possible to the surface helps reducing deviation. Unrolling was considered successful as long as the normal deviation between flat and 3D did not exceed fabrication tolerances (5/10 mm). For critical detail planes close to an edge of the surface, additional points in negative plane space are added to the process. All operations then simply follow a plane-to-plane transformation to be mapped onto the flat lamella. Detailed volumes were generated for both 3D and unrolled parts. Quality control involved visual checking of overlayed 3D and flat sibling parts to ensure that they were correctly oriented, and all cutting operations were properly transferred between the 3D and flat version.

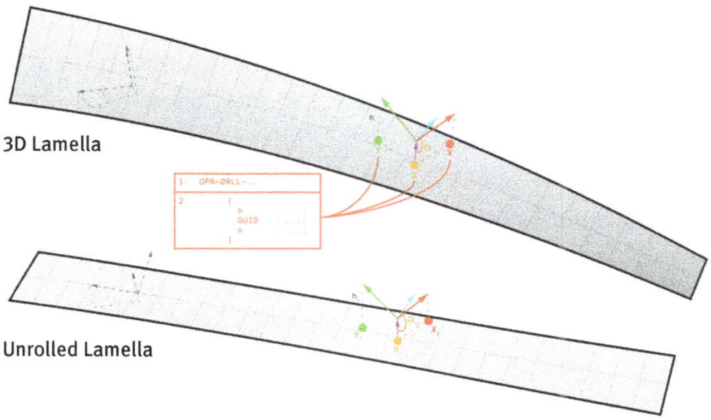

Fig. 11: Schematic unroll of detail planes.

As shown with the third demonstrator and by extension the final project, both approaches of unrolling operations proved to be valid solutions in terms of geometric representation. Although the second more pragmatic method is more relevant at an architectural scale in terms of fabrication and part size.

10 Conclusion

Fabrication and assembly optimization concerns were central during The Wisdome gridshell conception. The precise milling of unrolled lamellae out of flat LVL panels proved to be a viable (time and cost-wise) alternative to the glulam blanks approach with notable upside regarding fabrication (milling from a flat panel is easier than milling from a 3D blank) and shipping (transporting bundled flat lamellae is easier than formed 3D segments). On-site assembly was more labor intensive than a glulam gridshell but this was alleviated by precise and unambiguous pre-fabricated details, comprehensive naming system and assembly strategy. The entire project execution was immensely facilitated by the extensive digital methodology: front-loading the Material, Engineering, Fabrication, Assembly and Maintenance constraints allowed to keep potential issues in-check and pro-actively find solutions and plan for fabrication in a lean and efficient way, allowing the gridshell to be planned and ultimately assembled in time and on budget.

Acknowledgement

Many thanks to all people and parties involved in the project: Johan Oscarson and Arin Alia from Elding Oscarson; Jessika Szyber and Carl Humble from Stora Enso; Astrid

Stenberg and Frederik Eriksson from Tekniska Museet; David Riggenbach, Valentin Künzle, Martin Looser-Frey, Kai Strehlke and Simon Huber from Blumer-Lehmann AG; Hermann Blumer from Création Holz; Christoph Meier, Stefan Rick and Dominik Fischinger from SJB Kempter Fitze; Herbert Schmid and Sven Bill from Balteschwiler AG; Thomas Brühlman from Gebhard Müller AG; Matthias Hornung and Valeriia Vlasenko from Design-to-Production. Additional thanks to Jonas Van den Bulcke for his help during proof reading and Valentin Kubatta and Adrien Gesulfo for their contribution on this paper's illustrations.

References

Bechert, S., Aldinger, L., Wood, D., Knippers, J., and Menges, A. 2021. Urbach Tower: Integrative structural design of a lightweight structure made of self-shaped curved cross-laminated timber. *Structures,* (33): 3667–81.

Chilton, J. and Tang, G. ed. 2017. *Timber Gridshells: Architecture, Structure and Craft.* New York: Routledge.

Emy, A. R. 1837. *Traité de l'art de la charpenterie.* Paris: Carilian-Goeury (etc.).

Gliniorz, K.-U., Mosalam, K. M., and Natterer, J. 2002. Modeling of layered timber beams and ribbed shellframeworks. *Composits part B: engineering,* (33): 367–81.

Harris, R. and Roynon, J. 2008. The Savill Garden Gridshell Design and Construction. Paper presented at 10th *World Conference of Timber Engineering.*

Lara-Bocanegra, A. J., Roig, A., Majano-Majano, A., and Guaita, M. 2018. Innovative developments in the design and construction of a permanent elastic timber gridshell. *Word Conference of Timber Engineering.*

Lara-Bocanegra, A. J. 2022. Elastic timber gridshells, from material to construction. Doctoral diss., Polytechnic University of Madrid.

Miller, J. F. and Bulleit, W. M. 2011. Analysis of Mechanically Laminated Timber Beams Using Shear Keys. *Journal of Structural Engineering (ASCE),* (137): 124–32.

O'Ceallaigh, C., Harte, A. M., and McGetrick, P. J. 2022. Dowel Laminated Timber Elements Manufactured using Compressed Wood Dowels. *Civil Engineering Research in Ireland,* 222–27.

Quinn, G. 2018. Pneumatic erection of elastic gridshells, design, simulation and realization. PhD diss., Berlin University of the Arts.

Scheurer, F., Stehling, H., Tschümperlin, F., Antemann, M. 2013. Design for Assembly–Digital Prefabrication of Complex Timber Structures. *Proceedings of the IASS Annual Symposium.*

Scheurer, F. and Stehling, H. 2020. New Paradigms for Digital Prefabrication in Architecture. In *Design Transactions Rethinking Information Modelling for a New Material Age,* edited by Bob Sheil, 42–49. London: UCL Press.

Slabbinck, Evy, Koerner, Axel, and Knippers, Jan. 2017. Torsion as a design driver in plate-bending-active tensile structures. *Proceedings of the IASS Annual Symposium.*

Sotayo, A., Bradley, D., Bather, M., Sareh, P., Oudjene, M., El-Houjeyri, I., Harte, A. M., Mehra, S., O'Ceallaigh, C., Haller, P., Namari, S., Makradi, A., Belouettar, S., Bouhala, L., Deneufbourg, F., Guan, Z.. 2020. Review of state of the art of dowel laminated timber members and densified wood materials as sustainable engineered wood products for construction and building applications. *Developments in the Built Environment, (1).*

Tomei, V., Grande, E., and Imbimbo, M. 2022. Design optimization of gridshells equipped with pretensioned rods. *J. Build. Eng.,* (52).

Romain Mesnil, Takara Muto, Krittika Walia, Cyril Douthe, Olivier Baverel

Design and Fabrication of a Pseudo-Geodesic Gridshell

Abstract: The aim of this paper is to present the design and construction of a GFRP gridshell based on pseudo-geodesic curves. Pseudo-geodesic curves on a surface have a constant angle between their normal vector and the surface normal. They are thus a generalization of asymptotic curves (where this angle is equal to 90°) or geodesic curves (where this angle is 0° or 180°), which have already been used to construct elastic gridshells. Like asymptotic or geodesic gridshells, pseudo-geodesic gridshells benefit from a flat unrolling property as well as a standard connection detail. The paper presents the design strategy of the first pseudo-geodesic gridshell. A parametric structural analysis shows that the gridshell outperforms geodesic gridshells significantly but also that it is sensitive to local lateral-torsional buckling. The fabrication stage confirmed that the construction of a unique type of connector is possible. In total, 160 connectors have been used for the prototype. The mathematical properties of pseudo-geodesic curves have thus been successfully turned into a technological solution to materialize doubly curved shapes.

1 Introduction

1.1 Context

Gridshell structures are efficient structural systems that enclose large volumes with little material. The rationalization of their construction is an active topic of research in the architectural geometry community, as it is deeply related to the fundamental properties of discrete networks and to the mechanical behavior of the structure. This article studies *elastic gridshells,* which are constructed through the elastic deformation of slender beams. Case studies have shown that a significant share of cost and engineering is spent on node design [1]. Following recent research on fabrication-aware design, we are thus interested in the design of a gridshell with *congruent nodes* [2,3].

1.2 Previous Work

Some gridshells were built with circular hollow sections and swivel connectors, following Frei Otto's experiments with rebars [4]. However, most elastic gridshells are built with anisotropic cross-sections, like planks. Two factors may explain this: first,

https://doi.org/10.1515/9783111162683-007

Fig. 1: Two gridshells with high node congruence and straight development: the Polydome at EPFL (left) and the Inside/Out Pavilion at TU Munich (middle) and the pseudo-geodesic gridshell (right).

connections between planks are easier to construct, and second, materials adapted to the materialization of gridshell structures are usually fibrous materials, like timber or GFRP, and can easily be produced as planks [5]. Two rationalization strategies have been proposed to guarantee two properties: *node repetition* and *straight-flattening*.

Julius Natterer proposed to build geodesic gridshells, where planks are laid along geodesic curves of the reference surface [6]. The *Polydome* at EPFL is a fine example of this approach. Schling et al. [2] later proposed to construct asymptotic gridshells, where the planks are laid perpendicular to the surface osculating plane. Asymptotic gridshells are limited to negative curvature, and Schling proposed to construct them with planks that have an initial geodesic curvature. The main drawback of this approach for fiber-reinforced materials is that the material anisotropy is disregarded, and remains limited to isotropic materials, like steel.

An extension of the two design approaches has been proposed with pseudo-geodesic gridshells [8]. The authors claim that this structural system improves structural efficiency compared to geodesic gridshells while offering more formal possibilities than asymptotic gridshells. However, no experimental validation has been made yet.

1.3 Problem Statement

Asymptotic and geodesic gridshells are well-known structural systems that allow cost-effective materialization of doubly curved shapes. However, both systems have practical limitations: either the need for a double layer or strong formal constraints. Pseudo-geodesic gridshells offer the potential to find a compromise between the two approaches. The geometrical properties of pseudo-geodesic curves seem to allow for the construction of nodes with high congruence.

The aim of this paper is to discuss the potential application of pseudo-geodesic gridshells through a practical case study. We first present some theoretical considerations on pseudo-geodesic gridshells and highlight their potential for the construction of efficient structures. We then discuss the design of the *Pseudo-Geodesic Pavilion*, with critical feedback on the design tools currently available. We conclude the paper by presenting the construction of the pavilion, with a special focus on constructive details and their link to geometry.

2 Pseudo-Geodesic Gridshells

2.1 On Straight Flattening

The straight flattening property is linked to an underlying assumption for the geometry of the strip that requires some comments, since previous publications [2] focus on the geometric formulation of the problem. Here, we consider a thin strip that is to be mapped to an unstressed (generally planar) configuration, as shown in Fig. 2. We adopt a beam model [9,10], where the beam is described by a curve Γ and the ruling direction **r** (material frame).

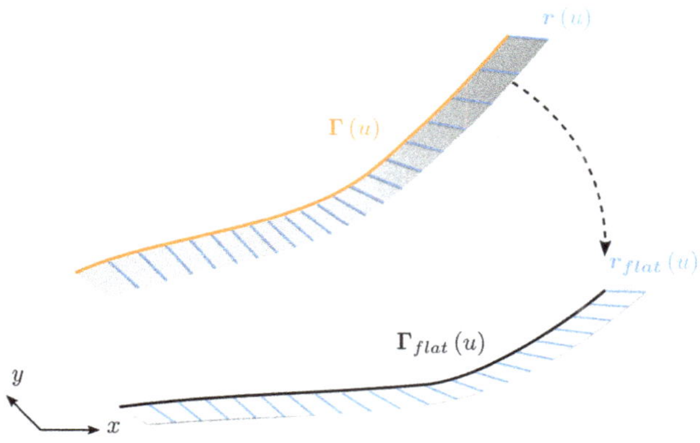

Fig. 2: A possible flattening of a ruled surface.

The strain energy of the strip, with the assumption of inextensibility, is given in Eq. (1):

$$\mathcal{E} = \frac{1}{2}\int_0^L \left[EI_1(\kappa_1 - \overline{\kappa}_1)^2 + EI_2(\kappa_2 - \overline{\kappa}_2)^2\right] ds + \frac{1}{2}\int_0^L GJ(\tau - \overline{\tau})^2 ds \quad (1)$$

where κ_i are curvatures in the material frame and τ is the torsion (with respect to a rotation-minimizing frame on the final configuration). Symbols with overlines correspond to the curvatures and torsions of the unstrained configuration. For an initially straight rod, they are null. The flattening problem consists of minimizing the strain energy with respect to the unstrained curvature and torsion.

When the ruled surface is also developable, it is admitted that the flat unrolling problem has a solution, thanks to *Theorema Egregium*. The computation of isometric

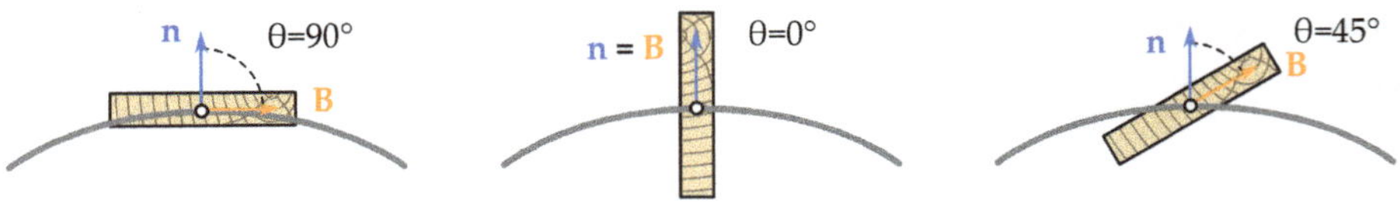

Fig. 3: Orientation of planks for geodesic gridshells (left), asymptotic gridshells (middle), and pseudo-geodesic gridshells (right) [6].

transformations corresponds to an inextensibility constraint, which imposes the torsional strain energy to be zero (the material frame follows the rotation-minimizing frame). It is also admitted that the unrolling is straight and planar when the ruling direction (or material frame) is along the binormal vector **B** of the curve: this is precisely the case for geodesic and asymptotic gridshells. In such cases, it is easy to verify that the strain energy is the sum of bending energy along the weak axis and torsional energy, as detailed in eq. (2):

$$\mathcal{E} = \int_0^L EI_2\kappa^2 ds + \int_0^L GJ\tau^2 ds \tag{2}$$

It is therefore assumed that the contribution of bending along the strong beam axis to the total strain energy is much higher than the two others and that the laths tend to avoid this flexural mode. Indeed, when the ratio between lath height h and width b goes to zero, the bending stiffness along the strong axis $Ehb^3/12$ becomes much larger than the two other stiffnesses $Ebh^3/12$ and $Gbh^3/4$.

2.2 Pseudo-Geodesic Curves

Pseudo-geodesic curves were first studied by Wunderlich [11] and later applied in architectural geometry [7]. They are curves traced on surfaces so that the angle between the curve binormal **B** and the surface normal **n**, written θ is a constant. Another definition is that the ratio between total curvature and normal curvature is constant, thanks to Meusnier's theorem:

$$\frac{\kappa_n}{\kappa} = \sin\theta \tag{3}$$

The ruled surface generated by the envelope of binormal vectors has a straight flattening property. It also has a constant angle with the surface normal. This property can be used to construct congruent connectors. Geodesic and asymptotic curves are special cases of pseudo-geodesic curves with $\theta = 90°$ and $\theta = 0°$ respectively (see Fig. 3).

One of the natural questions that one may ask is whether it is possible to cover a shape with a pseudo-geodesic net. Indeed, it is known that geodesics can be constructed through any point in any direction, while asymptotic curves only exist on anticlastic surfaces. Mesnil and Baverel have constructed a lower bound for the total curvature of pseudo-geodesic curves on synclastic surfaces by combining the Meusnier theorem and Euler's formula, recalled in eq. (4).

$$|\kappa| > \frac{|\kappa_2|}{\sin\theta} \tag{4}$$

This equation has practical implications for the construction of elastic gridshells since the bending stress in the laths is proportional to curvature and should remain below the design stress, as shown in eq. (5):

$$\sigma_{\max} = \frac{E\kappa h}{2} < \sigma_d \tag{5}$$

Therefore, the inequality constraint from eq. (4) becomes an inequality on the strip height h:

$$h < \frac{2\sigma_d \sin\theta}{E|\kappa_2|} \tag{6}$$

In practice, laths with high curvature have to be thinner, reducing the stiffness of the structure.

2.3 Mechanical Performance

Mechanical efficiency was one of Frei Otto's main motivations for constructing elastic gridshells. The construction of safe structures with as little material as possible is still highly relevant in the current context of dwindling resources and climate change. Pseudo-geodesic gridshells offer a compromise between geodesic and asymptotic gridshells and may mitigate some of their limitations.

First, we shall recall that gridshells are efficient because of their ability to transfer out-of-plane loads through axial forces in their members. This is only possible when the curves have *normal* curvature. Thus, asymptotic gridshells mainly work through bending because asymptotic curves have no normal curvature. This fact was hinted at by Panozzo and coauthors [15], who noticed that the orientation of thrust networks in negatively curved regions is severely constrained. In contrast, geodesic curves only have normal curvature, and are therefore more likely to work as a true gridshells.

Second, gridshells are known to be prone to buckling, which depends on their bending stiffness. A homogenized bending stiffness D has been proposed in [6] based on prior works:

$$D \sim \frac{E}{l}\left(\frac{hb^3}{12}\cos^2\theta + \frac{bh^3}{12}\sin^2\theta\right) \tag{7}$$

When factoring in the restriction on maximal strip height h imposed by eq. (6), it appears that the optimal angle θ is around 45°. However, a parametric study on toroidal caps in [6] shows that pseudo-geodesic gridshells may be sensitive to local lateral-torsional buckling, and that the optimal θ angle was around 30°.

3 Designing the Pseudo-Geodesic Pavilion

A full-scale pseudo-geodesic gridshell was constructed as a pavilion to investigate the feasibility of the concept (see Fig. 4). This section discusses practical design issues and strategies employed to materialize the pseudo-geodesic pavilion.

Fig. 4: The pseudo-geodesic pavilion, made of 100 mm GFRP lath at 60° with a unique Z-connector. (Picture: Charlene Yves)

3.1 Design Constraints

The pavilion spans a rectangular frame. Covering a fixed boundary is a common design problem in gridshells; notable examples include the courtyards of the British Museum in London, the Odeon in Munich, or Dresden Castle. The pavilion is covered with polycarbonate panels. The length of the rectangle is $L = 6$ m, while the short span is 4 m. Following [6], we choose an angle of 60°, the best compromise between bending stiffness and flexural torsional buckling.

A CAD-based approach [12,13] was used for the design because several constraints required adjustments for the production of shop drawings. The reference surface is a NURBS with globally positive gaussian curvature but locally negative curvature in the corners of the surface. Several design goals or constraints were at stake:
- Limit the distortion of the mesh, designed to be as uniform as possible;
- Minimize the number of rods while satisfying ULS design criteria, including buckling;
- Maintain planarity of panels within acceptable limits for a cover with thin polycarbonate sheets.

3.2 Grid Generation

Pseudo-geodesics on arbitrary surfaces are usually generated with a shooting method [7,8]. Indeed, once a seed point and a tangent vector have been chosen, a unique pseudo-geodesic curve can be traced (when $\theta \neq 0$). The main limitation with this approach is the difficulty of controlling the behavior of the pseudo-geodesic away from the seed point, as already noticed in [7]. The morphological strategy is represented in Fig. 5: a geodesic splitting the surface in two regions was traced, and tangent vectors along this curve can be generated as a list of scalar values. The second network was generated by symmetry. With this strategy, the parameters controlling the appearance of the grid are the number of points, the orientation of the pseudo-geodesics and the spacing between seed points on the geodesic. Choosing the spacing corresponds to a reparametrization of the curve which is here optimized to get a uniform meshing of the surface (see Fig. 6, 90 cm).

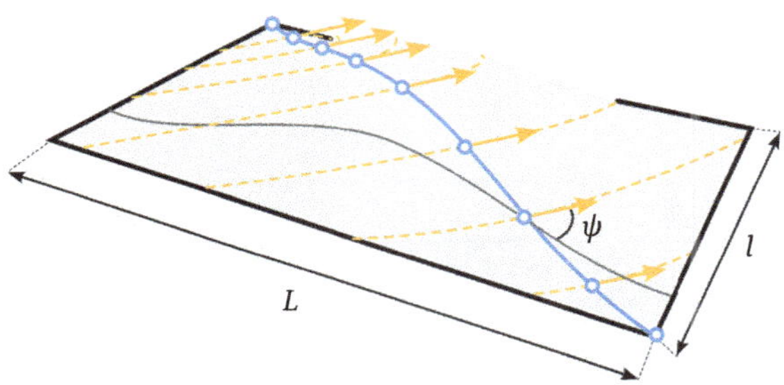

Fig. 5: Generation of the pseudo-geodesic net. A geodesic splitting the surface is chosen (blue). Each tangent vectors (orange) is described by the scalar ψ, which represents the angle between the pseudo-geodesic and the isoparametric curve going through the seed point (blue).

3.3 Shape Variations

A sensitivity study was performed to select the geometry that satisfied construction and mechanical constraints. Figure 6 shows the top view of five different grids. The higher the rise of the surface, the more positive Gaussian curvature there is, and the more irregular the pseudo-geodesic net becomes. The angles φ measuring deviation from perpendicularity for each connector are shown in Tab. 1 (smaller values are preferred). Since pseudo-geodesic curves have a tendency to have more total curvature on positively-curved surfaces, gridshells with lower rises are preferable.

Beyond deviation angles, the planarity of quad panels aligned with the pseudo-geodesic network is also evaluated for the 5 grids (see Tab. 1). It is measured as the ratio between the distance between the two diagonals and the length of the longest diagonal. As expected, the more total curvature in the surface, the larger the planarity

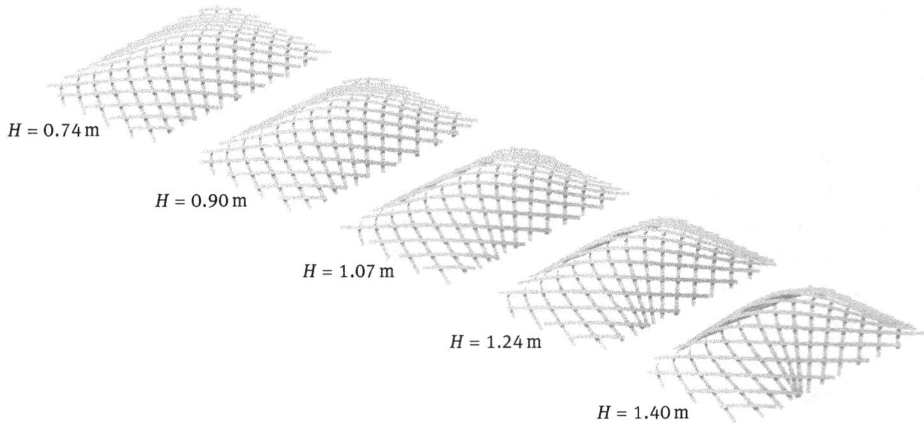

Fig. 6: Perspective view of the pavilion for pseudo-geodesics with $\theta = 60°$. The network becomes irregular for surfaces with a higher rise.

Tab. 1: Comparison of key features of the five grid configurations presented in Fig. 6. The grid with 90 cm height will be finally chosen as a good compromise between mechanical and geometrical constraints.

Gridshell height	74 cm	90 cm	107 cm	124 cm	140 cm
Min. deviation angle φ_{min}	−24°	−29°	−36°	−42°	−46°
Max. deviation angle φ_{max}	17°	19°	20°	24°	32°
Average planarity default	1.2 %	1.6 %	2.1 %	2.6 %	3.0 %
Max. planarity default	3.0 %	4.4 %	6.3 %	8.9 %	12.7 %
Max. lath curvature	0.43 m^{-1}	0.51 m^{-1}	0.59 m^{-1}	0.68 m^{-1}	0.76 m^{-1}
Max. vertical deflection	4.3 mm	4.2 mm	5.6 mm	5.8 mm	8.3 mm

default. For all configurations, the ratio exceeds the usual value accepted for glass panels (0.5 %) with only one quarter of the panels below this threshold.

Concerning mechanical performances, two criteria were selected during this feasibility study. The first one is linked with the prestress level of the laths, which is proportional to the curvature of the lines: the higher the curvature, the higher the permanent bending stresses, and by extension, the lowest the admissible lath thickness. The second one is linked to the grid stiffness, which is here evaluated as the maximum deflection of a grid made of 100 mm wide and 8 mm thick laths under self weight. Both are presented in Tab. 1. For the lath curvature, results follow intuition: values grow with the grid height but remain admissible for all grids (for the laths and material used for the prototype, the maximum allowable curvature is 0.83 m^{-1}). On the contrary, for the deflection, the highest surfaces show the lowest mechanical performances, which comes from the irregularity of grids on highly curved shapes (see Fig. 6).

The surface with $H = 90$ cm was selected in the end, as a good compromise between fabrication and mechanical constraints. The rise-over-span ratio is 22.5 %, which is still compatible with a pure shell behavior.

3.4 Structural Analysis

The CAD-Based design space exploration allowed the design of a grid structure satisfying SLS and ULS criteria with 30 GFRP laths, with a cross-section of 8 mm × 100 mm. The grid weighs 170 kg and 160 connectors cumulatively weigh 95 kg. The governing factor for design is the buckling: we restricted our study to a linear buckling analysis, and made sure that the buckling factor remained superior to 10, which is arguably a conservative approach. The first buckling mode was a lateral-torsional buckling.

Additionally, sensitivity analysis was conducted: the angle θ was sampled between 50° and 90° and finite element analysis was conducted on each geometry similarly to 6]. Some results are summed up in Fig. 7: first, they confirm Mesnil et al. founding [6]: 60° is the best angle on this surface and the mechanical efficiency of our design is doubled compared to geodesic gridshells. Second, they also show that the pre-stress due to bending (around 130 MPa) is much more than stress induced by external loads (around 15 MPa for critical wind load). Finally, the sensitivity to the addition of a triangulation was also studied: adding a triangulation doubles the stiffness, which is less than for other projects [4]. We explain this by the relatively small scale of the pavilion, by the

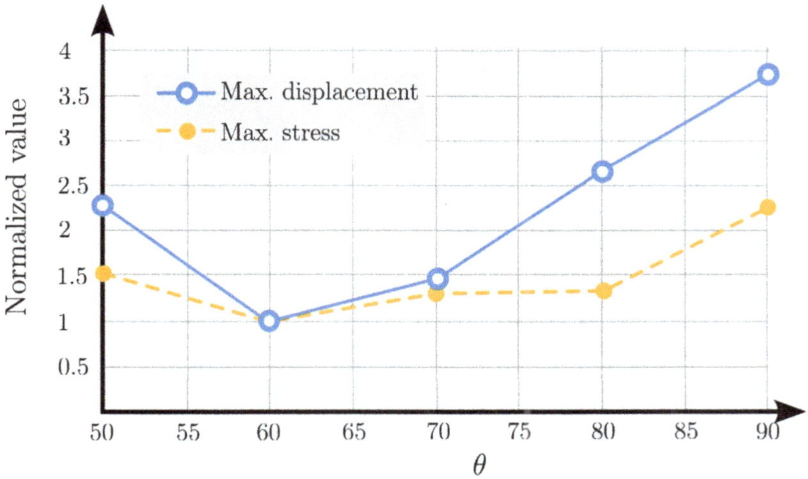

Fig. 7: Relative gridshell performance for various lath inclinations for a uniform load of 1 kPa applied to the structure.

fact that the grid is not a Chebysheff: the inclination of laths makes the in-plane shear stiffness rather high, which limits the need for additional bracing. The sensitivity of pseudo-geodesic gridshells to bracing may be an interesting research direction in the future.

4 Fabrication of the Pseudo-Geodesic Pavilion

4.1 High Node Congruence

One of the main contributions of this case-study is the joint design, which translates some geometrical properties of pseudo-geodesic curves into technological strategies. One of the main objectives was to construct congruent nodes that could be assembled from standard components. Additionally, the intersection of the laths is not allowed like it is done in a scissor joint, as the sensitivity of the composite material to crack or notch, and also to allow the re-usage of the laths after the structure is disassembled. Thus, the connectors are eccentric to the laths. In practice, this means that the neutral axes of the beams are not on the target surface; they are reconstructed from the ruling direction of the pseudo-geodesic strip.

Figure 8 shows a congruent connector satisfying the geometrical requirements. It mainly consists of two stainless steel Z plates and one long bolt passing through its center, allowing rotation in one direction. The Z plate serves to maintain the constant angle between two vectors, which was 60° in our case. The centerline of the laths is offset, and it is clamped between the Z plate and the small plate. Jappy bolts were used, which allowed the laths to be pressed easily at an accurate position. This pressing is essential to transmit shear force. Additionally, EPDM tape was used at the location of the connectors in order to increase the frictional force between the laths and the steel connector.

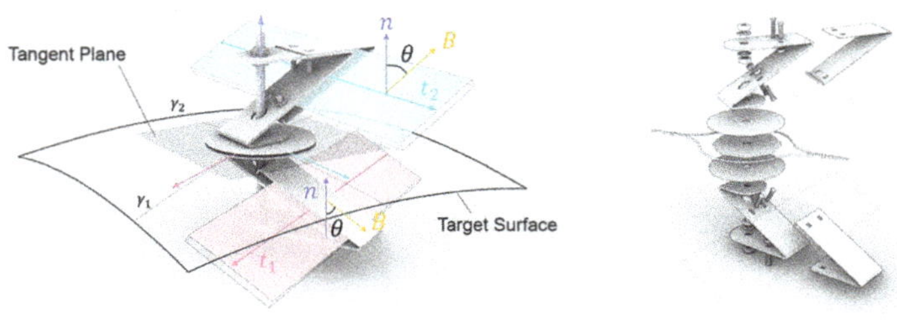

Fig. 8: Alignment of connector with respect to the given surface (left) and exploded view (right).

4.2 Grid Assembly

The network of the shell consists of the initially straight flat laths made of GFRP. Unlike most elastic gridshells, the structure was not deployed but assembled lath by lath. Indeed, the structure does not have repeating lengths. This property may be one of the factors explaining the stiffness of the structure and the relatively low influence of triangulation on the buckling capacity and stiffness of the gridshell. The assembly of all laths was done in one day. The inclination of the laths brings several architectural properties, such as semi-transparency of the roof and a unique edge of the eaves. Hence, the solar shading differs at different angles (see Fig. 9, the lower layer seems much wider than the upper one).

Fig. 9: The standard nodes developed in this study and bracing system. Laths cross-section: 100 × 8 mm.

4.3 Bracing

The circular plates and octagonal plate (Fig. 9) in the middle of the connectors serve as the fixture for a diagonal cable for bracing, allowing triangulation of the existing mesh of the pavilion. Cables were pulled inward thanks to zip ties. The geometry of the bracing, reminiscent of pre-stress strategies invented by Eladio Dieste [14], acts as a mechanical amplifier: a small pinching force results in a large tension in the cable. All the plates of the connector assemble together at one node, which gives semi-transparency to this grid shell.

4.4 Covering

The prototype was finally covered with 3 mm-thick polycarbonate quads aligned with the pseudo-geodesic network. As discussed in Section 3.4, the deviation from planarity is limited, and panels can be easily obtained from a flat sheet of material. Before fabricating the panels, it was decided to control the geometrical accuracy of the grid, whose form was only determined by the position and orientation of the anchorage and connectors. To this end, the length of every edge and diagonal was measured for each quad panel. An average accuracy of ±0.5 % on the diagonal length was found, corresponding to a very acceptable ±3 mm in the positions of the connector axes. As the panels are point-fixed, this accuracy can be easily handled through tolerances in panel openings. An overview of the covered gridshell is presented in Fig. 10: panels rest locally on the top surface of the connectors, which is parallel to the reference surface (see Fig. 8) and use the connector axes for clamping.

Fig. 10: Gridshell covered with quadrangular polycarbonate shingles.

5 Conclusion

This paper investigates the potential of pseudo-geodesic gridshells through the design and fabrication of a composite gridshell. The study confirmed that the straight flattening property can be achieved with high node congruence for pseudo-geodesic gridshells. A parametric study has also shown that the structure outperforms a similar geodesic gridshell on the same surface, confirming previous results [6].

Some questions are left open for future work. Regarding design, the tracing of pseudo-geodesics still relies on a shooting method that is difficult to control. An ad hoc design strategy was developed, but gaining a better understanding of the behavior of pseudo-geodesic nets would be a great help to designers. Regarding mechanical efficiency, lateral-torsional buckling is the critical failure mode; individual member stabilization is thus a new research topic worthy of exploration. Regarding the covering, a proposition was made using nearly-planar polycarbonate shingles; other solutions could be explored in combination with stabilizing systems against member lateral-torsional buckling. From an architectural point of view, the asymmetry of the grid due to the constant angle of the lath with the surface normal introduces a parameter that can be used to control daylight by varying the solar shading through the surface. This aspect is also worth further exploration. Finally, the question of deployability is an interesting challenge for future work. Although intellectually satisfying and convenient for prefabrication, non-deployability may be a plus for structural performance.

References

[1] Knippers, J., and T. Helbig. (2009). Recent developments in the design of glazed grid shells. *International Journal of Space Structures* 24(2), 111–26.

[2] Schling, E. Design and construction of curved support structures with repetitive parameters. 2018. *Advances in Architectural Geometry 2018.*

[3] Mesnil, R. et al. (2015). Isogonal moulding surfaces: a family of shapes for high node congruence in free-form structures. *Automation in Construction* 59, 38–47.

[4] Douthe, C., O. Baverel, and J.-F. Caron. (2006). Form-finding of a grid shell in composite materials. *Journal of the International Association for Shell and Spatial structures* 47(1), 53–62.

[5] Kotelnikova-Weiler, N. et al. (2013). Materials for actively-bent structures. *International Journal of Space Structures* 28(3–4), 229–40.

[6] Natterer, J. et al. (2000). Roof of the main hall at EXPO 2000 in Hanover, Germany. *Structural engineering international* 10(3), 167–69.

[7] Jiang, C. et al. Curve-pleated structures. (2019). *ACM Transactions on Graphics (TOG)* 38(6), 1–13.

[8] Mesnil, R., and O. Baverel. (2023). Pseudo-geodesic gridshells. *Eng. Structures* 279, 115558.

[9] Dias, M. A., and B. Audoly. (2015). "Wunderlich, meet Kirchhoff": A general and unified description of elastic ribbons and thin rods. *Journal of Elasticity*, 119, 49–66.

[10] Lefevre, B. et al. (2017). A 4-degree-of-freedom Kirchhoff beam model for the modeling of bending–torsion couplings in active-bending structures. *Int. Journal of Space Structures* 32(2), 69–83.

[11] Wunderlich, W. (1951). Raumkurven, die pseudogeodätische Linien eines Zylinders und eines Kegels sind. *Compositio Mathematica*, 8, 169–84.

[12] Hojjat, M., E. Stavropoulou, and K.-U. Bletzinger. The vertex morphing method for node-based shape optimization. (2014). *Computer Methods in Applied Mech. and Eng.* 268, 494–513.

[13] Mesnil, R. et al. (2018). Fabrication-aware shape parametrisation for the structural optimisation of shell structures. *Engineering Structures*, 176, 569–84.

[14] Pedreschi, R. (2004). Structural innovation in pre-stressed brickwork. *Construction and Building Materials*, 18(2), 99–109.

[15] Panozzo, D., P. Block, and O. Sorkine-Hornung. (2013). Designing unreinforced masonry models. *ACM Transactions on Graphics (TOG)*, 32(4) 1–12.

Eike Schling, Zongshuai Wan, Hui Wang, Pierluigi D'Acunto

Asymptotic Geodesic Hybrid Timber Gridshell

Abstract: This paper presents a strategy to design and build strained timber gridshells from exclusively straight timber planks, which are interwoven and elastically deployed into a doubly curved web. For this purpose, we combine asymptotic (A) and geodesic (G) curves into hybrid AAG-webs on curved surfaces. We present a digital method to design and geometrically optimize the timber AAG-webs to include equal intersection angles and geodesic boundaries. This new construction system benefits from the targeted use of the two differing bending axes of timber planks for flexibility and rigidity. The flat geodesic planks are interlaced at the midpoint of the asymptotic beams to create a tri-hex pattern, which lowers the buckling length and decisively increases the overall stiffness. As a proof of concept, a large-scale timber gridshell covering an area of 60 m^2 was designed and built. We document the construction process of manufacturing, prefabrication, elastic deformation, on-site assembly, and installation of the polycarbonate cover to verify constructive tolerances and feasibility. The structure is tested and simulated to validate our computational results.

Keywords: asymptotic curves, geodesic curves, timber gridshells, elastic bending and torsion, curvature analysis, timber AAG-webs, resistance through form, cradle-to-cradle

1 Introduction

The construction industry is one of the biggest contributors to environmental degradation and CO_2 emissions. Construction waste can be avoided by using standardized elements that can be disassembled and reused in the future. This cradle-to-cradle approach seems to be incompatible with the idea of freeform design, which usually demands highly bespoke elements and complex fabrication. Our research is looking to resolve this contradiction, by exploring the design and fabrication of doubly curved gridshells from exclusively flat and straight, off-the-shelf timber planks with repetitive joints. This offers substantial material savings compared to CNC-milled lattice gridshells. The flat packed planks can be transported easily, which in turn reduces CO_2 emissions, and assembled utilizing the kinetic behaviour of lamella grids.

1.1 Geometrical background

Architectural Geometry has introduced methods to control and optimize doubly curved surfaces (Pottmann et al. 2007). The curvature $\kappa = |c''(s)|$ of a curve $c(s)$ (with arc length parameter s) on a surface S is decomposed into two parts i. e. normal curvature κ_n and geodesic curvature κ_g, which are the two projections of the curvature vector

https://doi.org/10.1515/9783111162683-008

c'' onto the surface normal and tangent plane at a surface point (Fig. 1a). These three curvature elements have the relation $\kappa^2 = \kappa_n^2 + \kappa_g^2$. Let α be the angle between c'' and the surface normal, then $\kappa_n = \kappa\cos(\alpha)$, $\kappa_g = \kappa\sin(\alpha)$. The value of constant α throughout the curve has two special cases ($\pi/2$ and 0), making two special kinds of curves on the surface. When $\alpha = \pi/2$ along the curve, it characterizes an *asymptotic curve* with $\kappa_n = 0$, and the curvature vectors lie on the tangent planes. If $\alpha = 0$, then it is a *geodesic curve* with $\kappa_g = 0$, and the curvature vectors are in the planes spanned by the tangents and surface normals. By constraining either κ_n or κ_g to zero, we can design gridshells that can be built from *straight asymptotic* or geodesic planks (Fig. 1b,c).

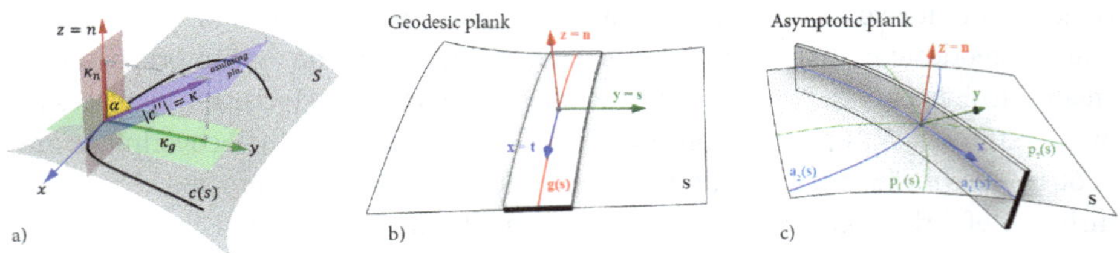

Fig. 1: Curvature of curves and planks: a) Relationship of κ, κ_n and κ_g of a curve $c(s)$ on a surface S. b) Plank along geodesic curve $g(s)$ on a surface S. c) Plank along asymptotic curve $a_1(s)$ on a surface S. Asymptotic curves $a_1(s)$ and $a_2(s)$ are symmetric with respect to the principal curvature lines $p_1(s)$ and $p_2(s)$.

1.2 Geodesic and asymptotic structures

Geodesic and asymptotic planks have, until now, been used independently for the design of gridshells. *Geodesic structures* have been studied (among others) for their construction simplicity and structural performance (Weinand and Pirazzi 2006). Julian Natterer developed Timber gridshells from flat geodesic planks (Natterer et al. 2000), using a technique of layering to create bespoke ribbed shells. Research in Architectural Geometry has investigated versatile geodesic patterns (Pottmann et al. 2010), their self-forming behavior from flat to curved grid (Soriano 2017) as well as their potential to reuse recycled, off-the shelf material (Haskell et al. 2021).

Asymptotic structures have been described through mathematical theory as early as (Finsterwalder 1897). Recently, a higher attention has been paid to asymptotic networks, their developable properties (Tang et al. 2016), and their discrete optimization as quad meshes (Jiang et al. 2020). The first asymptotic gridshell, the INSIDE/OUT pavilion was constructed in stainless steel including diagonal steel (Schling et al. 2018). Asymptotic grids can be assembled on flat ground and deformed into the global design shape without need for formwork, as their movement is restricted by the scissor joints, and the directional bending of planks. This kinetic mechanism has been investigated in depth by (Schikore et al. 2020) and used to design transformable structures, such as the

Kinetic Umbrella. Because the asymptotic curves are locally symmetric to the principal curvature directions, an isothermal asymptotic network can be covered with planar quad panels or developable strips in the diagonal direction (Schling and Wan 2022). The panel strips, however, are curved, i. e., do not naturally follow a geodesic path and can thus not easily be built from straight, off-the-shelf panels.

Structural behavior. For both geodesic and asymptotic structures, the slender planks have an effect on global and local stiffness (Schikore 2023). In geodesic gridshells, the weak axis of a plank lies tangential to the design shape, which significantly lowers its global stiffness and results in buckling out-of-plane of the gridshell. Usually, this is overcome by adding additional layers of planks. Asymptotic gridshell are globally stiffer, as their planks stand upright and can carry loads through their strong axis. However, the slender planks may buckle locally in case of high internal compression or bending (lateral-torsional buckling) (Wan and Schling 2022).

1.3 Contributions and Overview

We implement the design and construction of a triangulated timber gridshell by combining asymptotic (A) and geodesic (G) curves into *hybrid AAG webs* (Schling et al. 2022). The geodesic elements triangulate the grid, halve the buckling length, and offer simple façade solutions with standardized aluminum mullions and developable panels. The structure is assembled from only straight off-the shelf planks, without offcuts, or bespoke boundary elements, with the goal to design for disassembly and reuse. In relation to previous work, this paper presents the following contributions:
- an architectural design workflow to create a variety of initialization meshes
- a custom energy term in the optimization problem to fit specific design constraints
- constructive solutions, realized in a case study, the 60 m^2 Timber Vault

The remainder of this paper is structured as follows. In Sec. 2, we present the computational workflow to design and optimize AAG webs. We start with a negatively curved rotational surface to initialize the design routine. The complex AAG web is then generated through discrete optimization using planar constraints of vertex stars. Additional optimization goals, such as equiangular joints, planar and geodesic boundaries are added to simplify fabrication. In Sec. 3, we present the construction development of the AAG Timber Vault. We describe the curvature analysis and material choice, prefabrication, and assembly, including the installation of a semi-discrete façade from bent polycarbonate panels. In Sec. 4, we highlight the design impact on structural performance. We present a digital simulation and validate it through physical load tests. Finally, in Sec. 5, we reflect on the challenges and potentials of this technology and highlight ongoing research in this field.

2 Computational Design and Optimization

Finding a good starting mesh to optimize an AAG-web is not a trivial problem. It highly depends on the mesh parametrization. We adopt rotational surfaces, in which the principal curvature lines are the parallel circles and meridian curves, and the meridian curves are naturally geodesic. We extract mesh patches from negatively curved rotational surfaces and use them as initial meshes for our optimization approach. This enables quick testing of alternative shapes consisting of one or more rotational surfaces that are cropped, intersected, copied, or flipped to create meaningful shell designs (Fig. 2). These models function as preliminary design options to develop construction strategies, details, and material investigations to a high level (Fig. 3), without relying on the final refined mesh geometry.

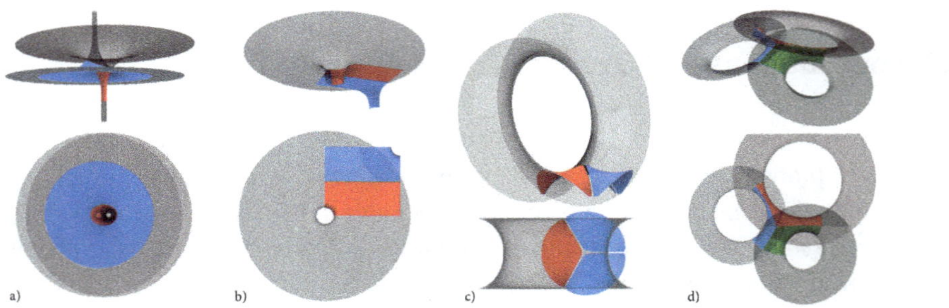

Fig. 2: Designing architectural alternatives with rotational surface (colors differentiate subsequent steps in creation, red, blue and green). a) Flipped rotational surface creating column and roof. b) Cropped rotational surface creating half of the symmetric roof. c) Cropped surface is rotated to create a triple-arched gridshell. d) Three individual rotational surfaces are intersected and cropped to create a vault with three entryways.

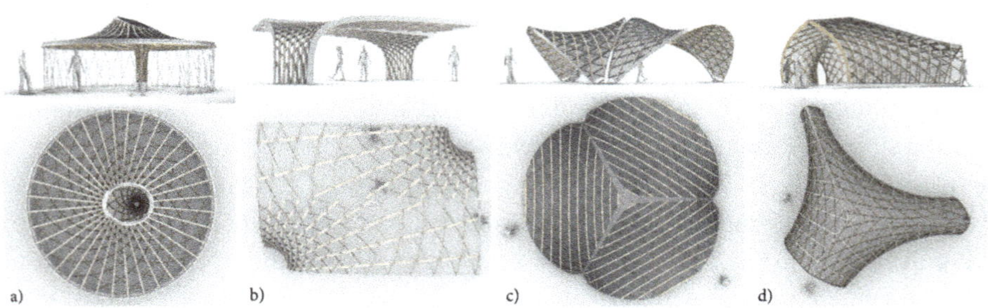

Fig. 3: AAG designs, that are based purely on rotational surfaces. In these cases, joints are not equiangular, and not all boundaries are geodesic. These designs are used to develop architectural ideas quickly without the need for optimization.

2.1 Discrete Nets and Webs

Discrete asymptotic nets (A-nets). The curvature vectors of asymptotic curves lie on the tangent plane of the surface. In a discrete network, three consecutive vertices of an asymptotic polyline define a discrete osculating plane. The mesh is optimized by enforcing that two osculating planes agree at each vertex star (Fig. 4 left). Therefore, a discrete A-net is defined by a quad mesh with planar vertex stars, in which five vertices $v_i, v_{ij} (j = 1, \dots, 4)$ are coplanar (Sauer 1937, 1970).

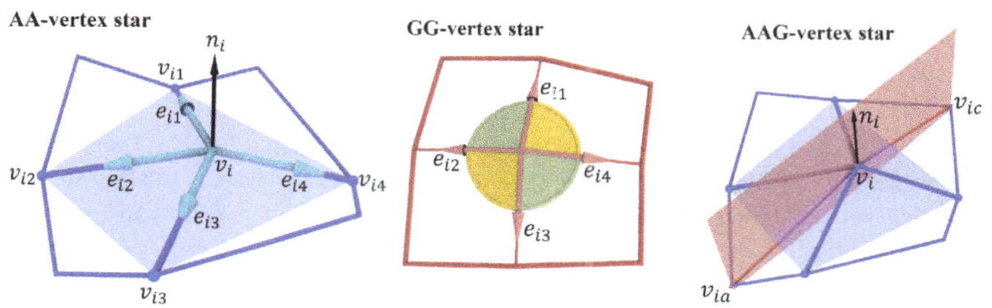

Fig. 4: Left: Planar vertex star of A-net. Middle: Vertex star of a conventional G-net, where two opposite angles are equal. Right: Vertex star of AAG-web, combining a planar vertex-star (AA) with a diagonal plane through the surface normal (G).

Discrete geodesic nets (G-nets) were first introduced by (Wunderlich 1951). A discrete G-net is defined as a quad mesh where two pairs of opposite angles at each vertex star are equal (Fig. 4 middle). However, this angle condition cannot help to get meshes with only one family of geodesic curves.

Discrete asymptotic geodesic nets (AAG-nets). We utilize the geometry property that the discrete osculating plane passes through the surface normals to get the discretization of diagonal geodesics in AAG-webs. The numerical model is achieved by optimization of A-nets with one family of diagonal polylines being geodesic. Depending on the design surface, this logic can be rotated to enforce the asymptotic vertices to be diagonal (see Fig. 4 right).

2.2 AAG optimization

The geometric properties of AAG-webs enable the use of planar rectangular timber strips, which are bent orthogonally on the surface along asymptotic curves and tangentially on the surface along geodesic curves. Numerical optimization is used to simplify the fabrication of the gridshells. To this end, we optimize for constant net angles between asymptotic curves to create repetitive joints. For our timber gridshell design (Fig. 5, 6), we also constrain the bottom boundary curve to be planar and geodesic to simplify the fabrication of supports. We optimize a symmetric mesh, which is then rotated to form a triple symmetric vault. Below, we list all the constraints as optimization energies we used for our designed shell. Notations refer to Fig. 4. The optimization problem is solved by regularized Gauss-Newton algorithm, where all the geometric conditions are represented by constraints with no more than quadratic equations (Tang et al. 2016).

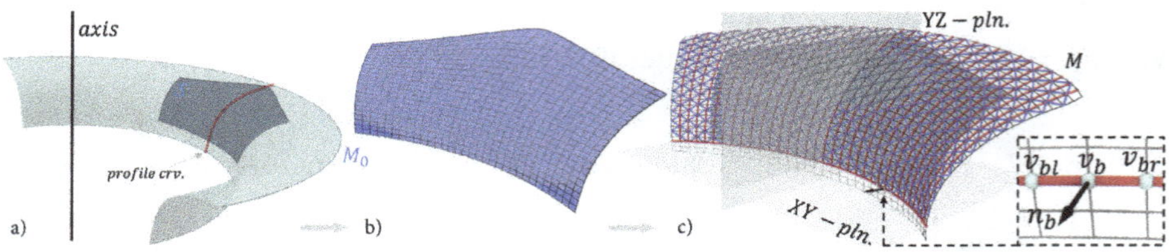

Fig. 5: a) A patch surface S is created from a rotational surface. b) A quad mesh M_0 is rebuilt from S and is used as the initial mesh for optimization. c) An AAG-web MM is achieved by solving the optimization problem with the target function Eq. (7), where one family of mesh polylines are geodesic curves (red), and two diagonal polylines (blue) are asymptotic.

A-net energy. The energy term of A-net is expressed with unit vertex normal n_i of the discrete surface as auxiliary variables. $|V|$ is the number of regular vertex stars, where vertices i are connected by 4 neighboring vertices ij:

$$E_{\text{Anet}} = \sum_{i=1}^{|V|} \sum_{j=1}^{4} (n_i \cdot (v_{ij} - v_i))^2 + \sum_{i=1}^{|V|} (n_i \cdot n_i - 1)^2 \tag{1}$$

Geodesic energy. To express the geodesic condition in one of the diagonal directions, we ask for three coplanar vectors $v_{ia} - v_i$, $v_{ic} - v_i$, n_i,

$$E_{\text{geo}} = \sum_{i=1}^{|V|} (n_i \cdot [(v_{ia} - v_i) \times (v_{ic} - v_i)])^2 \tag{2}$$

Angle energy. Given constant angle θ_0 between asymptotic curves, the isogonal condition at each vertex star can be expressed with the help of the unit tangent vectors $t_{i1} = \frac{e_{i1}-e_{i3}}{|e_{i1}-e_{i3}|}$, $t_{i2} = \frac{e_{i2}-e_{i4}}{|e_{i2}-e_{i4}|}$ by $t_{i1} \cdot t_{i2} = \cos\theta_0$, where $e_{ij} = \frac{v_{ij}-v_i}{|v_{ij}-v_i|}$ $(j = 1, \ldots 4)$ are unit edge vectors emitting from the vertex v_i. Norms $|v_{ij} - v_i|$ and $|e_{i1} - e_{i3}|$, $|e_{i2} - e_{i4}|$ can be computed in each iteration. Then the angle energy term is:

$$E_{\text{angle}} = \sum_{i=1}^{|V|} \left(\frac{e_{i1} - e_{i3}}{|e_{i1} - e_{i3}|} \cdot \frac{e_{i2} - e_{i4}}{|e_{i2} - e_{i4}|} - \cos\theta_0 \right)^2 + \sum_{i=1}^{|V|} \sum_{j=1}^{4} \left(e_{ij} - \frac{v_{ij} - v_i}{|v_{ij} - v_i|} \right)^2 \tag{3}$$

Bottom curve energy. The bottom curve is planar and geodesic. The curve vertices lie on the XY-plane.

$$\sum_{b \in \text{bottom\_polyline}} (v_b \cdot [0, 0, 1])^2 = 0$$

We apply the geodesic energy Eq. (2) to any three consecutive vertices v_{bl}, v_b, v_{br} of the bottom curve. However, since the valence of the boundary vertices is 3, the vertex normal n_b is not as clearly defined as any regular vertex stars in Eq. (1). We thus extend the bottom curve with additional rows of vertices so that the original boundary vertices are interior (Fig. 5c, red curve), and the planar vertex stars, vertex normal Eq. (2), and geodesic curves are well defined:

$$\sum_{b \in \text{bottom\_polyline}} (n_b \cdot [(v_{bl} - v_b) \times (v_{br} - v_b)])^2 = 0$$

After suitable convergence is achieved, these artificial rows of vertices can be removed, and the geodesic property of the assigned bottom boundary curve is preserved. This bottom curve energy term is combined as

$$E_{\text{bottom}} = \sum_{b \in \text{bottom\_polyline}} \left((v_b \cdot [0, 0, 1])^2 + (n_b \cdot [(v_{bl} - v_b) \times (v_{br} - v_b)])^2 \right) \tag{4}$$

Symmetry energy. The left and right areas of the AAG-web are symmetric with respect to the central YZ-plane, so the left vertices v_{left} and right vertices v_{right} form the energy term

$$E_{\text{sym}} = \sum \left((v_{\text{left}} - v_{\text{right}})^2 \times [1, 0, 0] \right)^2 \tag{5}$$

Fairness energy. A regularizer for any three consecutive vertices of a polyline allows for obtaining smoother polylines. It works as a weighted fairness term

$$E_{\text{fair}} = \sum_{i \in \text{poly}} (2v_i - v_{il} - v_{ir})^2 \tag{6}$$

Target function. The objective function for our AAG-web is a combination of (1)-(6) as

$$E = E_{\text{Anet}} + E_{\text{geo}} + E_{\text{angle}} + 10 \cdot E_{\text{bottom}} + E_{\text{sym}} + \omega E_{\text{fair}} \tag{7}$$

where $\omega(< 1)$ is a smaller weight enabled during the optimization and disabled in the final optimization steps to ensure convergence. To make sure the property of the bottom curve is highly satisfied, we choose a high weight (i. e. 10) for this term. We use the guided projection method (Tang et al. 2014), which is a Gauss-Newton method, to get the solver of Eq. (7). The algorithm is implemented in Python and tested on an Intel Xeon E5-2687W 3.0 GHz processor.

The densely optimized AAG-web (Fig. 6a) is converted into NURBS geometry through interpolation. Alternating curves of A and G are selected (b) to generate a tri-hex kagome pattern, which reduces density while improving constructive and load-bearing qualities (see Sec. 3.1 and 3.2). The grid is then cropped along geodesic boundaries (to simplify fabrication), which align with the grid on the bottom and sides but follow a misaligned geodesic path along the ridge, to closely offset to the symmetry plane. This NURBS geometry model is used to set up a parametric model (c) in Rhinoceros 3D – Grasshopper, and develop construction strategies, offsets, thicknesses, details, and materiality to a high level. The final design of the Asymptotic Geodesic Hybrid Timber Gridshell is based on three symmetric AAG webs that form a negatively curved vault.

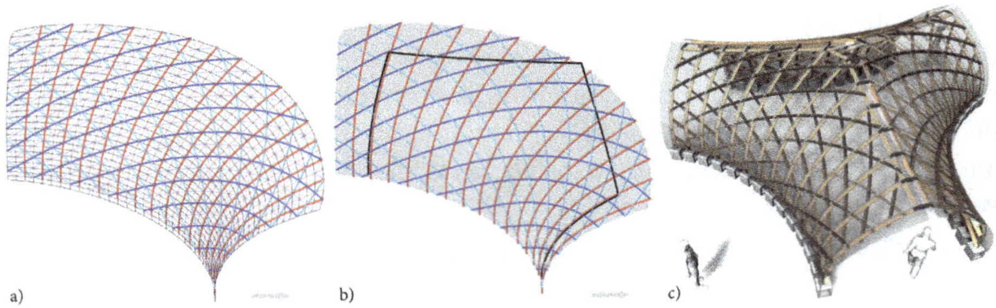

Fig. 6: The optimized mesh (a) is converted into a NURBS surface and curves through interpolation (b). Specific asymptotic and geodesic curves are selected to generate a tri-hex pattern. The curves and surface are cropped and processed, to create a detailed architectural model (c).

3 Construction Development

Finger jointed ash lamellas were chosen as the building material due to their outperforming properties of strength and elasticity (Kovryga et al. 2020). To avoid any waste of resources, the thinnest cross-sections supplied by the sawmill of 25 × 100 mm were purchased and simply split to create 2 × 12 × 100 mm planks. The choice of material and cross-section became a decisive driver for the design geometry, as it naturally limits the minimum radii for bending ($r_{min} = 1.7$ m) and maximum twist ($\beta_{max} = 60°/\text{m}$).

3.1 Timber Network and Joints

We create a tri-hex web in which geodesic planks connect to the midpoints of each asymptotic beam (Fig. 7i, 8d). This has three decisive advantages for construction and load-bearing behavior: 1) The geodesic planks can be assembled independently, either in the flat or curved stage of the erection process. They are connected simply with a steel screw to the coupling blocks of asymptotic beams. 2) The geodesic planks are used as locking mechanism of the kinetic erection process. During erection, the geodesic planks slide between the asymptotic layers and allow for a change in geometry. Once the final design shape is reached, the geodesics are fixed and brace the structure. 3) The geodesic planks are interwoven between top and bottom asymptotic beams at mid-points, and thus half the buckling length of both the asymptotic beams, as well as the geodesic planks themselves.

The timber is cut to length and drilled with minimal holes to allow reuse. This low-tech fabrication is achieved by assembling the three families of planks in three separate layers, with asymptotic beams on top and bottom and geodesic planks in between. Each asymptotic beam is constructed from two planks, which are individually bent and coupled with 24 mm blocks. These *coupled asymptotic beams* have three advantages:

Fig. 7: Top Row: Prefabrication. 12 × 100 mm planks (a) are marked (b), drilled (c), and coupled with 24 mm studs (d) into pre-bent (e) beams (f). This process embeds the geodesic curvature. Bottom Row: Assembly and Erection. The asymptotic beams and geodesic planks are interlaced into flat segments (g), and deformed into the doubly curved shape (h). The combination of asymptotic and geodesic planks (i) creates a stiff shell, that is transported (j) and assembled (k) on site.

1) The two sibling planks are pre-drilled individually, incorporating slight differences in the node-to-node distance. This allows us to embed the geodesic curvature during prefabrication (Fig. 7f). 2) The planks can be deformed individually but, after they are coupled, act as one strong beam with high lateral bending stiffness. 3) The parallel layout of planks at 24 mm distance allows a centric connection of top and bottom asymptotic beam using 22 cm long (24 × 24 mm) hexagonal studs (Fig. 7d).

3.2 Prefabrication and Assembly

The prefabrication is simple (Fig. 7). The ash planks (a) are first marked by hand (b) and cut to the right length. Groups of three identical planks (for three shells) are then aligned and tightly clamped (to overcome individual warping due to humidity and drying of the material) and pre-drilled using templates (c). The asymptotic beams are assembled each from two individual planks with 24 mm studs (d) in between. The planks are laid across two stands so that they naturally bend and sag into a curved shape, until the pre-drilled holes align. The two planks are then coupled at each stud with two steel bolts (e). With this method, 66 pre-curved asymptotic beams were prefabricated and stored to wait for further processing (f).

As a next step, the asymptotic beams and geodesic planks are *assembled* flat on the ground (g) into the three-layered tri-hex web. The top and bottom asymptotic beams are connected loosely (to still allow some rotation) through the long studs, while the geodesics are simply slid in between (without connections). The flat segment is then hoisted up (h) and elastically deformed into the design shape (i). During this process, the geodesic planks shift through the grid, the asymptotic beams are twisted, and the

Fig. 8: Completed timber gridshell. Top Row: a) Precast concrete blocks with steel supports. b,c) Birds-eye view of the timber gridshell without cover. d) Inside the Timber Vault. Bottom Row: Polycarbonate panels are tailored with concave edge (e), and installed using aluminum mullions along the geodesic directions (f), to create a smooth, transparent skin (g).

AA-joints slightly rotate into the final 60° angle. The final geometry is fixed by adding geodesic planks at upper and lower boundaries and fixing all bolts for asymptotic and geodesic planks. These stiff shell elements are transported to the building site (j) and combined along the roof ridge with steel blades (k). The timber gridshell (Fig. 8) is supported by 42 precast concrete blocks (14 for each shell) to resist wind suction and horizontal thrust. The shell elements are fixed with vertical steel bolts (a) that allow individual height adjustment to overcome construction tolerances. A *curved transparent cover* was installed (Fig. 8 e-f) to protect the wood from weathering and create a sheltered space. 13 strips of 10 mm thick polycarbonate multi-skin sheet are clamped and bent along the geodesic directions to form a semi-discrete skin. Each panel is tailored with a long concave edge (e) to accommodate the negative curvature. The AAG Timber Vault is the first architectural application that combines asymptotic and geodesic planks in a triangulated web. The structure consists of three shells covering an area of 60 m^2. Each shell weighs approx. 300 kg for timber, 30 kg for steel screws and plates, and 40 kg for façade panels and aluminum profiles. This equates to a total of 1100 kg for 100 m^2 roof area (approx. 11 kg/m^2).

4 Structural Performance

The design of the Timber Vault was driven by structural criteria. To activate a shell behavior, the grid was designed with a vaulted shape, linear supports, and was adjusted to its highest possible double curvature without exceeding the minimal bending radii of lamellas. The tri-hex grid with constant node angles ensures a homogenous edge length and additional bracing effect to increase the overall stiffness.

4.1 Structural Analysis

During the planning process, multiple load combinations were considered in detail, combining self-weight, snow load (full-span/half-span), wind load from three directions, as well as the residual stresses caused by the elastic deformation process during assembly (see Sec. 3.2). In this paper, we can only shortly highlight the full-span load and physical test results. For this article, the structural performance of the timber gridshell was evaluated using the Finite Element (FE) software Karamba 3D (Preisinger 2013), using a linear solver to simulate the structural behavior. All members are modelled using 2-node beam elements with 6 DOFs at each node. The connections are assumed to be all rigid, except for the geodesic connection, which allows rotation around local z. An elastic material model is used for the ash timber with Young's modulus of 14000 N/mm^2 and bending strength of 50 N/mm^2, obtained from the literature (DIN EN 338 2016) and confirmed by preliminary material testing. The boundary is fixed in three directions as hinge supports at every end of the geodesic lamellas.

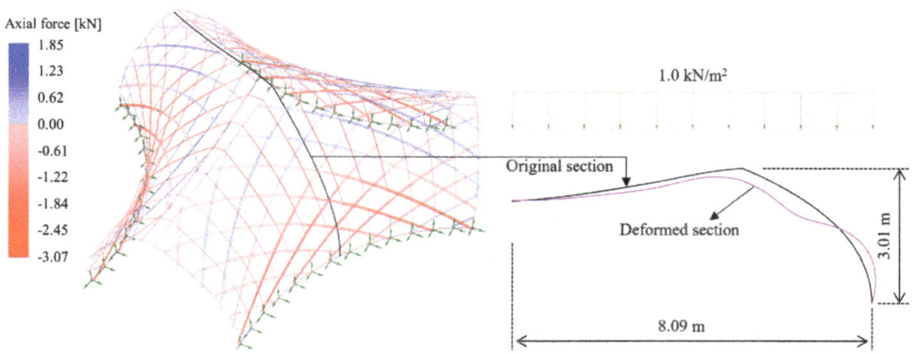

Fig. 9: Axial force (left) and section deformation mode (right) of the asymptotic-geodesic timber vault under self-weight and a full-span homogenous load of 1.0 kN/m^2; deformation is scaled by 80.

Full-span distributed load. To visualize the overall load-bearing behavior, we simulate the gridshell under its self-weight (without residual stress) and a full span homogenous load of 1.0 kN/m^2 on the horizontal projection plane. The external surface load is discretized and applied to the joints of the structure. The structural behavior (Fig. 9) can be described as an "inverted dome". The shell's top part deforms inwards. Due to the negative curvature, this deflection stretches the upper members in the horizontal direction and generates tension forces in the asymptotic beams. On the contrary, the

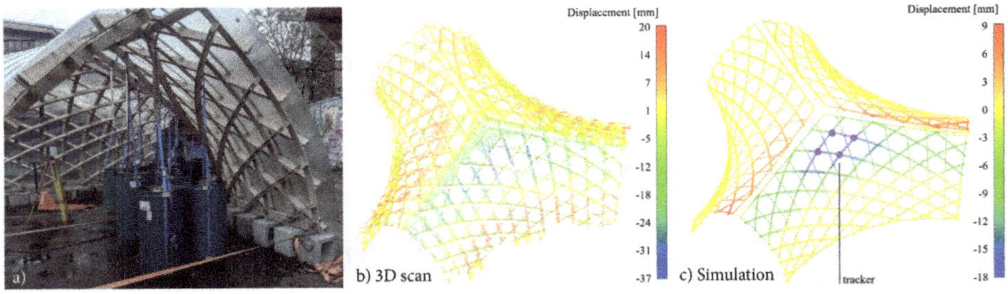

Fig. 10: Structural test under an asymmetric load of 26 kN. a) 13 barrels were filled with up to 200 l of water. b) The scanned point cloud illustrates the vertical deflections, which were compared with the simulation result (c).

shell's lower part deforms outwards, shortening the asymptotic lamellas and leading to compression. The compression in the geodesic planks decreases in the lower part, as they resist the scissor-distortion of the asymptotic grid. This arrangement allows the structure to transfer loads in a membrane-like force within the shell. The upright asymptotic planks provide out-of-plane stiffness, while the flat geodesic planks provide in-plane stiffness and reduce the buckling length.

4.2 Load Tests

To assess the loading capacity of the built structure, we conducted several physical loading tests on the constructed prototype, including one-point loading, local region loading, and one-shell asymmetric loading. The one-shell asymmetric loading is shown in Fig. 10. To avoid the supports sliding on the ground, we used six belts as horizontal ties between opposite supports. Gradually, 13 barrels were each filled with 200 liters of water, reaching a total of 26 kN vertical load distributed on 13 loading points (Fig. 10a). During this test, no failure, cracking or sudden movement of the structure was recorded. The structure was scanned using both a point cloud (Fig. 10b) and individual trackers. The deformation modes from the load test are well compatible with the results of the FE simulation. The average vertical deflection of four joints (marked with purple in Fig. 10c) is measured as 16 mm in the FE simulation and 31 mm in the scanning. Three factors may account for the magnitude difference. First, the material properties used in the FE simulation are only an estimation of the material properties of the real structure (Sec. 4.1). Second, unlike the FE simulation, the supports were not fully fixed in the load testing (as the horizontal belts were able to stretch). Third, the joints in the real structure are not completely rigid due to construction tolerance.

5 Conclusion

In this paper, we implemented the design and construction of triangulated doubly curved AAG gridshells from standardized planks of timber, with the goal to enable a circular use of building material. To enable an efficient computational workflow, we first designed negatively curved rotational surfaces to then initialize the discrete optimization of specific constraints that would simplify fabrication and improve structural performance. The dependency between geometric parameters and material properties played a decisive role in the design process, which is seeking a beneficial shape for structure and supports, a balance of curvature and elasticity, a clean arrangement of the 3-web to accommodate façade and boundaries, and well-resolved layering and detailing of planks. This iterative process yielded a triple symmetric negatively curved AAG-Vault with constant 60° nodes. The use of repetitive joints was significant to streamline fabrication and create a homogenous network with structural and aesthetic qualities. The construction process takes advantage of the global kinetic behavior of asymptotic grids, and thus poses challenges due to the residual stress, which limits the plank thickness, causes relaxation and creep, and affects the building accuracy. Physical tests and structural simulation were compared via 3D scan, to validate the load-bearing efficiency of this system. The accuracy of results, however, is dependent on the precise knowledge of specific material properties, as well as the stiffness of supports and joints. An official proof of structural stability for this structure has been completed following Euro-Code 5. The system will further be developed with industrial partners (Holzbau Amann and Erhard Brandl) to create economic and ecological prefabricated shells for modular timber construction. The research will address questions of relaxation and accuracy, supports and assembly on site, as well as climatic behavior and maintenance with changing humidity. An in-depth publication on structural behavior and applicability to larger spans is in progress. Finally, we plan to disassemble the gridshell after 2 years, survey the quality and creep of planks, and evaluate their reuse in a new construction cycle.

Acknowledgements

The research has been supported by the University Grants Committee (UGC) of Hong Kong, Early Career Scheme (RGC Ref No. 27604721), the Special Projects Fund, HKU DoA, the Professorship of Structural Design, TUM, as well as our industry partners Holzbau Amann and Metallbau Erhard Brandl. We thank the students of the course Structural Research at TUM and colleagues at TUM, and HKU for their joint effort in modelling, prefabricating and erecting the Timber prototype. We acknowledge Kayser+Böttges / Barthel+Maus, BIGA, Máté Péntek and Jörg Rehm for their consultations in structure, timber, wind and fire safety.

References

DIN EN 338 (2016). DIN EN 338: Bauholz für tragende Zwecke – Festigkeitsklassen. Berlin.

Finsterwalder, S. (1897). Mechanische Beziehungen bei der Flächen-Deformation. In *Jahresbericht der Deutschen Mathematiker-Vereinigung,* vol. 6., 43–90.

Haskell, C., Montagne, N., Douthe, C., Baverel, O., Fivet, C. (2021). Generation of elastic geodesic gridshells with anisotropic cross sections. *Int. J. Space Struct.* 36(4), 294–306.

Jiang, C., Wang, C., Schling, E., Pottmann, H. (2020). Computational design and optimization of quad meshes based on diagonal meshes. In *Advances in Architectural Geometry 2020.*

Kovryga, A., Stapel, P., van de Kuilen, J. W. G. (2020). Mechanical properties and their interrelationships for medium-density European hardwoods, focusing on ash and beech. *Wood Material Science & Engineering* 15(5), 289–302.

Natterer, J., Burger, N., Müller, A. (2000). Holzrippendächer in Brettstapelbauweise – Raumerlebnis durch filigrane Tragwerke. *Bautechnik* 77(11), 783–92.

Pottmann, H., Asperl, A., Hofer, M., Kilian, A. (2007). *Architectural geometry.* Bentley Institute Press.

Pottmann, H., Huang, Q., Deng, B., Schiftner, A., Kilian, M., Guibas, L., Wallner, J. (2010). Geodesic Patterns. *ACM Trans. Graphics* 29 (4).

Preisinger, C. (2013). Linking Structure and Parametric Geometry. *Archit. Design* 83(2), 110–13.

Sauer, R. (1937). *Projektive Liniengeometrie.* Walter de Gruyter&Co.

Sauer, R. (1970). *Differenzengeometrie.* Berlin, Heidelberg: Springer Berlin Heidelberg.

Schikore, J. (2023). Compliant Grids. Theory, Design and Realization. Doctoral Dissertation (to be published). Technical University of Munich, Munich. Chair of Structural Design.

Schikore, J., Schling, E., Oberbichler, T., Bauer, A. M. (2020). Kinetics and Design of Semi-Compliant Grid Mechanisms. In *Advances in Architectural Geometry 2020,* 108–29.

Schling, E. (2018). Repetitive Structures. Design and Construction of Curved Support Structures with Repetitive Parameters. Dissertation. Technical University of Munich, Chair of Structural Design. DOI: 10.14459/2018md1449869

Schling, E., Kilian, M., Wang, H., Schikore, J., Pottmann, H. (2018). Design and Construction of Curved Support Structures with Repetitive Parameters. In *Advances in Architectural Geometry 2018,* 140–65.

Schling, E. and Z. Wan. (2022). A geometry-based design approach and structural behaviour for an asymptotic curtain wall system. *J. Build. Eng.* 52, 104432. DOI: 10.1016/j.jobe.2022.104432

Schling, E., Wang, H., Hoyer, S., Pottmann, H. (2022). Designing asymptotic geodesic hybrid gridshells. *Computer-Aided Design* 152 (0010-4485), p. 103378. DOI: 10.1016/j.cad.2022.103378

Soriano, E. (2017). Low-Tech Geodesic Gridshell: Almond Pavilion. *Archi DOCT* 4(2), 29–40.

Tang, C., Kilian, M., Bo, P., Wallner, J., Pottmann, H. (2016). Analysis and design of curved support structures. In *Advances in Architectural Geometry 2016,* 8–23.

Tang, C., Sun, X., Gomes, A., Wallner, J., Pottmann, H. (2014). Form-finding with polyhedral meshes made simple. *ACM Trans. Graph.* 33(4), 1–9.

Wan, Z., Schling, E. (2022). Structural principles of an asymptotic lamella curtain wall. *Thin-Walled Structures* 180, p. 109772. DOI: 10.1016/j.tws.2022.109772

Wang, Hui and H. Pottmann. (2022). Characteristic parameterizations of surfaces with a constant ratio of principal curvatures. *Computer Aided Geometric Design* 93, p. 102074.

Weinand, Y. and C. Pirazzi. (2006). Geodesic Lines on Free-Form Surfaces. Optimized Grids for Timber Rib Shells. With assistance of IBOIS. In *World Conference in Timber Engineering WCTE.*

Wunderlich, W. (1951). Zur Differenzengeometrie der Flächen konstanter negativer Krümmung. In *Österreich. Akad. Wiss. Math.-Nat. Abt. IIa* 160, 39–77.

Daniel Chauhan, Shane Orme, Serena Gugliotta, Zachariah Wynne, Alex Black-Roberts, Diego Padilla Philipps

Daisy: A Data-Driven Multi-Objective Design Tool

Abstract: Meeting climate pledges demands that building performance for new construction is optimized in a multi-objective manner that minimizes operational energy and embodied carbon whilst balancing architectural objectives such as daylight. Daisy is a performance-driven, parametric design tool which harnesses machine learning to enable designers to navigate the design space for multi-objective optimization which has rarely been attempted at the concept stage. In this paper we explore a case study design for a 63-storey commercial building in central London. This paper highlights how early-staged design space exploration facilitates higher performing engineering and architectural performance providing optimized whole-life carbon design. The Daisy workflow allows the relative importance of early-stage design input parameters relating to the building's shape, size, orientation, structure, façade and building service systems to be estimated by harnessing data science techniques. Daisy was able to demonstrate the design improvement potential from the benchmark design with two clear trade-off strategies. First, by reducing the spatial daylight autonomy from 63.5 % to 52.9 %, the energy use intensity can be reduced by 4.1 % without increasing building embodied carbon. Secondly, by increasing the spatial daylight autonomy by 14.7 % and accepting an increase in building embodied carbon, the energy use intensity of the design can be reduced by 7.1 %. The application and benefits of surrogate modelling for interactive design exploration and design variable importance is discussed.

1 Introduction

To meet the challenge of the carbon emergency requires a step-change in how buildings are designed. Multi-objective optimization as part of early-stage design of buildings has the potential to significantly reduce building embodied carbon associated with material use and whole life operational carbon emissions associated with a building's energy use. Such reductions may greatly outweigh carbon savings which are potential later in the design process. However, despite an increased emphasis on building performance (Lützkendorf 2015; Swan et al. 2015; LETI 2020) little work has explored the potential for optimization in early-stage conceptual design. Effective early-stage design optimization requires balancing conflicting performance objectives including energy use associated with operation and maintenance of the building, the carbon emissions arising from the structural materials, and user-comfort criteria such as achieving a target illuminance level during operational hours. Balancing design trade-offs can

https://doi.org/10.1515/9783111162683-009

be achieved through multi-objective optimization (MOO) which simultaneously maximizes a set of performance objectives through searching a multi-dimensional design parameter space (Deb 2011; Evins 2012). Rather than identifying a single "globally optimal" design, MOO is used to identify the pareto front; designs which have optimal performance for one objective without negatively impacting the performance of any other objective (Deb 2011).

Despite advances in parametric design and MOO for building structures (Machairas et al. 2014; Samuelson et al. 2016; Brown and Mueller 2016; Gehry et al. 2020; Brown et al. 2020; Gauch et al. 2023), limited work has explored MOO for simultaneous optimization of building embodied carbon, operational energy use and user-comfort metrics such as spatial daylight autonomy.

1.1 Daisy

This work introduces Daisy, a performance MOO parametric design tool incorporating machine learning and data- driven computational workflows for designer-led optimization as part of early-stage building design. A workflow of Daisy is presented in Fig. 1, highlighting how it incorporates parametric design (Jabi 2013) and design cataloguing (Brown et al. 2020).

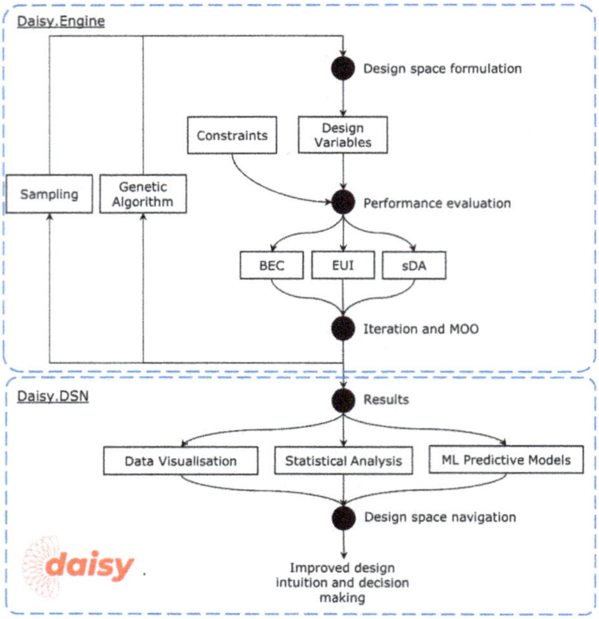

Fig. 1: Daisy design methodology flowchart. Chauhan et al. (2023).

Daisy.engine utilizes Rhinoceros 3D (Robert McNeel & Associates, 2020) and Grasshopper modelling tools (Akos et al. 2014) for performance-based parametric design, while Daisy Design Space Navigator (Daisy.DSN) utilizes surrogate models, statistical approximations of the non-linear design space, which enable users to intuitively navigate the design space and generate near real-time predictions of performance objectives. Daisy allows designers to simulate how changes in design parameters impact the global performance of the building and allows sets of designs which fall on the pareto front to be identified. In Daisy.engine, parametric design variables are selected through Latin Hypercube sampling (selecting sets of parameters from an n-dimensional grid, Loh 1996) and the NSGA-II genetic algorithm (Deb 2011), implemented using the Wallacei plugin (Makki et al. 2018).

For this study the three performance indicators for the MOO in Daisy are:

1. Building embodied carbon (BEC) per square meter of gross internal floor area
2. Energy use intensity (EUI) per square meter of gross internal floor area
3. Spatial daylight autonomy (sDA)

1.1.1 Daisy.DSN

Daisy is designed to seamlessly integrate with the existing user-orientated design process through tools such as a web-hosted dashboard, which allows designers and clients to explore the design space and form an intuitive understanding of the relationships between design variable and performance objectives.

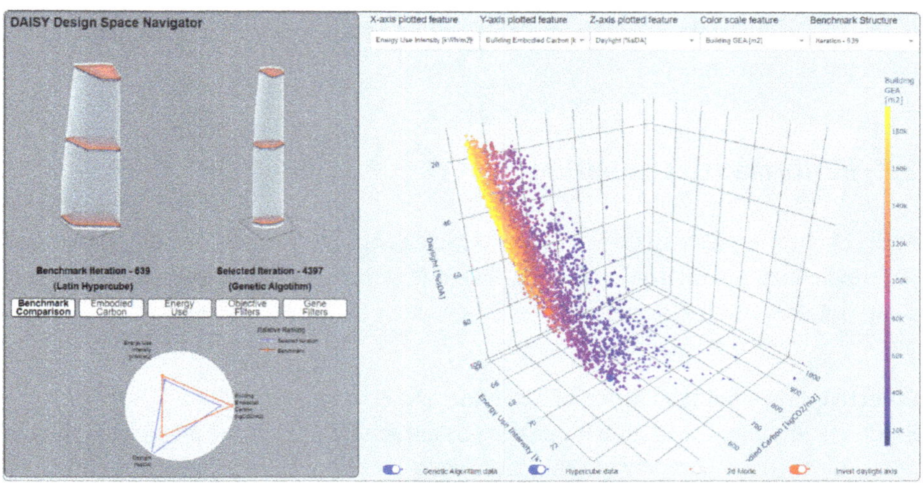

Fig. 2: Daisy.DSN web hosted dashboard.

To further strengthen designer understanding of the design space, Daisy.DSN implements surrogate models including quantile regression models (Fahrmeir et al. 2013) and random forests (Williams and Cremaschi 2019). These models enable informed navigation of the design space, highlighting to users the relative impact of changes on design variables on the performance of a benchmark structure, and allow approximated near real-time design simulation.

2 Case Study Methodology

In this paper Daisy is applied to the design of a real-world 63-storey commercial building. Surrogate modelling is demonstrated as an efficient way to aid user-led design exploration and communicate to clients the factors involved when weighing design decisions. The case study building, initially designed using a conventional design methodology, is a 63-storey, reinforced concrete framed, commercial building in central London and is typical of modern tall urban buildings. The results presented in this paper are anonymized with a benchmark design selected which has similar EUI, BEC and sDA selected for benchmarking relative performance.

2.1 Design space formulation

Histograms of the 13 building façade and shape variables used in the Daisy.engine design simulations and which define the design space for optimization are presented in Fig. 3. Further details of the implementation of the analysis in Daisy.engine are provided in Chauhan et al. (2023).

2.2 Performance evaluation

At the client's request, all design objectives used within the case study are normalized per square meter of internal floor area. This was carried out to investigate the effects of a varying floor area on design trade-offs between performance objectives.

Objective 1 – Embodied Carbon of Structure and Façade
Eurocode compliant structural frame designs are generated through an internal calculation engine, with accompanying embodied emissions calculated using BS EN 15978:2011 (BSI 2011). Emission coefficients, including construction and transport contributions, are derived from the ICE database (Hammond and Jones, 2008).

The façade embodied carbon is calculated individually for each façade orientation through a weighted average of a typical façade bay, adjusted for the design space

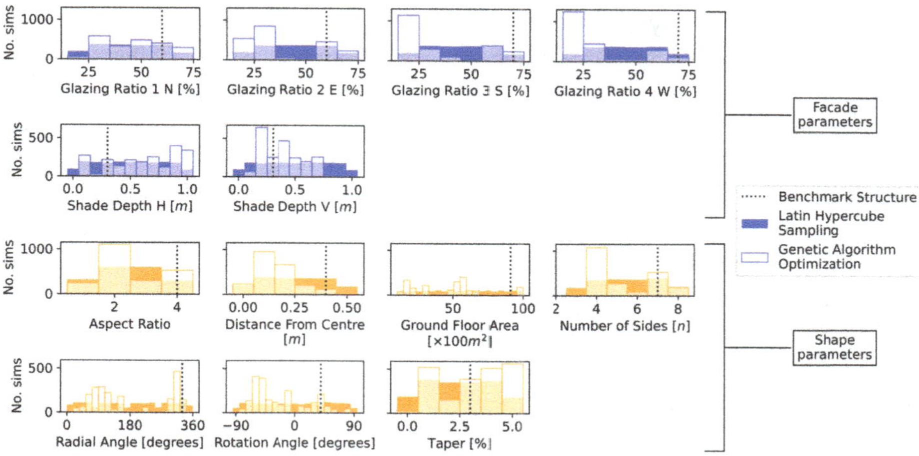

Fig. 3: Histograms of variables used in design optimization of case study building.

variables, following the standard for systemized building envelopes (Centre for Window and Cladding Technology 2006). Cradle-to-gate global warming potential of all façade components are assessed through environmental product declarations using the methodologies specified in BS EN 15804 (BSI 2012) and BS EN 15978 (BSI 2011).

Objective 2 – Operational Energy Use

Dynamic thermal models are implemented in Grasshopper using the Honeybee plugin (Roudsari et al. 2013). Within the models, ceiling and floors are modelled as adiabatic and each floor is separated into a core zone surrounded by a perimeter zone of 4.5 m depth. Heating and cooling units located at the edge of the perimeter zone are assumed to minimize mixing between perimeter and core zones, with an assumption of zero convective loss between regions. A closed cavity system with fixed U-values and G-values are used across the designs.

Simplified mass site context, derived from OpenStreetMap (OpenStreetMap contributors 2022), is used for modelling of shading provided by surrounding buildings. Lighting use is modelled as dynamically controlled. Operational energy use is calculated through a weighted average of the energy use for the ground floor, top floor, and a floor at mid-height of the building. For each floor, an area weighted operational energy use is calculated for each façade orientation. HVAC systems are modelled as air handling units with thermal wheel heat recovery which supplied demand controlled minimum fresh air to fan coil units. Heat pumps and air-cooled chillers, with dynamic efficiencies following ASHRAE limiting performance curves (RANSI 2020), are used for heating and cooling calculations.

Objective 3 – Spatial Daylight Autonomy

Spatial daylight autonomy (sDA), the percentage of floor area which achieves a target illuminance level exceeding 300 lux for 50 % of occupied hours, is used to measure access to daylight in the building. sDA is used in the WELL building standard (WELL Building Institute 2014) and the Leadership in Energy and Environmental Design (LEED) green building certification (Kubba 2016) and accounts for local weather data and site context building geometry. The sDA is calculated using a 2 m sensor grid, at height 1.2 m, with the perimeter core scaled by 25 % and internal partition zones removed to account for daylight within the central core area.

3 Case study results

Plotted in Fig. 4 are comparisons of the performance indicators calculated in Daisy.engine for the genetic algorithm (GA) optimization and the hypercube sampling, and data from the simulated benchmark structure. The pareto fronts for each combination of performance indicators are shown independently in the 2D subplots, with a global pareto front calculated from the combined GA and hypercube data shown in the 3D subplot.

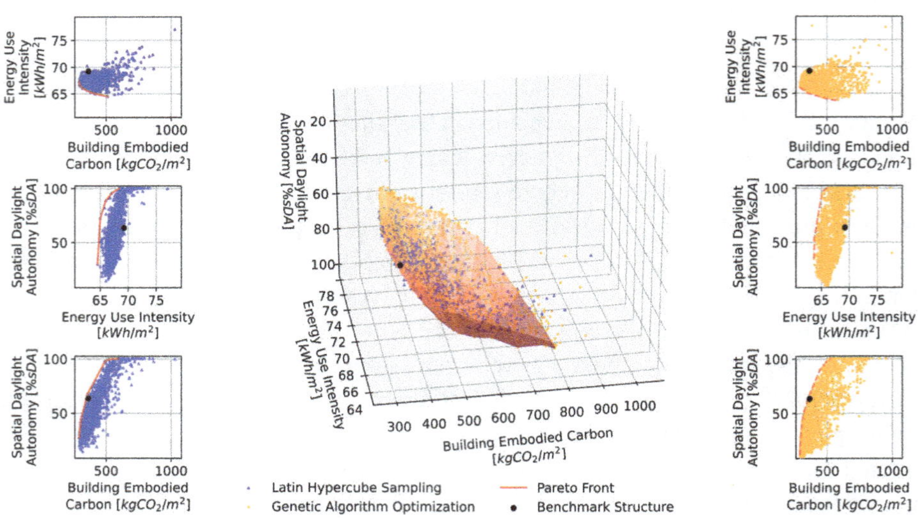

Fig. 4: BEC, EUI and sDA for hypercube sampling (blue triangles, left) and GA optimization (yellow dots, right). Performance of benchmark structure highlighted in red.

Figure 4 highlights that greater mapping of the solution space and higher average performance was achieved through GA optimization than hypercube sampling, as shown through comparison of the pareto fronts. This suggests that the GA successfully improved the performance of the building designs whilst maintaining diversity within the design parameters.

4 Discussion

For individual performance indicators, designs are identified with a BEC of 287.7 kg CO_2/m^2, 25.1 % lower than the median value, or an EUI of 63.6 kWh/m^2, 5.2 % lower than the median value. Numerous designs are identified with a sDA of 100 %, 2.17 % higher than the median. These designs are extremities of the pareto front identified by Daisy; while they maximize performance for one objective this is often at the expense of other performance indicators. Other designs, while not maximizing a single performance indicator, may offer more palatable design trade-offs that balance the competing needs of reducing EUI and BEC, whilst maximizing sDA. Note, potential reductions in EUI are limited due to a range of fixed energy uses within the building which reflect typical commercial buildings. This provides a lower bound EUI which can be achieved.

4.1 Benchmark structure performance

Relative to other designs, the benchmark structure has a high external window area and a low form factor, as demonstrated in Fig. 5. While the benchmark structure is pareto efficient, as shown in Fig. 4, with an sDA higher than 72.3 % of designs simulated and a BEC lower than 60.4 % of designs simulated, the benchmark structure's EUI is higher than 94.1 % of other designs.

4.2 Balancing performance objectives

Figure 6 presents the normalized performance indicators relative to the gross internal area (GIA) of the designs obtained from hypercube and GA sampling. The results highlight that, while a building which maximizes GIA allows for reductions in BEC through minimizing the area of carbon intensive façade required, this severely limits the maximum sDA which can be achieved. A lower sDA results is an increased EUI, due to the need for artificial lighting. To achieve a high sDA, the GIA must be smaller. However, as the GIA decreases, other design parameters must be holistically selected to avoid an increased EUI arising from higher peak cooling and heating requirements due to factors such as high glazing ratios or aspect ratios.

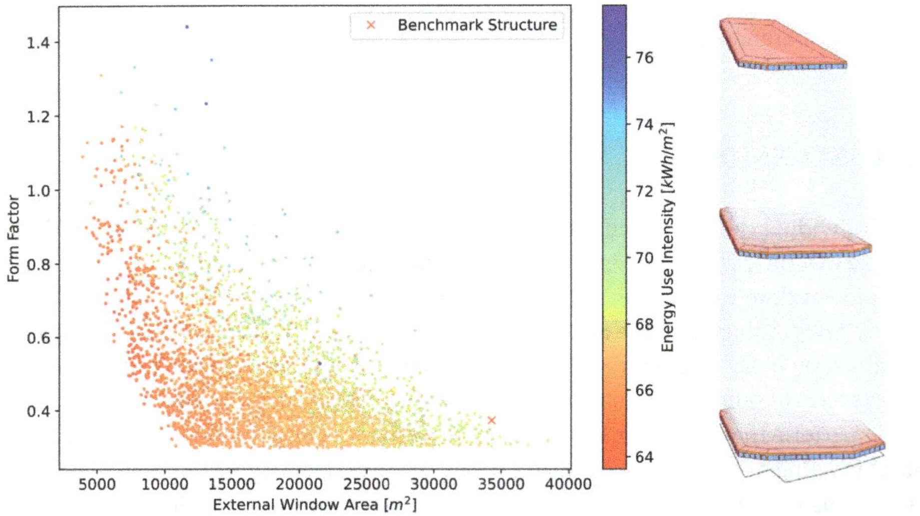

Fig. 5: Left: Comparison of EUI, form factor and external window area; Right: Simulated benchmark structure.

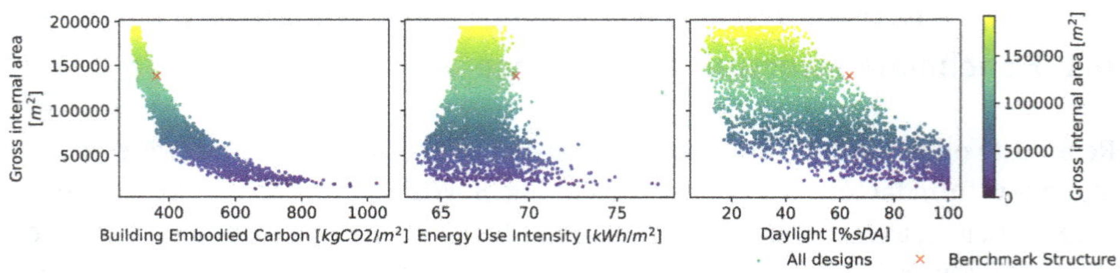

Fig. 6: GIA plotted against normalized performance indicators.

4.3 Random Forests: Pareto optimization and design compromises

Random forests (Williams and Cremaschi 2019) are used to identify roadmaps for improving global design efficiency, identify predictive variables for specific design performance, and enable informed navigation of the design space. Random forests of binary decision trees are implemented with the Scikit Learn Python package (Pedregosa et al. 2011), with 40 % of the data for model verification. Tree depth is increased sequentially to achieve 95 % training accuracy. No meaningful differences in classification accuracy between training and held back data was noted for the case study data, suggesting robustness of the fitted models.

The output from the random forests is the normalized feature importance's, the average total reduction in inaccurately classified designs arising from the use of the design parameter in the binary classification. These feature importance's provide a relative ranking that allow informed navigation of the design space.

Three design scenarios, shown in Fig. 7 and based on the real-world needs of the designers, are examined:

1. Characterizing designs with sDA < 40 %, the limit of good daylighting as defined in the LEED standard (Kubba 2016),
2. Identifying routes to reduce EUI of the benchmark structure through increasing sDA and BEC,
3. Identifying routes to reduce EUI of the benchmark structure through decreasing BEC and sDA whilst ensuring sDA > 40 % (Kubba 2016) or sDA > 55 % (WELL Building Institute 2014).

In relation to design scenarios 2 and 3, as the benchmark structure is already on the pareto front for BEC and sDA, to reduce the EUI compromises must be made in either the sDA or the BEC.

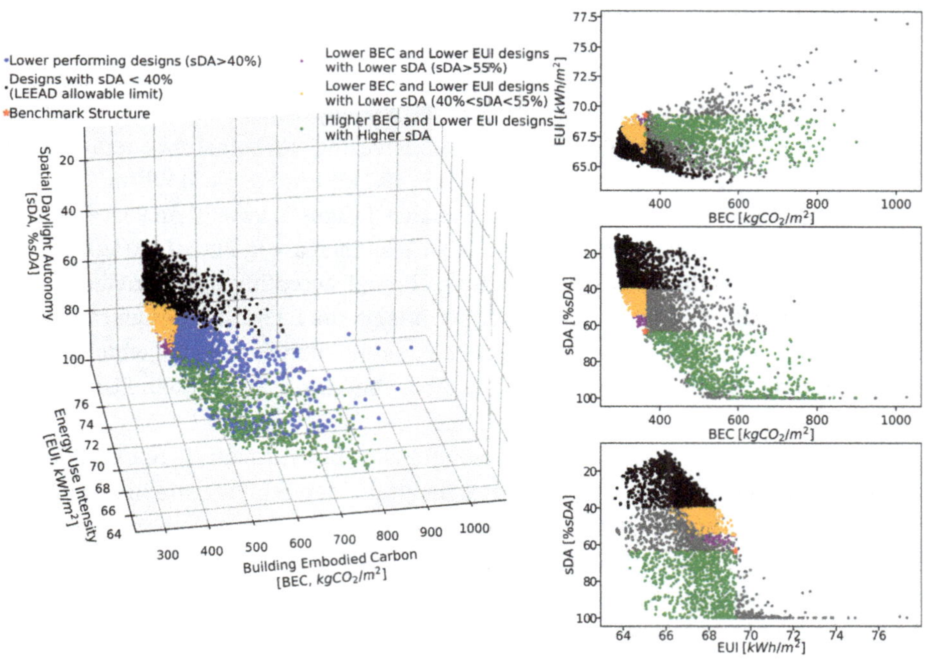

Fig. 7: Design scenarios explored through random forests; comparison of performance indicators.

Designs with sDA < 40 %

A random forest with a maximum depth six, identifies the external wall area as the greatest predictor of sDA < 40 % (feature importance = 0.810), as plotted in Fig. 8. This is followed by the minimum radius of the design (feature importance = 0.030) and a range of other parameters with smaller importance. Counterintuitively, the average window

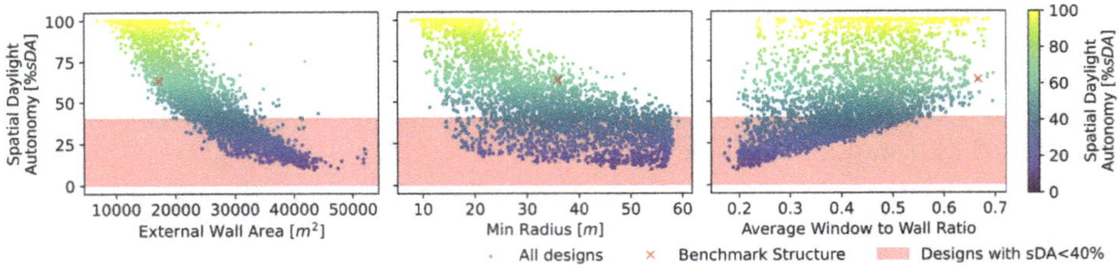

Fig. 8: Comparison of design parameters relative to sDA < 40 %.

to wall ratio of the building is a relatively poor predictor of designs with sDA < 40 % (feature importance = 0.003) as it fails to account for the combined effect of the building form and internal floor area.

Reducing EUI through increasing sDA and BEC

Increasing sDA of the design allows reductions in lighting and heating, potentially reducing the EUI of the design. However, due to the high BEC of the façade, there is a strong positive correlation between sDA and BEC and any increase in sDA is likely to increase BEC. The simulations demonstrate that increasing sDA can allow a reduction in EUI of the benchmark design of up to 7.1 %, from 69.2 kWh/m^2 to 64.3 kWh/m^2, moving the EUI from higher than 94.1 % of other simulated designs to lower than 98.8 % of simulated designs. The cost of this design decision is an increase in BEC of 220 kg CO_2/m^2, 60.9 % of the BEC of the benchmark structure. Through accepting smaller reductions in EUI more palatable designs can be identified through the Daisy.DSN design catalogue, an example of which is shown in Fig. 9. For example, there is a design which has an EUI lower than 70 % of other designs simulated, enabling a reduction in EUI of 3.7 %, but requires an increase in BEC of only 2.7 %.

Design parameters which enable reductions in EUI through increased sDA, as identified through a random forest with a maximum depth of five, are high external wall areas (feature importance 0.665), high minimum radii (feature importance 0.210), high window to wall ratios (feature importance 0.070) and a high surface area to volume ratio (feature importance 0.027), all plotted in Fig. 10. The key contribution of these features to the EUI efficiency are to significantly reduce the building's heating and lighting requirements.

Reducing EUI through decreasing sDA and BEC

Decreasing sDA improves the energy efficiency of the façade and decreases the cooling requirements for the building, enabling a lower EUI. In the context of carbon neutral buildings, this has the added advantage that, due to the strong correlation between sDA and BEC, reducing the sDA reduces the BEC. The limitation of this approach is that, to meet the LEED standard (40 % sDA), the maximum reduction in EUI which

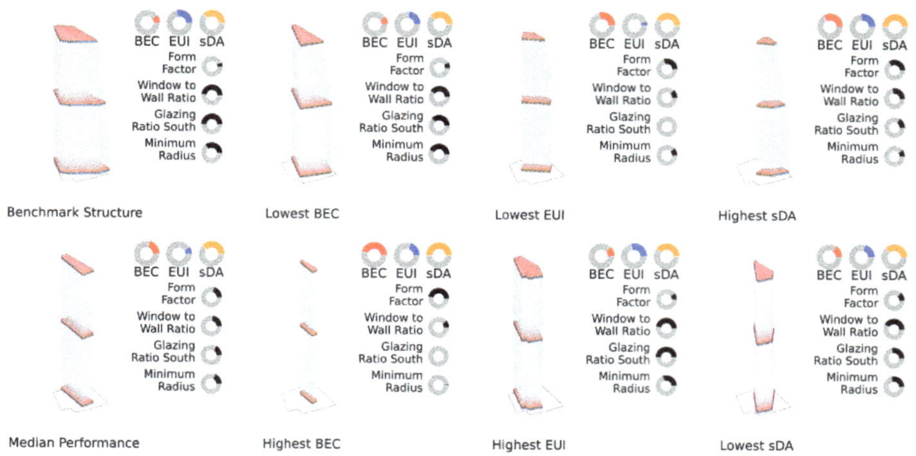

Fig. 9: Partial design catalogue for reducing EUI, through increasing sDA and/or increasing BEC. Radial dials show performance of designs relative to combined hypercube and GA datasets, scaled to have a minimum value of zero and a maximum value of one.

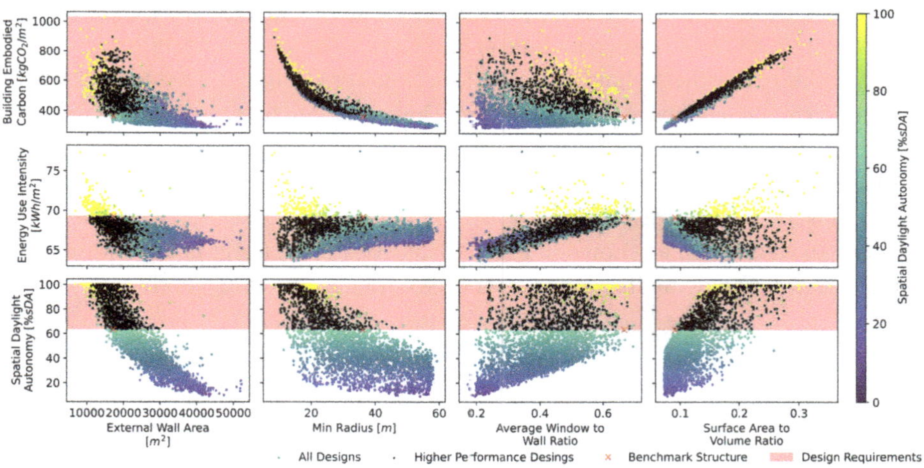

Fig. 10: Comparison of design parameters with a reduced EUI, increased sDA and increased BEC relative to the benchmark structure.

can be achieved is 4.4 % (69.2 kWh/m^2 to 66.2 kWh/m^2), lower than that which could be achieved when increasing the sDA. However, the EUI of this design remains lower than 76.3 % of other designs and allows for a BEC which is lower than 62.8 % of other designs simulated. As with reducing EUI through increasing sDA, there are a range of designs which enable a reduced EUI through decreased sDA, select examples of which are presented in Fig. 11.

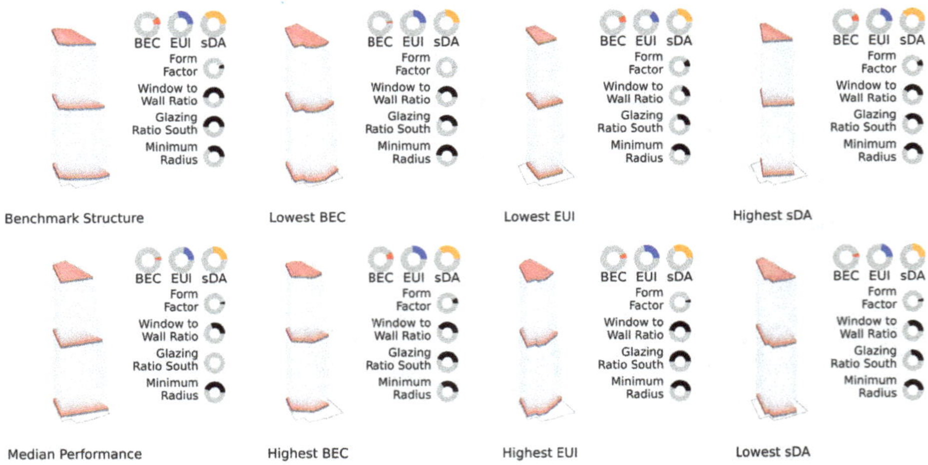

Fig. 11: Partial design catalogue for reducing EUI, through decreasing sDA and/or decreasing BEC.

Three features dominate the random forest for designs which enable reduced EUI, BEC and sDA, allowing the target classification accuracy of 95 % to be achieved with a maximum depth of 3. The key design parameter is external window area (feature importance = 0.526). Unexpectedly, designs which achieved a lower EUI through reduced sDA have a higher external window area than average. This arises due to the interaction with the other key design parameters of form factor (feature importance = 0.278) and external wall area (0.196). Designs with lower sDA and EUI have low form factors and moderate external wall areas, corresponding to stockier buildings with relatively high average window to wall areas, as shown in Fig. 12. However, these designs fall within local optima of EUI, balancing the competing needs of lighting, cooling, and heating.

5 Conclusion

The case study results have highlighted how Daisy, a multi-objective optimization (MOO) data-driven design methodology, enables generation of high-performance building designs that maximize spatial daylight autonomy (sDA), whilst minimizing energy use intensity (EUI) and building embodied carbon (BEC). Beyond this, surrogate modelling in the form of random forests, implemented as part of the Daisy design space navigator, allows designer-led optimization by quantifying relative feature importance, enabling informed navigation of the design space.

Through surrogate modelling, two roadmaps for reducing the EUI of the benchmark structure were identified, by increasing sDA and BEC, or reducing sDA and BEC. A broad range of designs fulfilling these criteria were identified allowing the creation of design catalogues whereby the user can weigh the trade-offs necessary to improve performance.

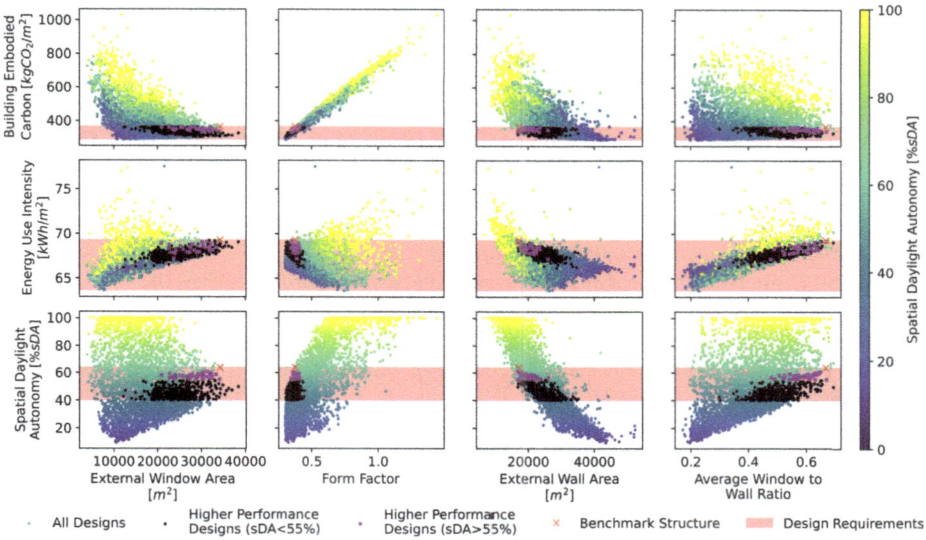

Fig. 12: Comparison of design parameters with a reduced EUI, decreased BEC relative to the benchmarkstructure and with sDA>40 % (LEED Standard) or sDA>55 % (WELL Building Standard).

In the context of the case study, this included whether a design which increased sDA and enabled a reduction of EUI of 3.7 %, to give an EUI lower than 70 % of other simulated designs, and required an associated increase in BEC of 2.1 %, was more attractive than a design which reduced EUI by 7.1 %, to give an EUI lower than 98.8 % of other simulated designs but required an increase of BEC in excess of 25 %. Alternatively, by reducing sDA, a reduction in EUI of 4.4 % relative to the benchmark could be achieved with a BEC which is lower than 61.8 % of other designs simulated. Quantifying these design trade-offs through multi-objective optimization allows for the user-informed, data driven decision making which is crucial in the move to a net-zero built environment.

The development of Daisy will continue to be driven by the needs of designers, for example by allowing limits on acceptable gross internal floor area to be defined by the user, allowing parallel simulation of multiple buildings within a single area, and allowing real-time approximated simulation of designs through surrogate modelling.

References

Akos, G., Parsons, R., Jamison, S., Reitz, A., Jahanshahi, A., Quinones, L., Katzenstein, E., Parsons, K., and Godinez, R. (2014). *The Grasshopper Primer*. Mode Lab, UK, 3rd edition.

BSI (2011). BS EN 15978:2011 Sustainability of construction works. Assessment of Environmental Performance of Buildings. Calculation method. BSI, UK.

BSI (2012). BS EN 15804:2012. Sustainability of construction works. Environmental product declarations. Core rules for the product category of construction products. BSI, UK.

Brown, N. C., and Mueller, C. T. (2016). Design for structural and energy performance of long span buildings using geometric multi-objective optimisation. *Energy Build.*, 127, 748–761.

Brown, N. C., Jusiega, V., and Mueller, C. T. (2020). Implementing data-driven parametric building design with a flexible toolbox. *Autom. Constr.*, 118-103252, DOI: 10.1016/j.autcon.2020.103252

Centre for Window and Cladding Technology (2006). *Standard for systemized building envelopes.* CWCT, UK.

Chauhan, D., Orme, S., Gugliotta, S., Wynne, Z., Black-Roberts, A., and Padilla Philipps, D. (2023). Daisy: A Multi-Objective Design Tool for decarbonising buildings at the concept stage. *Proc. Inst. Civ. Eng. Eng. Comput. Mech.* Under review.

Deb, K. (2011). *Multi-objective optimisation using evolutionary algorithms.* Wiley, UK.

Evins, R., Joyce, S. C., Pointer, P., Sharma, S., Vaidyanathan, R., and Williams, C. (2012). Multi-objective design optimisation: getting more for less. *Proc. Inst. Civ. Eng. Civ. Eng.*, 165(5), 5–10.

Fahrmeir, L., Kneib, T., Lang, S. and Marx, B. (2013). *Regression: Models, Methods, and Applications.* Springer, UK.

Gauch, H. L., Dunant, C. F., Hawkins, W., and Cabrera Serrenho, A. (2023). What really matters in multi-storey building design? *Appl. Energy,* 333-120585 DOI: 10.1016/j.apenergy.2022.120585

Gehry, F., Lloyd, M., and Shelden, D. (2020). Empowering design: Gehry partners, Gehry technologies and architect-led industry change. *Archit. Des.*, 90(2): 14–23, DOI: 10.1002/ad.2542

Hammond, G. P. and Jones C. I. (2008). Embodied energy and carbon in construction materials. *Proc. Inst. Civ. Eng. Energy,* 161(2), 87–98, DOI: 10.1680/ener.2008.161.2.87

Jabi, W. (2013). *Parametric design for architecture.* Hachette UK, UK.

Kubba, S. (2016). *LEED V4 practices, certification, and Accreditation Handbook.* Elsevier, Amsterdam.

LETI (2020). *LETI Climate Emergency Design Guide. How new buildings can meet UK climate change targets.* LETI, UK.

Loh, W. L. (1996). On Latin hypercube sampling. *Ann. Stat.*, 24(5), 2058–80.

Lützkendorf, T., Foliente, G., Balouktsi, M., and Wiberg, A. H. (2015). *Net-zero buildings: incorporating embodied impacts.* Build. Res. Inf., 43(1), 62-81, DOI: 10.1080/09613218.2014.935575

Machairas, V., Tsangrassoulis, A., and Axarli, K. (2014). Algorithms for optimisation of building design: A review. *Renewable Sustainable Energy Rev.*, 31, 101–12, DOI: 10.1016/j.rser.2013.11.036

Makki, M., Showkatbakhsh, M., and Song, Y. (2018). *Wallacei: An evolutionary and Analytic Engine for Grasshopper 3D.* Wallacei, UK.

OpenStreetMap contributors (2022). *OpenStreetMap database.* OpenStreetMap Foundation: UK. Available under the Open Database Licence from: openstreetmap.org.

Pedregosa, F., Varoquaux, G., Gramfort, A., Michel, V., Thirion, B., Grisel, O., Blondel, M., Prettenhofer, P., Weiss, R., Dubourg, V., and Vanderplas, J. (2011). Scikit-learn: Machine learning in Python. *J. Mach. Learn Res.*, 12, 2825–30, DOI: 10.48550/arXiv.1201.0490

RANSI (2020). *Thermal environmental conditions for human occupancy. American Society of Heating.* Refrigerating and Air-Conditioning Engineers, Atlanta, GA.

Robert McNeel and Associates (2020). *Rhinoceros 3D, Version 7.0.* Seattle, WA.

Roudsari, M. S., Pak, M., and Smith, A. (2013). Ladybug. In *Proceedings of the 13th international IBPSA conference.* Lyon, France, 3128–35.

Samuelson, H., Claussnitzer, S., Goyal, A., Chen, Y., and Romo-Castillo, A. (2016). Parametric energy simulation in early design: High-rise residential buildings in urban contexts. *Build. Environ.*, 101, 19–31, DOI: 10.1016/j.buildenv.2016.02.018

Swan, W., R. Fitton, and P. Brow. (2015). A UK practitioner view of domestic energy performance measurement. *Proc. Inst. Civ. Eng. Sustainability,* 168(3), 140–47, DOI: 10.1680/ensu.14.00056

WELL Building Institute (2014). *WELL Building Standard v1.* New York, USA.

Williams, B. A., and Cremaschi, S. (2019). Surrogate model selection for design space approximation and surrogate based optimisation. *Comput. Aided Chem. Eng.*, 47, 353–58.

Daria Dordina, Cyrill Milkau, Zlata Tošić, Daniel Lordick, Danilo Schneider

Point Cloud to True-to-Deformation Free-Form NURBS

Abstract: Built reality is always subject to tolerances. Existing buildings can only be represented to a limited extent by geometric primitives. With the help of Non-uniform B-Spline curves and surfaces (NURBS), complex structures can be represented as parameterized geometry, suitable for further usage in BIM applications. This research was conducted within the framework of a project FreeForm4BIM, a part of the ZIM innovation network twin4BIM, which aims to enable true-to-deformation creation of free-form elements of historic buildings based on point cloud data. The conducted free-form geometry, therefore, precisely describes objects, which have lost their initial shape characteristics over time. The considered approach consists of the following steps: in preparation, feature-based segmentation is performed on the cloud. The further processing analyzes the basic structure of a free-form segment and divides it to smaller auxiliary segments – patches. They correspond to the general architectural concept and satisfy the most convenient U- and V-spline arrangement for the NURBS surface. In the last step of the algorithm, NURBS surfaces are constructed from boundary geometry and parametrized curve network, extracted from point cloud data for each patch. This article demonstrates the implementation of the process on the case study of the Buddenbrookhaus scan data (Lübeck, Germany) and reflects its challenges: vault geometry, requiring precise true-to-deformation representation.

1 Introduction

The project aims to develop a workflow for the semi-automated, correct to form and deformation creation of BIM-compliant building models based on point clouds. To achieve this goal, point clouds have to be filtered, segmented, and geometry elements, with a particular focus on free-form elements, have to be modeled as NURBS surfaces. The choice of NURBS was done due to their adjustable level of accuracy and modeling logic that is well-suited to construction considerations. The workflow is intended for the building industry and focused on historical architectural elements (Germany, 18–19th century).

Consequently it was an important challenge, to reduce the manual part of an established processing approach used by the project partners. State of the art was that the building scans were segmented within the LupoScan program based on the normal deviation. After that, the NURBS-geometry was created in Rhinoceros 3D out of

https://doi.org/10.1515/9783111162683-010

manually drawn boundary curves and manually selected point cloud segments. This cumbersome process obviously offered potential for automation.

2 Backgrounds

The task of reconstruction of scanned buildings in the context of Architecture/ Engineering/Construction (AEC) applications has become very important in recent years. However, the specific task of representing free-form true-to-deformation geometry is relatively infrequent. In the majority of cases, authors work with mesh geometry. However, for AEC projects it is important to obtain parametrized architectural models, which is possible using NURBS.

Segmentation and classification of the underlying point cloud is an important part of the approach. Early segmentation approaches focused on finding edges or surfaces of points with similar attributes within the entire point cloud. A widely used and frequently adapted method is continuous region growing of neighboring points meeting certain growth criteria (Rabbani et al., 2006). When dealing with noisy data, statistical estimation methods such as Random Sample Consensus or Hough transformation have a high success rate for single object segmentation. Recently, there has been high interest in applying machine learning concepts to point clouds derived from imagery or lidar data.

The meaningful parameterization of the input point cloud segments is the main challenge in NURBS reconstruction. Several methods were investigated over the past two decades in order to meet this challenge. For example, a simple planar or predefined by user surface can be used as an initial guess. This was demonstrated in the commercial software RhinoRESURF. Banfi makes use of assuming user knowledge of common architectural geometry types in his method for the manual reconstruction of historical buildings (Banfi 2019). It lacks automatization of the geometry creation process, as the guide curves should be detected and manually drawn by the user. We propose an approach that takes the inner architectural logic of segments into consideration. The main property in this case are high curvature regions which represent the sharp features, as widely present in architectural structures.

3 Methods

3.1 Case study: Buddenbrookhaus

The article is based on the 3D scan data of the Buddenbrookhaus (1758) in Lübeck. The laser scan we worked with was done in 2022, after a partial reconstruction and reinforcement of the house. In this article we concentrate on the masonry vault geometry in the building basement. A cell segment including walls and floor has been selected as a benchmark (Fig. 1).

Fig. 1: Overview of basement scan in grayscale intensity with chosen room (red).

3.2 Explanation of the workflow and data exchange

The original scan data was provided in the E57 format, at 5 mm scanning resolution. The segmentation process was supported with methods from the PCL Point Cloud Library (Rusu and Cousins 2011). Adapting strategies from the PCL region growing implementation, we automatically detect initial seeding points using a minimization criterion on certain features at a subdivision level of the cloud. Manual selection of preferred seeds is optional. Adjacent seed clusters are glued together to form segments. The creation of the NURBS geometry succeeds automatically in the Grasshopper script with an E57 segment as an input. The user only starts the computation and defines curve interpolation and surface parameters (3.4, steps 3,5). For the algorithm developmentwere used Grasshopper plugins: Cockroach (Vestartas and Settimi 2020), Leafvein. Moreover, some parts are written in Python for Grasshopper. The choice of the programs was determined by the practice of the project partners and the objectives of the project.

3.3 Segmentation

We relied on attribute-based region growing approaches because they tend to not presume underlying models on segmented objects. The simplified workflow is shown in Fig. 2. When using spatial clustering algorithms for segmentation, the main goal is to find a good trade-off betweenaccuracy of results and computational efficiency. To meet the true-to-deformation requirement, we worked closely to the resolution of the original scan data, namely 20 mm, as we focused on only a small portion of the building.

Fig. 2: Left: Growth process of two clusters (growth iteration colored ascending from blue to red); center: two distinct cluster with border points emphasized; right: merged segment.

Custom region growing approach

The initial point cloud data contains coordinates and possibly intensity or color information. Using principal component analysis (PCA) (Maćkiewicz and Ratajczak 1993), geometric features are calculated pointwise, namely normals, eigenvalues and curvatures, since the segmentation algorithms used depend strongly on them.

The latter goal of modeling NURBS surfaces requires all segments to have reliable boundaries. We propose to use the natural behavior of the region growing to obtain them directly. Therefore, we implemented a custom region growing approach that tracks point status information throughout the growing process. The general approach to region growing is to select an initial set of points from the cloud as seed points that define individual cluster structures. By examining neighboring points, it is evaluated whether they belong to the same cluster as the seed by checking deviations from the seed of pre-computed features such as normal orientation or curvature. In addition, the presented implementation tracks not only cluster membership, but also violations of growth conditions in the form of stop points.

Since the growth starts simultaneously from different seeds (Fig. 2 left), an important task is to merge clusters that are likely to form a joint segment. This is evaluated by applying the same growth condition to the shared border points of neighboring clusters (Fig. 2 center). Borders appear naturally as growing is only allowed on points that do not yet belong to any cluster. Merging is repeated for all possible candidates and yields the point clusters for the next process steps (Fig. 2 right).

In addition to the point cloud and the seeds, the growth condition is passed as an input parameter. In our approach, we check the normal deviation of the points as the main criterion for clustering.

Boundary estimation and neighborhood relation

The resulting clusters are automatically merged into object-describing segments. Additionally the points, where the growth stops, are also returned. While these stop points can be viewed as a wireframe model of the segmentation (Fig. 3 center, red), subsets of them are also associated with individual segments such as the to be modeled vault segment S_0. The combination of stop points and S_0 results in a denser vault segment S_0 with a clearly identifiable boundary (Fig. 3 right, light blue). By examining points from multiple segments lying close to a shared boundary, one can also infer a simple neighborhood relationship between them.

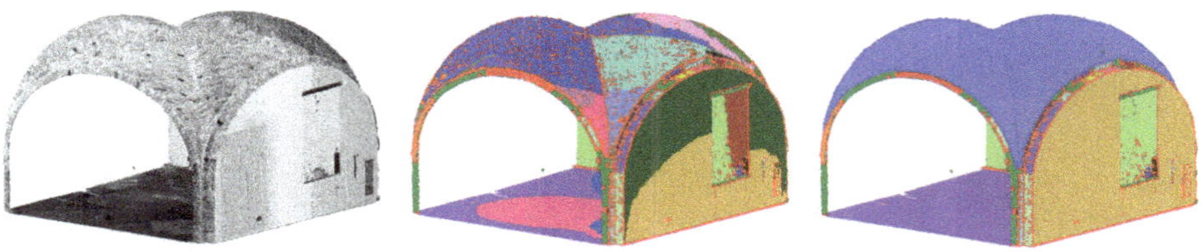

Fig. 3: Left: initial point cloud (grayscale intensity); center: resulting clusters from region growing (random coloring) and stop points of all clusters (red emphasized); right: final segment vault S_0 and re-filled with stop points belonging to merged clusters (darker blue in vault).

3.4 Surface generation

As the focus of the project is on free-form geometry, the geometry creation is now demonstrated in detail on the example of the groin vault segment S_0 (Fig. 4).

Fig. 4: Groin vault segment S_0 (grayscale intensity).

Step 1

The preliminary geometric analysis is conducted through a surface variation calculation. The surface variation value sv (Hackel, Wegner, and Schindler 2016) is set for each point in the segment. High curvature areas are isolated through a value threshold $sv_t = 0.1 \cdot \max(sv)$ and clustered. The result of this operation is segment S_1 (Fig. 5 center). Further structure extraction includes the following automatized steps: segment thinning and construction of a principal graph (Fig. 5 right).

Fig. 5: Left: visualization of the surface variation value in grayscale, segment S_0; center: cleaned segment S_1; right: graph describing S_1(cross), simplified.

The graph of the S_1 allows distinguishing intersections and rib ends.

Step 2

In the next step, we estimate the boundary data of segment S_0 and conduct the geometric analysis to get the boundary curves and corners. A clustering algorithm was developed for splitting point contours into constituent parts: lines or curves segments. The algorithm plots the slope of the tangents of all ordered boundary points with a global X axis and groups the segments of the point cloud according to the identified clusters. The outliers (Fig. 6 right, red) in this algorithm represent the corners of the surface and correlate with the results of step 1. As a result of step 2 we get the point sets for each guide curve and the exact position of corners.

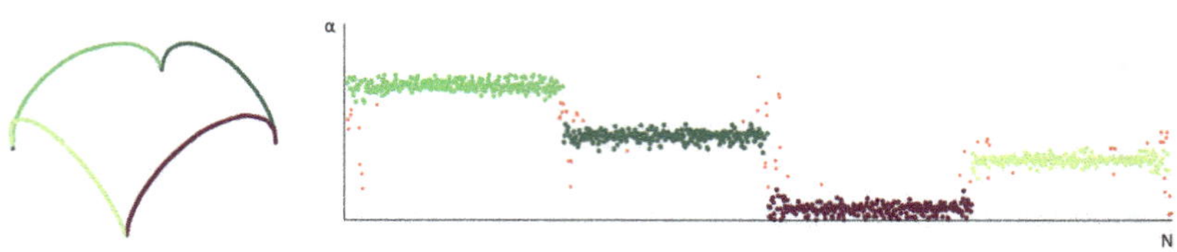

Fig. 6: Left: clustered boundary; right: clustered graph (outliers in red color). α – declination angle, N – index number of each point.

Step 3

Based on the first two steps we associate the graph structure with the S_0 and find consistent point sets on the regions with high curvature (S_1). The inputs for the algorithm are point cloud S_1, start and end points from the graph from step 1. We start the path finding algorithm from thecorners and analyze the surrounding points by different criteria such as curvature value, the distance between points, angle estimation, etc. The values for each point will be summed and based on this final score the next point will be chosen. This algorithm allows us to estimate the complex deformed geometry with high precision. Further we conduct the interpolation (parameter - curve degree) and get a set of ordered rib curves as a result (Fig. 7).

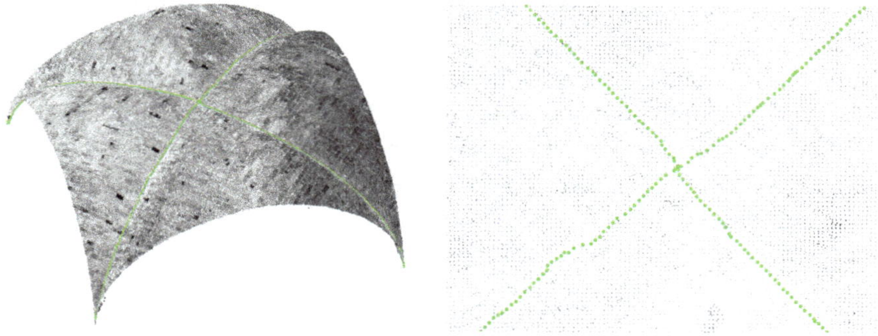

Fig. 7: Left: vault segment S_0 with estimated ribs position; right: detail in top projection: the deformations and roughness of the masonry vault are preserved.

Step 4

Next we prepare the surface division into patches (S_2). A patch is defined with two dominant directions U, V and four boundary curves. The patch structure definition is done in 2D projection by processing boundaries obtained in steps 2 and 3. If the polygon is triangle-like (Fig. 8 left), in order to satisfy the basic patch condition, this segment has to be divided into three patches. On the basis of this structure the point cloud sub segmentation through polyline cropping is conducted (Fig. 8 center).

During this subdivision process spatial relationships between subsegments are preserved: edges shared by different subsegments (for this case, four triangular vault subsegments) are defined as crease edges. Edges belonging to patches inside one subsegment are classified as smooth edges for later curve network interpolation. At the end of this step the boundary curves for every patch are interpolated (Fig. 8 right).

Fig. 8: Left: triangular subsegments derived from vault structure; center: point cloud subdivision; right: boundary curves for each patch S_2.

Step 5

We define a U- and V-network for each patch and conduct sections through the patches accordingly. Control parameters are the U- and V- amount of curves for each direction. Intermediate and boundary curves will together create the UV-net for the NURBS geometry (Fig. 9 left). By the interpolation of intermediate curves we use the information about edge segments from the previous step: for the curves on smooth boundaries we additionally rewrite the end vectors as an average from two neighboring vectors meeting in this point).

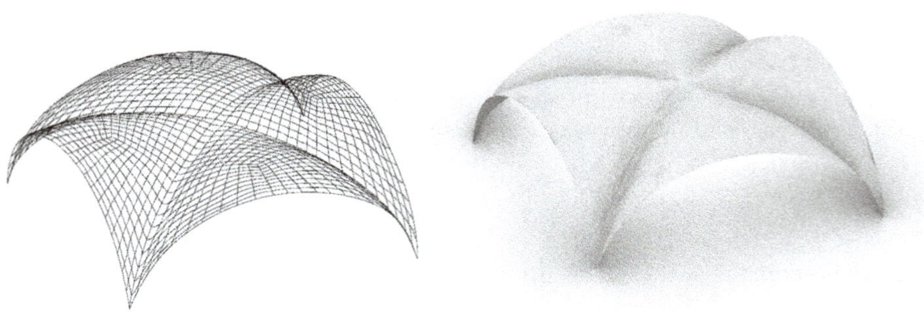

Fig. 9: Left: U- and V-network; right: final NURBS surface.

4 Results and reflection

4.1 Segmentation

For evaluation, we compare our segmented vault S_0 with a reference segmentation derived from a projection-based approach in LupoScan. The main purpose of the presented segmentation approach is to separate the vault as accurately as possible for the point cloud-to-NURBS process while preserving meaningful boundaries. We compute the cloud-to-cloud distance to find the nearest neighbor points between the two clouds. Since the segments originate from the same cloud, identical points should overlap, resulting in a point distance of zero. Any distance above this indicates missing points, which is highlighted by the green to red color scale in Fig. 10. Only deviations greater than 1 cm are shown. We can see that by using our segmentation we approach and include more points in the corners of the arch.

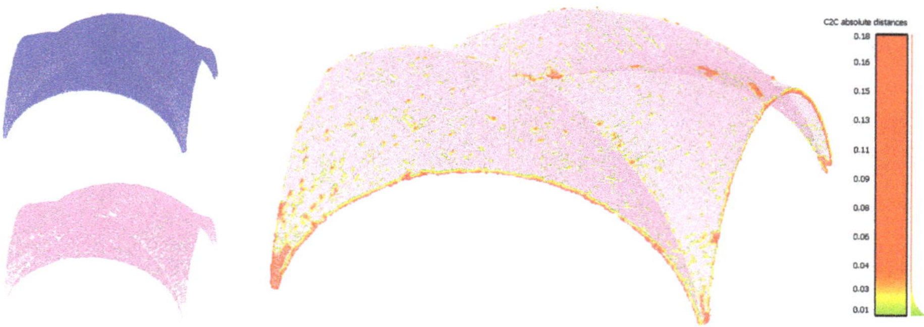

Fig. 10: Left: vault segments gained with proposed segmentation method (blue) and reference data from manual segmentation (pink); right: colored point to point distance of our vault segmentation to reference segmentation.

4.2 NURBS

The resulting NURBS geometry is evaluated based on the different values of the input parameters in the surface reconstruction algorithm such as curve network density (D), with the goal of not exceeding the aimed tolerance between point cloud data and free-form NURBS surface (0.02 m according to the AEC-requirements for construction drawings).

The distances are calculated between the sample points and their matching "closest surface point" of the gained NURBS surface. By parameter $D = 10$ 98.9 % of points have a deviation equal or smaller than 0.015 m (Fig. 11), 96.4 % smaller than 0.01 m. By increasing the parameter D to 15 the result is improved by a few percent. Decreasing this parameter creates more generalized surfaces with bigger deviation. In any case, there are a few points that lie relatively far (0.15 m) from the resulting surface, these are scanning defects.

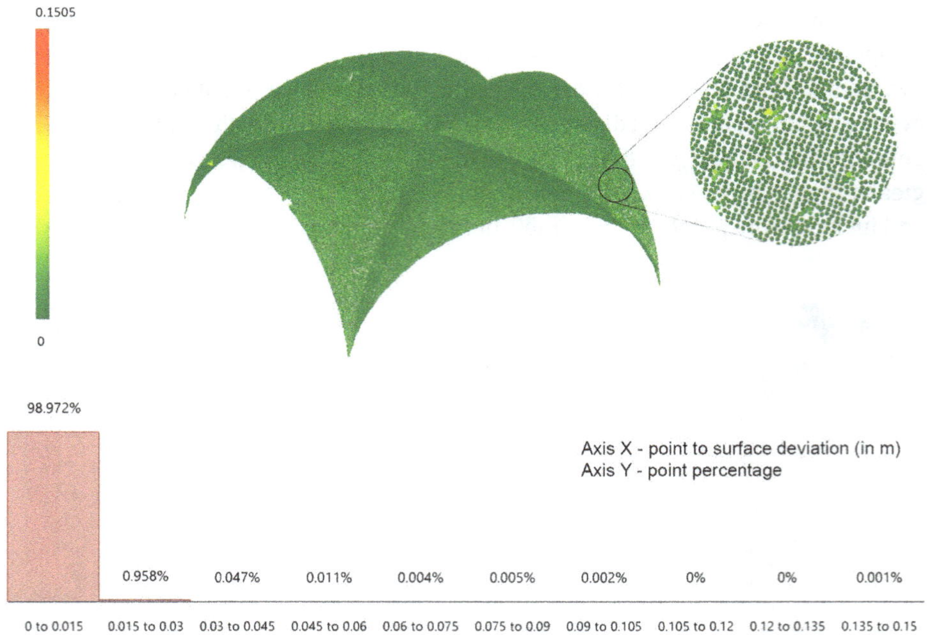

Fig. 11: Left: Point-to-surface deviation visualization; right: deviation percentage bar. Parameter *D*: 10 curves in *U*- and *V*-direction pro patch.

4.3 Limitations and future work

In general, the proposed method strictly depends on the quality of the input data: it should be complete scans without too many additional objects in the scene. Moreover, the tests were conducted on a data set with the resolution of 20 mm, as the most applicable resolution in case of architectural scans. The use of PCA methods is dependable on point cloud density, as in some cases local curvature will overcome the segment structure. Such datasets should be downsampled.

The main issue when dealing with segmentation algorithms is the high dependency on user given input which leads mostly to over- or undersegmentation of the scene. Even small changes in the growing conditions may lead to non-predictable output of the resulting segmentation. Also the (automatic) choice of seed points can affect the result, although the merging approach tries to manage it as best as possible. Though boundaries can be inferred almost automatically, there is yet no way of deducing them at open segment ends or distinguish between inner and outer boundaries. We consider the most important steps in the further development to be in the direction of classification and semantic labeling. This can be supported by expanding the segment relation deduction described in Sec. 3.3 to promote associations between building components.

To tackle a variety of architectural geometries, more surface types should be detected in the early stage of classification and the classification label should then trigger different algorithms for different architectural typologies. As freeform surfaces include a huge variety of geometries demanding different strategies, we have to estimate the geometry structure first and to continue with the fitting strategy over it. The proposed workflow was developed as a touchstone using high-curvature regions as the most important input and therefore applicable for more complex vaults or other geometries with sharp features. For different geometry types, for instance, sweeps or composite surfaces, other methods of underlying structure definition such as skeleton calculation or directional fields should be used. Steps 3, 4 and 5 will be adjusted to the new structure type.

5 Conclusion

The proposed automated workflow solves the praxis-oriented problem of precise creation of NURBS geometry from point cloud data of architectural objects. It is already possible to automatically achieve true-to-deformation and exact in detail groin vault geometry from a pre-segmented point cloud.

While defining custom region growing conditions is already common practice, tracking the region growing has not yet been investigated in these fields of application. Our approach yields the ability to either adapt growing parameters at borders or deducing adjacency implicitly without having to analyze the cloud a posteriori.

Although the topic of NURBS recreation of historical architectural objects was widely studied by Banfi (2019), the full potential of automatisation was merely explored in his works. The main contribution of the proposed paper is the automatisation of routine time-consuming processes such as manual guide curves drawing for further geometry creation. Another important point is an idea to analyze the geometrical structure of elements for a smart strategy choice. Moreover, the workflow is divided into algorithmical steps with an idea to flexibly use them during further extension of the project.

Acknowledgements

This research was funded by the Federal Ministry for Economic Affairs and Climate Action based on a resolution of the German Bundestag. We thank our industrial partner Scan3D GmbH and Fokus GmbH Leipzig for providing us with access to point cloud data and their engineers for preparing data and generating ground truth segmentations and NURBS surfaces used in our experiments.

References

Banfi, F. 2019. HBIM generation: extending geometric primitives and BIM modelling tools for heritage structures and complex vaulted systems. *International Archives of the Photogrammetry, Remote Sensing and Spatial Information Sciences* XLII-2/W15, 139–48. DOI: 10.5194/isprs-archives-XLII-2-W15-139-2019

Deschaud, J.-E., and F. Goulette. 2010. *A Fast and Accurate Plane Detection Algorithm for Large Noisy Point Clouds Using Filtered Normals and Voxel Growing*, 3DPVT. Paris, France.

Hackel, T., J. D. Wegner, and K. Schindler. 2016. *Contour Detection in Unstructured 3D Point Clouds*, IEEE Conference on Computer Vision and Pattern Recognition (CVPR).10.1109/CVPR.2016.178

Maćkiewicz, A., and W. Ratajczak. 1993. Principal components analysis (PCA). *Computers & Geosciences* 19, no. 3 (3): 303–42. DOI: 10.1016/0098-3004(93)90090-R

Rabbani, T., F. van den Heuvel, and G. Vosselman. 2006. Segmentation Of Point Clouds Using Smootheness Constraint. *International Archives of the Photogrammetry, Remote Sensing and Spatial Information Sciences – ISPRS Archives*, S. 248–253.

Rusu, R. B., and S. Cousins. 2011. 3D is here: Point Cloud Library (PCL), In *Proceedings of the IEEE International Conference on Robotics Automation*. Shanghai, China.

Vestartas, Pe., and A. Settimi. 2020. *Cockroach: A Plug-in for Point Cloud Post-Processing and Meshing in Rhino Environment*, EPFL ENAC ICC IBOIS.

Keyan Rahimzadeh, Evan Levelle, John Douglas

Beyond the Hypar: Predicting Buckled Shapes in Bent Glass with Machine Learning

Abstract: The cold-bending of rectangular glass panels into anticlastic shapes is a well-established practice. A recently built project consists of four towers featuring thousands of individually unique glass panels which are non-rectangular, cold-bent to an extreme degree, and sometimes curved. The depth of bending in the panels approaches the elastic instability limit, which prompts several challenges: 1) edges no longer remain linear, thereby becoming incompatible with linear frames, 2) surfaces produced by CAD packages do not account for secondary nonlinear effects, and therefore are not accurate enough to fabricate, 3) some panels combine curving and bending, which, along with the parallelogram shape, place the problem outside the bulk of related literature which uses flat, rectangular shapes as a basis. To enable fabrication within an accelerated timeline, a Machine Learning model was trained on 3,500 material simulations that are parametrically generated from the Building Information Model. The model was then deployed to predict the 3-dimensional surface of each unique panel while accounting for non-linear effects, and ensuring straight, linear edges after deformation. The paper describes the geometric problem, including how buckling effects manifest differently in non-rectangular shapes; the development and deployment of the Machine Learning model; and comparisons to physical specimens, including the final construction of some 11,000 unique panels.

1 Introduction

In the pursuit of realizing architectural projects with free-form geometries, the technique of cold-bending glass has seen growing adoption. Among the many challenges that the technique presents is the intractability of predicting the surface geometry that the glass will have after its deformation.

As we discuss in detail, the further you bend the glass, the more the real-world surfaces of glass diverge from the theoretical surfaces created in CAD packages. Predicting the final geometry is useful for many reasons, such as understanding visual reflections. Here we tackle the challenge of reverse-engineering the flat shape of these warped panels, such that they can be fabricated accurately.

There is an ever-growing body of research regarding cold-bent glass; the work presented here has a unique mix of characteristics:
- The panels are quadrilaterals but highly non-rectangular.
- The depth of bending is substantial enough to approach the elastic buckling limit, thereby exhibiting nonlinear behavior.

https://doi.org/10.1515/9783111162683-011

- Every single panel is different, therefore requiring an approach for solving the geometry of each one.
- Techniques were needed that were consistent and compatible with contemporary engineering practice (e. g. Finite Element Analysis), and could be verified.

2 Background and Methodology

2.1 Project Description

The project motivating this research consists of four high rise towers located in Qatar. The facades have a 50/50 mix of vision and opaque panels. Each building can be represented as a morph between two different curved shapes, with a spiraling subdivision pattern; each panel takes the shape of a parallelogram that is non-planar and unique.

The schedule was compressed, moving from architectural concept to first fabricated panel in less than two years, with 23,000 panels fabricated and installed within another two years.

Fig. 1: Construction Project

2.2 Methodology

The research detailed herein was carried out in the process of executing a construction project, and therefore subject to constraints of schedule, cost, and the need to be implemented immediately, and at scale. The investigation and the solution had to be developed simultaneously, an approach commonly called "action research".

We drew upon our practical knowledge of the literature available and the state of the art, and prioritized development that we knew could be implemented with a predictable – and short – timeline, with a reliable degree of quality. We therefore favored an approach grounded in methods that were familiar to the industry (e. g. finite elements, parametric modeling, etc.), and could be easily interpreted or even replicated by 3rd party reviewers.

2.3 Facade System

The buildings contain 11,716 glazed panels with bending depths ranging from 10mm to 442 mm. Some panels also needed to be curved at the top edge, bottom edge, or both, to avoid clashes with the slab. Panels with less than 180 mm[1] of deformation are fabricated flat and deformed out of plane when they are mounted to the frame, which were assembled in their final 3-dimensional shape after being fabricated digitally. The development of the 3D framing is discussed in detail by the authors in a forthcoming publication.

For all panels that were curved and/or above the 180 mm threshold, the glass is produced using a different method: both layers of the IGU are made of laminated glass. The glass is deformed out of plane in the factory before going into the autoclave and being laminated "in-place". Once laminated, the structural interlayer' stiffness helps retain the deformed shape. This is sometimes called "Lamination Bending"[2] or "Warm Bending"; we include it here within the "Cold Bending" category as the process never raises the temperature enough to alter the properties of the glass itself.

3 Geometry of Cold Bent Surfaces

Design of cold-bent glazing must consider a range of technical concerns regarding structural mechanics, engineering, and system design (Datsiou 2017, Beer 2015, Nardini 2018). Here we focus on the geometric aspects of the problem.

It is a known phenomenon that as glass is deformed elastically out of plane, a secondary mode emerges that diverges from the idealized shape of a hyperbolic paraboloid (Bensend 2015, 2016). While a hyperbolic paraboloid is a double-ruled surface, it is anticlastic and not developable; it follows therefore that glass, being rigid, cannot be forced into such a shape.

It is also established that above a certain level of distortion, "linear Kirchhoff–Love theory of plates [does not apply, so] the plate bulges into an asymmetric configuration and the edges are not any more straight" (Galuppi 2014). In this project, we found that the panels' edges deviated from in-plane straightness by as much as 6.5 mm. This

1 This threshold is derived in more detail in later sections.

2 This technique is known to exhibit behavior wherein the final shape "varies in time due to the viscosity of the polymeric interlayer" (Galuppi 2015). The panels in this project however are adhered to a rigid structural frame, and the long edges further have an exterior cap, which together work to retain the perimeter shape of the panel.

is consistent with in-plane edge deviations found in other studies (Eversmann et al. 2016). Other related work also indicates that optimal shape finding must “reduce the tangential force exerted at the boundary” (Gavriil et al. 2020).

The suggestion is that edges should not be assumed straight on the fabricated shape; rather, to develop a methodology that accounts for the distortions such that panels have straight edges *when in their final position*. This is important because the vertical framing elements are also straight; any deviation in the glass edge straightness could compromise the quality of the structural bond between the glass and the frame (Fig. 2). Through discussions with the fabricator, we set a target deviation of 1.0 mm.

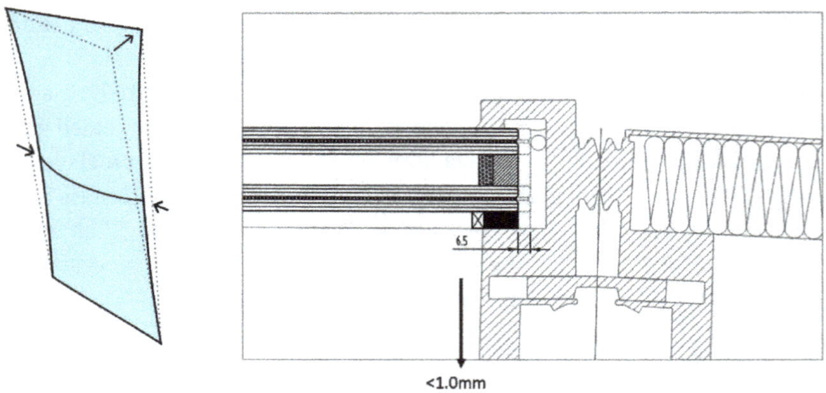

Fig. 2: As the edge deviates from straight, it affects the structural interface with the frame.

Some existing work puts an emphasis on pursuing shapes that conform to architectural design surfaces (Gavriil et al. 2020, Beer 2015) or an assessment of “cold bending distortion... and the resulting optical quality” (Datsiou 2016), but here priority is given to the geometrical conformance between panel edges and the framing.

3.1 Effects of Non-Rectangularity

As mentioned above, the bulk of existing literature assumes panels that are rectangular. Rectangles alone cannot tile most complex surfaces. As the shape deviates from rectangular, the buckling behavior of glass panels changes. We include here a parametric study of panels with skewed, parallelogram shapes, in line with the panels of the project. The study indicates that between 50 mm and 100 mm of skew, there is a bifurcation in behavior in the buckling behavior (Fig. 3).

Bensend establishes a non-dimensional twist ratio, denoted k, and suggests that for rectangular panels the moment of buckling usually begins when $k = 13.5$. The

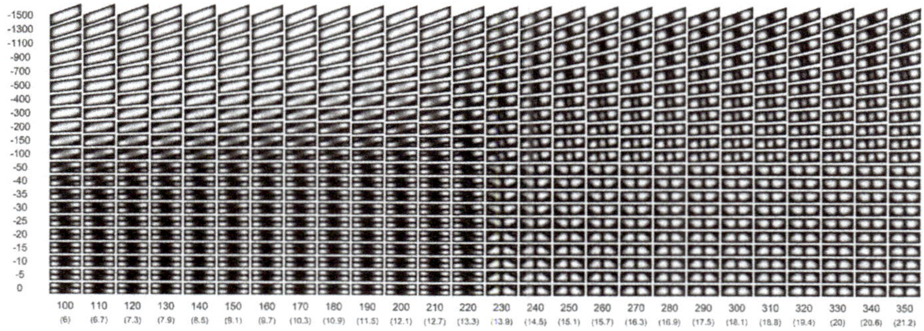

Fig. 3: Plot of surface deviations relative to idealized hypar surface. Above image shows plots normalized to the maximum of each individual panel. Below image is normalized to the maximum across all samples. The horizontal axis shows depth of bending in real dimensions and non-dimensional *k*-factor (in parentheses).

parametric study above indicates a bifurcation in buckling behavior between k_{buckling} values of 13.3 and 13.9, which is consistent. Plots below show that as the skew increases, the deviation from a hypar is greater, even at small amounts of bending. As bending approaches the buckling limit, rectangular panels show a sharp increase in shape changes, while the skewed panels have a more subtle transition (Fig. 4 and Fig. 5).

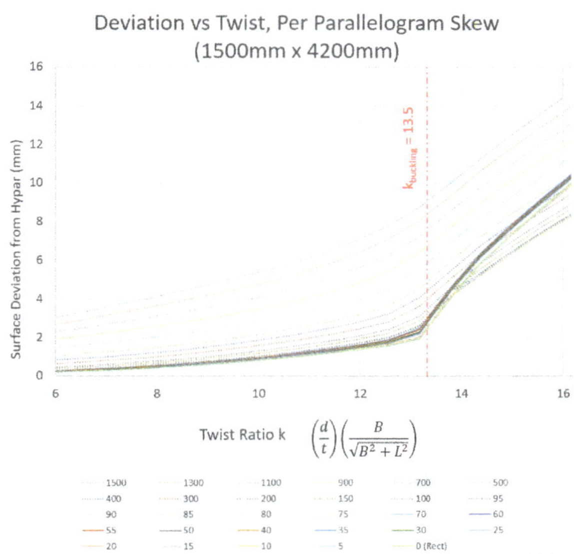

Fig. 4: Comparison of surface deviation and panel twist for different dimensions of panel skew.

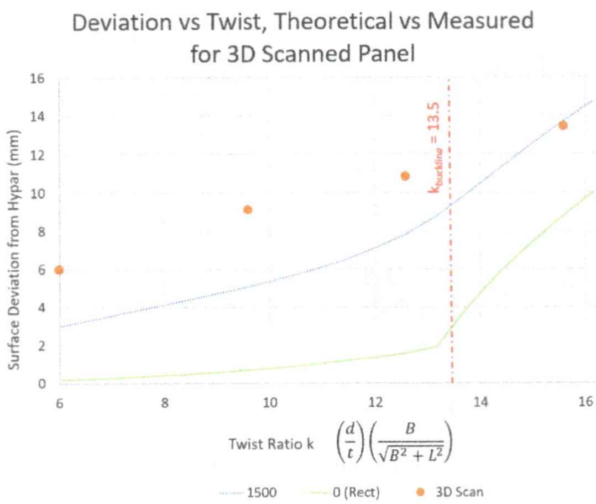

Fig. 5: Surface deviation vs panel twist for 3D Scanned panel, compared to theoretical for skewed and rectangular equivalent.

3.2 Preliminary Physical Testing

Prototype panels were fabricated, with 3D scans performed for comparison with the structural simulations from Finite Element Analysis (FEA) and deviations from the theoretical surfaces. The prototypes clearly exhibited the secondary nonlinear effects consistent with the FEA, though the results are not a perfect match. While broadly consistent, with two local "bulges" of deviation, there are distinct ripples in the pattern (Fig. 6). These are artifacts from the plastic bag draped across the back of the glass, the shadows of which may affect the registration of the scanner. The scanning results were only available after the prototype testing had concluded, and the timeline did not allow for further testing. In future studies, it would be preferable to mask the panel with tape or otherwise opaque material.

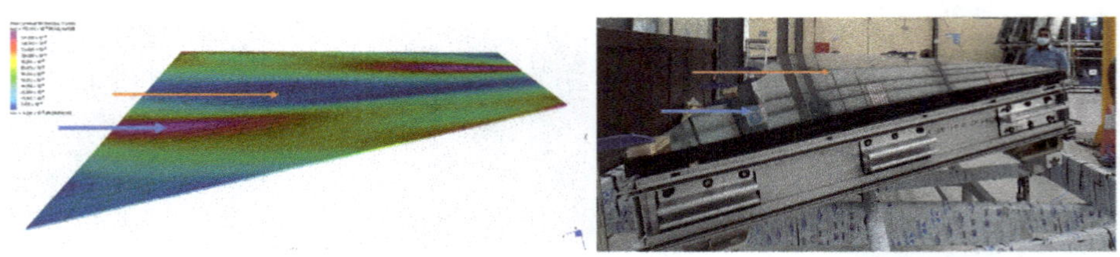

Fig. 6: Correlation of FEA simulations and fabricated panels.

Having built, bent, and scanned multiple prototypes, the scans were post-processed to measure the point clouds against the theoretical hyperbolic paraboloid, for different degrees of bending/twist. Due to the inherent noise in the sampling, as well as errant

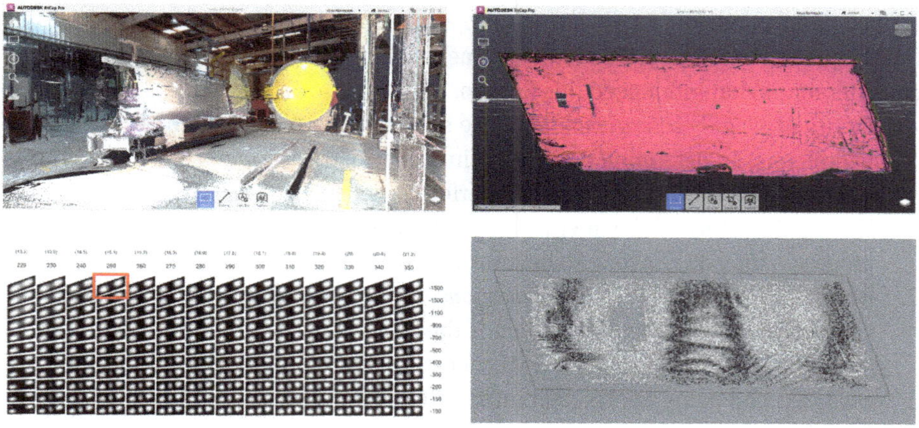

Fig. 7: Top left: color view of full 3D scan; top right: selected area of 3D scan, indicating normals as measured by scanner; bottom right: calculated deviation of 3D scan from theoretical surface; bottom left: corresponding plot of expected deviation according to FEA.

scan points due to things like clamps and other hardware, statistically distinct outliers were removed from the data.

As the panel approaches and passes the buckling limit, the scans converge with the expected simulation results. Below this level, the scans show greater deviation from the theoretical than we would expect from the FE analysis. We believe this is due to the realities of fabrication and construction, for example imperfect boundary conditions: FEA can represent infinitely stiff edge conditions, while in reality, materials flex and bend, and the restraints on the panel will have some additional give, thereby introducing greater deviations.

While a tighter match between the FEA and the scans would be preferred, the overall buckled shape of the glass was consistent with our simulations (Fig. 6), and the further the panel was bent, the tighter the numerical correlation.

Bensend suggests a knock down factor of 0.6 to the twist factor if buckling is to be all together avoided. This sets the threshold at 180 mm as discussed previously. When accounting for the thinner plies used in the lamination bending, that 180 mm threshold is equivalent to a k-factor of 10.8, as shown in Fig. 4. Further confidence is provided because the correlation between scans and simulations are stronger above this level.

4 Machine Learning

To accurately unroll the glass surface, one must accurately predict the 3-dimensional surface, accounting for the material properties and nonlinear buckling effects described in the previous section. For this project we chose to pursue a Machine Learning (ML) approach with integrated Finite Element Analysis. Listed below are the primary considerations that influenced this decision.

1. *Inputs are highly interdependent.* The shape of the panel is defined by several parameters. To determine the relative influence of these inputs, one must perform many simulations and/or tests and perform a sensitivity analysis. One strength of ML algorithms is that the interaction of inputs is computed as part of the regression process – they do not have to be known in advance.
2. *Requirement for a large set of simulations.* Any analytical technique would require simulations or tests for validation. If the large dataset is required regardless of the technique, then opting for a ML model sidesteps the need for heavy analytical work, instead using analysis for intermittent checks to ensure the model outputs are reasonable.
3. *Integration with industry standard workflows.* Ultimately a mathematical description of the surface would need to be translated into a CAD package. Previous work that aims to predict glass panel outcomes from definition of its shape is primarily aimed at providing singular numerical values that are relevant to the engineer, such as peak deflection or peak stress. The more sophisticated of those techniques rely on a large array of tabulated data along that would require the practitioner to do several interpolations (Galuppi 2020). We instead pursued an approach that could exist "natively" inside of our 3D modeling environment.
   The process we deployed generates simulations directly from the model, trains the ML model, and returns data points in a model-friendly format (this is expanded upon below). This avoids having to translate data between different contexts (e. g. CAD to mathematical formulations and back), and leverages the tried and tested surfacing algorithms in the 3D CAD environment for representation. Being able to use verified, trusted FEA packages also gave us greater confidence in the approach, because each database entry could be retrieved, opened, interrogated, or even shared with other engineers. Strand7 was used for our simulations, which is an industry standard package, such that most 3rd party reviewers would be able to review or even replicate any simulation in a software package that they know and trust.
4. *Flexibility and Generalization.* The ML model employed here simply analyzed three parameters (described below), whereas the largest machine learning models in use today have tens of billions of parameters, so the principles of this approach could be used to encompass a more general set of panel geometries.

4.1 Automated Generation of FEA models

To produce a ML model, one must first compile the data on which the model is trained. Since it is important that the database accounts for the physical behavior of the plates, the database is comprised of structural simulations that encompass the range of panel geometries in the project.

To establish the training set, each panel starts as a NURBS surface from Rhino that generates a quad mesh that is suitable for FEA. The twisted surface is then "pulled flat" in a simulation process that constrains its edges to the ground plane, allowing the center of the panel to deform and buckle as necessary within that flat boundary. The aim in this step is to approximate what the final flat shape might be. The perimeter of the "flattened" panel is then extracted, and a new mesh, truly flat, is created from it. That new mesh is then pushed back to 3D, again with edges constrained. This simulated shape will now feature the characteristic "bulges" as indicated in the prototypes.

4.2 Shape Functions and Parameterization

When designing a ML model, one of the most important decisions is the selection of inputs and outputs. For this study, inputs are selected to provide a numerical representation of the panel perimeter, and outputs constitute a numerical representation of the surface.

For the inputs, the panel perimeter is known in advance. The known parameters that vary from panel-to-panel include: Out-of-plane bending dimension, top edge curvature, bottom edge curvature, internal angles of the parallelogram, top edge length, and bottom edge length. For the outputs, we needed to define discrete numerical values that can define a continuous surface. To do this, we borrowed a concept from Finite Element methods called a Shape Function. In FEA, this is a function that allows for computation of a value within the field of an element, based on the values at its nodes.

A similar approach is used in Galuppi (2020) to determine peak stress or deflection, where ours is for recreating surface geometry. We selected 7 locations within the field of the quadrilateral, and measured their deviation from the idealized CAD geometry. The locations were selected by a process of trial and error but followed an underlying logic: if/when the top and bottom edges of the panel are curved, three points provide sufficient definition of the resulting nearby curvature (P1–P3 for the top, P5–P7 for the bottom).

We determined that the panel's central area could be defined with a single sample point – the curvature across the center was always symmetrical and sufficiently described by a single interpolation point. Through development of the mapping functions, we also found that some input parameters had weak correlations, and could be ignored. For example, the internal angle of the panel appears to have very little correlation with sample point P3, whereas the depth of bending has a clear correlative relationship (Fig. 8).

After analysis, the only three inputs required were the first three listed above: depth of bending, and curvature of the top and bottom edges.

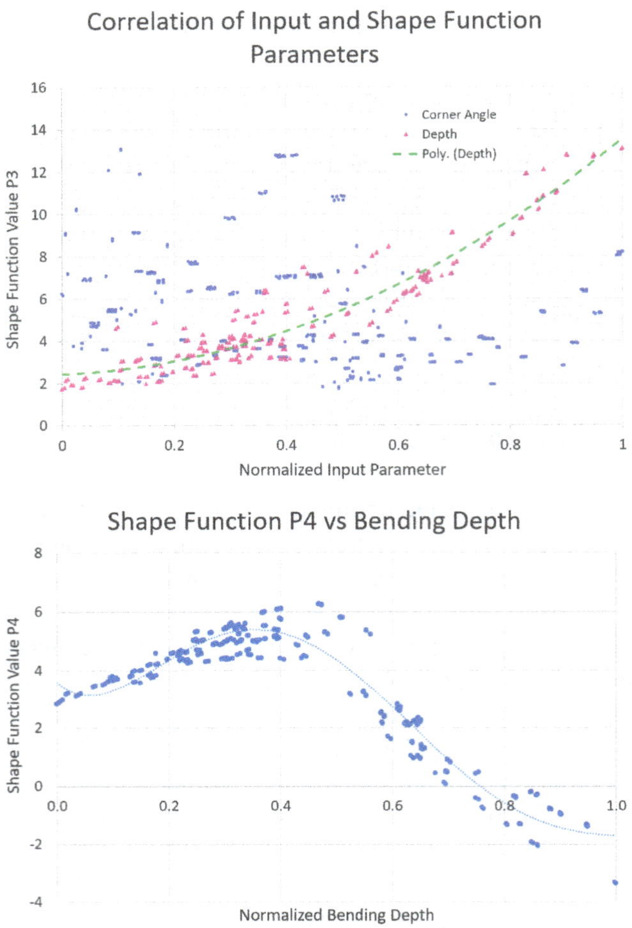

Fig. 8: Correlations of shape parameters relative to normalized inputs.

4.3 Training

With the establishment of the I/O parameters complete (Fig. 9), the process was then executed over 3,500 panels to create the training set. As suggested by the plots in Fig. 9, the relationships between parameters are not linear, so polynomial regression was used. These polynomial regression results can be abstractly represented in 3D visuals, as shown in Fig. 10 and 11. The result of the training is 7 polynomial expressions, one for each sample point, each having the three inputs as the expression variables.

In many ML approaches, the model receives an input database and maps them directly to outputs, with weights and relationships concealed within: the classic "black box" problem. Further, if you query the model with a set of inputs that matches an entry in the database, it may simply return the outputs that were originally paired with

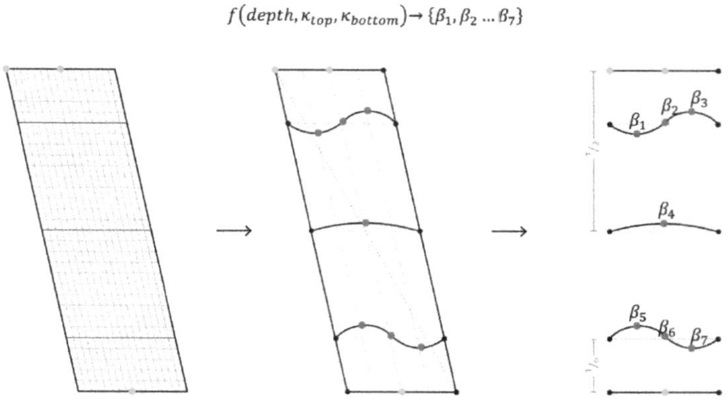

Fig. 9: Function mapping from 3 inputs to 7 output Shape parameters.

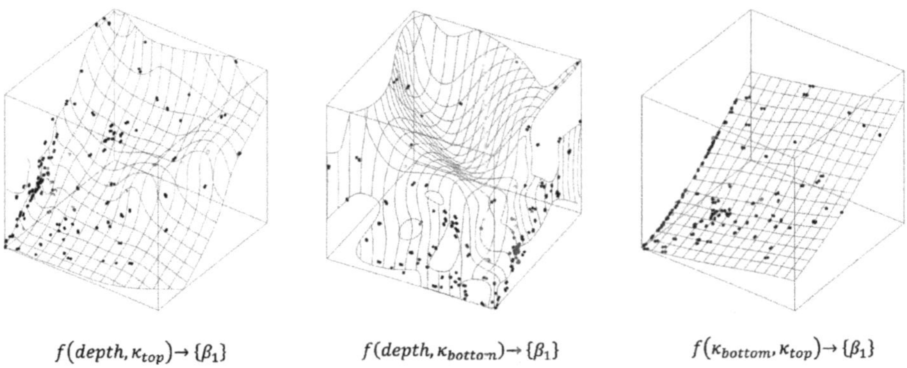

Fig. 10: 3D representations of the ML model, showing two inputs on the horizonal axes, and the resulting shape parameter B1 in vertical.

it. In our case, this is insufficient since the partnered outputs were only approximations. We want to account for the accumulated contours of the sample space, i. e. to "average" across shapes that were similar. By producing a polynomial function, we solve both problems: an explicit description of how the calculation is made, and the function serves as an interpolation of the sample space.

If the predicted 3D surface is found to diverge from the 3D perimeter beyond the agreed tolerance, this indicates there is a gap in the dataset. However, while that simulation was out of tolerance, the deviations were on the order of < 10 mm, meaning the surface was still a decent approximation of the surface's contours. It can therefore be added to the database of approximations, the and model/polynomials retrained. This results in the pseudo-self-improving nature of the ML model.

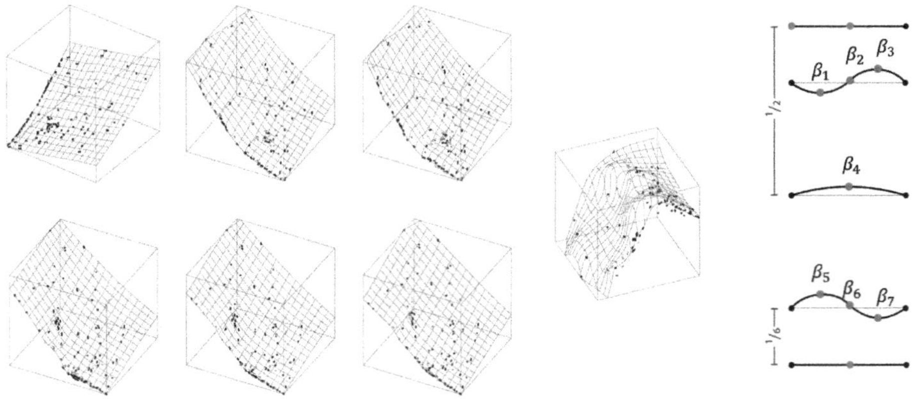

Fig. 11: 3D Representations of all 7 shape functions.

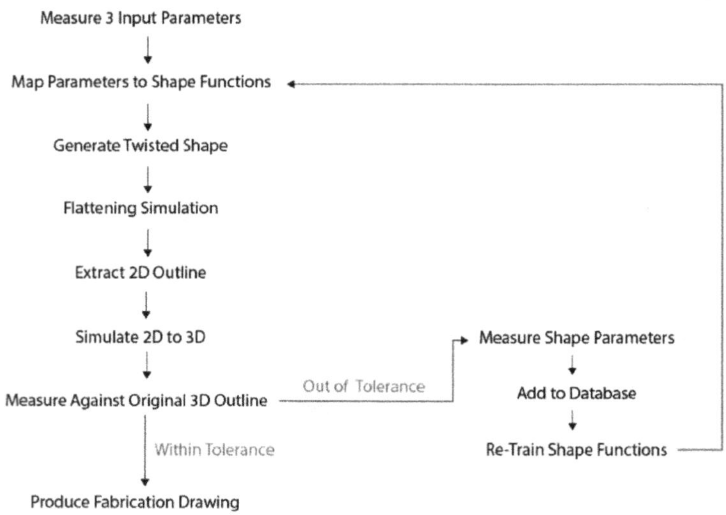

Fig. 12: Flow chart of complete training and production routine.

5 Results and Reflection

By using the ML model to predict the surface geometry of the cold bent, curved, semi-buckled shapes, the edge deviations were reduced from 6.5 mm to a max of 0.5mm, which is well within the allowable threshold. In this regard, the process was useful and constructive. The technique enabled the fabrication of thousands of unique, cold-bend pieces of glass, and to realize the project even with an extremely ambitious timeline.

Fig. 13: Image of some of the unique, highly twisted panels.

We have also shown that current guidance for estimating the buckling threshold appears consistent even with panels that are not rectangular, though the transition is smoother in panels that are skewed.

This study was still centered around a narrow set of geometries and is far from comprehensive. We believe this provides an entry point for a potential larger study across a wider variety of shapes, and the use of ML could provide an efficient way of simulating the "real" surfaces that glass panels will manifest.

The primary motivation for these exercises was fabrication and assembly. However, it would certainly be valuable for architects to be able to assess the aesthetics of surfaces that would arise from given panelizations, a possible extension of this work.

In the context of this building it proved to be critical that any technical solution could fit naturally into a 3D production pipeline. The pace of the project would not have tolerated workflows that require high-friction translations of information across different software packages. While not our central focus, this proved essential for the realities of the project, including time demands and verifiability.

6 Conclusion

As cold bent glazing proliferates as a strategy for realizing complex geometries, one should consider the translation of surfaces represented in CAD environments into reality. Non-planar quadrilaterals are often represented as double-ruled hyperbolic paraboloids. However, we have demonstrated that glass panels appropriate for construction do not adhere to such surfaces, and we present one technique that successfully mapped 3D outlines to a 3-dimensional surface that accounts for material behavior.

Edge deviations were reduced from 6.5 mm to less than 0.5 mm, and the workflow was implemented directly into a 3D CAD production pipeline. Future work should include an expansion of the range of shapes, and a more interactive interface that can be useful for designers.

References

Beer, B. 2015. Structural silicone sealed sold-sent slass–sigh-sise projects experience leading to a new design concept. In *Glass Performance Days 2015 Conference Proceedings*, Tampere, 2015

Bensend, A.: Beneath the surface: buckling of cold formed glass. In: *Glass Performance Days 2015 Conference Proceedings*, Tampere, 2015

Bensend, A. 2016. Maximizing the twist of cold formed glazing. *Challenging Glass Conference Proceedings* 5: 65–80. DOI: 10.7480/cgc.5.2230

Bensend, A. 2018. The Effects of Cold Warping on Glass Stiffness. *Challenging Glass Conference Proceedings* 6: 85–96. DOI: 10.7480/cgc.6.2119

Datsiou, K. C. 2017. Design and Performance of Cold Bent Glass. Thesis, University of Cambridge. DOI: 10.17863/CAM.15628

Datsiou, K. C., and M. Overend. 2016. The Mechanical Response of Cold Bent Monolithic Glass Plates during the Bending Process. *Engineering Structures* 117: 575–90. https://doi.org/10.1016/j.engstruct.2016.03.019.

Eversmann, P., A. Ihde, and C. Louter. 2016. Low Cost Double Curvature – Exploratory Computational Modelling, FE-Analysis and Prototyping of Cold-Bent Glass. *Challenging Glass Conference Proceedings* 5: 81–92. DOI: 10.7480/cgc.5.2233

Galuppi, Laura, Simone Massimiani, and Gianni Royer Carfagni. 2014. Buckling Phenomena in Double Curved Cold-Bent Glass. *International Journal of Non-Linear Mechanics* 64. https://doi.org/10.1016/j.ijnonlinmec.2014.03.015.

Galuppi, Laura, and Gianni Royer Carfagni. 2019. Betti's Analytical Method for the Load Sharing in Double Glazed Units. *Composite Structures* 235: 111765. DOI: 10.1016/j.compstruct.2019.111765

Galuppi, L., G. F. M. Royer-Carfagni, L. Barbieri, and M. Maffeis. 2020. Sharing of general loading in double glazed units. The BAM analytical approach. *Challenging Glass Conference Proceedings* 7. DOI: 10.7480/cgc.7.4484

Gavriil, K., R. Guseinov, J. Pérez, D. Pellis, P. Henderson, F. Rist, H. Pottmann, and B. Bickel. 2020. Computational Design of Cold Bent Glass Facades. *ACM Transactions on Graphics* 39: 1–16. DOI: 10.1145/3414685.3417843

Nardini, V., and J. Hilcken. 2018. Mistral Tower: Value of System Design, Manufacturing and Installation in Cold Bent SSG Units. *Challenging Glass Conference Proceedings* 6: 117–34. DOI: 10.7480/cgc.6.2128

Samim Mehdizadeh and Oliver Tessmann

Roll-Locks: A Fabrication to Self-Assembly Design-Framework for Reusable Discrete Concrete Elements

Abstract: *Roll-Lock* is a research project in which building elements are informed for motion and future reuse. This paper gives technical insight into systematically investigating lightweight, hollow, and kinetic concrete elements' geometrical and physical properties. The massive-looking yet lightweight, hollow, robotically materialized concrete elements are designed to join together, but not permanently. A single person can move and assemble them with their dry-joint, interlocking connection and charged center of mass (CoM) (Fig. 1). The elements are programmed for guided self-assembly, disassembly, and re-assembly through embedded geometrical and physical attributes (Fig. 2). The reconfiguration allows for a more circular approach and reuse of concrete elements in the architectural context. Furthermore, this research seeks to provide novel strategies to save building material resources by rotoforming (Tessmann and Mehdizadeh 2020), thereby minimizing material consumption and significantly reducing the weight of hollow concrete elements (Fig. 4).

Keywords: design for reuse, digital concrete, RotoForm, programmable material, sequential dynamic casting, interlocking assembly, sequential assembly

Fig. 1: Human interaction with the 1:1 hollow concrete element for guided self-assembly.

https://doi.org/10.1515/9783111162683-012

1 Introduction

Current construction technologies extensively rely on non-reversible mechanical fasteners or adhesives. Even precast elements are mostly connected with mortar and grout (Elliott 2002), preventing non-destructive disassembly and reversibility. In this paper, we offer a design to materialization framework that allows for reconfiguration and (re)assembly with interlocking hollow elements. We regard reconfiguration as a crucial strategy for the long-term use of building elements and reusability. Most importantly, it allows for more circularity in design and construction. We investigate geometrical and physical principles that enable and ease humans' lifting, rolling, assembling, and disassembling of building elements without any heavy equipment. Thus, reversibility reduces material consumption, allowing new forms of interaction between human and architectural elements. The presented work is based on the robotic fabrication method RotoForm (Tessmann and Mehdizadeh 2019), which saves material of both concrete and formwork by producing hollow elements.

Fig. 2: Human interaction: The process of assembly, disassembly, and lifting of the concrete element through rotational interlocking and manipulated center of mass.

2 Research background

2.1 Self-assembly

The research on self-assembly is relevant for this paper because it includes strategies of inscribing assembly information into the geometry of parts instead of using the architectural strategy of plans, drawings, and details as abstract and formalized assembly instructions for human or machine agents. The studies of Georges Popescu (Popescu n.d.) at MIT on digital materials and digital fabrication describe the programmability of parts with their simplicity and discrete logic. All parts of a self-assembly system are geometrically simple, should be able to have at least two states (e. g., on/off or attached/detached), and should be able to respond to the instruction sequence. Based on Popescu's research Skylar Tibbits suggests in his master thesis, "Logic Matter", parts for self-assembly and guided self-assembly are enabled through assembly information inscribed in their geometry and embedded digital logic (Tibbits 2012). Tibbits describes the following requirements for self-assembly systems.: 1) a simple assembly sequence, 2) programable parts, 3) force and energy for activation, and 4) redundancy. The research "growing machines" (Griffith 2004) at MIT explores the idea of self-assembly and guided self-assembly through self-replicating machines (Penrose and Penrose 1957) and identifies a discrete system that consists of different programable parts; rather than arrays of similar components. He describes the hierarchy of assemblies in biological models as a technique for building complex matter assemblies.

2.2 Interlocking

Research on interlocking originates from various fields, such as computer graphics, material science, and structural engineering. Computer graphics research defines interlocking in two categories: either the voxel-based approach assembling parts along orthogonal vectors in a particular sequence (Song et al. 2015), or the catalog-based approach, in which small pieces form a puzzle difficult to solve since the assembly must follow specific rules. While the combinatorial complexity of the puzzle's piece arrangements is exceptionally high (Xin et al. 2011). Luo proposes a digital framework, "Chopper" (Luo et al. 2012), for partitioning large geometries into interlocking parts with male/female joinery systems. Wang et al. introduce a method to cluster voxel-based geometries for a broader range of assemblies with many different shapes (Wang et al. 2018). In summary, the computer graphics community focuses on chopping the geometry into smaller buildable (printable) pieces – a top-down approach for solving the fabrication issue of large parts.

The geometrical concept of topological interlocking can be traced back to the baroque era of stereotomy (Fallacara 2009) and resurfaced through research in material science (Dyskin et al. 2003). The application potentials of the topological interlocking in the field of computational have been explored by Tessmann (2012), Gata et al. (2019), and Weizmann et al. (2017). The researcher sought to increase the geometrical repertory of topological interlockings and integrated generative and analytical capacities into a computational design framework. The flexural performance of interlocking flat vaults made from topologically interlocking concrete blocks demonstrates good performance compared to the monolithic concrete plates (Rezaee Javan et al. 2017). Moreover, Baghdadi et al. (2023) developed analytical digital models to evaluate the structural strength of dry joint interlocking interfaces between the concrete elements. The geometrical interface as a dry join possibility has been applied and explored by Oval et al. (2023) as well as Rippmann et al. (2018) as a dry joint connection between concrete, discrete elements. The architectural practice AAU-Anastas has built a couple of architectural projects using interlocking elements as flat and curved vaults.

2.3 Robotic fabrication and digital concrete

Over the last decade, a vast range of research in computational design and robotic fabrication sought to change the conventions in construction. Robotic and concrete researchers propose novel materialization methods for formwork fabrication (Wangler et al. 2019) and concrete deployment. Besides 3D printing of concrete, a technology that made the leap from research labs to construction sites (Tessmann et al. 2022), material-robot systems for concrete processing offer great potential for exploration and innovation. This research field explores a more substantial involvement of robotic kinematics in material placement, such as slip casting (Shahab et al. 2013) or rotoforming (Tessmann and Mehdizadeh 2019). Material-intensive formwork systems are replaced by robotic movement during concrete processing. Rotoforming is used in this research to materialize the building elements.

The technology utilizes the concrete flow and robotic trajectory to materialize hollow concrete elements with variable cavities inside the elements. A minimum amount of liquid concrete is poured in a form gently rotated by a machine. The material is just enough to form a thin hollow shell as the rotation makes it flow along the formwork. RotoForm allows casting in several layers of concrete with different thicknesses from outside to inside. The differentiated material and weight distribution inside the hollow elements shift the elements' center of mass (CoM).

2.4 Assembly through weight

As architectural elements are mostly meant to be static as part of a larger tectonic system, another research field focusing on dynamic objects needs to be explored. Bächer et al. developed an algorithm that generates spinning objects by optimizing rotational dynamics properties. A solid 3D model with a random, asymmetric shape and a desired axis of rotation forms the input. The algorithm then modifies the mass distribution such that the principal directions of the moment of inertia align with the target rotation frame (Bächer et al. 2014). The 3D-printed objects ideally rotate around a vertical axis while being asymmetric in their outer shape. Invisible inner cavities achieve mass symmetry.

The design research project "Walking assembly" by matter design (Brandon Clifford) addresses different aspects which are also present in this research, such as assembling the concrete Blocks through weight distribution (Swingle et al. 2020). Differently, Clifford manipulates the CoM by gradually changing the concrete density. Clifford also presents studies on Geometry and force design, which allows humans to lift and move megalithic heavy Concrete objects without cranes (Clifford 2016). A design to build a framework concerning the weight distribution in the global assembly by (Wibranek et al. 2020) illustrates the ability to assemble modular building components without the need for support structures.

3 Method

This proposal aims for objects that can roll into various assembly configurations. All elements are based on a 3D voxel grid. The catalog of elements is shown in Fig. 3.

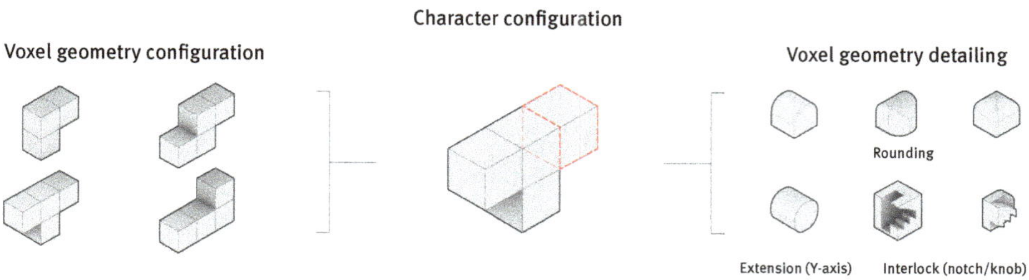

Fig. 3: The topology catalog of elements; Left: the chosen voxel-based shapes (L, T, and S), Right: the replacement voxels.

Every voxel of an element can be formed differently to gain rotational or interlocking properties (Fig. 3 right). The digital workflow in this design to simulation and fabrication process starts with developing forms that enable rotational movement. Motion and geometry are closely linked. The joints that enable rotational interlocking are derived from the rotational trajectories of the objects (Fig. 4). The interlocking performance is subsequently tested in rigid body simulations as well as the placement of the CoM through designing internal cavities and varying wall thickness. Finally, the robotic fabrication data that controls the material distribution during rotoforming is generated. We conducted several iterations of numerical and experimental experiments and produced a series of physical prototypes before evaluating the results. The following paragraphs describe the relationship between the different topics.

3.1 Motion path and rotational interlocking

The target is to generate shapes that roll and interlock into a solid connection (Gilibert et al. 2022). We characterize all part's motion paths and interlocking procedures with three transitional and three rotational degrees of freedom. The voxelated shape of

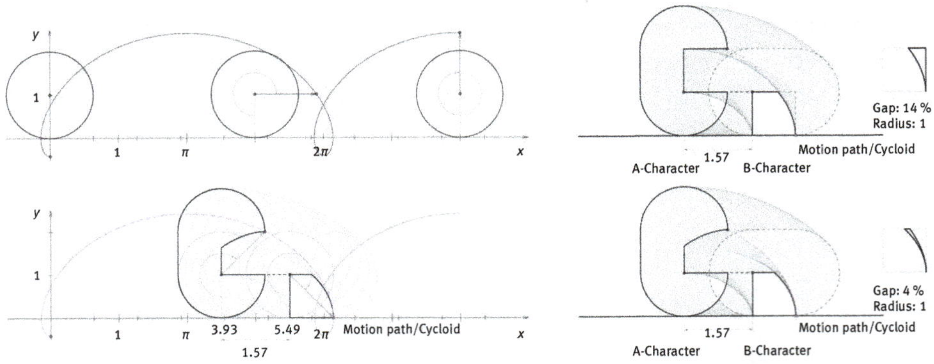

Fig. 4: A cycloid, form finding and research to increase the contact area in the locked state.

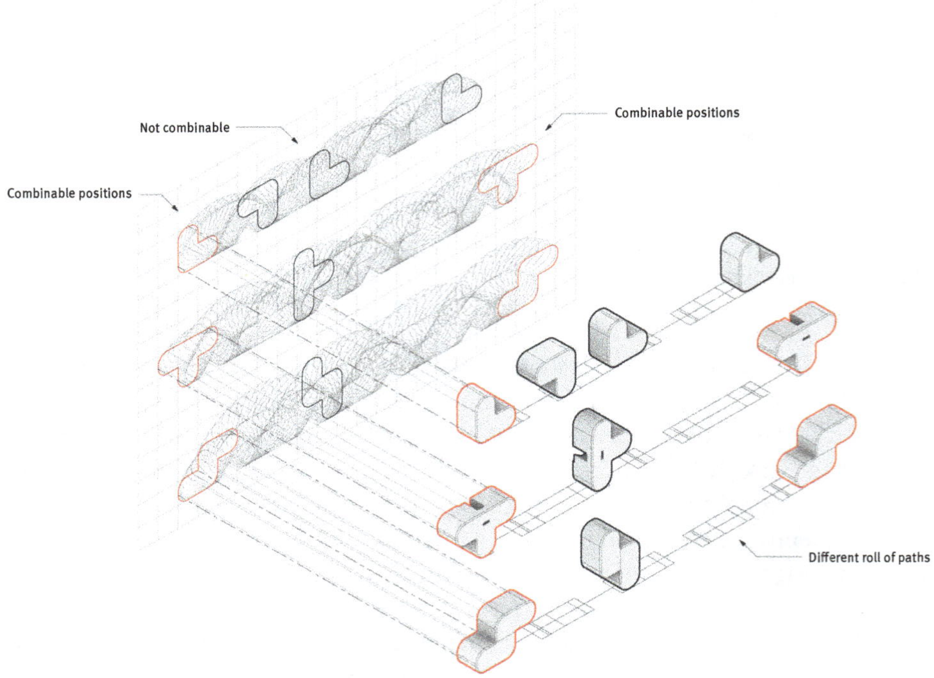

Fig. 5: Moving (rolling) paths regard to the shape of elements in 2D projection address the initial and target position for assembly.

elements and the rotational interlocking system allows for solid connections in all transitional directions (*X*, *Y*, *Z*) (Fig. 6).

To define a (rolling) motion path in relation to rotational interlocking, we use the geometrical principle of the cycloid. "A cycloid is a curve traced by a point on the circumference of a circle which rolls without slipping along a straight line" (Barra

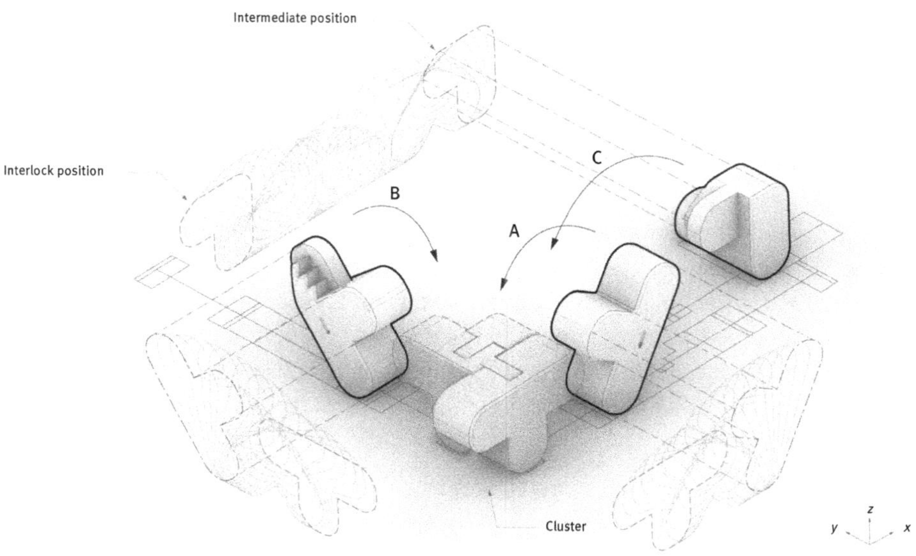

Fig. 6: Assembly sequence of the lowest hierarchy with three elements (T, T, L).

1975). A generative digital model (form finding algorithm) based on the cycloid allows for defining all the possible motion paths of the convex shapes based on their curvature (Fig. 4 left, Fig. 5). The generated motion paths define the shape of interlocking joints made from notches and knobs (male/female connection). In order to increase the contact area in the locked state between two elements, the form-finding algorithm minimizes the geometrical deviation of curvature between the male and female parts (Fig. 4 right). The number of rotational interlocking joints in the physical prototypes is increased with different radii to increase the contact pressure on the side surfaces.

3.2 Motion and assembly sequences

This scope focuses on geometry as a tool for generating assembly sequences. Designing the recursive assembly refers to the issue of finding the sequence for full (dis)assembly of all the parts. Throughout this research, we have developed and tested two main approaches to recursive assembly sequences with different shape types: a) clustering assembly, b) sequential assembly. The assembly sequence of both approaches consists of interlocking voxelated shapes with a single-key property. When the assembly is completed, all the elements in each cluster are notched and locked in all transitional directions. Disassembly is only possible through the rotational motion of the single key element. We use the mathematical method of Non-Directional Blocking Graphs.

(NDBG) proposed by (Gilibert et al. 2022) to define the assembly sequence concerning the interlocking.

3.2.1 Clustering assembly approach

This approach determines a catalog of discretized building elements in three Tetris-like shapes (L, T, and S). In order to decrease the complexity of assembly, each shape is defined with three to five voxels. Every voxel can incorporate a notch/knob joint or become a curved voxel according to the desired function of motion or interlock (Fig. 4.). Thus, each element consists of a regular, notch, and curved voxels in different orders. This catalog-based geometry generation allows for a wide range of roll-lock scenarios (Fig. 7).

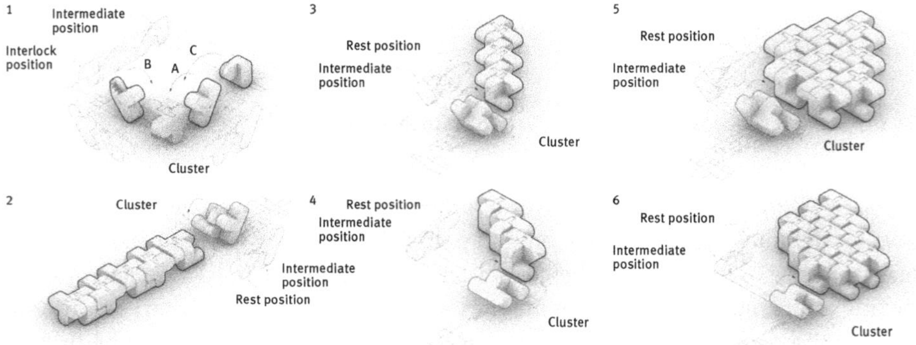

Fig. 7: The catalog of hierarchy models for one-directional (linear strings) and two-directional orders (weaved surfaces).

3.2.2 Sequential assembly approach

A second generation of rolling assembly elements is based on a sequential assembly logic through which elements join together in a specific order one after each other. With every new element added to the assembly, new rotational directions are unlocked while others are deactivated. The sequential assembly of different shapes is locked and released by one single key responsible for immobilizing all possible translation directions (Wangler et al. 2019). Each element has two states: attached and detached (on/off). The curved corners of the elements allow for rolling movement in certain positions. The topology of the element and the assembly sequence control the interlocking. The elements are different. The assemblies are designed to be finally locked in a solid composition by adding the last keyelement (Fig. 8). The sequential (dis)assembly has only one single fixed order. In a digital simulation, weights could be assigned to (solid/hollow status) to test the motion behavior (Fig. 10). (See also Sec. 3.4.)

Fig. 8: The assembly sequence of elements. After adding the key element, the topological-interlocking principle holds all the parts together during the motion.

3.3 RotoForm, robotic sequential casting

The rotoforming process is iterative for several material layers, the material quantities and qualities allow for a differentiation of the layers. Using a six-axis robotic arm as a RotoForm-machine, the robotic trajectory and rotation velocity enable varying thickness of a layer in different positions (Mehdizadeh et al. 2022) (Fig. 9). We use these two parameters for programming the material (weight) distribution inside the hollow elements. Subsequently, we can manipulate the CoM (Mehdizadeh 2023) and activate the kinetic behavior of the elements through this technique.

3.4 Weight distribution inside Hollow elements

Initial activation energy is required for the guided self-assembly with hollow-concrete elements. This proposal conceives the activation energy applied by a human agent, as it is meant to offer physical affordance for reconfiguring architectural assemblage through its users. To minimize the initial activation energy for assembly, we aim for a specific position of CoM and optimize the weight distribution inside the elements. For the digital optimization algorithm, we voxelize the designed element and define each of the voxels in two categories, solid and hollow, and assign a certain weight to each

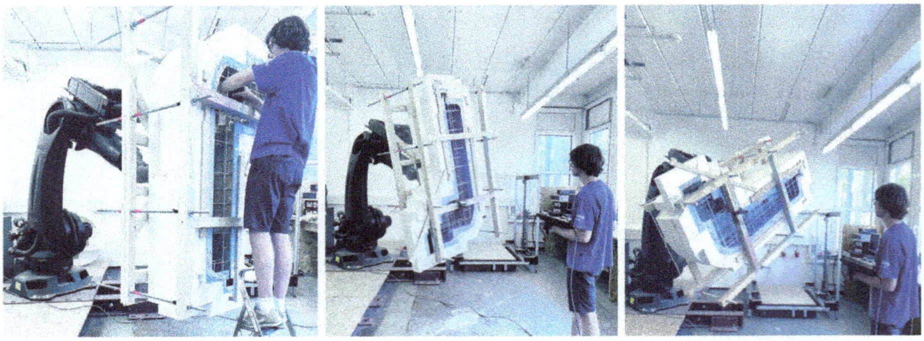

Fig. 9: Robotic sequential casting (RotoForm) of 1:1 hollow concrete element. Left: adding the calculated amunt of casting material to the formwork. Right: the robotic arm rotates the formwork to distribute the material alng the desired trajectory.

voxel. Increasing the resolution of voxels in the digital simulation model increases the simulation accuracy. The average Vector in the middle of the touching surface on the ground is the equilibrium axis (EA). To have a stable element status (in equilibrium), the total weight amount multiplied by its distances to EA should be equal on both sides of the EA. The optimization algorithm aims for the equilibrium of the element in the desired position and assigns the weight to voxels to reach the goal. After the optimization phase, we run a rigid-body simulation to track the motion of elements from any position to the stable status using Nvidia-PhysX inside the CAD environment. Rhino/Grasshopper.

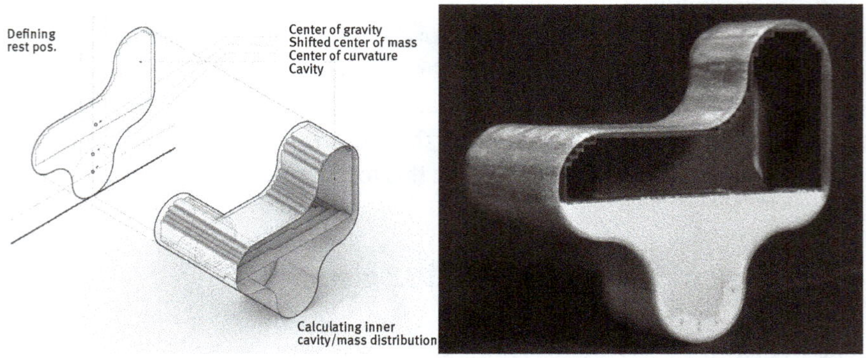

Fig. 10: Left: voxel-based optimization of weight inside the elements, right: physical prototype with 3D printed cavity.

4 Results

The continuous digital workflow from assembly motion design to robotic fabrication, allows for fabrication-aware design. Recursive interlocking architectural assemblies could become a principle for reversible, dry-joint, and circular constructions. The rotational interlocking mathematical model is the optimal solution for dry joinery of the elements that assemble through rolling (not the transitional directions) and results in a solid connection between the elements (Fig. 11). Rotoforming allows materializing the prototypes in 1:1. To evaluate the results; we used a four-dimensional survey method to track the motion of the physical prototypes and compare them with the digital simulation (Mehdizadeh 2023).

The comparison between the motion track data and rigid-body simulation illustrates the following points:

(a) Rotoforming with a robotic arm allows for precise positioning of the CoM inside the concrete elements.
(b) Manipulating the CoM allows for programming the motion behavior of elements as in simulation.
(c) Adjusting the global equilibrium of assembly clusters allows for levering and moving more elementsonce they are attached.

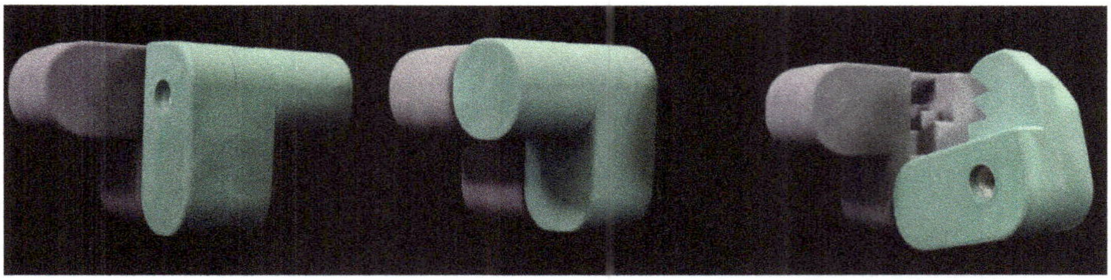

Fig. 11: Physical Prototypes of Hollow element with a rotational interlocking system in 1:10 Scale.

5 Conclusion

The research "Animate Concrete" contributes to circular approaches in design and "design for reuse" by providing a design framework for guided self-assembly and reassembly (Fig. 2). Moreover, using the robotic sequential casting method (RotoForm) reduces material consumption significantly. This study demonstrates the potential of embedding kinetic behavior and assembly information in hollow concrete elements by manipulating the physical and geometrical attributes of elements. This study shows that large-scale hollow concrete elements with particular weight distribution can be easily moved, lifted, and mobilized through a gentle touch of a human. The simulation results address the great potential of applying Nvidia-PhysX on different digital programs within the CAD environment Rhino/Grasshopper to increase the calculation efficiency. Designing assembly motion is uncommon in architecture and construction, as architects and structural engineers have rigid and stable tectonics.

6 Future works and outlook

Future studies will integrate structural analysis to analyze the load-bearing capacity of reinforced hollow concrete elements and the dry-joint solid connection between the concrete elements. The hollow concrete elements are slightly sliding on the ground when moving them. This issue decreases the accuracy of assemblies and should be solved in the following steps. Therefore, Digital material characterization and a finite element model analysis are necessary for the high-precise simulation of contact-rich and friction-fit connections. A study on environmental impact, reusability and a life cycle analysis (LCA) model for assessing the hollow concrete elements is our targeted next step.

A significant finding in this research was the ability to adjust the cluster's global CoM, which enables it to leverage the assembly's weight and move larger assemblies from horizontal to vertical status (Fig. 8. bottom right) (much like self-erection Tower cranes) (Dehlsen and Mikhail 2005). We illustrate this possibility as our next stage to lift gigantic assemblies through the same technique (Fig. 12).

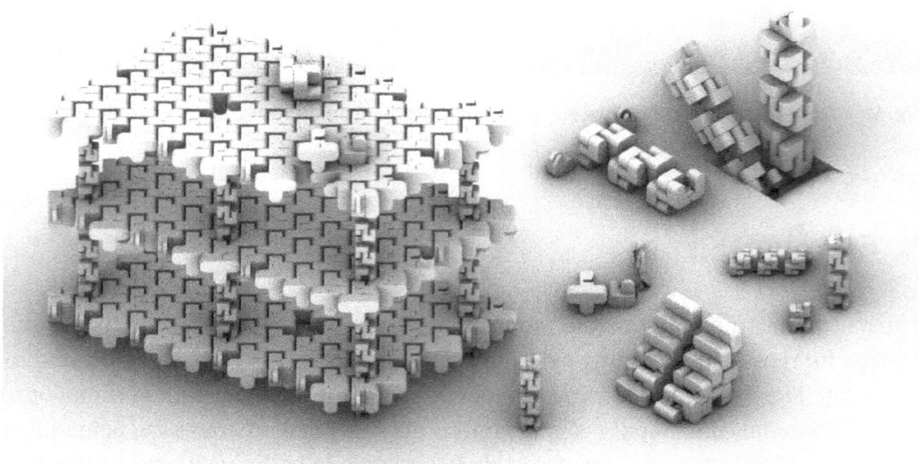

Fig. 12: A visionary visualization of various assemblies as a building system.

Acknowledgment

The research project "Animate Concrete" is developed and materialized with architecture students' outstanding contribution at Digital Design Unit, TU Darmstadt, in S.2022 under the supervisory of Authors. The initial Idea of Roll-Lock developed with the students: Joshua Schäfer, Mirko Duschke, Sara Elif Yildiz, Cristina Calin, Daniel Schinkels, Luck Schüßler, and Ye Hung. Among all the work of Joshua Schäfer impacted the development of the design simulation and motion tracking framework. The support of the student assistant Danial Ahmad and Leon Witschorke specially assisted our team with the robotic fabrication. The motion design assistance of student Annika Beissner assisted us with designing the interaction concepts with the concrete elements. The motion tracking part of this project is developed in collaboration with IAS TU Darmstadt, Dr. Boris Belousov, and Timm Schneider. Special thanks to Prof. Dr. Anna-Maria Meister. Prof. Moritz Dörstelmann, Prof. Dr. Timo Carl, and Prof. Mirko Becker for their amazing insights as a jury member of the design studio "Animate Concrete".

References

AAU ANASTAS News - AAU ANASTAS. Accessed March 18, 2023. http://aauanastas.com/.

Bächer, M., Whiting, E., Bickel, B., Sorkine-Hornung, O., 2014. Spin-it: optimizing moment of inertia for spinnable objects. *ACM Trans. Graph.* 33, 96:1-96:10. DOI: 10.1145/2601097.2601157

Baghdadi, A., Ledderose, L., Ameri, S., Kloft, H., 2023. Experimental and numerical assessments of new concrete dry connections concerning potentials of robotic CNC manufacturing technique. *Engineering Structures* 280, 115605. DOI: 10.1016/j.engstruct.2023.115605

Barra, M., 1975. The Cycloid. *Educational Studies in Mathematics* 6, 93–98.

Clifford, B., 2016. The McKnelly Megalith: A Method of Organic Modeling Feedback. https://doi.org/10.52842/conf.acadia.2016.440

Dehlsen, J. G. P., Mikhail, A. S., 2005. Self-erecting tower and method for raising the tower. US6955025B2.

Dyskin, A. V., Estrin, Y., Kanel-Belov, A. J., Pasternak, E., 2003. Topological interlocking of platonic solids: A way to new materials and structures. *Philosophical Magazine Letters* 83, 197–203. DOI: 10.1080/0950083031000065226

Elliott, K. S., 2002. *Precast concrete structures.* Butterworth-Heinemann, Oxford; Boston, MA.

Fallacara, G., 2009. Toward a Stereotomic Design: Experimental Constructions and Didactic Experiences. In *Proceedings of the Third International Congress on Construction History, Cottbus.*

Gata, K. M., Mueller, C., Valiente, E. E., 2019. Designing strategies for Topological Interlocking Assemblies in architecture. Flat Vaults. In *Proceedings of the IASS Annual Symposium, Barcelona.*

Gilibert, P., Mesnil, R., Baverel, O., 2022. Rule-based generative design of translational and rotational interlocking assemblies. *Autom. Constr.* 135, 104142. DOI: 10.1016/j.autcon.2022.104142

Griffith, S. T., 2004. Growing machines (Thesis). Massachusetts Institute of Technology.

Luo, L., Baran, I., Rusinkiewicz, S., Matusik, W., 2012. Chopper: partitioning models into 3D-printable parts. *ACM Trans. Graph.* 31, 1–9. DOI: 10.1145/2366145.2366148

Mehdizadeh, S., Zimmermann, A., Tessmann, O., 2022. RotoColumn: A Continuous Digital Fabrication Framework for Casting Large-Scale Linear Concrete Hollow Elements, In *Towards Radical Regeneration: Design Modelling Symposium Berlin 2022.* Springer, pp. 476–486.

Mehdizadeh, S. and Tessmann, O., 2023. Animate Concrete (Materialization of Discrete Kinetic Element Assemblies) (No. 10115). EasyChair.

Oval, R., Nuh, M., Costa, E., Madyan, O. A., Orr, J., and Shepherd, P., 2023. A prototype low-carbon segmented concrete shell building floor system. *Structures* 49, 124–38. https://doi.org/10.1016/j.istruc.2023.01.063

Penrose, L. S., and Penrose, R., 1957. A Self-reproducing Analogue. *Nature* 179, 1183–1183. DOI: 10.1038/1791183a0

Popescu, G. A., n.d. Digital Materials for Digital Fabrication.

Rezaee Javan, A., Seifi, H., Xu, S., Ruan, D., Xie, Y. M., 2017. The impact behaviour of plate-like assemblies made of new interlocking bricks: An experimental study. *Materials & Design* 134, 361–73. DOI: 10.1016/j.matdes.2017.08.056

Rippmann, M., Liew, A., Van Mele, T., and Block, P., 2018. Design, fabrication and testing of discrete 3D sand-printed floor prototypes. *Materials Today Communications* 15, 254–59. DOI: 10.1016/j.mtcomm.2018.03.005

Shahab, A. R., Lloret Kristensen, E., Fischer, P., Gramazio, F., Kohler, M., Flatt, R. J., 2013. Smart dynamic casting or how to exploit the liquid to solid transition in cementitious materials. Presented at the 7th RILEM International Conference on Self-Compacting Concrete and 1st RILEM International Conference on Rheology and Processing of Construction Materials.

Song, P., Fu, Z., Liu, L., Fu, C.-W., 2015. Printing 3D objects with interlocking parts. *Computer Aided Geometric Design* 35–36, 137–48. DOI: 10.1016/j.cagd.2015.03.020

Swingle, T., Zampini, D., Clifford, B., 2020. Walking Assembly: A Method for Craneless Tilt-Up Construction. In *Impact: Design With All Senses.* Proceedings of the Design Modelling Symposium, Berlin 2019. Springer, pp. 237–49.

Tessmann, O., Mehdizadeh, S., 2020. Hollow-Crete: Prestressed Membranes as Formwork for Material Efficient Hollow Concrete Building Elements. *In Impact: Design With All Senses.* Proceedings of the Design Modelling Symposium, Berlin 2019. Springer, pp. 474–86.

Tessmann, O., Mehdizadeh, S., 2019. Rotoform: realization of hollow construction elements through roto-forming with hyper-elastic membrane formwork. In *Proceedings of the Symposium on Simulation for Architecture and Urban Design,* pp. 1–7.

Tibbits, S., 2012. Design to Self-Assembly. *Architectural Design* 82, 68–73.

Wang, Z., Song, P., Pauly, M., 2018. DESIA: a general framework for designing interlocking assemblies. *ACM Trans. Graph.* 37, 1–14. DOI: 10.1145/3272127.3275034

Wangler, T., Roussel, N., Bos, F. P., Salet, T. A. M., Flatt, R.J., 2019. Digital Concrete: A Review. Cement and Concrete Research 123, 105780. DOI: 10.1016/j.cemconres.2019.105780

Weizmann, M., Amir, O., Grobman, Y. J., 2017. Topological interlocking in architecture: A new design method and computational tool for designing building floors. *International Journal of Architectural Computing* 15, 107–118. DOI: 10.1177/1478077117714913

Wibranek, B., Wietschorke, L., Glaetzer, T., and Tessmann, O., 2020. Sequential Modular Assembly – Robotic Assembly of Cantilevering Structures through Differentiated Load Modules. DOI: 10.52842/conf.caadria.2020.2.373

Xin, S., Lai, C.-F., Fu, C.-W., Wong, T.-T., He, Y., Cohen-Or, D., 2011. Making burr puzzles from 3D models. *ACM Trans. Graph.* 30, 97:1-97:8. DOI: 10.1145/2010324.1964992

Aly Abdelmagid, Zlata Tošić, Anahita Mirani, Ahmed Hussein, Ahmed Elshafei

Design Model for Block-Based Structures from Triply Orthogonal Systems of Surfaces

Abstract: A design model for structures constituted by a system of planar blocks based on the geometry of Triply Orthogonal Systems of Surfaces (TOS) is presented. This model goes from the different TOS types generated together with preserving Möbius transformation to the blocks' assembly and sizing. Based on the design model, a prototype for a TOS version of the topological interlocking system is explored.

1 Introduction

Triply Orthogonal Systems of Surfaces (TOS) take the geometric advantages of principal patches, in particular, generating planar faces (up to architectural tolerance) cf. (Bobenko and Tsarev 2007), (Liu et al. 2006) and establishing them in a three dimensional setting cf. (Abdelmagid et al. 2022). A TOS can be parameterized by three mutually-orthogonal families of principal patches, giving rise to a 3D array of points generating a 3D array of blocks, as seen in the Figure 1. The system of TOS blocks in question is useful in the fabrication of structures made by assembling blocks in an array of one, two, or three dimensions. In this context, we present a design model for block-based structures arising from the geometry of TOS. More precisely, we use three ways of generating TOS by differential-geometric methods, as well as constructing their Möbius transformations which preserve their TOS structures. Furthermore, we show how the TOS can generate a version of the topological interlocking system (based on regular tessellations of space) for stereotomy of stone vaults cf. (Brocato and Mondardini 2010) to an interlocking system based on irregular tessellations of space with TOS blocks made out of planar faces while maintaining their construction principles and the planarity of the contact faces between the blocks.

2 Geometry

To begin let us briefly recall some results from differential geometry, for more extensive reading check (Eisenhart 1909) and (Gray et al. 2006).

https://doi.org/10.1515/9783111162683-013

2.1 TOS generating three-dimensional array of points

A triply orthogonal system of surface on an open set $\mathcal{U} \subset \mathbb{R}^3$ consists of three families $\mathcal{R}, \mathcal{G}, \mathcal{B}$ of surfaces that intersect each other orthogonally. More concretely, we think of a TOS as a curvilinear coordinates system arising from a smooth three-dimensional patch $X(u, v, w)$ satisfying the mutual orthogonality condition (with respect to the Euclidean scalar product $\langle , \rangle$ on $\mathbb{R}^3$) given by

$$\langle X_u, X_v \rangle = 0 \quad , \quad \langle X_v, X_w \rangle = 0 \quad , \quad \langle X_w, X_u \rangle = 0. \tag{1}$$

We recall that by the theorem of Dupin, the curves of intersections of the surfaces in a TOS patch X are principal curves on the surfaces in question. In other words, for fixed parameters u_o, v_o, w_o the two-dimensional patches R, G, B defined by

$$R(v, w) = X(u_o, v, w) \quad , \quad G(w, u) = X(u, v_o, w) \quad , \quad B(u, v) = X(u, v, w_o)$$

are all the so-called principal patches. These are known to be the continuous smooth analogues of circular planar quadrilateral meshes cf. (Liu et al., 2006). This means that by evaluating the continuous smooth TOS patch $X(u, v, w)$ at discrete values of u, v, w we obtain a three-dimensional array of discrete points whose (quadrilateral) faces are all planar (up to architectural tolerance) cf. (Bobenko and Tsarev 2007), as seen in Fig. 1.

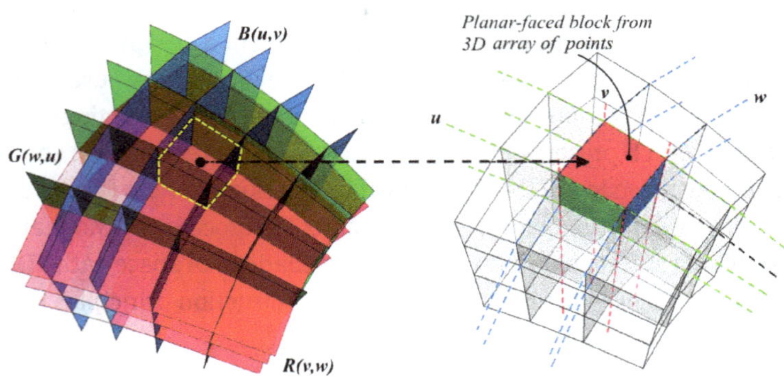

Fig. 1: R, G, B principal patches inducing a 3D array of points forming blocks with planar faces.

2.2 Generating TOS

A known result of Darboux states that the determination of a general TOS amounts to solving partial differential equations of the third order called the Lamé system (Eisenhart 1909). A solution consists of the (unitary) tangent vector fields to the TOS which are further integrated to obtain the patch X. However, we use here three differential-geometric constructions of TOS bypassing the Lamé system and giving us explicit TOS patch parameterizations as follows.

TOS from Quadrics: This TOS type is explained by the authors (Abdelmagid et al. 2022).

TOS from DOS: The second TOS type's construction is based on revolving or extruding a Doubly Orthogonal System of Curves (DOS) which is given as a principal patch $B(u, v)$ in the plane or space. In the revolving case, B is a portion of the plane $\mathbb{R}^2$ which is then rotated about an axis l with angle w giving the third coordinate. In the extrusion case, B defines a surface in $\mathbb{R}^3$, which is then extruded along its normals N with offset distance w giving the third coordinate. Denoting by ρ the rotation matrix in question, then the TOS patches are given by

$$\text{Revolving:}\quad X(u, v, w) = \rho(l, w) \cdot B(u, v) \tag{2}$$

$$\text{Extruding:}\quad X(u, v, w) = B(u, v) + wN(u, v). \tag{3}$$

The TOS from revolving DOS has its $\mathcal{B}$ family planar and its $\mathcal{R}$, $\mathcal{G}$ families surfaces of revolution, as seen in Fig. 2a, while the TOS from extruding DOS, has its $\mathcal{B}$ family all offsets of B and its $\mathcal{R}$, $\mathcal{G}$ families all developable surfaces (and in some cases planar), as seen in Fig. 2b.

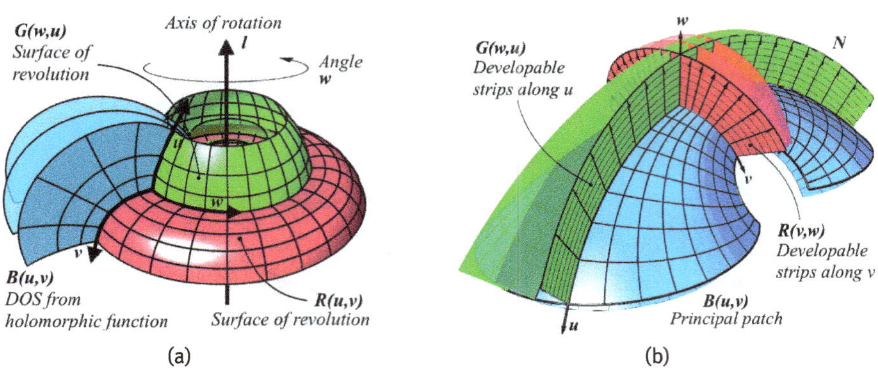

Fig. 2: TOS arising from DOS: (a) TOS from revolving DOS, (b) TOS from extruding DOS.

We will focus in particular on planar DOS patches $B(u, v)$ arising from (regular) holomorphic functions $f(z)$ defined on (an open subset of) the complex plane. This is following from the fact that such a function f is conformal (i. e. preserves angles). This is seen in the planar DOS patch $B(u, v)$ in Fig. 2a which is defined by the real and imaginary parts of a holomorphic function $f(z)$ that is

$$B(u, v) = (\Re f(z), \Im f(z)) \quad , \quad z = u + iv \in \mathcal{U} \subset \mathbb{C} \simeq \mathbb{R}^2. \tag{4}$$

TOS from Bianchi Transforms: In this TOS type, the $\mathcal{B}$ family are surfaces of negative constant Gaussian curvature (CGC) $-1/\rho^2$ with $\rho > 0$, also known as the TOS of Ribaucour cf. (Eisenhart 1909). This follows from a theorem of Ribaucour stating that if $F(u, v)$ is a surface of negative CGC $-1/\rho^2$ then the circles in the tangent planes at points $p \in F$ of radius ρ are the orthogonal trajectories of a family $\mathcal{B}$ of surfaces of negative CGC $-1/\rho^2$, as seen in Fig. 3a. It turns out that family $\mathcal{B}$ can be realized from the Bianchi transformation of the negative CGC surface F. To see this, we assume that $F(u, v)$ is a Tchebyshef principal patch of angle function $\theta(u, v)$ cf. (Gray et al. 2006). Next, finding a solution $\theta^*(u, v, w)$ to the Bianchi-Darboux equations

$$\begin{cases} \theta_u^* + \theta_v = \sin\theta^* \cos\theta \\ \theta_v^* + \theta_u = -\cos\theta^* \sin\theta \end{cases} \tag{5}$$

where the constant of integration denoted w will provide the third coordinate in the TOS, parameterizing the circles orthogonal to the $\mathcal{B}$ family, as seen in Fig. 3a. Finally, the TOS patch X in question is defined by the Bianchi transformation

$$X(u, v, w) = F(u, v) + \left(\frac{\cos(\theta^*(u, v, w))}{\cos(\theta(u, v))}\right) F_u + \left(\frac{\sin(\theta^*(u, v, w))}{\sin(\theta(u, v))}\right) F_v \tag{6}$$

where for every fixed w_o we have a negative CGC surface $B(u, v) = X(u, v, w_o)$ which is a Bianchi transform of $F(u, v)$, while F itself is not part of the TOS.

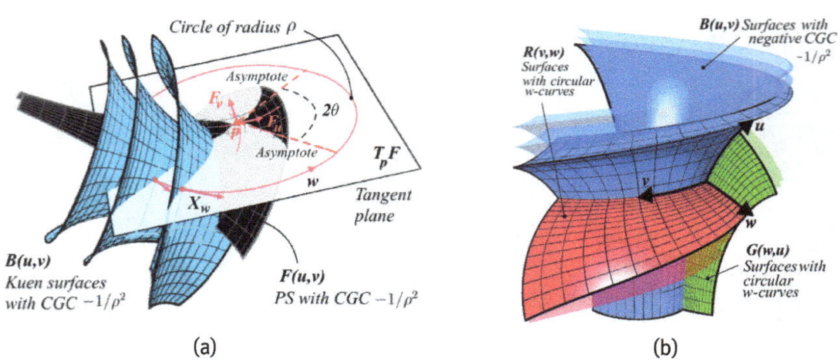

Fig. 3: TOS of Ribaucour from Bianchi transformation: (a) $\mathcal{B}$ family orthogonal to circles, (b) $\mathcal{B}$ surfaces of negative CGC.

2.3 Möbius transformations of the TOS

We will now show how to transform the TOS patches to obtain other TOS patches, using the so-called Möbius transformation. The fact that Möbius transformations preserve TOS patches follows from their preservation of principal patches, a result due to Darboux cf. (Eisenhart 1909). Intuitively, Möbius transformations are constructed from compositions of reflections and inversions. However, for convenience of application, albeit less intuitive we construct them as Lorentz transformations preserving a light cone in Minkowski space $\mathbb{R}^{4,1}$ and composed with the stereographic projection cf. (Cecil 2018), as seen in Fig. 4.

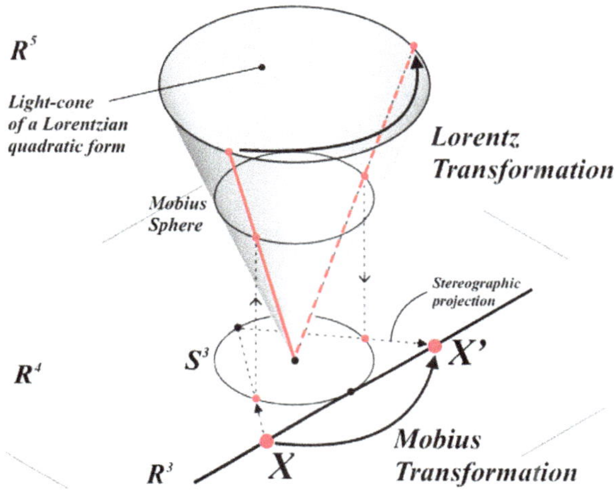

Fig. 4: Möbius transformation as Lorentz transformation.

Thanks to this construction, a Möbius transformation is given by a 5 × 5 matrix

$$\mu = A^k(\mathfrak{a}) \cdot B^l(\mathfrak{b}) \cdot C^m(\mathfrak{c}) \cdot D^n(\mathfrak{d}) \tag{7}$$

$$A(\mathfrak{a}) = \begin{pmatrix} \frac{1}{\sqrt{1-\mathfrak{a}^2}} & 0 & 0 & 0 & -\frac{\mathfrak{a}}{\sqrt{1-\mathfrak{a}^2}} \\ 0 & 1 & 0 & 0 & 0 \\ 0 & 0 & 1 & 0 & 0 \\ 0 & 0 & 0 & 1 & 0 \\ -\frac{\mathfrak{a}}{\sqrt{1-\mathfrak{a}^2}} & 0 & 0 & 0 & \frac{1}{\sqrt{1-\mathfrak{a}^2}} \end{pmatrix} \quad B(\mathfrak{b}) = \begin{pmatrix} 1 & 0 & 0 & 0 & 0 \\ 0 & \frac{1}{\sqrt{1-\mathfrak{b}^2}} & 0 & 0 & -\frac{\mathfrak{b}}{\sqrt{1-\mathfrak{b}^2}} \\ 0 & 0 & 1 & 0 & 0 \\ 0 & 0 & 0 & 1 & 0 \\ 0 & -\frac{\mathfrak{b}}{\sqrt{1-\mathfrak{b}^2}} & 0 & 0 & \frac{1}{\sqrt{1-\mathfrak{b}^2}} \end{pmatrix}$$

$$C(\mathfrak{c}) = \begin{pmatrix} 1 & 0 & 0 & 0 & 0 \\ 0 & 1 & 0 & 0 & 0 \\ 0 & 0 & \frac{1}{\sqrt{1-\mathfrak{c}^2}} & 0 & -\frac{\mathfrak{c}}{\sqrt{1-\mathfrak{c}^2}} \\ 0 & 0 & 0 & 1 & 0 \\ 0 & 0 & -\frac{\mathfrak{c}}{\sqrt{1-\mathfrak{c}^2}} & 0 & \frac{1}{\sqrt{1-\mathfrak{c}^2}} \end{pmatrix} \quad D(\mathfrak{d}) = \begin{pmatrix} 1 & 0 & 0 & 0 & 0 \\ 0 & 1 & 0 & 0 & 0 \\ 0 & 0 & 1 & 0 & 0 \\ 0 & 0 & 0 & \frac{1}{\sqrt{1-\mathfrak{d}^2}} & -\frac{\mathfrak{d}}{\sqrt{1-\mathfrak{d}^2}} \\ 0 & 0 & 0 & -\frac{\mathfrak{d}}{\sqrt{1-\mathfrak{d}^2}} & \frac{1}{\sqrt{1-\mathfrak{d}^2}} \end{pmatrix}$$

with k, l, m, n integers and parameters $\mathfrak{a}, \mathfrak{b}, \mathfrak{c}, \mathfrak{d}$ of norm less than one, applied to a TOS patch X seen as an element of the light cone, as seen in Fig. 4.

3 Design model

Now that the geometric constructions are defined, the following part establishes a design model based on these constructions. In essence, the model takes the form of a series of choices induced by the geometric constructions, that we will call Degrees of Design Freedom (DF). Through the five DF, the designer goes from defining the global shape to its detailing on the block level.

3.1 DF-1 TOS-type

The TOS-types are interpreted as morphological choices. And they are classified as follows:

- **A01 and A02:** TOS Quadrics (Elliptic) and TOS Quadrics (Parabolic)
- **B01 and B02-(a,b):** TOS from DOS (Revolving) and TOS from DOS (Extruding)
- **C01:** TOS from Kuen (Bianchi transform).

3.2 DF-2 Variants

Each TOS-type contains within it a spectrum of variants that are controlled firstly by the coordinates u, v, w and secondly by parameters a, b, c, .. involved in the expression of the TOS patch X of each TOS-type. The coordinates u, v, w determine the surface choice in the families $\mathcal{R}$, $\mathcal{G}$, $\mathcal{B}$, while the parameters a, b, c, .. determine the shape of the TOS-type. Below, is a selection of examples.

B01: TOS-type is Revolving DOS obtained from considering the holomorphic function $f(z)$ (on an appropriate part of the complex plane) of the form

$$f(z) = \frac{(a_1 + ia_2)z + (b_1 + ib_2)}{(c_1 + ic_2)z + (d_1 + id_2)}$$

and inducing the DOS $B(u, v) = (r(u, v), s(u, v))$ given by

$$r = \frac{a_2d_1u - a_1d_2u + a_2c_1u^2 - a_1c2u^2 + a_1d_1v + a_2d_2v + a_2c_1v^2 - a_1c_2v^2 - b_1(d_2+c_2u+c_1v) + b_2(d_1+c_1u-c_2v)}{\left(d_1^2 + d_2^2 + 2d_2(c_2u + c_1v) + 2d_1(c_1u - c_2v) + (c_1^2 + c_2^2)(u^2 + v^2)\right)}$$

$$s = \frac{a_1d_1u + a_2d_2u + a_1c_1u^2 + a_2c_2u^2 - a_2d_1v + a_1d_2v + a_1c_1v^2 + a_2c_2v^2 + b_2(d_2+c_2u+c_1v) + b_1(d_1+c_1u-c_2v)}{\left(d_1^2 + d_2^2 + 2d_2(c_2u + c_1v) + 2d_1(c_1u - c_2v) + (c_1^2 + c_2^2)(u^2 + v^2)\right)}$$

The DOS is then revolved around an axis l with rotation parameter w and the variants are obtained by varying $a_1, a_2, b_1, b_2, c_1, c_2, d_1, d_2$ as seen in Fig. 5.

B02-b: TOS-type is Extruding DOS obtained from considering a principal patch which is a Dupin cyclide (elliptic type) given by the parameterization

$$\left(\frac{b\sin(2\pi u)(a - d\cos(2\pi v))}{a - c\cos(2\pi u)\cos(2\pi v)}, \frac{b\sin(2\pi v)(c\cos(2\pi u) - d)}{a - c\cos(2\pi u)\cos(2\pi v)}, -\frac{d(c - a\cos(2\pi u)\cos(2\pi v)) + b^2\cos(2\pi u)}{a - c\cos(2\pi u)\cos(2\pi v)}\right)$$

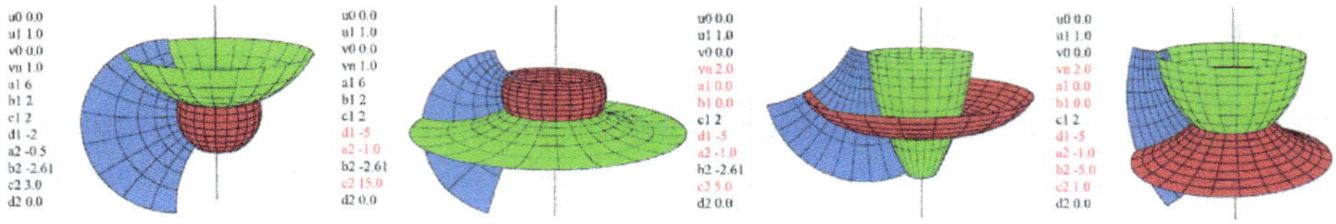

Fig. 5: Variants of TOS-type B01 (Revolving DOS).

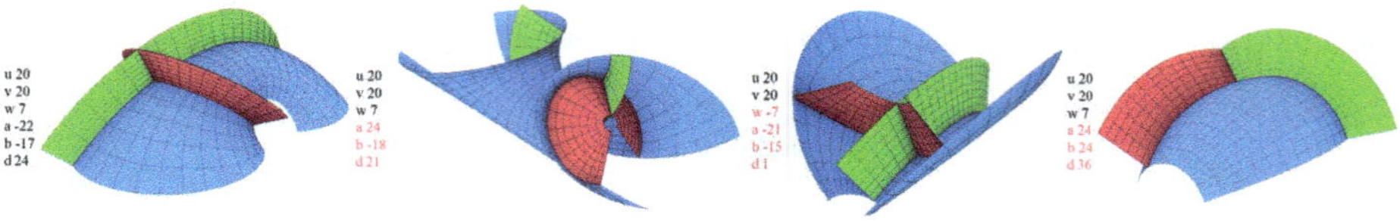

Fig. 6: Variants of B02: TOS from DOS (Extruding).

with $0 < b \leq a$ and $c^2 = a^2 - b^2$. The TOS is then generated by extruding this DOS along the normals with extrusion parameter w and the variants are obtained by varying the parameters a, b, c, d as seen in Fig. 6.

C01: TOS-type is Bianchi transform obtained from considering a Pseudo-Sphere (PS) that is parameterized by a Tchebyshef principal patch $F(u, v)$ with its angle function $\theta(u, v)$ given by

$$F(u, v) = \left(\frac{\cos v}{\cosh u}, \frac{\sin v}{\cosh u}, u - \tanh u\right) \quad , \quad \theta(u, v) = 2 \arctan\left(e^u\right).$$

Next, solving the Bianchi-Darboux Equations (5) we obtain a one-parameter family of solutions $\theta^*(u, v, w)$ with parameter w (constant of integration) given by

$$\theta^*(u, v, w) = 2 \arctan\left(\frac{-v - w}{\cosh u}\right).$$

Finally, putting everything in Bianchi transform Expression (6) we obtain a family of Kuen surfaces $B(u, v)$ forming a TOS patch $X(u, v, w)$ and where the w-curves shared by the $\mathcal{R}$, $\mathcal{G}$ surfaces are circular arcs as seen in Fig. 7

$$X = \rho\left(\frac{2(\cos v + (v + w)\sin v)\cosh u}{(v + w)^2 + \cosh^2 u}, \frac{2(\sin v - (v + w)\cos v)\cosh u}{(v + w)^2 + \cosh^2 u}, u - \frac{\sinh(2u)}{(v + w)^2 + \cosh^2 u}\right).$$

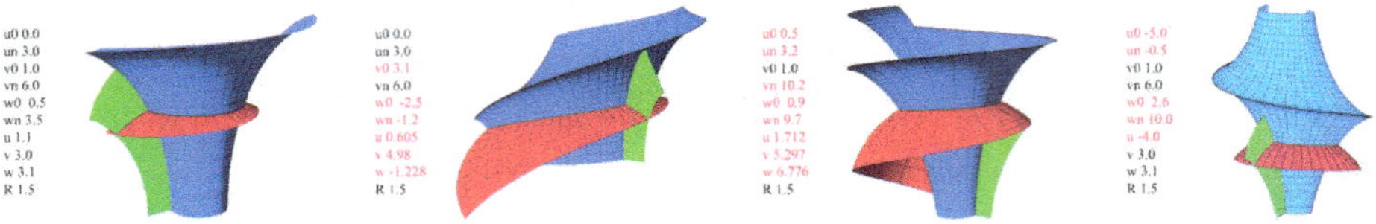

Fig. 7: TOS-type C01 (Bianchi transform).

3.3 DF-3 Möbius transformations

Möbius transformations expand even further the variety of morphological solutions while maintaining the structure of their associated TOS. A Möbius transformation μ given by the Expression (7) allows two-level manipulations. First, by different combinations of the generators A, B, C, D to obtain words of the form $AABCACD\ldots$ and second, by varying the parameters $\mathfrak{a}$, $\mathfrak{b}$, $\mathfrak{c}$, $\mathfrak{d}$. Selected examples of the application of Möbius transformations on some TOS-types are seen in Fig. 8.

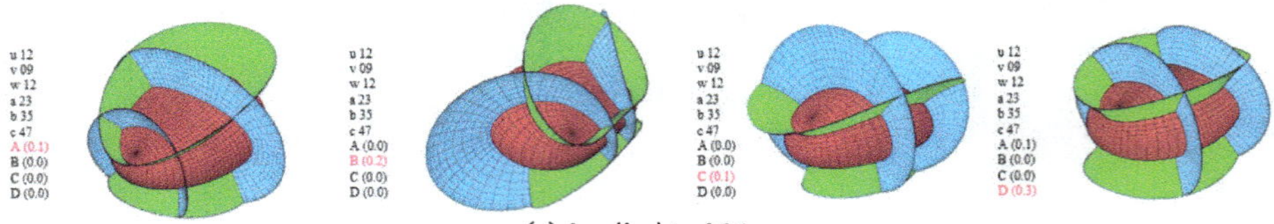

(a) Applied to A01

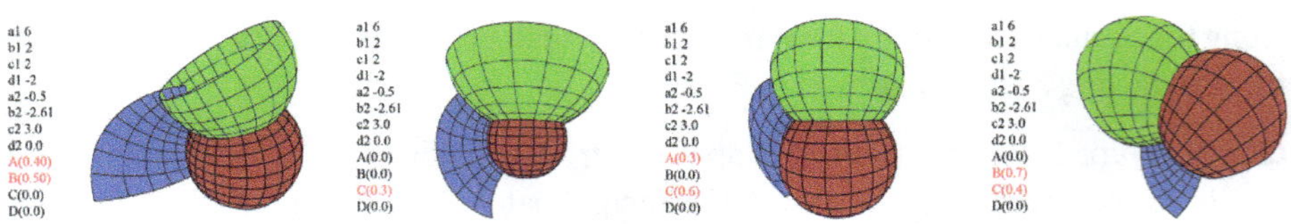

(b) Applied to B01

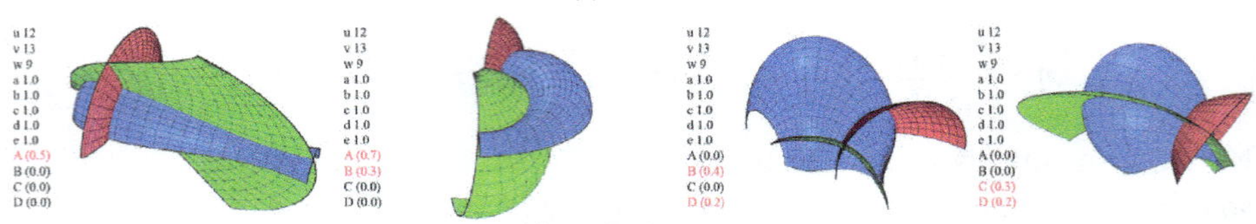

(c) Applied to B02-a

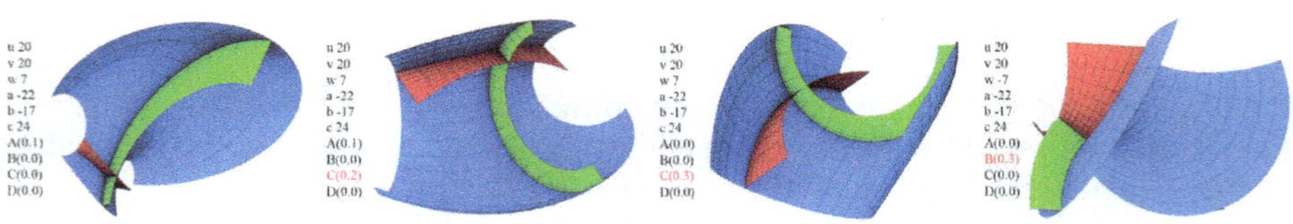

(c) Applied to B02-b

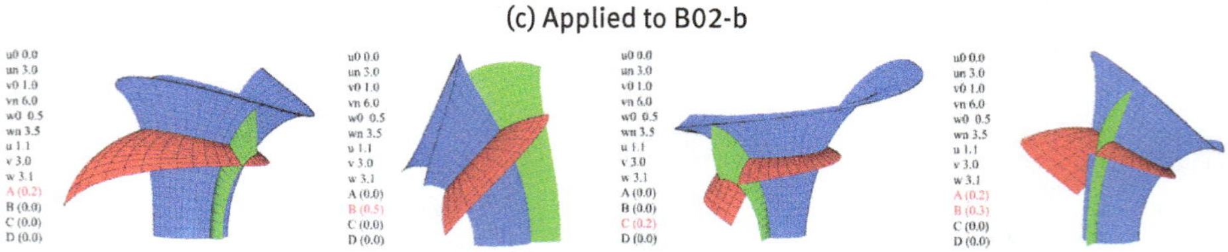

(d) Applied to C01

Fig. 8: Selected examples of Möbius transformation to some TOS-types.

3.4 DF-4 Blocks assembly

As explained above, any of the TOS-types creates a 3D array of points by evaluating the TOS patch X at discrete values of u, v, w, thus creating blocks with planar faces (see Fig. 1). It then follows naturally that DF-4 is the manner of assembling these blocks, namely, in arrays of one, two or three dimensions (Fig. 9). In particular, we observe that the 3D array offers a (non-regular) tessellation of (part of) $\mathbb{R}^3$, as seen in Fig. 9c.

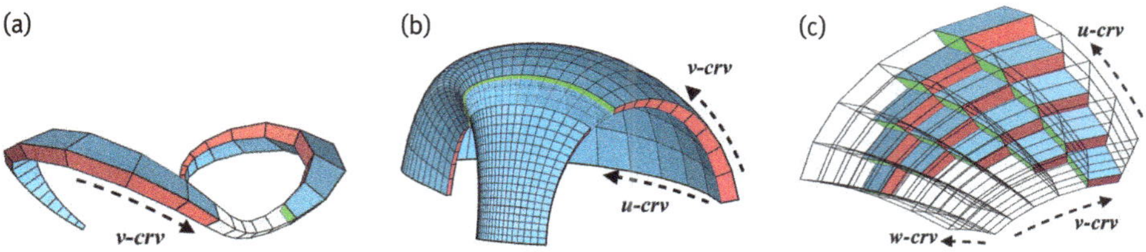

Fig. 9: Three fundamental types of blocks assembling: (a) 1D array, (b) 2D array and (c) 3D array.

3.5 DF-5 Blocks Sizing

This DF provides a way to control the sizes of the blocks arising from the TOS patch X using a reparameterization of the form $Y(x, y, z) = X(u(x), v(y), z(w))$ where $u(x)$, $v(y)$, $w(z)$ are the inverses of the functions $x(u)$, $y(v)$, $z(w)$. In Fig. 10, we can see how these functions control the block sizes by altering both the lengths of the u, v, w-domains and the step sizes of their discrete values. The option to size the blocks while maintaining their geometric properties provides the designer with a rationalized way to optimize material use and fabrication time.

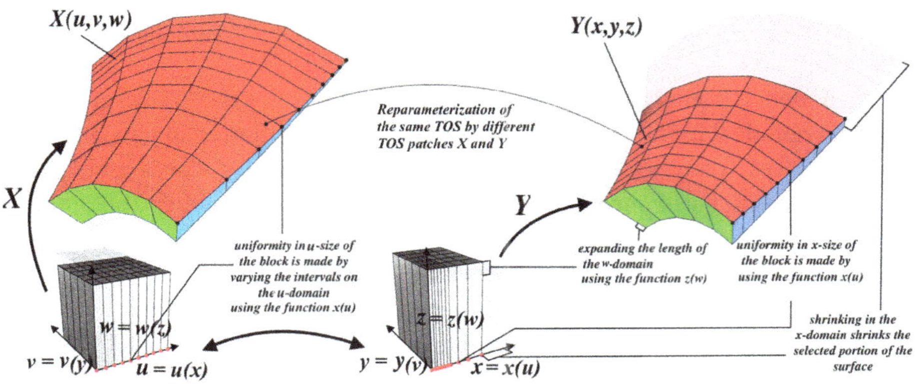

Fig. 10: Controlling blocks size through reparameterization.

4 Design application (TOS stone vault)

Next, we present a stone vault prototype based on an interlocking system of blocks arising from the TOS-type A01 3D array of points. The continuous nature of the TOS patch $X(u, v, w)$ allows us to generate an "intermediate" step size for the discrete blocks providing an overlapping version of the TOS 3D array which ensures planar contact faces between the blocks.

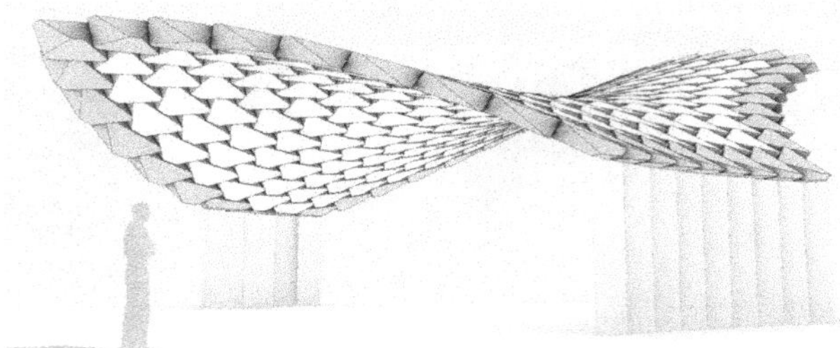

Fig. 11: Model of the TOS stone vault

Recall that TOS 3D points-array generate an irregular aperiodic tessellation of $\mathbb{R}^3$ where, the TOS blocks are neither platonic solids nor repeating. Nevertheless, they follow the structural principles of the classical interlocking system cf. (Brocato and Mondardini, 2010). More precisely, the mutual bearing between the blocks that are held together by a peripheral constraint as seen in Fig. 11.

TOS interlocking system

Observe that the overlapping of the planar contact faces between the blocks is achieved by using two-step sizes, namely, unit TOS blocks and big TOS blocks (clusters of 8 units). That means that each big TOS block is made out of 24 peripheral planar faces as seen in Fig. 12.

Even though having as many faces in a big TOS block is structurally sound, it is generally not ideal for fabrication, where it is naturally preferred to have the least amount of faces to cut. Notice that the 24 faces can be reduced to 6 faces per block by selecting one unit face from every cluster of four and intersecting them, obtaining a "modified" TOS block, as seen in Fig. 13.

To obtain hexagonal blocks, we consider the diagonal on each of the 6 new contact faces on a block, then define a plane that passes approximately through their midpoints (mid-plane). Finally, we offset this plane to obtain two parallel truncation planes defining the intrados and extrados of the block, as seen in Fig. 14. The offset distances of the truncation planes vary so as to reduce the overall weight and material waste while maintaining the overall stability of the structure.

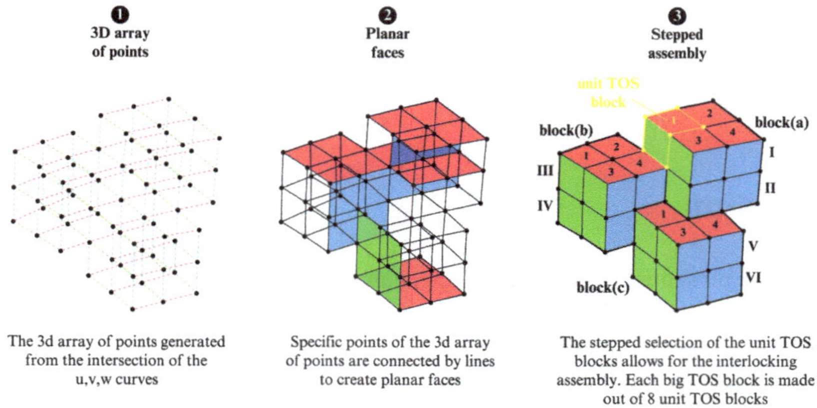

Fig. 12: Big TOS blocks (clusters of 8 units).

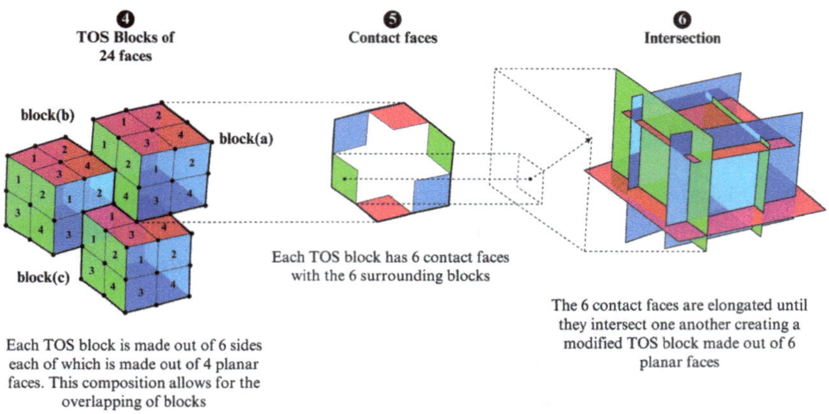

Fig. 13: Generating TOS blocks of 6 faces.

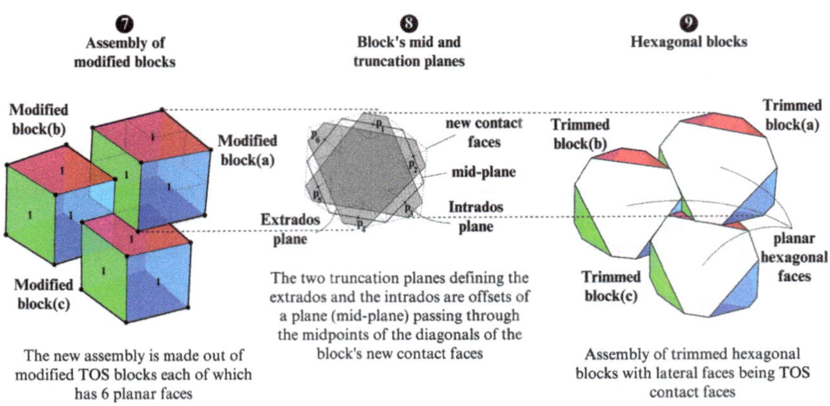

Fig. 14: Obtaining hexagonal blocks.

The version of the TOS interlocking system presented above was developed jointly with Maurizio Brocato and Paul Nougayrède of the GSA lab as a pre-rationalized approach to the design and construction of doubly curved stone structures.

5 Conclusion

In the article, we presented a geometric setting, namely the TOS, with clear advantages for architectural design and fabrication. Next, we established a design model which articulated viable ways of creating a search space for morphological options and parameters to manipulate it at the block level while remaining true to the initial geometry. Finally, we showed how once a morphological option is chosen though the 5 DF, it is translated into a design prototype, namely a stone vault from an interlocking system arising from the TOS. Hence, setting the ground for the next phase of this research, where the parameters that went into the conception of this prototype are highlighted.

Acknowledgments

The research of Elshafei, A. was partially financed by Portuguese Funds through FCT (Fundação para a Ciência e a Tecnologia) within the Projects UIDB/00013/2020 and UIDP/00013/2020. The research of Tošić, Z. is a part of the priority program SPP 2187: Adaptive Modular Construction with Flow Production Methods – Precision High-Speed Construction of the Future in the subproject Formwork-free Flow Production of Adaptive Supporting Structures from Variable Frame Elements – Adaptive Concrete Diamond Construction (ACDC) funded by the German Research Foundation (DFG).

References

Abdelmagid, A., A. Elshafei, M. Mansouri, and A. Hussein. (2022). A design model for a (grid)shell based on a triply orthogonal system of surfaces. In *Towards Radical Regeneration: DMS Berlin.*

Bobenko, A. and S. Tsarev. (2007). Curvature line parameterization from circle patterns. https://arxiv.org/abs/0706.3221

Brocato, M. and L. Mondardini. (2010). Geometric methods and computational mechanics for the design of stone domes based on abeille's bond. In *Advances in Architectural Geometry 2010.*

Cecil, T. E. (2018). Lie sphere geometry and dupin hypersurfaces. *Mathematics Department Faculty Scholarship.*

Eisenhart, L. (1909). *A treatise on differential geometry of curves and surfaces*. Ginn and Company, Boston.

Gray, A., E. Abbena, and S. Salamon. (2006). Modern differential geometry of curves and surfaces with Mathematica. 3rd Edition. Chapman & Hall/CRC.

Liu, Y., H. Pottmann, J. Wallner, Y. Yang, and W. Wang. (2006). Geometric modeling with conical meshes and developable surfaces. *ACM Tr., Proc. SIGGRAPH,* 25(3): 681–89.

Tobias Schwinn, Lasath Siriwardena, Achim Menges

Integrative Agent-Based Architectural Design Modelling for Segmented Timber Shells

Abstract: Methods of planar re-meshing are an intensively studied topic in the fields of Architectural Geometry and Computer Graphics, which allow the design and eventual fabrication of double-curved shell structures from flat stock material. Planar hex-dominant (PH) meshes have vertex valence 3, a property that is important for the structural behavior of segmented timber shells. These shells are not only lightweight structures, but also offer additional benefits such as resource efficiency, low embodied carbon, and ease of prefabrication. Nevertheless, integrating architectural, structural, and fabrication constraints in the construction of PH-meshes remains a challenge, especially when transitioning from synclastic to anticlastic regions on a single design surface. Furthermore, state-of-the-art segmentation methods are largely inaccessible during early-stage architectural design. This paper presents an agent-based approach that addresses both the complex synclastic-anticlastic case and early-stage design usability. The model offers a novel conceptualization of timber shell segments as agents consisting of sub-agents representing segment vertices. We show how sub-agent-level decision-making and segment-level evaluation enables greater design flexibility and deeper integration of interdisciplinary constraints resulting in fairer segmentations than previous approaches. Real-world constraints from the realization of a multi-shell, large-scale segmented timber roof have informed method development thereby ensuring its applicability in design practice.

1 Introduction

Design, engineering, fabrication, and construction of segmented shells is a multi-facetted challenge addressed from multiple disciplinary perspectives. Recent research covers such diverse topics as design for assembly [1], segmentation methodology [2, 3], structural analysis [4], prefabrication and building system development [5] (Fig. 1), or extending the existing building stock through lightweight timber shells [6]. However, to successfully realize segmented shells, integrating disciplinary constraints in the design process is crucial. This paper consequently proposes a novel integrative approach for segmented shell design.

Shells are curved enclosing structures that are thin in relation to their span relying on 'membrane action' for structural stability [7]. The architectural relevance of shells lies in their ability to cover large areas column-free while using a minimal amount of material, and in their lightweight aesthetics where form and forces are inextricably

https://doi.org/10.1515/9783111162683-014

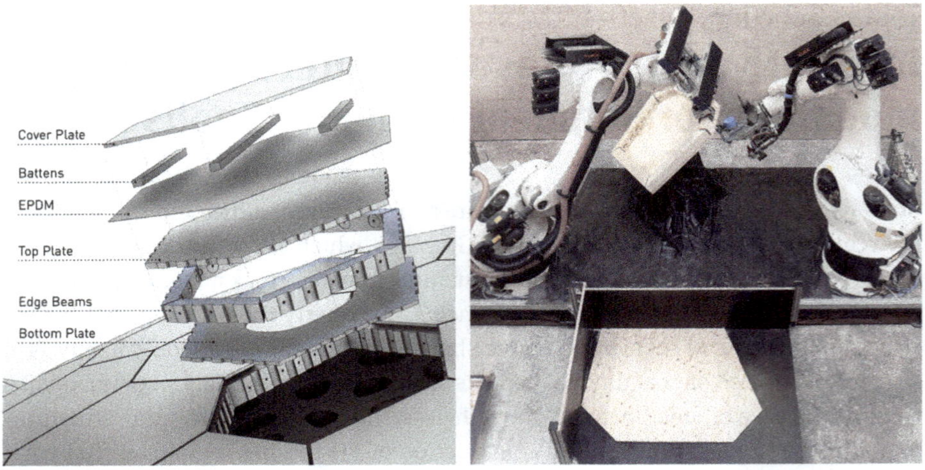

Fig. 1: Segmented timber shell building system of BUGA Wood Pavilion. Left: Exploded view of hollow cassette building system. Right: Robotic prefabrication of building system. Images: ©2019 University of Stuttgart.

linked. Segmented timber shells, besides being lightweight, can serve as carbon sinks, if designed for longevity and circularity.

Planar approximation methods for double-curved surfaces play an essential role in realizing segmented shells economically and ecologically. These methods allow construction from standard planar stock material, avoiding the need for custom molds or time-intensive milling. Furthermore, planar hex-dominant (PH) meshes, with most faces having 6-sides, exhibit architecturally relevant properties, such as vertex valence 3, a structurally stable configuration, where forces are translated between segments as normal and shear forces [8]; and, in contrast to triangular meshes, PH-meshes can be offset with constant thickness, where edges and faces remain parallel [9].

Constructing PH-meshes in architectural contexts requires considering geometric, structural, fabrication, assembly, aesthetic, and budgetary concerns. Constraints include the size of the available stock material in fabrication; size and weight limitations, when handling segments in assembly; and the joint pattern's structural and design implications. What makes the design of segmented shells a complex problem is that most of these evaluation criteria are intertwined, and changing one aspect will inevitably affect others. Despite substantial research into segmentation methods, constructing PH-meshes that integrate multi-disciplinary constraints remains a challenge—especially when transitioning from synclastic to anticlastic regions on a single design surface.

The contribution of this paper is an architectural design-oriented approach for free-form segmented timber shells that integrates constraints from all stakeholders involved in design, engineering, fabrication, and construction. The approach combines PH-meshing and agent-based modelling and offers a novel conceptualization of timber shells as a system of agent systems. A case study validates its applicability in design practice in designing of a multi-shell, large-scale segmented timber roof under real-world constraints. The research offers valuable insights for the scientific community and industry by 1) proposing a systematic way for co-designing the segmentation under different disciplinary constraints; 2) providing direct control of segment shape and sizes in contrast to previous agent-based approaches; and 3) extending the accessibility of PH-meshing to the architectural design stages.

2 Related Work

PH meshing. Planar approximation of an arbitrary surface is a special case of remeshing based on selected criteria, which is an established research area in Computer Graphics (CG). It has specific applications in Architectural Geometry, including planar quad (PQ) meshing and planar hex-dominant (PH) meshing [9]. PH-meshing can be approached through Clustering, Tangent-Plane-Intersection (TPI), or "Dupin" methods.

Clustering methods, such as the ones proposed by Cutler and Whiting [10], generalize well, but result in unordered segmentations with heterogeneous segment sizes, edge lengths, and angles.

The simplest TPI method—also known as the 'apple-slicing' method [11]—addresses only the synclastic case. TPI methods based on the plate-lattice dualism [12], such as the one proposed by Troche [13], apply to both synclastic and anticlastic curvature cases and construct PH-meshes from the intersection points of tangent planes at the mesh vertices. While geometrically planar, the only available design variables are the distribution of vertices on the surface and the topology of the mesh, which limits the number of design criteria that can be applied to the segmentation. Zimmer et al. [14] consequently relaxed the tangency constraint, allowing more design freedom, and subjected the segmentation to optimization to achieve selected design goals. The main limitations of the TPI approach are the indirect control of plate properties, such as dimensions and edge lengths, as well as the challenge of transitioning from areas of positive to negative Gaussian curvature, the synclastic-anticlastic case, where intersection planes become almost co-planar.

Dupin methods share with the TPI methods the idea of a duality between triangle mesh and PH-mesh. Instead of generating planar plates directly, an intermediate non-planar dual mesh approximates the desired PH-mesh, which is optimized to minimize non-planarity. In this approach, "planar" consequently means "planar within tolerance", not truly planar as in the case of TPI. In the synclastic case, a non-planar barycentric dual of a triangular mesh suffices [15]. For anticlastic cases, the Dupin-indicatrix shapes the dual mesh based on surface curvature to produce an 'informed' dual mesh [16]. Further development uses the conjugate curve network of principal curvature lines to generate a shifted quad layout as the basis for optimization [17].

The most computationally advanced methods for creating PH-meshes fall into the Dupin category, resulting in ordered, principal curvature-aligned segmentations that 'naturally' follow the principal curvatures in the shell surface. However, the non-linear optimization involved goes beyond industry-standard Computer-Aided Geometric Design (CAGD) software. The first research objective therefore is to develop accessible methods for the design of segmented shells based on the duality approach and to validate them in a case study.

Agent-based Design of Segmented Timber Shells. Agent-based modelling and simulation (ABMS) is a numerical method for modelling and simulating complex systems that is used in various fields including architectural design, engineering, and construction (AEC) [18]. In ABMS, a segmented shell is considered a system of interacting agents that implement segment generation while pursuing their own objectives. Since 2014, two of the above methods have been implemented through ABMS and used to design and fabricate two segmented timber shells.

In the Landesgartenschau Exhibition Hall (Fig. 2, left), which spans 11 m, agents represent solid Beech plywood plates generated through TPI on a design surface with both synclastic and anticlastic regions [19]. In the BUGA Wood Pavilion for the Bundesgartenschau 2019 (Fig. 2, right), agents represent hollow cassettes generated through the 'apple-slicing' method on a synclastic design surface [20]. These cassettes consist of top and bottom plates connected by edge beams (Fig. 1), allowing for a span of close to 30 m with a similar weight-area ratio as the Landesgartenschau shell. The related ABMS innovation is the conceptualization of plate edges as sub-agents that steer their plate's tangent plane.

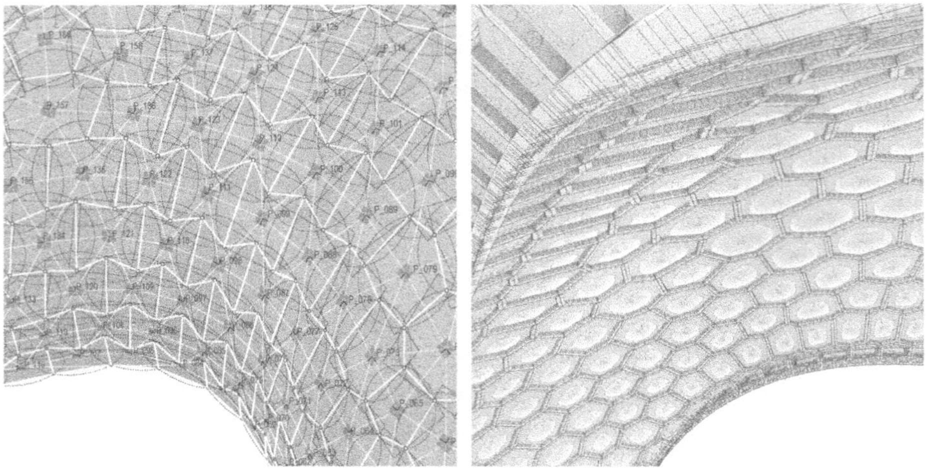

Fig. 2: Agent-based design of segmented shells. Left: Single-layer solid timber plates of Landesgartenschau Exhibition Hall 2014. Right: Double-layer hollow cassettes of BUGA Wood Pavilion 2019. Images: ©2014, 2019 University of Stuttgart.

These projects showed that ABMS effectively addresses the problem of indirectly controlling plate outlines by managing the complexity of iterative repositioning of tangent planes and subsequent remodeling of plate outlines for large segmentations. They also showed that ABMS enables the integration of later-stage AEC constraints into the early design process, significantly influencing segmentation. The BUGA project showcased that segments composed of sub-elements can lead to much higher performing structures despite increased fabrication complexity. However, current agent-based implementation rely on TPI methods and their associated limitations. The second research objective therefore is to develop a Dupin-based agent model to avoid the strict tangency constraint thereby expanding the capabilities of ABMS in segmented shell design.

3 Methods

To address the second research objective, we conceptualize a segmented timber shell as a system of agents representing a planar segments called plates. Each plate is defined by a group of sub-agents representing its vertices. In line with the Dupin approaches mentioned above, the segmentation of the design surface S is based on an "ideal" triangulation M, which is aligned with the principal curvature directions. The minimum principal curvature lines running through the umbilical points of S act as spines from which the two sets of principal curvature (PC) lines are constructed using the Runge-Kutta (RK4) method [21] (Fig. 3, left). The intersections of these PC lines form the vertices of the primal mesh M. The dual mesh D is generated for each vertex in M using barycenters of the adjacent faces (Fig. 3, center). D then initializes the agent system, which computes planar segments meeting given design constraints (Fig. 3, right).

3.1 Agent System

Agent systems typically consist of three main system constructs: agents, their behaviors, and the environment in which the agents interact [18]. Each plate agent represents a mesh face of the dual mesh D, with coincident vertices as vertex agents (Fig. 4, left). Edges in D indicate interaction topology, whereby every vertex agent is connected to at least two, typically three, others. Adjacent closed loops indicate segment membership, with every vertex agent being a member of up to three plate agents (Fig. 4, right). Vertex agents at the perimeter of a shell, i. e., the 'naked' vertices of D are considered anchors by default. Plate agents that accommodate a column head or technical equipment, such as smoke vents, at system initialization can be flagged accordingly. Figure 4 illustrates the specific attributes defining agent types and states.

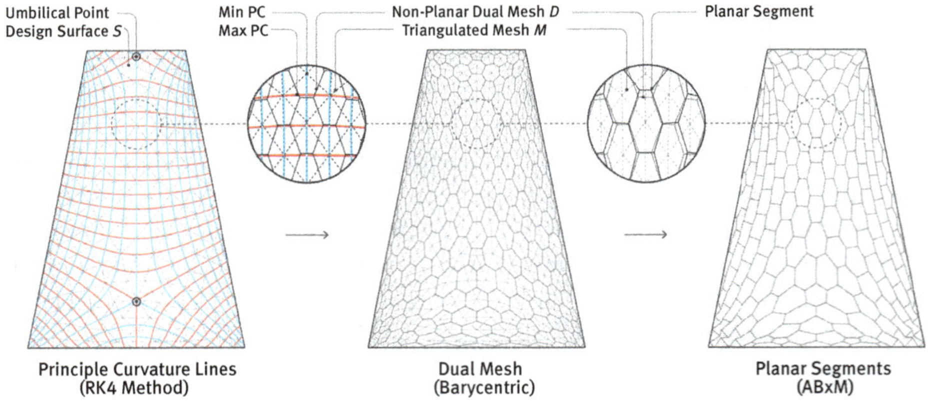

Fig. 3: Steps of segmentation process. Left: Generation of primal triangle mesh. Center: Generation of polygonal dual mesh. Right: Constraint-based planarization of segments.

Consequently, a vertex agent's environment is defined by the states of neighboring vertex agents, its parent agents, as well as base surface properties such as local curvature. Vertex agents iteratively update their states based on their internal states and the state of their environment, using goal-oriented behaviors. The resulting model is a hierarchical network-based agent system, with decision-making and action occurring at the of sub-agent level and evaluation at the segment-level.

3.2 Agent Behaviors

According to Levitis et al. [22], behaviors are the internally coordinated response (action or inaction) of whole entities to internal or external stimuli. Defining a behavior from a system modeler's point of view thus means providing the rules for the internal computation that leads, in our case, to a change of location of vertex agents such that the plate agent approaches its desired state. This translation **v** w.r.t. the current location of a vertex agent takes the fundamental form of

$$\mathbf{v} = \alpha \cdot \frac{1}{N} \cdot \sum_{i=1}^{N} (P_i - P_a), \text{ with } 0 < \alpha \leq 1 \tag{1}$$

where α is a user-defined scaling factor controlling the resultant step size, N is the number of neighbor agents, P_i the target point computed by the current behavior w.r.t the neighbor agent i, and P_a is the current position of the vertex agent. In the following, we describe the rules to achieve six design objectives related to geometry, architecture, fabrication, technical systems, and structure.

Planarity behavior (PL). This geometrical behavior aims to achieve planarity of the plate polyline by minimizing the cumulative distance of the plate vertices to the fitplane. The target point P_i is the closest point on the frame of adjacent plate agent i.

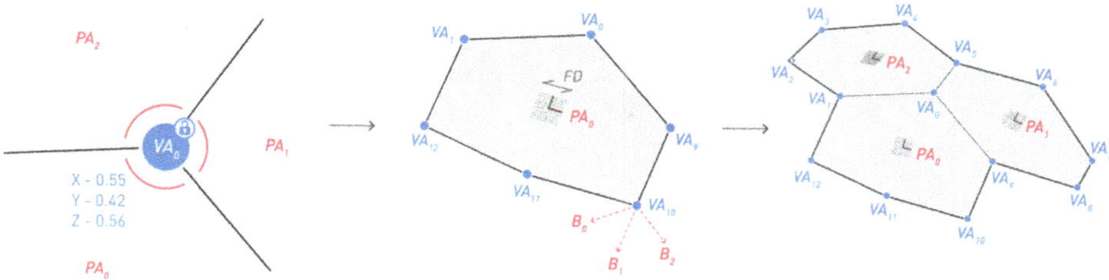

VertexAgent	
Position	Cartesian coordinates
IsAnchor	Anchor Flag
IsMemberOf	List of PolyAgent IDs
HasUpdated	Modification Flag

PolyAgent	
VertexAgents	List of VertexAgent IDs
Frame	Fit-Plane of Vertices
Polyline	Outline of Plate
Centroid	Vertex Average
Fibre Vector	Grain Direction
Area	Project Area on Frame
IsColumn	Column Flag
IsVent	Contains Vent Flag
Behaviors	Behaviors for vertex agents to update states to reach a specified goal

PolyAgent System	
PolyAgents	List of PolyAgent IDs
Total Displacement	Cumulative Translation of all Vertices
Threshold	System Convergence Value based on Total Displacement

Fig. 4: Agent system constructs. Left: Vertex Agent. Center: Plate Agent. Right: Agent system.

Every vertex agent will compute up to 3 different target points P_i, one for each plate that it belongs to, and will average its resultant translation vectors (Fig. 5, top left).

Surface Approximation behavior (SA). This design behavior aims to closely approximate the base surface by minimizing 1) the distance between the plate agent's centroid and the closest point on the surface P_c and 2) the angular distance between the agent's Frame orientation and the surface normal at P_c. First, a proxy frame $F_{px,i}$ of plate agent i is calculated:

$$F_{px,i}(P, N) = \alpha \cdot (T_{pc,i} - F) \tag{2}$$

where $T_{pc,i}$ is the tangent plane at P_c and F_i is the i-th plate agent's frame. Then, the target point P_i is the closest point on the proxy frame $F_{px,i}$. Like PL, every vertex agent will compute up to 3 different target points P_i (Fig. 5, top right).

Plate width behavior (PW). This fabricational behavior aims to limit the width of the bounding rectangle, which encloses the plate polyline, to a maximum user-defined value to ensure fabricability. This value will usually correspond to the dimensions of a chosen stock material, such as 3-ply or CLT. If the bounding rectangle is larger than the user-defined value, the target point Pi of the current vertex agent is the centroid of plate agent i, effectively shrinking the plate (Fig. 5, center left).

Plate area behavior (PA). This technical behavior aims to ensure a minimum surface area of the segment, e. g, to allow technical systems to be embedded in selected plates. This behavior operates in the opposite way of the PW behavior in that the target point P_i

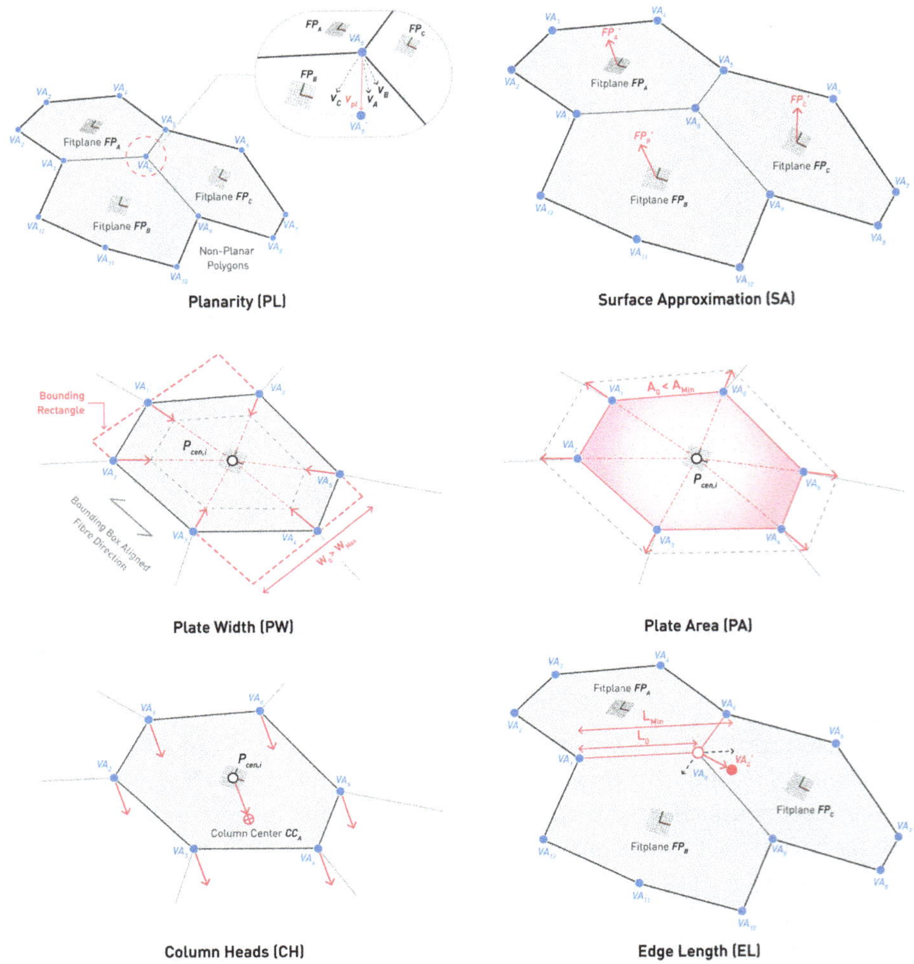

Fig. 5: Plate Agent Behaviors. Planarity, Surface Approximation, Plate Width, Plate Area, Plate Column Head, and Plate Edge Length.

of the translation lies in the opposite direction to the centroid of plate agent i effectively increasing the size and area of the plate up until a user-defined minimum area target value is met (Fig. 5, center right).

Column heads behavior (CH). This structural behavior aims to ensure that a segment flagged as having a column will be roughly centered on the column head to avoid eccentricity. The target point P_i is the column head associated with plate agent i (Fig. 5, bottom left).

Edge length behavior (EL). This structural behavior aims to enforce a given minimum edge length to make sure loads can be transferred across the edge via a minimum amount of crossing screw pairs. If the current vertex agent is closer than a user-defined minimum distance value to its neighbor vertex agent i, then the target point P_i of the translation lies in the opposite direction to vertex neighbor i (Fig. 5, bottom right).

Resulting translation. The behaviors described above will compute up to six individual translation vectors for each vertex agent. The resultant translation $\mathbf{v}_t$ of a vertex agent is the weighted average of all its behaviors:

$$\mathbf{v}_t = \frac{w_{\mathrm{pl}} \cdot \mathbf{v}_{\mathrm{pl}} + w_{\mathrm{sa}} \cdot \mathbf{v}_{\mathrm{sa}} + w_{\mathrm{pw}} \cdot \mathbf{v}_{\mathrm{pw}} + w_{\mathrm{pa}} \cdot \mathbf{v}_{\mathrm{pa}} + w_{\mathrm{ch}} \cdot \mathbf{v}_{\mathrm{ch}} + w_{\mathrm{el}} \cdot \mathbf{v}_{\mathrm{el}}}{w_{\mathrm{pl}} + w_{\mathrm{sa}} + w_{\mathrm{pw}} + w_{\mathrm{pa}} + w_{\mathrm{ch}} + w_{\mathrm{el}}} \tag{3}$$

where w_{pl}, w_{sa}, w_{pw}, w_{pa}, w_{ch}, and w_{el} are user-defined weight parameters. As demonstrated in previous work, those parameters can be subject to parameter optimization to minimize a certain score, such as the number of iterations to convergence or a measure of the segmentation quality [3].

3.3 Implementation

The model described in Sec. 3.1 and 3.2 is implemented in C# using an object-oriented approach as an extension to the agent-based modelling framework ABxM [23]. Custom agent and behavior classes are derived from the framework's base classes. The vertex agent class is derived from the Cartesian agent class, which allows for position-based modelling and faster convergence than previous force-based approaches. System updates are performed iteratively and synchronously to ensure consistency between agent updates within each iteration.

To address the first research objective, 1) the source code and UML diagram of this implementation is published open access [24]; and 2) ABxM is an addon to the popular CAGD software environment Rhinoceros 3d and is available through its package manager. Implementing the model in a widely used software environment increases accessibility and ensures that the approach can be part of a larger design-to-fabrication workflow as demonstrated in the following case study.

4 Large-Scale Construction Robotics Laboratory

The research described above was conducted as part of the design of the Large-Scale Construction Robotics Laboratory (LCRL) for the Cluster of Excellence “Integrative Computational Design and Construction for Architecture” (IntCDC) in Stuttgart, Germany. The roof design of the LCRL building thus serves as a case study and validation of the modelling approach.

4.1 Project Description

The LCRL is a three-story laboratory building serves as a demonstrator for IntCDC's research. It features a column-free 1415 m^2 testing hall and a 3200 m^2 timber roof covering the entire structure, including offices and workshops (Fig. 6). Unlike prior segmented timber shells, the shell is integrated into a full-scale building, hovering 10 m above ground, and is only point supported. The roof consists of seven trapezoidal shells, spanning 28.8 m, that roughly form barrel vaults and transition between synclastic and anticlastic curvature regions at their corners (Fig. 7). These shells address additional constraints such as interfacing with timber columns, the facade, solar panels, fire vents, HVAC systems, and a fibre-reinforced skylight above the office section. The roof is expected to be completed in Q3 2025.

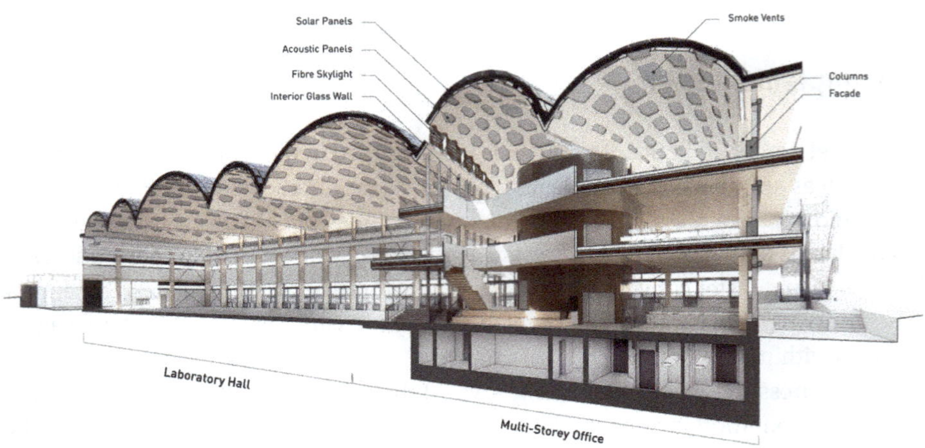

Fig. 6: Section view of the case study Large-scale Construction Robotics Laboratory (LCRL).

4.2 The Workflow

The segmented shells of the LCRL are designed using a workflow that includes 1) modelling of a design surface, 2) segmentation, and 3) evaluation of the segmentation based on AEC criteria.

Surface Design. The goal is to create a shell surface S that adapts to boundary constraints such as support positions. Lofting five G2 curves yields a NURBS surface, which acts as an interim surface. This interim surface is then 'inflated' via vertex normal displacement to increase mean curvature, optimizing structural and material efficiency. A parameter study facilitates preliminary structural optimization, leading to a structurally informed free-form design surface S.

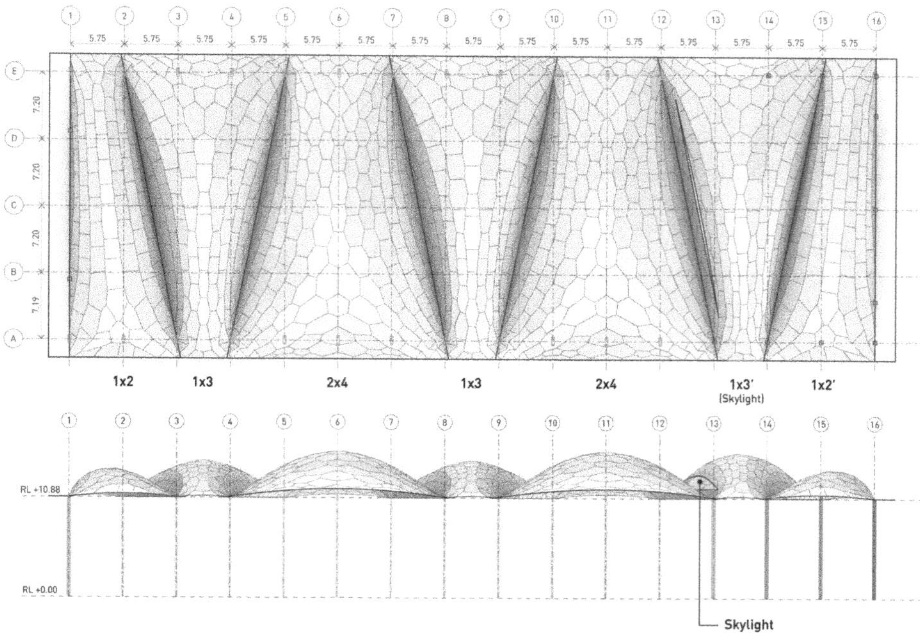

Fig. 7: Plan and Elevation of LCRL Roof.

Segmentation. In this step, the agent-based segmentation method described above is used (Fig. 3) including the creation of a PC-aligned mesh M from S and its transformation into a non-planar dual mesh D initializing the agent system.

The project-specific constraints applied in the case study include achieving segment planarity within a tolerance of 0.1 mm, approximating surface S, limiting the fiber-aligned plate width to 2.35 m (based on sheet sizes of 2.5×6.0 m), ensuring a minimum surface area of 6 m^2 for vent segments, centering column segments around the column head, and maintaining edge lengths larger than 0.3 m. M is produced using a custom mesher incorporating additional vertex points near the umbilical points that deviate from the PC lines, to avoid large changes in segment sizes between neighboring plates. Considering the anisotropic properties of timber, the fiber direction of the segments is aligned with the principal stress direction in the segments.

Evaluation. Each segment is continuously evaluated at runtime w.r.t. constraints that have been defined locally in the agent behaviors. After convergence, the resultant segmentation is also evaluated w.r.t. to global design criteria (Fig. 8). These include:

Structural evaluation to compute deflection, utilization, segment type (massive vs. hollow), material selection, and fibre direction. Massive cassettes are necessary in areas of high stress concentrations around the column heads. Structural criteria generally aim for reducing the number of segments, i. e., having bigger segments to

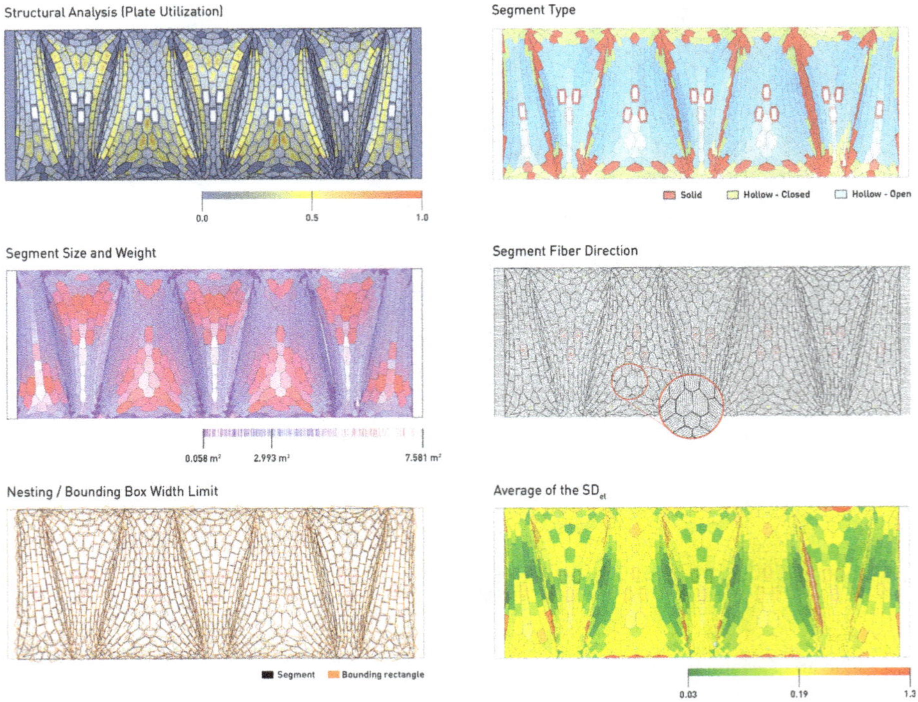

Fig. 8: Evaluation of LCRL Roof. Left-to-right, top-to-bottom: structural, solid vs hollow, size and weight, fibre direction, nesting/bounding box width limit, Distribution of fairness values of the SD_{el} centered on median value of 0.19 (yellow).

reduce global deflection. Limits are imposed by stock size availability and producibility (robot payload and reach).

Fabrication analysis evaluating robotic producibility w.r.t. fibre-aligned segment sizes (nesting), segment weight (payload), number of elements, fabrication steps, and expected fabrication time.

Cost analysis considering number and types of segments, build-up, and nesting efficiency using unit prices provided by the manufacturer as a reference. These factors heavily influence material usage and assembly time on-site and off-site and therefore cost estimates.

Interfacing with column heads and smoke vents is evaluated at run-time, while interfacing with the facade, is evaluated after convergence. During concept design, the evaluation is based on qualitative criteria such as connection complexity and presumed ease of production.

Architectural evaluation assessing the fairness of the segmentation. Fairness is defined by the angle between plates (avoiding sharp kinks for visual reasons and to minimize stress concentrations), the interior angle of plates, and the standard deviation of edge lengths SD_{el} to indicate regularity [3].

4.3 Results

The evaluation of different segmentations and comparing design options in terms of cost and structural performance are crucial for decision-making (Tab. 1). During concept and schematic design stages (HOAI stages 2 and 3), four of the many design iterations of the roof have been evaluated accordingly in collaboration with a manufacturer as part of a pre-construction agreement. Option 1, with a total of 1175 segments, is the first segmentation that has been fully analyzed and priced. It has a wide range of segment sizes, some exceeding the available stock sizes. It also exceeded the permissible structural utilization. Consequently, Option 2 was generated, specifically considering segment widths and adjusting the segment buildup. Options 2 and 3 are identical segmentations with 1128 segments but differ in terms of fibre orientation and consequently the distribution of massive and hollow cassettes. Option 4, with 1262 segments, aimed to reduce plate sizes further but increased deflections and localized stresses due to the increased number of segments and joints.

Tab. 1: Design options. Material options are Cross-Laminated-Timber (CLT), 3-ply (3P), Glulam beams (GL), and Laminated Veneer Lumber (LVL). Unit price of hollow cassettes includes costs for buildup and pressing, but no insulation, waterproofing, and cladding. (*) Max. deflection value in this option is not comparable due to different segment stiffness assumptions.

	Opt. 1	Opt. 2	Opt. 3	Opt. 4
No. segments	1175	1128	1128	1262
No. of massive	463	440	380	398
No. of hollow	712	688	748	864
Total Area	3240.1	3274.3	3216.4	3357.8
Massive cassettes				
Massive area	701.2	826.4	718.7	616.7
Buildup massive	CLT	CLT	CLT	CLT
Max. width (m)	2.9	2.4	2.4	2.35
Max. weight massive (kg)	855.6	694.0	694.0	600.0
Unit price change per m^2 (%)	100.0	88.9	88.9	88.9
Hollow cassettes				
Hollow area	2538.9	2549.9	2657.6	2741.1
Buildup hollow	3P/GL/3P	3P/GL/LVL	3P/GL/LVL	3P/GL/LVL
Max. width (m)	2.9	2.4	2.4	2.35
Max. weight hollow (kg)	348.0	326.3	326.6	298.00
Unit price change per m^2 (%)	100.0	106.5	105.7	107.4
Total len. edge beams (m)	5440.3	5802.6	5802.6	6277.3
Total len. screw edge (m)	3698.2	3684.4	3684.4	3934.00
Nesting efficiency (%)	58.4	56.6	59.2	60.2
Max. deflection (mm)	–*	−102	−94	−106
Structural utilization (top plate)	>1.0	<1.0	<1.0	<1.0
Average SD_{el}	0.299	0.319	0.319	0.306

Fig. 9: Interior view of the LCRL hall covered by the segmented timber roof.

Within the evaluation process, nesting efficiency emerges as a critical metric. Considering the size of the roof, the 2.6 % increase in nesting efficiency between Options 2 and 3 corresponds to a material saving of 262.2 m^2. To further increase material utilization, we aim to use offcut not suitable for roof plates as façade material. Based on these findings, Option 3 was chosen for further development in detail design (Fig. 9).

5 Discussion and Conclusion

In the case study, we successfully showed how 'real-world' constraints of the design of a large spanning roof can be integrated into the segmentation process and inform design decisions, thereby addressing our first research objective. Our main finding relates to our second objective: a plate system where plate vertices are conceptualized as sub-agents of plate agents allows for more detailed control of the final segmentation with respect to defined performance criteria than previous approaches—most notably TPI and black-box algorithms.

While the conceptualization of segments as agents makes it easier to formulate modelling goals, it still requires specialist training to be able conceptualize and model a system from the bottom up. Furthermore, the behaviors that we have implemented are only a subset of rules that can be imagined and implemented. This means that we cannot guarantee that we have identified the optimal behaviors to achieve the desired goals, which is a main limitation of metaheuristic approaches. Reinforcement learning has shown promising results for defining behavioral rulesets without preconceived rules of thumb [3].

A limitation of our approach is that the triangulation is not part of the agent-based approach, which means that changing the interaction topology by the agents themselves is currently not possible. In further work, we will therefore aim to integrate the curvature-aligned triangulation into the agent-based architectural modelling approach for segmented timber shells. Furthermore, additional evaluation criteria like carbon modeling need to be integrated into the post-convergence evaluation of the agent system.

Acknowledgements

This research has been supported by the Deutsche Forschungsgemeinschaft (DFG) under Germany's Excellence Strategy – EXC 2120/1 – 390831618.

We would also like to thank the architects and engineers of the IntCDC Planungs-GmbH, particularly Moritz Münzenmaier and Daniel Sonntag, for the design, analysis, and coordination of the LCRL building into which we can integrate our research.

References

[1] G. T.-C. Kao, A. Iannuzzo, B. Thomaszewski, S. Coros, T. Van Mele, and P. Block. 2022. Coupled rigid-block analysis: stability-aware design of complex discrete-element assemblies, *Computer-Aided Design,* 146, 103216, DOI: 10.1016/j.cad.2022.103216

[2] K. Pluta, M. Edelstein, A. Vaxman, and M. Ben-Chen. 2021. PH-CPF: Planar Hexagonal Meshing using Coordinate Power Fields, *ACM Trans. Graph.* 40(4), DOI: 10.1145/3450626.3459770

[3] T. Schwinn. 2021. A Systematic Approach for Developing Agent-based Architectural Design Models of Segmented Shells: Towards Autonomously Learned Goal-oriented Agent Behaviors, PhD, University of Stuttgart, DOI: 10.18419/opus-11633

[4] S. Bechert, D. Sonntag, L. Aldinger, and J. Knippers. 2021. Integrative structural design and engineering methods for segmented timber shells – BUGA Wood Pavilion, *Structures,* 34, 4814–33, DOI: 10.1016/j.istruc.2021.10.032

[5] H. J. Wagner, M. Alvarez, A. Groenewolt, and A. Menges. 2020. Towards digital automation flexibility in large-scale timber construction, *Constr. Robot.,* 4(3–4), 187–204, DOI: 10.1007/s41693-020-00038-5

[6] A. Groenewolt. 2023. Timber Plate Structures as a Roof Construction System, Doctoral Thesis, Institute for Computational Design and Construction, Stuttgart.

[7] Art & Architecture Thesaurus Online, Shell Structures. http://vocab.getty.edu/aat/300001276

[8] R. La Magna, F. Waimer, and J. Knippers. 2012. Nature-inspired generation scheme for shell structures, In *Proceedings of the International Association for Shell and Spatial Stuctures (IASS) Symposium,* Seoul, Korea. http://d-nb.info/1051812577/34

[9] H. Pottmann, M. Eigensatz, A. Vaxman, and J. Wallner. 2015. Architectural geometry, *Comput Graph* 47, 145–64, DOI: 10.1016/j.cag.2014.11.002

[10] B. Cutler and E. Whiting. 2007. Constrained planar remeshing for architecture, In *Proceedings of Graphics Interface 2007,* ACM Press, p. 11. DOI: 10.1145/1268517.1268522

[11] A. Bagger. 2010. Plate shell structures of glass Studies leading to guidelines for structural design, Dissertation, Technical University of Denmark.

[12] T. Wester. 1989. Design of Plate and Lattice Structures Based on Structural Dualism, In *Proceedings of the IASS Annual Symposium 1989,* Madrid.
[13] C. Troche. 2008. Planar Hexagonal Meshes by Tangent Plane Intersection, In *Advances in Architectural Geometry 2008,* Vienna, pp. 57–60.
[14] H. Zimmer, M. Campen, R. Herkrath, and L. Kobbelt. 2012. Variational Tangent Plane Intersection for Planar Polygonal Meshing, In *Advances in Architectural Geometry 2012,* pp. 319–32. DOI: 10.1007/978-3-7091-1251-9_26
[15] C. Robeller and N. Von Haaren. 2020. Recycleshell: Wood-only Shell Structures Made From Cross-Laminated Timber (CLT) Production Waste, *Journal of the International Association for Shell and Spatial Structures,* 61(2), 125–39, DOI: 10.20898/j.iass.2020.204.045
[16] W. Wang and Y. Liu. 2009. A Note on Planar Hexagonal Meshes, In *Nonlinear Computational Geometry,* I. Z. Emiris et al., Eds., The IMA Volumes in Mathematics and its Applications, vol. 151. New York, NY: Springer, pp. 221–33. DOI: 10.1007/978-1-4419-0999-2_9
[17] Y. Li, Y. Liu, and W. Wang. 2015. Planar Hexagonal Meshing for Architecture, *IEEE Trans. Vis. Comput. Graph.,* 21(1), 95–106, DOI: 10.1109/TVCG.2014.2322367
[18] D. Stieler, T. Schwinn, S. Leder, M. Maierhofer, F. Kannenberg, and A. Menges.2022. Agent-based modeling and simulation in architecture, *Automation in Construction* 141, 104426, DOI: 10.1016/j.autcon.2022.104426
[19] T. Schwinn, O. D. Krieg, and A. Menges. 2014. Behavioral Strategies: Synthesizing design computation and robotic fabrication of lightweight timber plate structures, In *Design Agency,* Los Angeles, pp. 177–88. DOI: 10.52842/conf.acadia.2014.177
[20] A. Groenewolt, T. Schwinn, L. Nguyen, and A. Menges. 2018. An interactive agent-based framework for materialization-informed architectural design, *Swarm Intelligence,* 12(2), 155–86, DOI: 10.1007/s11721-017-0151-8
[21] D. Rutten and S. Gregson. 2011. Principal curvature lines on surfaces, *Grasshopper3d Forum.* https://www.grasshopper3d.com/xn/detail/2985220:Comment:1486232 (May 25, 2023).
[22] D. A. Levitis, W. Z. Lidicker, and G. Freund. 2009. Behavioural biologists do not agree on what constitutes behaviour, *Anim. Behav.* 78(1), 103–10, DOI: 10.1016/j.anbehav.2009.03.018
[23] L. Nguyen et al. 2022. ABxM.Core: The Core Libraries of the ABxM Framework, *DaRUS,* DOI: 10.18419/darus-2994
[24] T. Schwinn et al. 2023. ABxM.PlateStructures: Agent-based Architectural Design of Plate Structures, *DaRUS,* DOI: 10.18419/darus-3438

Paul Nicholas, Ayoub Lharchi, Martin Tamke, Hasti Valipour Goudarzi, Carl Eppinger, Konrad Sonne, Gabriella Rossi, Mette Ramsgaard Thomsen

A Design Modeling Framework for Multi-Material Biopolymer 3D Printing

Abstract: This paper presents a novel design-to-fabrication pipeline for multi-material 3D printing. The pipeline is based on a multi-resolution grid-based approach that allows multiple data values to be stored and computed locally. Integrating contextual data, generative algorithms and print path generation, our pipeline incorporates efficient data storage and manipulation, seamless integration with existing digital workflows, advanced volumetric operations, and scalability to accommodate different scales of volumetric data. We demonstrate and assess the application of this workflow in the design and fabrication of Radicant, a biopolymer-composite wall system, exemplifying the possibilities of utilizing diverse materials and efficient modeling and fabrication techniques.

Keywords: volumetric modeling, adaptive design, context-aware modeling, growth algorithms, dynamic density adjustment, multi-material printing

1 Introduction

This paper introduces a digital workflow for multi material 3D printed bio-polymers and describes the application of this workflow to a panel-based interior wall cladding. Biopolymers are polymers that are produced by or derived from living organisms such as plants instead of petroleum. They can be 100 % biodegradable under natural conditions and produced at large scale relatively cheaply. Biopolymer composites are created by reinforcing a biopolymer with bio-fillers and fibers. By varying the choice and ratio of filler, differing mechanical, absorptive, and expressive properties can be achieved (Aaliya et al. 2021). It is the potential of combining this ability to locally differentiate and grade performance with the geometric possibilities of 3D printing that motivates this paper.

Robotic printing of bio-polymer composites offers the opportunity to create multi-material 3D printed bio-polymer elements at architectural scale, in which material properties can be varied in response to different performance criteria related to a design and its environmental and boundary conditions. However, to design and fabricate these elements, material specifications need to be closely coupled to local geometry descriptions as well as robotic control code. This interdependence means that methods typically associated with 3D printing such as slicing, and boundary representation-based modelling cannot be applied. Instead, there is a need for approaches that enable

https://doi.org/10.1515/9783111162683-015

design, material, and contextual information to be locally and volumetrically stored, queried, and manipulated via generative, parametric and fabrication workflows.

In this paper we present a novel workflow that seamlessly integrates contextual data, generative operations, and print path generation across multiple scales of resolution. We describe the application of this workflow to the case of a customizable 3D-printed biopolymer-composite paneling system, Radicant, and evaluate its performance. By showcasing the capabilities of this workflow in Radicant, we aim to highlight its potential for future improvements and extensions in the realm of modeling for multi-material 3D printing.

2 State of Art

The potential of multi-material 3D printing to produce elements in which material and geometric specifications are tightly coupled has been identified early on (Palz 2009; Oxman 2010; Richards and Amos 2014; Grigoriadis 2015). Multi and graded materials (Bever and Duwez 1972) are studied and applied within fields where it is crucial to provide great material performance, as in aeronautics (Saleh et al. 2020), for producing biological tissues (Watari et al. 2004) or mechanical engineering (Maalawi and Badr 2009), as well for architectural construction (Sinke et al. 2022, Nicholas and Tamke 2012, Pajonk et al. 2022). Attempts for software capable to describe complex variable material distributions have been undertaken within the traditional CAD framework of boundary representations (Luu et al 2022) or more radically, outside of it, such as with voxel-based representation (Michalatos and Payne 2016).

The generation of fabrication code for 3D printing was originally developed on the assumption of 3D boundary models and mono-material prints, with code generated from the slicing of a geometric model. Most current printing applications retain this approach or its extension into multi- filament printing, in which a geometric model is subdivided into parts that are each assigned to specific materials. With more advanced printing technologies such as the Stratasys Polyjet and its predecessors, which can selectively deposit layers of multiple acrylic-based photopolymers to predefined geometries via inkjet printing, it becomes possible to achieve more complex material descriptions that can be varied within a geometric part description. This approach assumes a linear process that applies material assignments onto a pre-existing geometric definition.

The work within the field of soft robotics and especially that of Hod Lipson (Cheney et al. 2014) on soft robots with multiple materials and a powerful generative encoding, demonstrate a yet unachieved potential of designing with multi material composites, and raises the question of how to link geometry design to material assignment to the fabrication process in a more integrated and less linear way. The locomotion that Lipson's multi-material robots achieve (in simulation) is accomplished by connecting design to materialization. A compositional pattern-producing network (CPPN) generates geometries, these are tested for their performance and the results are fed back to a new iteration of the design process. This interdependencies between performance, overall geometry and varying local material properties has been subject to research projects, such as Complex Modelling (Thomsen et al. 2017) and can be tackled with multi-scalar modelling strategies (Weinan et al. 2011, Nicholas et al. 2015, Faircloth et al. 2018). At the core of these approaches is a digital design pipeline which connects an ecology of models for different scales and functions, such as analysis and generation of geometries. The geometry of the design is herein a variable and not static as in the current modelling approaches for multi material 3d Print. The question is therefore whether a design to fabrication workflow can be established, which spans very different modelling paradigms and simultaneously combines the design-level information from site and architectural concept up to the generation of highly specific fabrication code for multi material robotic 3D printing?

3 Project Context

The workflow we describe in this paper is developed at the intersection of three related research projects which provide technologies as well as research directions:

- Predicting Response: development of the base material recipes,
- EcoMetabolistic Architecture: development of the custom robotic print heads, and
- Living Prototypes: provision of the case to develop a novel type of interior wall cladding.

We demonstrate the developed workflow in a specific design case: Radicant – a built demonstrator of 7 × 5 m, which was exhibited in the Aedes Gallery in Berlin in 2022/23. The Radicant collagen glue wall is a bespoke wall-paneling system created from a biopolymer composite reinforced with various types of cellulose from waste products. In the exhibition context it demonstrates the design, fabrication, and creation of a bio-based architecture. Each tile was 3D-printed using a different composition of biopolymer composites (Fig. 1), with the resulting variations arranged to create a branched interwoven form extending over six meters that is materially graded from bottom to top. With a design inspired by the silk tree, the biomaterial tiles are printed more densely where they are fixed at their trunks and become more open towards their leafy edges.

To establish a design-to-fabrication framework capable of spanning various modelling paradigms and incorporating high-level information on site and architectural concepts, a comprehensive data pipeline is necessary. This pipeline consists of several components, integrating both frontend and backend tools to achieve the desired multi-material 3D printing capabilities. The frontend was developed using the popular

Fig. 1: Use case: Bespoke wall-paneling system "Radicant", Aedes Gallery Berlin 2022.

modeling software Rhinoceros 3D and Grasshopper. The backend is built on the top of OpenVDB (Museth, 2013), which is an open-source volumetric data structure and toolkit designed for efficient storage and manipulation of sparse volumetric data. It is particularly useful for representing complex, graded material compositions in a voxel-based format. It enables the backend component of the pipeline to process and handle the complex geometric processes. Svilans et al has demonstrated the application of OpenVDB for the analysis and specification of biomaterials (Svilans et al. 2022). In this research we aim to extend this trajectory to explore how volumetric data can play a significant role in integrating design, material, and fabrication data within digitally driven design workflows. In this context, our framework possesses four main advantages:

- Efficient data storage and manipulation: The data structure enables the efficient storage of large volumes of heterogeneous material data, such as density, porosity, or mechanical properties.
- Integration with digital design workflows: It can be integrated into existing digital design environments and can be used as a base for further geometric algorithms.
- Advanced volumetric operations: The framework provides a range of operations for processing volumetric data, such as filtering, smoothing, and resampling.

- Scalability: The flexible data structure can accommodate different scales of volumetric data, from micro-scale material properties to macro-scale architectural elements, and the transfer of information between these scales.

4 Design Pipeline

The design pipeline for Radicant (Fig. 2) begins with a 3D scan to create an accurate digital model of the wall to which the paneling system will be attached. This model is then used to create a volumetric representation using a high-resolution voxel grid, which serves as the foundation for the creation of information layers, including local height, structural requirements, and pre-existing wall features. After the voxel-based model is populated with this underlying contextual information, it is then utilized as the basis for geometry and property generation that is informed by the material and fabrication properties, geometrical constraints, and design considerations. Instead of relying on an intermediary model to derive the print path, we directly create the print path from the volumetric model and associated information layers. This streamlined process eliminates the need for additional conversions or translations between models.

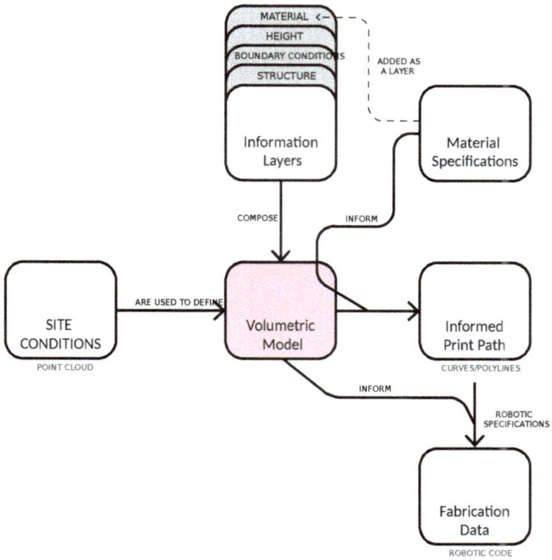

Fig. 2: Overview of the design pipeline.

The computational model follows a non-linear approach, allowing for dynamic updates and adjustments throughout the entire process. This adaptive framework enables a continuous refinement and modification of the design in response to changes in contextual data, material properties, or design goals. If new information becomes available or project requirements evolve, the computational model is able to seamlessly integrate these updates.

4.1 Volumetric Model

The volumetric model used here is a 3D model that represents an object or environment as a collection of voxels. Each voxel in the model can hold custom attributes, which makes it a useful tool for data storage, organization, and query. The base of the volumetric model is a 3D scan. The derived point-cloud provides accurate size and dimensions of the space, as well as information about existing elements on the site, such as walls, columns, and other architectural features. This data was used to establish the constraints and context for the design process, ensuring that the resulting architectural elements would integrate seamlessly with the existing site conditions. Once the model is created, we can store all the necessary parameters in the voxels, which can be queried at any level.

To store attributes in the voxel model, each voxel can be associated with a set of data values or properties. For example, a voxel might store information about its color, transparency, texture, or other relevant properties. The data can be stored directly in the voxel, or it can be stored in an external data structure that is indexed by the voxel's position. Each voxel in the grid is assigned a specific position in 3D space, typically represented by its XYZ coordinates. There are two approaches possible for storing data in a voxel model: to use a regular grid or a sparse representation, where only the voxels that contain relevant data are stored. While this approach can be useful for models that contain a large amount of empty space or for cases where high levels of detail are required, we have used the first approach. This storage method is combined with a coarsening and refining approach to adjust the level of detail in the model as needed (Fig. 3).

Grid coarsening or refinement can be applied to optimize the model for specific purposes, such as reducing memory requirements, increasing processing speed, or improving the visual fidelity of the model. To coarsen our volumetric model, the level of detail in the model is reduced by decreasing the number of voxels used to represent the object or environment. This is achieved by merging adjacent voxels together or by removing voxels that do not contribute significantly to the overall structure of the model. To refine the model, the level of detail in a grid is locally increased by adding more voxels or by subdividing existing voxels into smaller units. This can be done to improve the visual fidelity of the model, increase the level of detail in specific areas of interest, or to provide more accurate representations of complex shapes or structures. While it is possible to refine the computational model indefinitely to achieve a higher level of detail and precision, practical considerations related to the fabrication process must be considered. In our case, the smallest size of the 3D printing nozzle was 3 mm, which determined the minimum resolution achievable in the final printed structure.

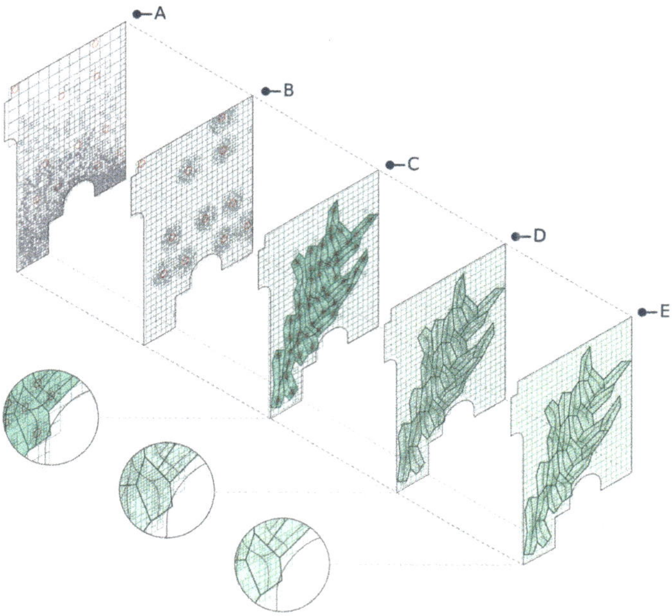

Fig. 3: llustration of the different information layers of the volumetric in variable resolution. A: Height; B: Distance to existing features on the wall; C: Distance to attachment points; D: Base panel geometry (layers 4–8); E: Base panel geometry (layers 8–12).

By incorporating this limitation and using 3mm as the smallest value in our model, we strike a balance between achieving intricate detail and ensuring the feasibility of the fabrication process.

4.2 Panelization

We utilized the information layers within the volumetric model to discretize the space into multiple panels of varying sizes. These layers provided essential data about the height, structural components, and elements of interest in the environment. We considered the constraints imposed by the existing site, such as connections to the wall and immovable elements, using them as drivers to inform the segmentation process. Additionally, we employed a dynamic relaxation model to determine the optimal wall connections for each panel.

By considering these factors, we were able to generate a suitable arrangement of panels that accommodated the site's unique conditions while maintaining the desired design intent. The discretization resulted in a total of 24 panels (Fig. 4), each tailored to their specific location and function while taking into consideration its neighbors to allow a small overlap for visual continuity.

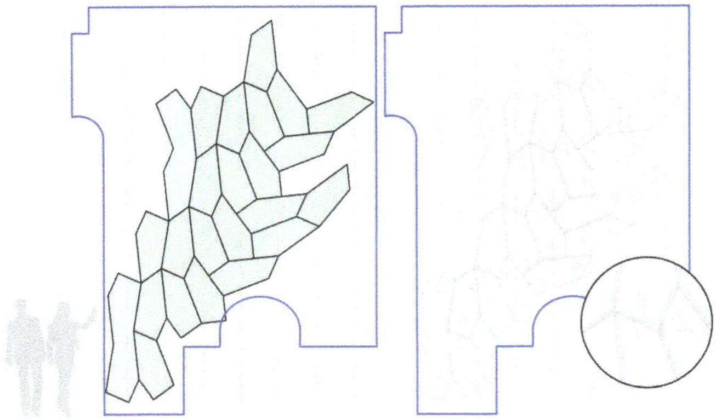

Fig. 4: Panelization strategy outcome.

4.3 Informed Print Path Generation

Our modeling approach for the panels was based on two distinct strategies. First, we focused on the "trunk", a strong and continuous design element that runs through all panels, serving as the backbone of the structure and providing a coherent visual narrative across the entire installation. The trunk connects the panels and reinforces the overall unity of the project, giving it a consistent aesthetic and structural identity. We utilized a natural growth algorithm called Space Colonization (Runions et al. 2007). This algorithm mimics the way that trees and other plants grow in nature, by simulating the growth of branches from a central trunk. In our case, we used a modified version of the Space Colonization algorithm to generate a set of points on each panel, representing the locations where branches would grow. Typically, the growth direction is calculated based on the distance to the points within a configurable field of view. However, for our purpose, the points were weighted based on the volumetric model, ensuring that the branching structure would be informed by the underlying contextual data and adapt to the specific requirements (Fig. 5).

We have also considered the smaller "branches" that extend from the trunk and exhibit unique characteristics based on their specific context within the installation. These branches were designed with varying parameters, such as thickness, length, and material behavior, allowing them to adapt to different functional requirements. To generate these branches, we employed an agent-based modeling algorithm, which simulates the growth of branches from the trunk while ensuring that they avoid collisions with each other. This algorithm considers the spatial constraints and specific characteristics of each panel and the different material recipes, allowing the branches to grow in a controlled and organized manner.

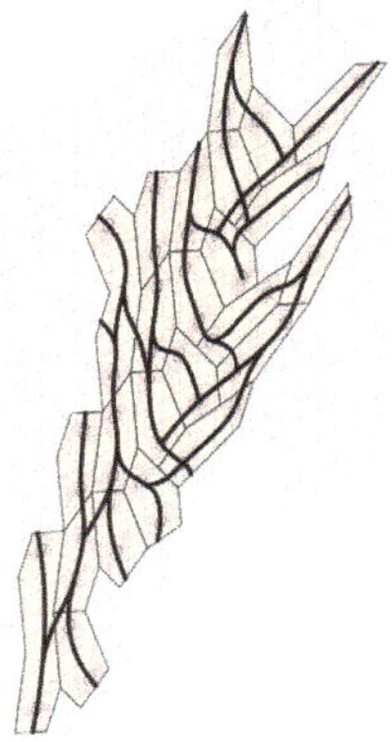

Fig. 5: "Trunk" on the design composition.

4.4 Material Specification

In the above sections, we have detailed how we used our modeling approach allows for a responsive toolpath geometry generation. Another benefit is its ability to carry information that drives the material aspects of fabrication process, therefore interfacing geometric description, material specification and robotic control code. The material specification of the print needs to respond to the geometry and structural considerations of the design, while incorporating aesthetical considerations. For example, the toolpath density and structural performance of a panel dictate its load-bearing capacity, while the choice of material, will affect its stiffness and weight.

In Radicant we have worked with a collagen-glue material mixed with waste-stream cellulose fillers: Bark flour, Wood flour, Cotton, and Seagrass. The choice of fibers affects not only the structural properties of the mix, but also its texture, color, and smell. For example, glue matrix mixed with 30 % wood flour produces a material with 0.7 g/cm^3 density, 1.35 GPa Youngs Modulus and a yellow sandy finish, whereas mixing the glue matrix with 30 % cotton results in a satin blue material with 3.15 GPa Youngs modulus, and 1.1 g/cm^3 density. A large number of permutations and interactions between different fillers and their percentages offers a customizable recipe space that can be leveraged by the print to respond to its functional and aesthetic requirements at different regions. This can be further empowered using a Machine Learning model that would return required material composition for a multi-objective specification.

By leveraging the suggested framework, we are able to assign specific material values to the individual voxels, therefore resulting in a comprehensive representation of the spatial distribution and characteristics of the installation. For example, we employ the stiffest and lightest materials in the base layers of the panels using a vertical gradient based on height blending from seagrass at the bottom to wood flour at the top. The top layer employs materials that are stronger in tension, creating a blending gradient from cotton to bark flour symmetrical to the trunk. This combined information is then translated into a vector field, where the material choices are discretized to the panel (Fig. 6), and which serves as the foundation for generating the final printing path with the correct layering sequence.

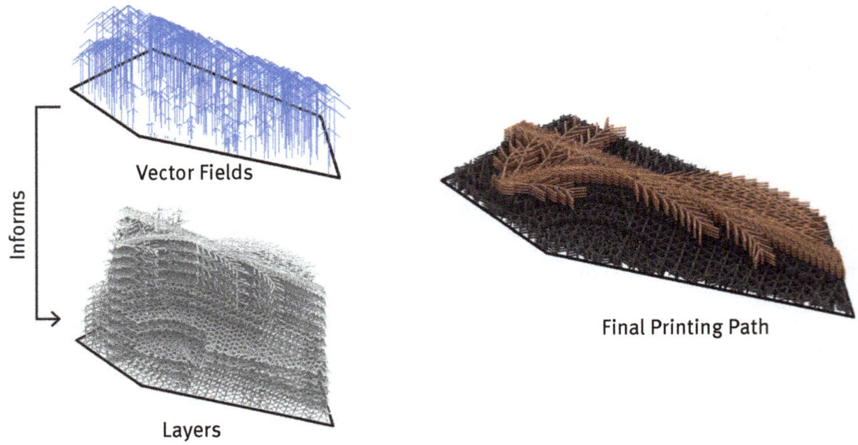

Fig. 6: Density, height, and material properties are used to generate the final print path.

4.5 Generation of Fabrication Information

The generation of fabrication data for the robotic printing process was an essential aspect of our design pipeline. This step involved transforming the geometric information and material properties derived from the volumetric model into a set of commands that the robotic printer could execute. To ensure a smooth and efficient printing process, we also considered the robot's reachability, making certain that the designed elements were within the operational range of the robot. In order to achieve a continuous and seamless print, we utilized an informed path generation algorithm that processed the curves and aligned their start and end points. This approach optimized the print path, minimizing the need for the robot to reposition itself, thus reducing overall printing time.

This fabrication information is input to two Universal Robots (UR16e) equipped with custom print heads. 24 panels with dimensions 2100×600 mm and 1200×600 mm were printed over a period of 3 weeks. In this setup, each Universal Robot printed with one material recipe, and multiple robots were used to print each panel (Fig. 7). Subsequent to this project we have developed a single multi-material print head that allows a panel to be printed by a single robot receiving multiple material feeds.

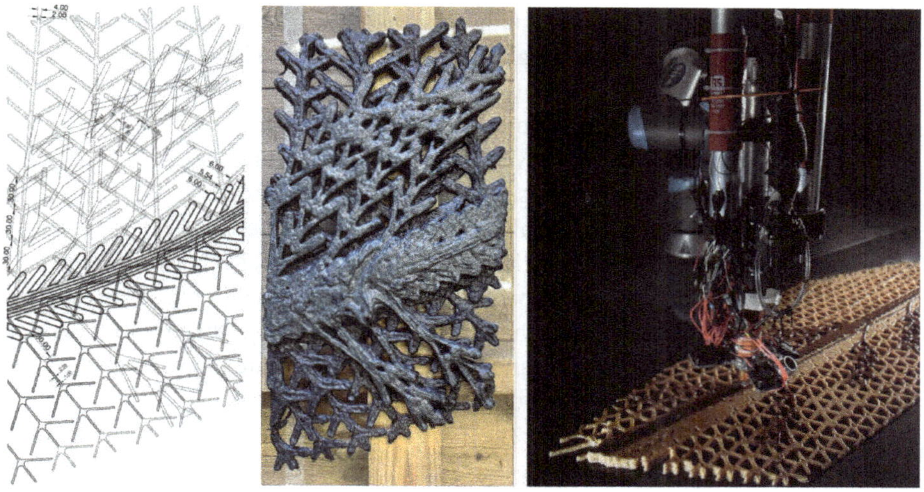

Fig. 7: 3D printing – Generated print path, test of Biomaterial with recycled cotton, and final print with seagrass and bark fibrous filler with multimaterial Biopolymer printhead.

5 Process Evaluations and Results

The development of the design to fabrication pipeline took place in parallel to the development of the protocols and hardware for large scale printing, drying and post-processing of collagen-based materials, the adaptation and optimization of the material for 3D printing with a range of fillers as well as the development and refinement of design concept and interfaces between the different parts of the structure. This process enforced a high level of modularity and adaptability into the developed framework, as feedback on conceptual design level, from small- and large-scale test prints and finally full-scale mockups needed to be integrated at any moment. The development of the framework followed the same linear order as the development of the design, from conceptual to full scale material production level. It was however the ability of the pipeline to connect these very different layers and their model representation, that was tested constantly and with positive results. Changes on any model could and have been executed at any moment until the end of the production phase. Here a measurement of the printed collagen panels was undertaken and showed as expected an uneven shrinkage of every panel of up to 20 % in area. The geometric model of the pipeline could be updated, and the model of the timber substructure was automatically updated – just in time for production (Lharchi et al. 2023).

The project allows us to reflect on the initial four aims for the framework:

- Efficient data storage and manipulation: The framework has been able to connect models of various datatypes such as point clouds from 3D scanning, voxel-based volumetric data for material properties, mesh models for structural analysis, and curve-based data for fabrication paths with the aim to produce finally information that steers the printing of 24 large panels, with over 18000 lines of robotic code for small panels, and the double for the larger ones. However, the performance could be improved, as each regeneration took several hours.

- Integration with digital design workflows: Our framework was integrated seamlessly into a Rhinoceros 3D/Grasshopper workflow and could ingest other data, as from point clouds, structure analysis, and material properties. The framework's potential for incorporating additional information layers, such as material aging and degradation or acoustic properties, highlights its versatility in addressing various design considerations. Furthermore, by selectively increasing the resolution where needed, the framework can provide a more accurate representation of the design, while maintaining computational efficiency.
- Advanced volumetric operations: The framework range of operations for processing volumetric data, such as filtering, smoothing, and resampling have been crucial for providing information on site, and other high- and low-level information in the pipeline.
- Scalability: The flexible data structure could indeed accommodate different scales of volumetric data, from micro-scale material properties to macro-scale architectural elements, and the transfer of information between these scales. It was especially important that this adaptation could take place at any time of the process, so that incrementally gained insights gained in process could be accommodated, as well more radical updates, as the required adaptation to the large shrinkage of the panels.

6 Conclusion

In this paper we have presented a grid-based workflow that integrates contextual data, generative operations, and print path generation across multiple scales of resolution. The workflow has been motivated by the architectural and sustainable potentials of multi-material 3D printed biopolymers. We have described the components and application of this workflow to the design and fabrication of an interior wall paneling system. Evaluating this specific case indicates the potential for a larger set of applications and extensions, based on incorporating more of the specificity of the printing process itself, the possibility for grading material more continuously and at a finer resolution, and through generalization of the pipeline to materials beyond biopolymers. More fundamentally, this research points towards the need to move from boundary to volumetric representations to better enable the full potential of design and fabrication with emerging biomaterials.

Author Contributions

The manuscript was written with the contribution of all authors. All authors have approved the final version of the manuscript. Nicholas, P. project conceptualization, methodology, design concept, writing – original draft, reviewing and editing, supervision, Lharchi, A., design concept, computational modeling framework development, 3D print strategy, fabrication, installation, writing – original draft, reviewing and editing, Tamke, M., project conceptualization, funding acquisition, methodology, writing – original draft, reviewing and editing, installation, supervision, funding acquisition, Valipour Goudarzi H., design concepts, visualization, prototyping, fabrication, installation, Eppinger, C., design concept, 3D print hardware development, prototyping, fabrication, installation, Sonne, K., prototyping, fabrication, installation, Rossi, G., material specification strategy, Ramsgaard Thomsen, M. project conceptualization, methodology, design concept, writing – review and editing, lab infrastructure, supervision, funding acquisition.

Acknowledgements

Radicant was funded through *EMA – Eco-Metabolistic Architecture* (3D printing strategies and steering), Predicting Response (workflow and intelligent adaptive fabrication) and Living Prototypes (dissemination and biobased materials for architecture).

It has received funding from the European Research Council (ERC) under the European Union's Horizon 2020 research and innovation program (grant agreement No 101019693). Predicting Response: Independent Research Fund Denmark (DFF) – Project Number 9131-00034B. Living Prototypes – Digital Fabrication Processes and Prototypes for Resource-Conscious Living Spaces. Zukunft Bau, Federal Institute for Research on Building, Urban Affairs and Spatial Development, German Ministry for Housing, Urban Development and Building. Project number 10.08.18.7-21.02

References

Aaliya, B., K. V. Sunooj, and M. Lackner. Biopolymer composites: A review. *International Journal of Biobased Plastics* 3, no. 1 (2021): 40-84.

Bever, M. B., and P. E. Duwez. *Gradients in composite materials.* Materials Science and Engineering 10 (1972): 1-8.

Cheney, N., R. MacCurdy, J. Clune, and H. Lipson. Unshackling evolution: evolving soft robots with multiple materials and a powerful generative encoding. *ACM SIGEVOlution* 7, no. 1 (2014): 11–23.

Faircloth, B., R. Welch, M. Tamke, P. Nicholas, P. Ayres, Y. Sinke, B. Cuffy, and M. Ramsgaard Thomsen. Multiscale modeling frameworks for architecture: Designing the unseen and invisible with phase change materials. *International Journal of Architectural Computing* 16, no. 2 (2018): 104–22.

Grigoriadis, K. Material fusion: a research into the simulated blending of materials using particle systems. *International Journal of Architectural Computing* 13, no. 3–4 (2015): 335–52.

Lharchi, A., Tamke, M., Nicholas, P., Eppinger, C., Sonne, K., Valipour Goudarzi, H. Rossi, G., and Ramsgaard Thomsen, M., 2023, September. Sensing and Augmenting for Adaptive Assembly Strategies (In Press) In *41th eCAADe Conference 2023: Digital Design Reconsidered*. eCAADe.

Luu, T. H., C. Altenhofen, A. Stork, and D. Fellner. GraMMaCAD: Interactively Defining Spatially Varying FGMs on BRep CAD Models. In *International Design Engineering Technical Conferences and Computers and Information in Engineering Conference*, vol. 86212, p. V002T02A014. American Society of Mechanical Engineers, 2022.

Maalawi, K. Y., and M. A. Badr. Design optimization of mechanical elements and structures: A review with application. *Journal of Applied Sciences Research* 5, no. 2 (2009): 221231.

Michalatos, P., A. Payne, A. Herneoja, T. Österlund, and P. Markkanen. Monolith: the biomedical paradigm and the inner complexity of hierarchical material design. In *Complexity & Simplicity-Proceedings of the 34th eCAADe Conference*, vol. 1, pp. 445–54. Oulu, Finland: eCAADe, University of Oulu, 2016.

Museth, K. VDB: High-resolution sparse volumes with dynamic topology. *ACM transactions on graphics (TOG)* 32, no. 3 (2013): 1–22.

Nicholas, P., D. Stasiuk, E. Clausen Nørgaard, C. Hutchinson, and M. Ramsgaard Thomsen. A multi-scale adaptive mesh refinement approach to architectured steel specification in the design of a frameless stressed skin structure. In *Modelling Behaviour: Design Modelling Symposium* 2015, pp. 17–34. Springer International Publishing, 2015.

Nicholas, P., and M. Tamke. Composite Territories: Engaging a bespoke material practice in digitally designed materials. In *Physical Digitality: Proceedings of the 30th eCAADe Conference*. 2012.

Oxman, N. Structuring materiality: design fabrication of heterogeneous materials. *Architectural Design* 80, no. 4 (2010): 78–85.

Pajonk, A., A. Prieto, U. Blum, and U. Knaack, 2022. Multi-material additive manufacturing in architecture and construction: A review. *Journal of Building Engineering* 45, p.103603.

Palz, N. Programmed matter. *Strukturalismus in Architektur und Städtebau Reloaded,* 20–21 November 2009 (2009).

Ramsgaard Thomsen, M., M. Tamke, P. Nicholas, A. H. Deleuran, P. Ayres, R. La Magna, and C. Gengnagel. Simulation in complex modelling. In *Proceedings of the Symposium on Simulation for Architecture and Urban Design,* pp. 1–8. 2017.

Richards, D. and A., Martyn (2014) Evolving Morphologies with CPPN-NEAT and a Dynamic Substrate. In *ALIFE 2014 – Fourteenth International Conference on the Synthesis and Simulation of Living Systems,* 30th July – 2nd August 2014, Manhattan, New York.

Runions, A., B. Lane, and P. Prusinkiewicz. Modeling trees with a space colonization algorithm. *NPH* 7, no. 63–70 (2007): 6.

Saleh, B., J. Jiang, R. Fathi, T. Al-hababi, Q. Xu, L. Wang, D. Song, and A. Ma. 30 Years of functionally graded materials: An overview of manufacturing methods, Applications and Future Challenges. *Composites Part B: Engineering* 201 (2020): 108376.

Sinke, Y., M. Ramsgaard Thomsen, M. Tamke, and M. Seskas. Strategies for Encoding Multi-dimensional Grading of Architectural Knitted Membranes. In *Towards Radical Regeneration: Design Modelling Symposium Berlin* 2022, pp. 528–42. Cham: Springer International Publishing, 2022.

Svilans, T., S. Gatz, G. Tyse, M. Ramsgaard Thomsen, P. Ayres, and M. Tamke. Deep Sight-A Toolkit for Design-Focused Analysis of Volumetric Datasets. In *Towards Radical Regeneration: Design Modelling Symposium Berlin* 2022, pp. 543–55. Cham: Springer International Publishing, 2022.

Watari, F., A. Yokoyama, M. Omori, T. Hirai, H. Kondo, M. Uo, and T. Kawasaki. Biocompatibility of materials and development to functionally graded implant for bio-medical application. *Composites Science and Technology* 64, no. 6 (2004): 893–908.

Weinan, E., 2011. *Principles of multiscale modeling.* Cambridge University Press.

Seiichi Suzuki, Alison Martin, Yingying Ren, Tzu-Ying Chen, Stefana Parascho, Mark Pauly

BamX: Rethinking Deployability in Architecture through Weaving

Abstract: Deployable gridshells are a class of planar-to-spatial structures that achieve a 3D curved geometry by inducing bending on a flat grid of elastic beams. However, the slender nature of these beams often conflicts with the structure's load-bearing capacity. To address this issue, multiple layers are typically stacked to enhance out-of-plane stiffness and prevent stability issues. The primary challenge then lies in deploying such multi-layered systems globally, as it requires significant shaping forces for actuation. This paper presents an alternative design approach that involves strategically connecting compact-to-volumetric gridshell components using weaving principles to shape a thick segmented shell. This innovative approach allows for an incremental construction process based entirely on deployable modules with volumetric configurations that locally provide the necessary structural depth for the entire system. To demonstrate this principle, we present the realization of BamX, a research pavilion constructed using deployable cylindrical components made from raw bamboo slats. These components are interconnected at carefully optimized interlocking woven nodes, resulting in a bending-active structural frame that is both strong and exceptionally lightweight. To determine the optimal topology and geometry of the pavilion, we employ an integrative computational approach that leverages advanced numerical optimization techniques. Our method incorporates a physics-based simulation of the bending and twisting behavior of the bamboo ribbons. By finding the ideal locations for ribbon crossings, we ensure that all external and internal forces are in global equilibrium while minimizing the mechanical stress experienced by each ribbon. BamX exemplifies how a symbiosis of refined weaving craft and advanced computational modeling enables fascinating new opportunities for rethinking deployability in architecture.

1 Introduction

Elastic gridshells are material- and structure-efficient systems that achieve 3D curved geometries by globally inducing bending and twisting on a flat grid of elastic beams. Based on the magnitude of shaping forces, external mechanisms such as cranes or inflatable cushions are needed to transform the flat grid as a whole into its final 3D shape. Slender beam profiles and materials with high strength and low bending stiffness are required to reduce the magnitude of shaping and residual forces. In practice, these characteristics of slenderness and flexibility generate stability problems even after finding a suitable geometric configuration. To counterbalance this condition, additional

https://doi.org/10.1515/9783111162683-016

grid layers need to be superimposed to create sufficient out-of-plane stiffness together with bracing elements to diagonalize the grid and increase in-plane shear strength.

The research presented in this paper showcases an alternative approach for elastic gridshells in architecture. The aim is to develop an integrative computational design framework for double-layered segmented shells built from deployable units. This approach was formed by the adaptation of elastic gridshell principles at the local level of components and from the application of basket weaving principals at the global level of the segmented shell. The design framework built the basis for the development of the BamX Research demonstrator.

2 Related work

2.1 Deployable Gridshells

Elastic gridshells are quad-grid structures shaped from continuous beams connected at nodes by revolute joints. During erection, beams deform elastically and rotate around the joints to achieve the desired geometry. The shape is then locked by diagonalizing the grid, blocking joints' DoFs, adding restraining elements, or a combination thereof. Gridshells are typically classified according to the lengths of their edges within the grid. Traditionally, elastic gridshells are considered to be regular grids that bend as a whole from a flat state (Lienhard and Gengnagel 2018). Irregular gridshells, in turn, require an incremental assembly process since elements need to be individually bent and connected. Recent studies are leveraging the geometrical incompatibilities of irregular grids for deployability and shaping purposes. X-shells (Panetta et al. 2019) are generalized types of irregular gridshells in which the key concept is to encode the 3D curved geometry into the planar state of the irregular grid. Similar studies were conducted by (Soriano et al. 2019; Pillwein and Musialski 2021) based on the design space of geodesic grids. In contrast to previous studies that have mainly concentrated on global deployment, our work addresses the notion of local deployability at the component level.

2.2 Bamboo

Bamboo has emerged as a viable and sustainable alternative to composite materials in the ongoing research on elastic gridshell structures. Previous studies by (Lienhard 2014) have indicated that bamboo exhibits a comparable ratio between Young's Modulus and permissible bending stress to that of Glass Fiber Reinforced Polymers (GFRP), thereby positioning it as a suitable material for both static and kinetic applications involving active bending. In particular, bamboo poles have been utilized in several built examples such as the ZCS Bamboo pavilion (Crolla 2017) or the Pantographic Bamboo Hybrid Amphitheater Structure (Seixas et al. 2021). In contrast to complete poles, our approach necessitated enhanced material flexibility, which we achieved by using thin slats obtained by splitting bamboo poles into thin slats.

2.3 Segmented shells

Segmented shells are a different type of discrete shell offering a simpler construction process than elastic gridshells. The fundamental principle is to connect prefabricated components in a step-by-step process to create a unified shell structure. The connection between components is crucial since any compromise in material continuity can reduce the structure's stiffness. Significant efforts have been conducted on segmented timber shells based on double-layered plate components with finger joint connections to increase in-plane stiffness (Li and Knippers 2015; Alvarez et al. 2019). Elastic bending has also been used to create double-layered components using custom-laminated beech plywood and sewing techniques to establish the connection between components (Sonntag et al. 2017). Other works deal with modular tubular fibrous morphologies produced through coreless filament winding technology (Pérez et al. 2020). Our research aligns with this study concerning the cylindrical typology of the component. However, the primary differentiation lies in the conceptualization of individual components as elastic gridshells and their interconnection through weaving principles.

2.4 Weaving

Basket weaving involves interlacing thin flexible ribbons in regular patterns to create a stable lattice structure. By alternating over- and under-crossings, the interlacing principle creates the necessary friction for holding ribbons in place (Martin 2015). Woven structures commonly utilize bi- and tri-axial arrangements of straight ribbons. In such cases, introducing topological singularities becomes crucial for inducing curvature Ayres et al. (2018). However, the combination of straight ribbons and topological singularities tends to produce kinks in the woven structure. Ren et al. (2021) presented an alternative approach to attain smooth free-form woven surfaces based on injecting in-plane curvature into the ribbons. An optimization-driven process is employed to accurately calculate the curved planar geometry of each ribbon, enabling the interwoven structure to approximate a specified target surface closely. In our work, we rely on straight bamboo ribbons but apply a similar optimization scheme to determine the ideal lengths and crossing locations to best approximate the desired target shape while minimizing stress.

3 Design principle

The presented study was conducted under the assumption that combining fun- damental principles from segmented shells, elastic gridshells, and basket weaving can lead to large-scale architectural applications. The realization of the BamX research demonstrator serves as a proof-of-concept of the above. The structure is an ultra-lightweight dome shaped from 36 deployable cylindrical components. When deployed, the cylinders connect seamlessly at 21 woven nodes to globally form a double-layered segmented shell (Fig. 1). Cylindrical components are designed as compact-to-volumetric gridshell units made from thin flexible bamboo slats. While deployable components can be highly compacted for transport and storage, they also facilitate the construction process of the entire structure. By sequentially deploying cylindrical units and connecting them using weaving principles, the global geometry of the structure progressively emerges. The introduction of weaving principles at the level of nodes allows the cylinder's scissor mechanism to be locked, gradually stabilizing the entire structure without adding external elements.

Fig. 1: BamX research demonstrator.

4 Computational framework

The design, modeling, and simulation of segmented shells composed of deployable gridshell modules entail considerable complexity. The kinematics of deployable components and the overall structural behavior of the shell are significantly affected by

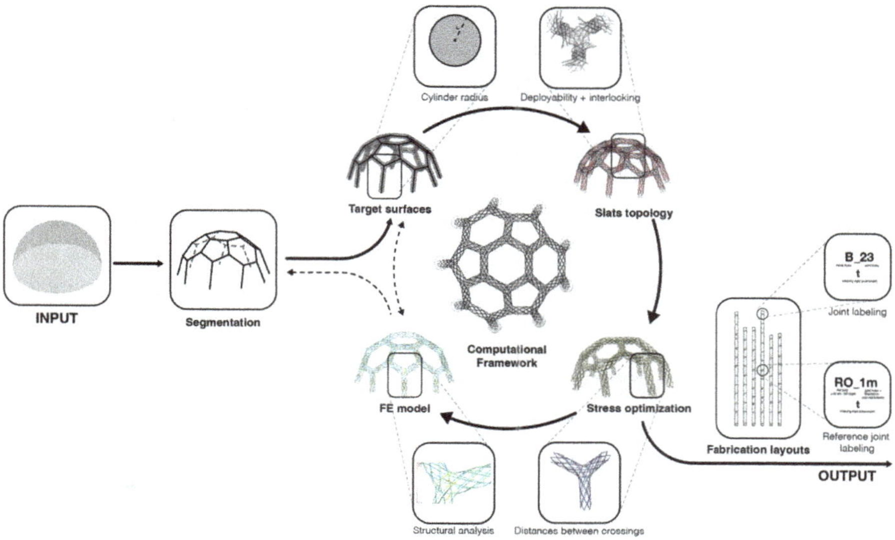

Fig. 2: Computational framework combining several models with different levels of detail.

geometric and topologic features at different scales. In addition, it is also necessary to incorporate materialization principles at conceptual design stages to ensure the viability of the design. A consistent digital approach was formulated, combining an iterative design process with diverse computational models that vary in detail (Fig. 2). These stages comprise simplified global models of segmented surfaces and detailed models encompassing all deployable grid layouts.

4.1 Modeling

A *segmented shell* is modeled using a valence 3 n-gon topology created from a structural free-form surface representing a desired geometry. The edges of this topology are used to create cylindrical surfaces that are interpolated at incident vertices to develop the surface of nodes. A simplified FE model based on a strut representation is then used to calibrate the diameter of cylinders and define material quantities. The results obtained from this model are in turn utilized to create the curve network of cylindrical components.

A *cylindrical gridshell component* is developed from a curve network consisting of two families with an equal and even number of spiral curves. Two construction planes positioned at the ends of the corresponding edge and aligned with the normals to the global surface are used to create spiral curves with precise control over their starting/ending points and tangent directions. The number of turns around the edge determines the number of crossings along the cylinder and its deployment.

A *woven node* is shaped from the connection of adjacent cylindrical gridshell components, wherein spiral curves are carefully paired and linked by alternating family types (Fig. 3). The connection is geometrically modeled by extending one of the curves using tangent directions and incorporating an overlap between connected curves. Materialization constraints based on the dimensions of the available bamboo rods are enforced at this point to restrict the maximum length achievable by those curves. Consequently, a tessellation containing triangular tiles is generated on the surface of the node through the connection of spiral curves. Triangular tiles are a desired feature to lock the kinematic behavior of the scissor linkage of the deployable cylinders. Since the resulting topology contains vertices with a valence greater than 4, interlacing principles need to be carefully defined to connect these curves at the level of the node.

Fig. 3: Woven node created from the connection of deployable cylinders.

For the design of the BamX research demonstrator (Fig. 4), the geometry of a spherical cap was approximated by using part of a truncated icosahedron. The structure spans 10.3 m and has a height of 4.60 m. Deployable gridshell cylinders have a diameter of 40 cm. Curve networks are created with 8 spiral curves per family and one turn around the corresponding edge-axis. In total, 432 bamboo slats were used to build the structure, with a total weight of less than 160 kg.

4.2 Optimization

To model the equilibrium state of the entire structure, it is necessary to simulate the bending and twisting behavior of flexible beams shaping the cylindrical gridshells and woven nodes as well as the forces occurring at the joints. Using volumetric FEM with continuous collision detection to model the beams is a non-trivial and computationally expensive approach that can compromise design exploration. Instead, we adopt the Discrete Elastic Rods model (Bergou et al. 2008) to simulate the physical behavior of beams. We employ a reduced joint model to constrain such beams at their intersections (Panetta et al. 2019).

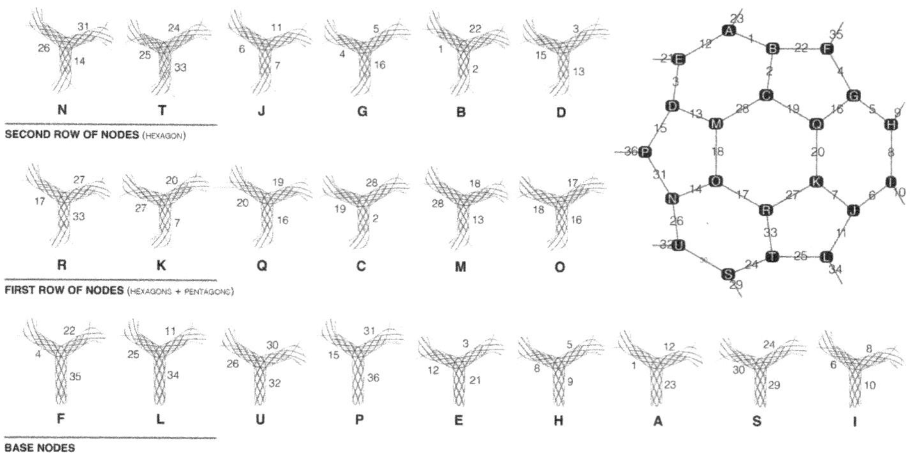

Fig. 4: Typologies of woven nodes and cylindrical gridshells in the structure.

The numerical model is constructed from the curve networks generated for each cylindrical unit. The required inputs are the topology defined by the crossings originated from the intersection of curves and the spacings between these crossings. While crossings with valence 4 are modeled as revolute joints, crossings with higher valences are modeled as fixed joints. Additional information regarding the offsets along the normal direction at crossings is also required to define the under-and-over logic for interweaving. On this basis, the 3D equilibrium state of the entire assembly can be computed. To better approximate the interpolated surface generated from cylinders and nodes, it is important for the simulation to keep joint positions close to the surface while also permitting them to glide along it. This objective can be formulated using closest point projection from a point to a 3D surface. Since the input topology is fixed to respect the weaving pattern on nodes and the deployability of cylindrical modules, the main design parameters are the spacings or "rest-lengths" between joints. Given a set of length parameters, the forward simulation can be run to compute the 3D joint positions and then evaluate the objective function. Using the adjoint state method and sensitivity analysis, a gradient-based optimization method is implemented to find the optimal length parameters while tracking the equilibrium state of the structure at each iteration. This design optimization permits to minimize stresses on the beams. More details about the method can be found in Panetta et al. (2019); Ren et al. (2022).

4.3 Analysis

Structural analysis is integrated into the framework at different stages, with models varying in their level of detail. Physical tests on cylindrical gridshell components were conducted to construct an FE representation of screw connections between flexible beams. The testing result of the cylinder prototypes also serves to inform the global design with an accumulated stiffness of the components. This step largely simplifies the initial FE analysis into a strut-represented model for rapid global topology iteration. Once the global topology optimization is accomplished, the characterized mechanical properties from material samples allow for a more sophisticated simulation of the detailed woven geometry.

An important consideration for developing the BamX research demonstrator was the utilization of raw bamboo slats. Up-scaling a structure built out of raw bamboo requires a calculation process that considers the material properties derived from test results. This is because raw bamboo has substantially varying mechanical properties and cross-section dimension along the axis, unlike more standardized and graded building materials such as steel, concrete, or timber. Hence, integrating structural information into the design loop requires the characterization and evaluation of the acquired batch of material. Multi-scale calibration on the material, component and global levels provides an understanding of the range of scalability the structure can achieve.

Calibrations are done on the component scale and the bamboo slat scale. 4-point flexural tests were carried out on 70 cm slat specimens. Wide and thin specimens were extracted respectively from the pole's root-end and tip-end. In total, 12 air-dried specimens were tested, and 2 sets of mechanical properties were characterized. In the final FE model, both stiffnesses were applied to check the maximum deformation under the lowest stiffness scenario and the maximum stress state under the highest stiffness scenario.

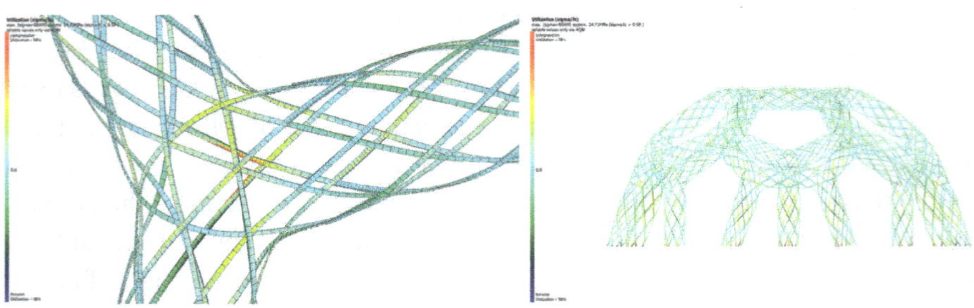

Fig. 5: Calibrated FE model. Left: Beams close up; Right: Detailed model with deformations.

5 Materialization

The computational process simplified construction instructions, making them independent of advanced fabrication technologies and industrial materials. The con- struction was divided into two phases: preparing bamboo slats and pre-fabricating cylinders in the first phase and transporting and connecting the cylinders at the construction site to form the structure in the second phase.

5.1 Off-site pre-fabrication

The final structure relies solely on raw bamboo slats and basic hand tools for construction. The bamboo was sourced from the Tuscany region of Italy, belonging to the Phyllostachys family, with a maturity of 3–4 years. The bamboo poles utilized on this project had a cross-section diameter of 5 cm from which 6 slats, each with a width of 2.5 cm and a thickness of 7 mm, were split using a traditional bamboo splitting device. To the greatest extent possible, it was important to maintain a consistent cross-section along the splits to ensure surface smoothness and material continuity. For this reason, only the central part of the plant was utilized to minimize the difference in diameters between the upper and lower part of the poles. The total amount of material came from 50 bamboo poles of 5 meters and 20 poles of 3 meters.

During pre-fabrication, a peer review system was instituted to ensure that manual measurement of the joints and label information precisely aligned with the data of the digital model. A custom-designed labeling system was adopted to include, in each bamboo slat, the required weaving information for assembly. Holes are then manually drilled and slats are cut to length with a hand saw. Finally, slats are connected with 3 mm screws to create the cylindrical components, which were stored in their compact state.

5.2 On-site assembly

Once pre-fabrication was done, cylinders were transported to the construction site to assemble the entire structure (Fig. 6). The general assembly principle involves the interweaving and connection of slats from three deployable cylinders that converge at a node. After weaving the slats through the node, the connection is finalized by overlapping slats of different cylinders, as defined in the computational model. The overlap between slats is always fastened by one pin joint and two cable ties attached at both ends (Fig. 7). Consequently, cylinders are joined seamlessly at nodes since the overlap between slats permits maintaining material continuity.

The assembly sequence of the entire structure was carefully planned by taking into account the maneuverability of cylinders and avoiding scenarios where slats can be over-stressed while weaving. A top-to-bottom strategy (Fig. 8) was adopted, which involved assembling the top hexagon on the ground and then lifting it up using tension cables. This enabled adding the remaining parts of the structure and weaving the nodes always at the floor level. From the initial hexagon, the remaining cylinders of the nodes

Fig. 6: Undeployed cylinders on the construction site.

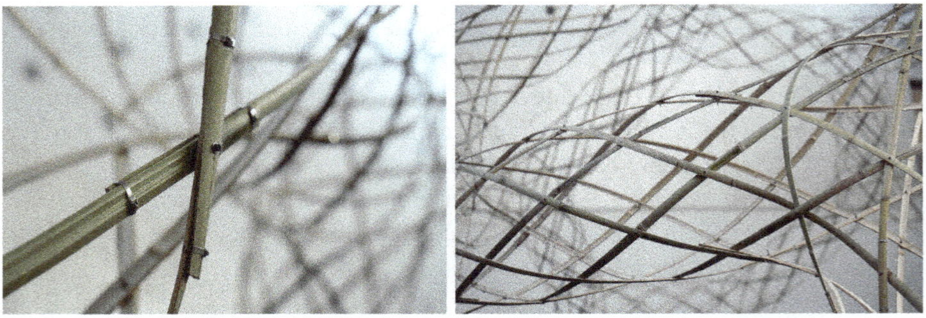

Fig. 7: Detail of the overlapping between bamboo slats (left); Connections occurring when cylinders start to shape nodes (right).

were added to form the first row of woven nodes. Note that the structure was raised gradually as deployed cylinders were added. The next row of woven nodes was created by adding the remaining two cylinders per node, which also served to close the first level of pentagons and hexagons. The third row of woven nodes was completed after adding the cylinders in contact with the floor. The structure was then carefully released from the tension lines with the end cylinders attached to the base, thereby becoming load-bearing.

6 Limitations and future work

The design and construction of the final pavilion posed a significant challenge for the proposed structural system and design framework. Several digital models with different

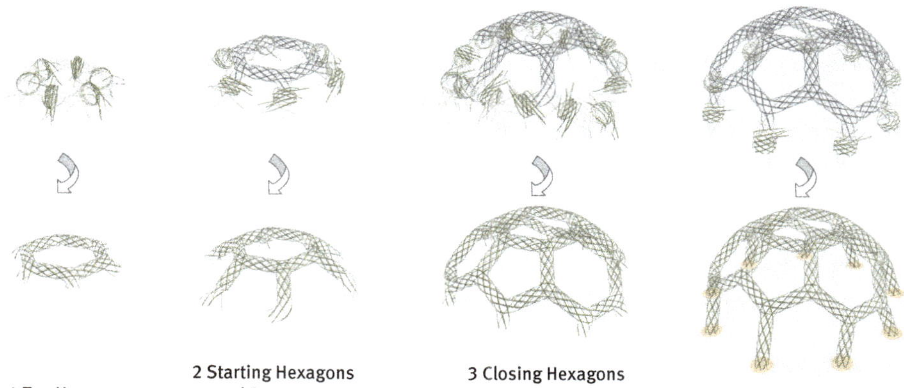

Fig. 8: Assembly sequence.

levels of detail incorporating physical behavior and materialization constraints have heavily informed the entire design process. The segmentation of the spherical cap modeled as a part of a truncated icosahedron was intended to facilitate our large-scale testing. However, the design framework can use different types of structurally stable free-from surfaces but some special features are always required. The segmentation of the target surface needs to only contain interior vertices of valence 3, with nearly symmetrical edges. At this stage, symmetrical edges secure the creation of triangular tiles on the surface of the node while maintaining the continuity of curves connected from two different cylinders. Further studies are required to include nodes with higher valence and accommodate more variation between the angles of adjacent edges.

Currently, cylinders showed a straight deployment path which forces curvature to be accumulated at the level of the woven nodes. While it is possible to develop positive and negative curvature with this deployment, smoother results on the target surface could be obtained by incorporating cylinders exhibiting a curved deployment path. As shown in preliminary studies, it is possible to obtain deployed curved cylinders by enforcing geometric incompatibilities on the curve network. Curvature can then be distributed at the level of woven nodes and along edges. However, this raises further problems regarding the woven connection of cylinders and the generation of triangular tiles. Future studies are needed to explore computational strategies for incorporating curved cylinders.

7 Conclusion

This paper introduces a computational framework to design, optimize, and fabri- cate double-layered segmented shells made of elastic gridshell components. The fundamental principle is to create rigid woven nodes by connecting compact-to- volumetric cylindrical components. The use of volumetric components creates sufficient out-of-plane stiffness to stabilize the entire structure with a similar effect of a double-layer grid. Triangular tiles are created on the surface of woven nodes by carefully interweaving curves from adjacent cylinders. This topologic feature permits to lock cylinders' kinematic behavior and stabilize the entire structure without adding external elements.

To demonstrate the effectiveness of the design framework, a dome structure with a span of 10.3 m was constructed using raw bamboo slats. The structure shows the enormous potential that the combination of segmented shells, deployable units, and natural materials has in architecture. Deployable cylindrical components can be pre-fabricated off-site and kept in their compact state for transportation and storage. Based on an incremental assembly process, components can then be deployed and interconnected without requiring external mechanisms due to their small weight and magnitude of shaping forces. The optimized geometric and topologic features are the only factors determining the stability of the entire structure. Additionally, the strategy presented here combines high-tech and low-tech strategies by using computation for the design and simulation of the material behaviour but relying on an intuitive manual fabrication process that takes advantage of traditional bamboo processing knowledge.

Acknowledgments

The project was made possible by funding from the Swiss National Science Foundation (Grant FNS 514543 / CF 1156). The authors would like to thank Filip Goč, Florin Isvoranu and Davide Pellis for their valuable contribution to the project; as well as to student assistants: Ruben Bento, Antoine Brunner, Guillaume Hueber, Alix Magnaguémabé, Christian Meier, Sepehr Mousavi, Maximilian Paulsen, Xinran Tao, Sixiao Xu; and volunteers: James Pauly and Hannah Pauly.

References

Alvarez, M., H. Wagner, A. Groenewolt, O. Krieg, O. Kyjanek, L. Scheder-Bieschin, S. Bechert, D. Sonntag, A. Menges, and J. Knippers (2019). The Buga wood pavilion – integrative interdisciplinary advancements of digital timber architecture. In *Proceedings of the ACADIA 39th Annual Conference*, Austin, Texas.

Ayres, P., A. Martin, and M. Zwierzycki (2018). Beyond the basket case: A principled approach to the modelling of kagome weave patterns for the fabrication of interlaced lattice structures using straight strips. In *Advances in Architectural Geometry.*

Bergou, M., M. Wardetzky, S. Robinson, B. Audoly, and E. Grinspun (2008). Discrete elastic rods. In *ACM SIGGRAPH 2008 Papers*, SIGGRAPH '08, New York, NY, USA. Association for Computing Machinery.

Crolla, K. (2017). Building indeterminacy modelling – the 'zcb bamboo pavilion' as a case study on nonstandard construction from natural materials. *Visualization in Engineering* 5 (1), 1–12.

Li, J.-M. and J. Knippers (2015). Segmental timber plate shell for the Landesgartenschau exhibition hall in Schwäbisch Gmünd. *International Journal of Space Structures Vol. 30.*

Lienhard, J. (2014). Bending-active structures: form-finding strategies using elastic deformation in static and kinetic systems and the structural potentials therein. Universität Stuttgart.

Lienhard, J. and C. Gengnagel (2018). Recent developments in bending-active structures. In *Proceedings of the IASS Annual Symposium 2018: Creativity in Structural Design.*

Martin, A. G. (2015). A basketmakers approach to structural morphology. *Proceedings of IASS Annual Symposia 2015 (29),* 1–8.

Panetta, J., M. Konaković-Luković, F. Isvoranu, E. Bouleau, and M. Pauly (2019). X-shells: A new class of deployable beam structures. *ACM Transactions on Graphics (TOG) 38* (4), 1–15.

Pillwein, S. and P. Musialski (2021). Generalized deployable elastic geodesic grids. *ACM Trans. Graph. 40* (6).

Pérez, M. G., B. Rongen, V. Koslowski, and J. Knippers (2020). Structural design, optimization and detailing of the buga fibre pavilion. *International Journal of Space Structures 35* (4), 147–59.

Ren, Y., U. Kusupati, J. Panetta, F. Isvoranu, D. Pellis, T. Chen, and M. Pauly (2022). Umbrella meshes: Elastic mechanisms for freeform shape deployment. *ACM Trans. Graph. 41* (4).

Ren, Y., J. Panetta, T. Chen, F. Isvoranu, S. Poincloux, C. Brandt, A. Martin, and M. Pauly (2021, jul). 3d weaving with curved ribbons. *ACM Trans. Graph. 40* (4).

Seixas, M., L. Eustaquio Moreira, P. Stoffel, and J. Bina (2021). Form finding and analysis of an active bending-pantographic bamboo space structure. *Journal of the International Association for Shell and Spatial Structures 62* (3), 206–22.

Sonntag, D., S. Bechert, and J. Knippers (2017). Biomimetic timber shells made of bending-active segments. *International Journal of Space Structures 32* (3/4), 149–59.

Soriano, E., R. Sastre, and D. Boixader (2019). G-shells:flat collapsible geodesic mechanisms for gridshells. In *Proceedings of the IASS Annual Symposium 2019: Form and Force.*

Seri Nishimoto and Tomohiro Tachi

Transformable Surface Mechanisms by Assembly of Geodesic Grid Mechanisms

Abstract: The scissors-like grid shells composed of members bent in the out-of-plane direction have structural advantages: the ease of deployment as a one-degree-of-freedom system and the stability as it is modeled as a shell once the scissors' transformation is fixed. However, it is impossible to generate a freeform doubly curved surface from a single grid while maintaining the geodesics. In this paper, we propose an approach based on assembling multiple units of grid systems to obtain a variety of transformable surfaces. We first provide a geometric model of bending active scissors system along the geodesics by interpreting the structures as surfaces that change the intrinsic metric according to the scissors' transformation. Then, we clarify the compatibility conditions between multiple patches and show approaches for designing patterns that satisfy compatibility. Based on the model, we show a simulation of surface transformation using a bar-and-hinge model. In particular, we propose a method for inversely generating a mechanism that deploys to a given target surface.

1 Introduction

Mechanisms that combine scissors' transformation of the grid with the elastic bending of material are used in various fields. Bending active grid shells are an example of such systems and has been used in the architecture as the efficient construction and deployment method [5]. However, the application of grid shells is not restricted to static structures. For example, the tubular braided structures used in McKibben-type artificial muscles, stents for expanding blood vessels, finger traps, and grips for pulling cables can be regarded as a similar grid system on a cylindrical surface and serve dynamic functions. Inspired by these transformable grid shell structures, we are interested in realizing an architectural scale mechanism that exploits the dynamic behavior of transforming surfaces. In this research, we aim to model and simulate the transformation of such bending-active grid shells and freely design the transformation of such mechanisms.

Transformable bending active grids are classified according to the bending directions of the members (Fig. 2b). (1) The most general case is when we allow the member to bend in two directions. In this case, the grid can reproduce part of the given free surface by parameterizing it with an isometric mesh called a Chebyshev net [3,1]. However, the obtained structure is not kinematically determinate, and it is difficult to control the motion of the deployment. Also, they require additional support to make the structure stable. (2) Asymptotic grid shells [10] use members that cannot bend in the normal di-

https://doi.org/10.1515/9783111162683-017

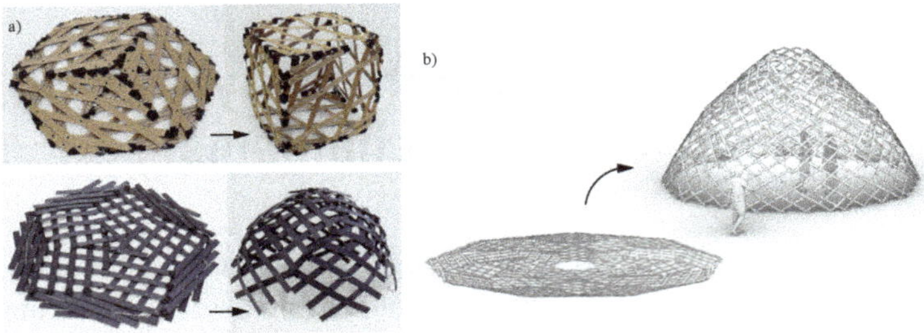

Fig. 1: Mechanisms using transformation of geodesic grids. a) Physical model. b) Architectural scale mechanism.

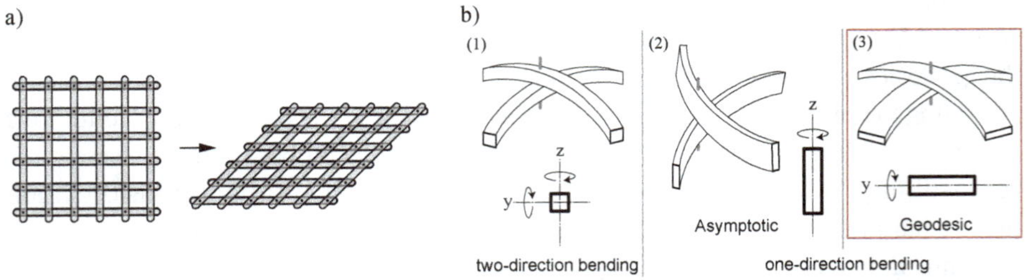

Fig. 2: a) Scissors' transformation of a grid. b) Classification of transformable bending active grids.

rection of the curved surface. Asymptotic grid shells still have a high degree of freedom in transformation and require the control of diagonal braces to make the form into the target shape. Also, the shape is limited to surfaces with negative Gaussian curvature. (3) Geodesic grid shells [7] use members that can only bend in the normal direction of the surface. They have one degree of freedom (DOF) in-plane transformation, allowing for easy deployment control. Therefore, the in-plane shape is expected to be rigid by bracing only a part of the scissors, and the whole mechanism behaves like a shell structure. In this research, we focus on such a grid shell based on geodesic lines.

However, grid shells with geodesic lines cannot generate arbitrarily curved surfaces from a planar state because the degree of freedom of transformation is constrained. Pillwein et al. [7] proposed methods for designing surfaces by introducing the play at the pivots of the scissors by converting holes into slits. This approach would again introduce excessive degrees of freedom due to the sliding that must be fixed after the deployment. Ono and Tachi [2] used an approach by segmenting the curved surface into multiple grids that are kinematically connected without play, so they produce one-DOF in-plane transformation. They generated surfaces with constant negative curvature, but the method was limited to rotationally symmetric shapes. The limitation of the shapes comes from the fact that the transformation is not compatible when multiple grids are combined in general. The conditions of compatibility between grids have not been explored.

This research aims to create a freeform surface as a geodesic grid shell by combining multiple grids so the entire mechanism is one-DOF. We first provide a geometric model of the mechanism based on macroscopic in-plane transformation and clarify the compatibility conditions of the mechanism (Sec. 2). Then, we propose a mechanism deployment simulation method using this geometric model (Sec. 3). We present a mechanism design approach that satisfies the compatibility conditions and applies them to more general curved surfaces. In particular, we propose a method for designing a mechanism that deploys to a given target surface (Sec. 4).

This paper focuses mainly on the system's geometric principles and programmability. However, we intend to extend our research to architectural applications and provide structural verification in the future. Potential applications of this research include building structures that can be constructed by assembling a grid-like shape in the plane and changing only the boundary shape, as well as the development of environmentally controllable adaptive facades, curved formwork, or interior equipment (Fig. 1).

2 Geometry of the Scissors' Transformation

2.1 Grid Transformation of Single Unit

Vector Notation and Scale Factor. When a square grid is transformed in-plane, the direction along the grid does not expand or contract, while the diagonal directions of the grid expand and contract at the maximum and minimum scale factors, respectively. Since the diagonal direction remains orthogonal before and after transformation, this transformation is expressed as a linear transformation with the two orthogonal diagonals as the principal directions. Therefore, the scale factor in any direction can be computed from these two scale factors. Here, we use these diagonal directions as the coordinate system.

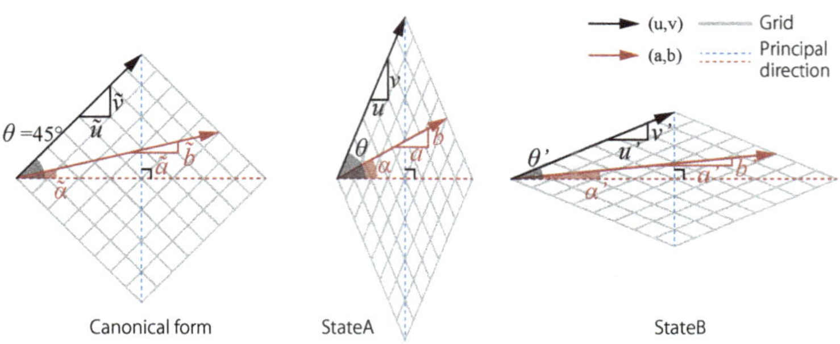

Fig. 3: Transformation of a grid using the principal directions as a coordinate system.

Let σ_1 and σ_2 be the scale factors in the principal directions from the initial state A to a transformed state B. A vector (u, v) along the grid in state A is transformed to (u', v') in state B as

$$\begin{pmatrix} u' \\ v' \end{pmatrix} = \begin{pmatrix} \sigma_1 & 0 \\ 0 & \sigma_2 \end{pmatrix} \begin{pmatrix} u \\ v \end{pmatrix}. \tag{1}$$

Since the direction along the grid does not expand or contract,

$$\sigma_1^2 u^2 + \sigma_2^2 v^2 = u^2 + v^2. \tag{2}$$

Next, let (a, b) be a vector in an arbitrary direction in state A, and the *direction angle* $\alpha = \operatorname{atan2}(b, a)$ be the angle between the principal direction and (a, b). Then, the transformed vector (a', b') can be described as

$$\begin{pmatrix} a' \\ b' \end{pmatrix} = \begin{pmatrix} \sigma_1 & 0 \\ 0 & \sigma_2 \end{pmatrix} \begin{pmatrix} a \\ b \end{pmatrix}. \tag{3}$$

Then, the scale factor S in (a, b) direction before and after transformation is given by

$$S^2 = \frac{\sigma_1^2 a^2 + \sigma_2^2 b^2}{a^2 + b^2}. \tag{4}$$

Notation Using Canonical Form. Because (u, v) is a length-preserving vector, we express it as $(u, v) = (\cos\theta, \sin\theta)$ using the *grid angle* θ. To handle a direction (a, b) independent of the grid angle, we define the *canonical form* of the direction angle $\tilde{\alpha}$ and the direction $(\tilde{a}, \tilde{b}) = (\cos\tilde{\alpha}, \sin\tilde{\alpha})$ to be the direction angle and the direction when $\theta = 45°$.

When the grid angle changes from 45° to θ, the principal scale factor is $(\sigma_1, \sigma_2) = (\sqrt{2}\cos\theta, \sqrt{2}\sin\theta)$. From (3), the direction (a, b) can be represented using the canonical direction $(\tilde{a}, \tilde{b})$ and the grid angle $\tilde{\alpha}$ as

$$\begin{pmatrix} a \\ b \end{pmatrix} = \begin{pmatrix} \sqrt{2}\cos\theta & 0 \\ 0 & \sqrt{2}\cos\theta \end{pmatrix} \begin{pmatrix} \cos\tilde{\alpha} \\ \sin\tilde{\alpha} \end{pmatrix} = \sqrt{2} \begin{pmatrix} \cos\theta\cos\tilde{\alpha} \\ \sin\theta\sin\tilde{\alpha} \end{pmatrix}. \tag{5}$$

Therefore, the direction angle α is

$$\tan\alpha = \frac{\sin\theta\sin\tilde{\alpha}}{\cos\theta\cos\tilde{\alpha}} = \tan\theta\tan\tilde{\alpha}. \tag{6}$$

Now we consider the transformation from state A with grid angle θ to B with grid angle θ'. Then, the direction angle at state B is

$$\tan\alpha' = \tan\theta'\tan\tilde{\alpha}. \tag{7}$$

The scale factor S in the direction angle $\tilde{\alpha}$ when transforming from state A to state B can be expressed as follows:

$$S^2 = \frac{1 + \cos 2\tilde{\alpha}\cos 2\theta'}{1 + \cos 2\tilde{\alpha}\cos 2\theta}. \tag{8}$$

2.2 Compatibility Between Multiple Modules

We call a polygon obtained by cutting a single grid into polygons a *unit*. The edges of a unit expand or contract as the grid angle changes. When two or more units are serially connected by shared edges, they form a one-DOF mechanism because the scale factor of the shared edge need to match, except for the case when the scale factor of one of the edges is 1. If a state for one unit is determined, all states for the units in the series are determined.

However, if we connect the units with interior vertices, the overall structure becomes over-constrained because it forms a closed loop in this chain. Therefore, the motion is generically incompatible, while the whole mechanism is compatible and transformable under special conditions. In the following, we discuss the compatibility condition around such a loop for making the mechanisms work together to form a one-DOF system. When they work together, the sum of α increases or decreases around the vertex; accordingly, a flat pattern transforms into a three-dimensional surface.

Consider the case where n units are connected around a vertex (Fig.4), and let α_i, β_i be the angles between the edge and the principal direction in the i-th unit ($i = 0, 1, 2, \ldots, n-1$, modulo n).

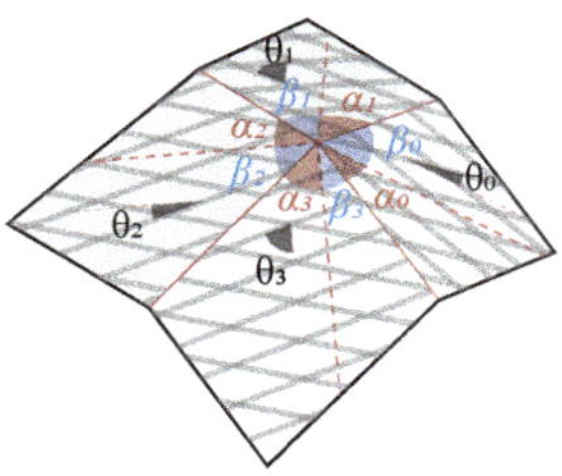

Fig. 4: Unit connection around an interior vertex.

Connection of Two Units. If the parameters of the two units (θ_i, θ_{i+1} and $\tilde{\alpha}_{i+1}, \tilde{\beta}_i$) are given, the i-th transformed grid angle θ'_i determines the $i+1$-th grid angle θ'_{i+1} as

$$x_{i+1} = r_i x_i + s_i, \tag{9}$$

where

$$x_i = \cos 2\theta'_i,\; r_i = \frac{\cos 2\tilde{\beta}_i \left(1 + \cos 2\tilde{\alpha}_{i+1} \cos 2\theta_{i+1}\right)}{\cos 2\tilde{\alpha}_{i+1} \left(1 + \cos 2\tilde{\beta}_i \cos 2\theta_i\right)},\; s_i = \frac{\cos 2\tilde{\alpha}_{i+1} \cos 2\theta_{i+1} - \cos 2\tilde{\beta}_i \cos 2\theta_i}{\cos 2\tilde{\alpha}_{i+1} \left(1 + \cos 2\tilde{\beta}_i \cos 2\theta_i\right)}. \tag{10}$$

Compatibility Around Vertex. When further connecting units, the i-th grid angle θ'_i is obtained from θ'_0 using (9) as

$$x_i = A_i x_0 + B_i, \quad \text{where} \quad A_i = \prod_{k=0}^{i-1} r_k, \quad B_i = A_i \sum_{k=0}^{i-1} \frac{1}{A_{k+1}} s_k. \tag{11}$$

For n units to be compatible around interior vertices, x_0 and x_n must be identical. Therefore, the compatibility condition around vertices is represented using grid angles at the initial state θ_i, and the canonical direction angles $\widetilde{\alpha}_i$ and $\widetilde{\beta}_i$ as

$$A_n = 1, \quad B_n = 0. \tag{12}$$

2.3 Special Mechanisms

It is not straightforward to find and design a solution that satisfies the identity equation (12) at each internal vertex. For multiple units to produce a synchronized motion, the scale factor function of the shared edges must be identified during the motion. Therefore, as a special case that satisfies the condition, we introduce mechanisms with *rhombic units* and *single units*.

Mechanism Using Rhombic Unit. A *rhombic unit* is a unit with a rhombic shape such that the principal direction of the grid and the diagonal of the unit coincide (Fig. 5). In this unit, the scale factors of the four edges are always equal and thus (11) holds around every vertex. The mechanism connecting rhombic units form a one-DOF mechanism (in its in-plane motion) which can be controlled by the common edge scale factor S. The exception is when the unit edge and the grid direction coincide (when $\theta = \alpha$) when the scale factor of the edge is 1, so the adjacent units cannot synchronize.

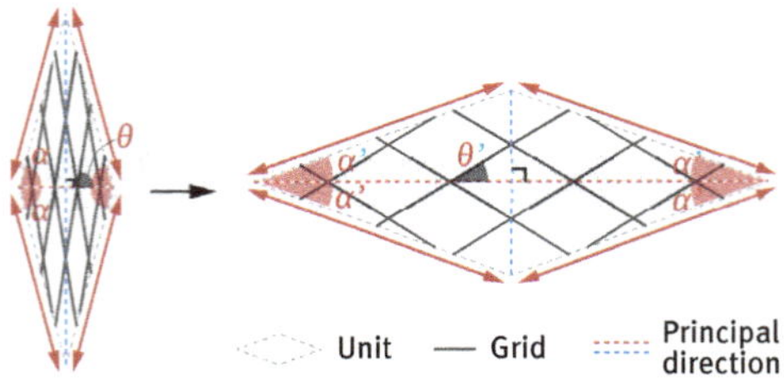

Fig. 5: Transformation of a rhombic unit. All four edges of the unit have an equal scale factor.

Mechanism with Single Unit. An alternative method is to construct a mechanism by repeating the same unit, so adjacent units share the edge of the same scale factor function. Since all units share the same direction angle of the edges during the motion, the grid angle is also the same. Therefore, it forms a one-DOF mechanism that can be controlled by the shared grid angle θ as a parameter.

Two universal ways to connect two units sharing the same edge are (1) connecting with 180° rotation around the midpoint of the edge and (2) mirror reflecting at the edge (Fig. 6). Since the grid has 180° rotational symmetry, units connected by 180° rotation can be regarded as a merged unit in a macroscopic viewpoint. This does not produce

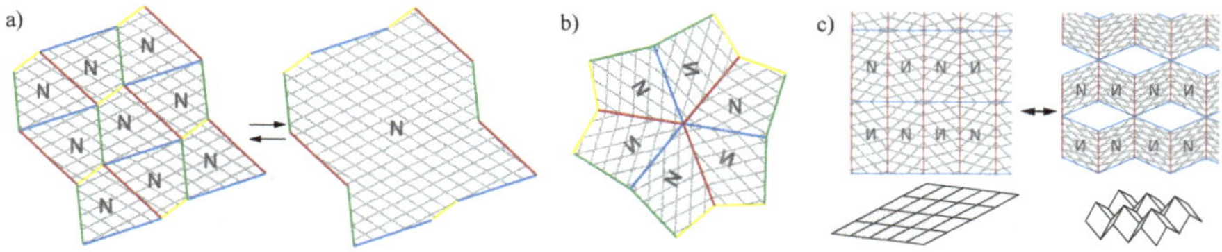

Fig. 6: Tiling of a single unit. Edges of the same color have the same scale factor throughout the transformation. a) Tiling by rotating 180° at the edge midpoint. b) Tiling by mirror reflection. c) A design example that can transform from a sheet to an eggbox surface.

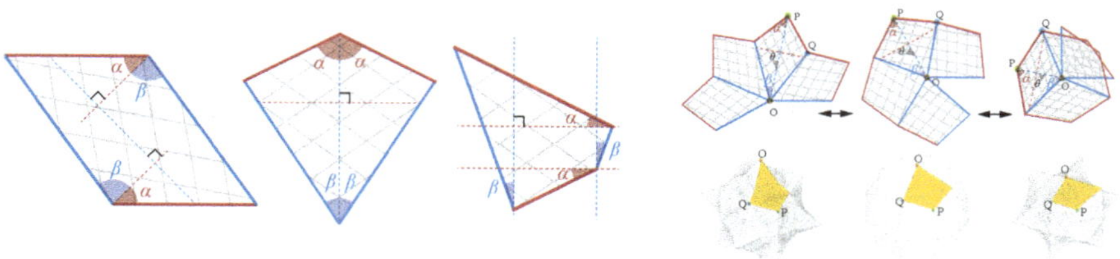

Fig. 7: Examples of units with symmetry. Parallel edges or edges with equal angles from the principal direction have the same scale factor.

angle defects or excess around interior vertices (if starting from a flat state, it keeps 360°) even if the unit is deformed (Fig. 6a). On the other hand, connecting by mirror reflection can generate a mechanism in which the angle around the vertex changes according to the scissors' transformation (Fig. 6b). Figure 6c shows a design example that can continuously change from a sheet to an eggbox surface [9].

Although mirror reflections are limited to the tiling of even-valency vertices, we can extend the single-unit tiling by introducing the symmetry of the tile. For example, using the property that parallel edges or edges with equal angles $\tilde{\alpha}$ from the principal direction have the same scale factor, various types of tiling are possible (Fig. 7). Figure 7 right shows an example using a mirror-symmetric kite shape that can produce polyhedral deformation from a deltoidal icositetrahedron (that requires degree-3 vertex). In this case, the mechanism does not have a planar state; it transforms from a polyhedron to another polyhedra.

3 Simulation

We propose a simulation method for each case (rhombic or single unit system) using the macroscopic model proposed in Sec. 2. This allowed us to confirm the transformation behavior of the entire mechanism before deciding on the specific component layout. We used the Grasshopper [8] and Kangaroo2 [6] for the implementation on 3D-CAD software Rhinoceros.

3.1 Simulation Method

For both unit cases, we first triangulate the unit and the total surface to make a triangular mesh. Then, we construct a dynamic relaxation model by defining several springs at the edges of triangles and minimizing the potential energy of the system as follows. Let $S_{\tilde{\alpha}}$ be the scale factor in the $\tilde{\alpha}$ direction. Here [] represents the goals name of Kangaroo2.

Linear Spring Make the length of the edges the original length times $S_{\tilde{\alpha}}$. [Length (line)]
Angle Spring 1 Keep the fold angle in the unit 0 (represents the out-of-plane stiffness). [Hinge]
Angle Spring 2 Keep the fold angle between units 0 (represents the out-of-plane stiffness). [Hinge]

Here, we use Angle Spring 2 only when adjacent units are joined to transmit the bending moment as in Fig. 16 left.

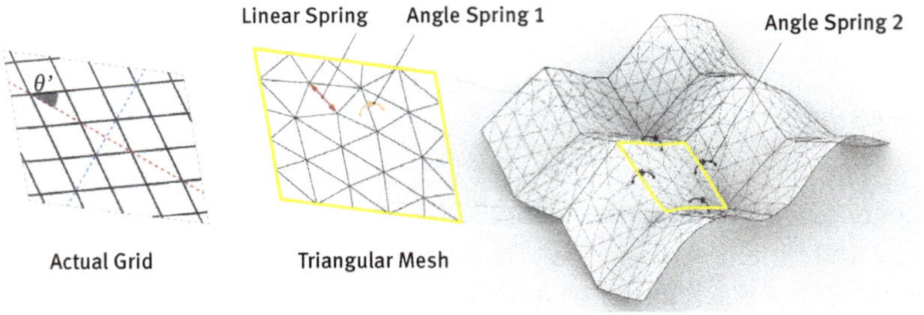

Fig. 8: Simulation model combining linear springs and two types of angular springs.

When simulating mechanisms with rhombic units, we use the scale factor S of all the unit edges as the parameter to control the transformation. When simulating mechanisms with a single unit, we use the common grid angle θ' as the parameter to control the transformation. The scale factor $S_{\tilde{\alpha}}$ for the length constraint of the mesh edges was obtained from θ', θ and $\tilde{\alpha}$ using equation (8) .

3.2 Effect of Triangulation

We observed that the method of dividing the triangular mesh may affect the accuracy of the simulation. Depending on the division, wrinkles may occur at the ridges when the units are deformed out-of-plane. For more accurate reproduction of transformation behavior, it is desirable to use a division pattern that can be easily bent in any direction, i. e., one with high angular resolution. For this reason, pattern E in Fig. 9 was adopted for models with large out-of-plane transformations.

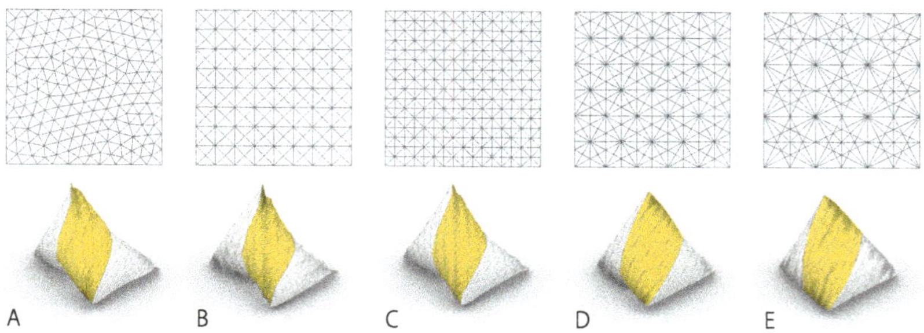

Fig. 9: Comparison of simulation results for different mesh patterns.

3.3 Simulation Examples

Cylinder. Cylindrical mechanisms can be created by connecting the opposite sides of a parallelogram-shaped unit. The transformation is coupled transformation of a twist, squeeze (diameter change), and elongation. If the principal direction and the unit edge are parallel, changing the grid angle θ will not cause twisting because the unit edges remain orthogonal to each other, the behavior observed in braided cords and finger traps. If the grid and unit edge directions are coincident, the diameter of the cylinder is unchanged, and only twisting and elongation occur.

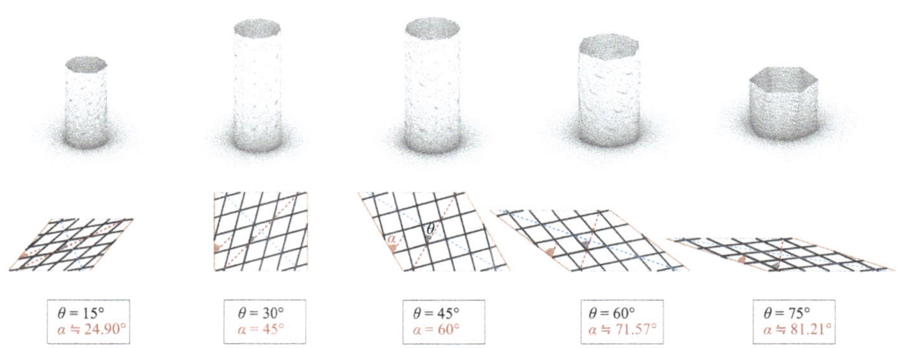

Fig. 10: Simulation result and unfolding of the cylinder model that can twist, squeeze, and elongate.

Tetrahedron. We created a model that transforms from a flat dihedral (double-sided rectangle) to a tetramonohedron (tetrahedron with equal faces). This model was made by connecting the four edges of two units that transform from a square to a parallelogram (Fig. 11). This mechanism has 90° rotational symmetry in the flat state and can be twisted in left and right directions to form tetrahedra in both cases.

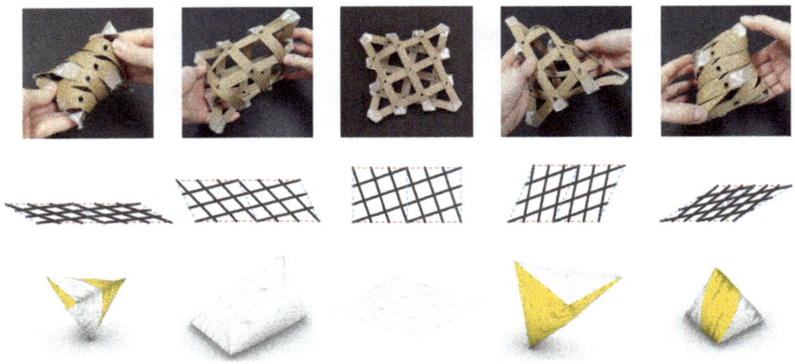

Fig. 11: Tetrahedron design using a single unit. Top: physical model, Middle: grid patterns, and Bottom: simulation results.

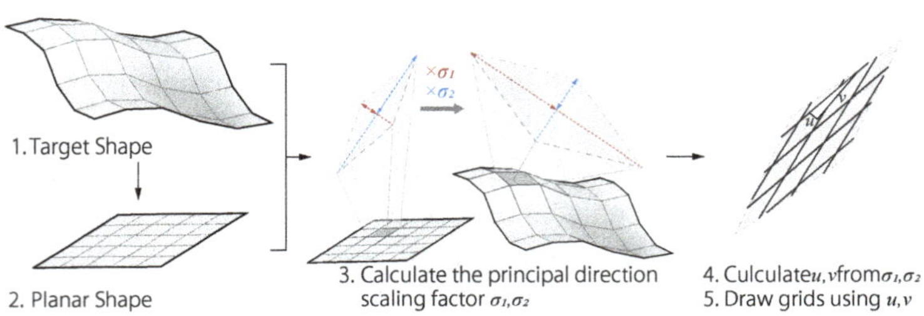

Fig. 12: Design flow of mechanisms that transform to target surfaces using rhombic units.

4 Designing Mechanisms Deploying to a Target Surface

Section 2.3 describes a method to generate compatible mechanisms using the symmetry of units. However, the shape after deployment cannot be predicted and can only be confirmed by simulation. In this section, we propose a method to design a mechanism that can be deployed to a freeform target shape by inversely computing from the target shape. To make interior vertices compatible, the design uses the rhombus unit described in Sec. 2.3. The design flow is as follows (Fig. 12).

1. Define the input target shape consisting of a quadrilateral with equal edge lengths (this may contain non-planar quadrangles).
2. Create a rhombic flat tile with the same connectivity to the target shape.
3. Determine the principal scale factors σ_1, σ_2 from the diagonal lengths of the units of the planar and target shapes.
4. Computing grid direction (u, v) in a planar state from σ_1, σ_2.
5. Draw the actual grid from (u, v) (Sec. 5.1).

4.1 Details of the Steps

Defining Target Shape. First, we define the final target shape. The target shape is given as a surface composed of quadrilaterals of equal edge length. The quadrilateral does not need to be planar, since out-of-plane bending of the final unit is allowed. Each quadrilateral represents a rhombic unit. Such an equal-length quadrangle mesh is characterized by the Chebyshev net on a surface. Existing construction methods of constructing an equal-length grid based on Chebyshev net [3,4] can be used for generating input shapes in our method.

Generating Planar Tile. We generate planar rhombus tiles as unit shapes in a planar state. First, we prepare a planar tile with the same connectivity to the target shape.

When we can find a planar tile with the same connectivity, we can use them as the initial mapping with proper scaling. If tiles with the same connectivity are not trivial, they can be generated by projecting the target shape onto a plane and constraining the length of the edges. After the projection, we need to adjust the edge lengths and proportions of the tiles, so the principal directional scale factor σ_1, σ_2 of each unit satisfies the following equation:

$$\sigma_1 < 1 < \sigma_2 \quad \text{or} \quad \sigma_2 < 1 < \sigma_1. \tag{13}$$

This condition indicates that one of the two principal directions contracts and the other extends. If $S = 1$, the conditional equation (13) is trivially satisfied but the mechanisms do not link together, so this needs to be avoided. Therefore, we need to find tiles where S is reasonably away from 1 but satisfies (13).

Because of this constraint, it is not always possible to construct a mechanism for an arbitrary shape using this method. In a general case, we use constraint-based form finding by solving (1) lengths fixed to the scaled length of the edges by $(1/S)$ and (2) diagonal inequalities through the dynamic relaxation method using springs that represent the constraints.

Computing Grid Direction. The grid inside the unit is calculated from the planar shape and the target shape. First, from the ratio of the diagonal lengths of the units before and after transformation, the principal direction expansion ratios σ_1, σ_2 are obtained. Next, using the property that the vector (u, v) in the grid direction does not expand or contract before and after transformation, (u, v) in the plane can be obtained from sigma by the formula

$$u = \sqrt{\frac{\sigma_2^2 - 1}{\sigma_2^2 - \sigma_1^2}}, \quad v = \sqrt{\frac{1 - \sigma_1^2}{\sigma_2^2 - \sigma_1^2}} \tag{14}$$

The necessary and sufficient condition for this expression is (13).

4.2 Design Example

Examples of mechanisms designed using the above methods are shown in Fig. 13. The target shape of the example A (Fig. 13) uses a catenary surface generated by applying an upward load to a three-fold symmetric rhombic mesh and constraining the length of the edges. Planar tiles were generated by scaling the rhombic mesh used as the basis for generating the target shape. The scale factor S is $S \simeq 1.21$ in this model. Examples B and C are a surface near a translational surface and a surface near a polar zonohedron, respectively. Figure 13 shows the grid and the deployment behavior of the mechanism obtained by the above design method. As a result of the deployment simulation, we were able to reproduce a shape close to the given target shape.

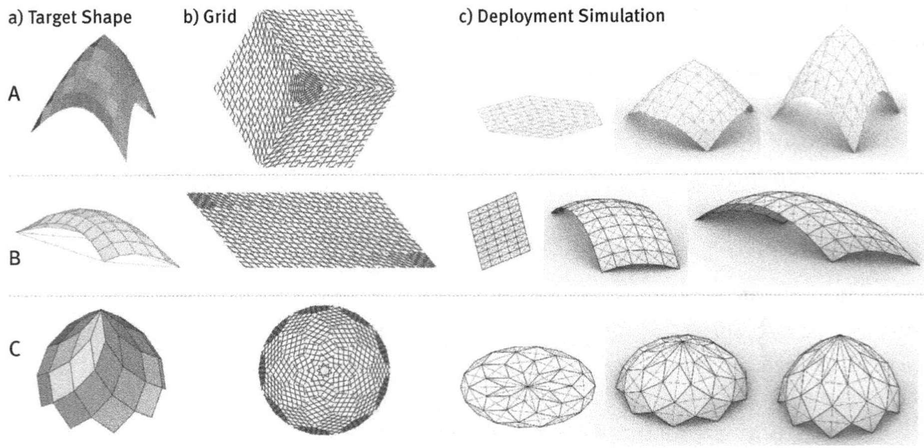

Fig. 13: Design examples using rhombic tiles. A: catenary surface, B: a translational surface, C: polar zonohedron. a) target mesh, b) grid obtained by the proposed method, c) result of the deployment simulation.

5 Joint Design and Fabrication

5.1 Connection of Grid Elements

After the grid angles are determined for units, the actual locations of the discrete grid need to be determined in order to fabricate the mechanisms with actual material. There are two necessary conditions for grid components: a) intersecting within the unit and b) connecting with adjacent units at the ends (Fig. 14).

Intersection within the unit. Depending on the grid layout, the grid elements may not intersect within the unit (Fig. 14a), when the transformation cannot be propa-

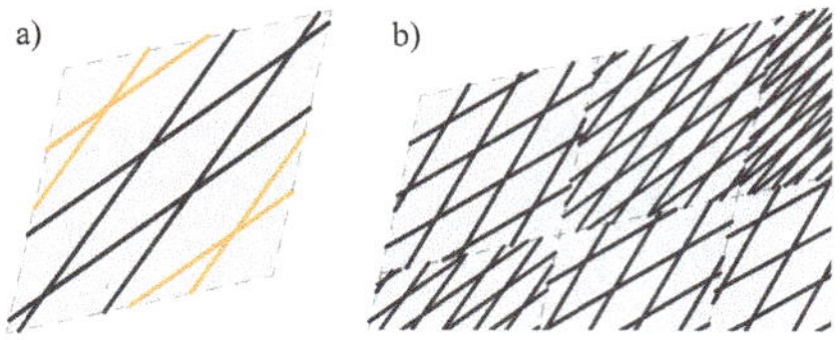

Fig. 14: Problems with grid connections. a) Grid elements not intersecting. b) Endpoints not connected.

gated within the unit. This problem can be solved by increasing the grid density or by changing the unit parameters.

Connection of endpoints. In general, if the grid is defined for each rhombic unit, the members of adjacent units are not connected to each other (Fig. 14b). One solution is to use the common locations of endpoints of the members at all shared edges. Specifically, we can locate the endpoints symmetric around the midpoints of each edge (Fig. 15).

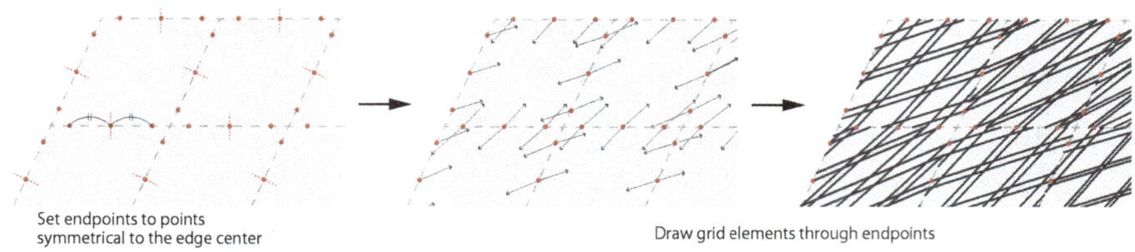

Fig. 15: Adjustment of grid member end point positions.

In the case of a single unit with mirror inversion, the members can be connected regardless of the grid arrangement. However, when the units are connected by parallel translation, the connection of the components must be adjusted.

5.2 Fabrication

The members within a unit are connected by pivots to form a scissors-like mechanism. Between the units, pivot hinges are used when out-of-plane bending was to be transmitted and pin joints are used when the units were to be folded (Fig. 16). We get a bending-active smooth surface when using pivot hinges and creased surfaces when using pin joints. For the model in Fig. 16, the first step is to determine the length, number, and hole locations of the members. Next, the members were cut out, holes were created at the hinge points using a single-hole punch, the members were joined, and finally, the ends of the members were connected. To manufacture the mechanism, 15 mm wide paper bands (craft bands) were used as scissors' members, tag pins or 5 mm diameter eyelets were used as pivot hinges, and the pin joints were realized using tapes.

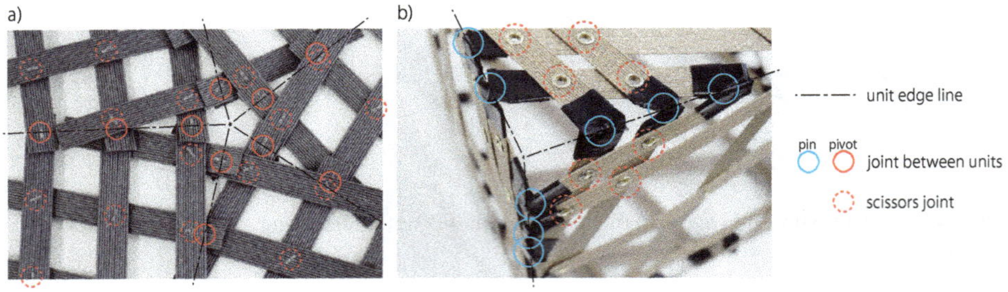

Fig. 16: a) Model with pivots between the units. The joints transfer out-of-plane bending moment, resulting in a smooth curved surface. b) Model with pin joints between units. The joints do not transfer bending moment, resulting in a polyhedral shape with folded hinges between the units.

We showed the conditions for mechanisms to be compatible around a single vertex. However, we have not yet fully solved the problem of how to obtain a general solution with multiple interior vertices, which is left as future work. In addition, the design method shown in Sec. 4 is limited to some families of surfaces, but its characterization is still open. Finally, detailed structural studies under loading conditions are a future requirement for this method to be applied to architecture.

Acknowledgment. This work was supported by JSPS Grants-in-Aid for Scientific Research 22H04954 and JST AdCORP JPMJKB2302.

References

[1] Baek, C., A. O. Sageman-Furnas, M. K. Jawed, and P. M. Reis. 2018. Form finding in elastic gridshells. *Proceedings of the National Academy of Sciences* 115(1), 75–80.

[2] Ono, F., and T. Tachi 2022. Growth deformation of surface with constant negative curvature by bending-active scissors structure. *Proceedings of IASS 2022.*

[3] Garg, A., A. O. Sageman-Furnas, B. Deng, Y. Yue, E. Grinspun, M. Pauly, and M. Wardetzky. 2014. Wire mesh design. *ACM Trans. Graph.* 33(4), 66:1–12. DOI: 10.1145/2601097.2601106

[4] Montagne, N., C. Douthe, X. Tellier, C. Fivet, and O. Bavarel. 2020. Voss surfaces: A design space for geodesic gridshells. In: *Proceedings of IASS Annual Symposium 2020.*

[5] Otto, F., E. Schauer, J. Hennicke, and T. Hasegawa. 1974. IL 10 Gitterschalen–Gridshells. Institute for Lightweight Materials, Stuttgart, Germany.

[6] Piker, D.: K2Goals. https://github.com/Dan-Piker/K2Goals

[7] Pillwein, S., K. Leimer, M. Birsak, and P. Musialski. 2020. On elastic geodesic grids and their planar to spatial deployment. *ACM Trans. Graph.* 39, 125. DOI: 10.1145/3386569.3392490

[8] Robert McNeel & Associates: Grasshopper 3d. https://www.grasshopper3d.com/

[9] Schenk, M., and S. D. Guest. 2010. Origami folding: a structural engineering approach. In *Fifth International Meeting of Origami Science, Mathematics, and Education.*

[10] Schling, E., D. Hitrec, and R. Barthel. 2018. Designing grid structures using asymptotic curve networks. In *Humanizing Digital Reality,* pp. 125–40. Springer

Christoph Schlopschnat, Marta Gil Pérez, Christoph Zechmeister, Rebeca Duque Estrada, Fabian Kannenberg, Katja Rinderspacher, Jan Knippers, Achim Menges

Co-Design of Fibrous Walls for Multi-Story Buildings

Abstract: Coreless filament winding (CFW) is an advancement of industrial filament winding for architectural applications. In this process, the formwork is reduced to an absolute minimum, allowing the fibers to span freely in space between anchor points. Using carbon and glass fibers with a resin matrix, it exhibits high potential for lightweight, material efficient building elements. While previous research demonstrated its applicability in shell, roof and long-span structures, the potential in using this method for multi-story wall and slab systems has not been thoroughly investigated. This paper elaborates on methods to develop structural wall components built entirely of carbon and glass fiber composite, which are specifically tailored to meet the requirements of multi-story construction in architecture. A computational design method based on tangent-based approximation was developed to generate bespoke fiber patterns and openings, allowing the wall components to act as load-bearing elements. This facilitates the generation of a multitude of pattern variations which can be adapted to axisymmetric and asymmetric boundary conditions. Structural performance of the building elements is evaluated throughout the design process by means of finite element analysis establishing a feedback loop between design, robotic fabrication and structural evaluation and informing the optimisation of the wall geometry and fiber layup. The developed methods were successfully applied in the design and fabrication of a multistory fiber installation exhibited at the 17th Architectural Biennale in Venice. It demonstrates the potential of coreless wound load-adapted fibrous walls as architectural building components leveraging integrative computational design, structural engineering and robotic prefabrication.

1 Introduction

1.1 Coreless filament winding in architectural design

Coreless filament wound structures (CFW) are lightweight fiber composite elements characterized by their low intrinsic weight and material- fficient fabrication method (Menges and Knippers 2015). Using industrial robots and an external rotation axis, fibers are robotically placed on a winding frame at structurally relevant locations during fabrication (Gil Pérez et al. 2021). This can be achieved by winding the fibers around anchor points on the frame, which secure their final position in the structure (La Magna

https://doi.org/10.1515/9783111162683-018

et al. 2016), as well as by digitally planning the placement order of the structurally informed fibers along these anchor points and developing a matrix that ensures as many fiber iterations as possible. The fiber layup is designed based on the definition of the sequence of connection points which is referred to as fiber syntax. In essence, the alignment of the fibers to the direction of the loads as well as their interaction between each other makes the fibers structurally efficient. The component geometry also plays a crucial role in the performance and material efficiency of the structure; they are directly linked to the finished form of the physical component, which in turn means that deviations and tolerances from the digital model must be kept as low as possible (Reichert et al. 2014). Computational design simulations and form-finding methods make it possible to develop genuinely digital building systems (Knippers et al. 2021) and to digitally simulate and predict the final shape, the fiber interactions, and the corresponding physical influences of the winding process (Menges et al. 2022).

Fig. 1: Overview of the full installation of Maison Fibre © ICD/ITKE/IntCDC, University of Stuttgart

The coreless filament winding method builds on more than a decade of research in fiber construction by ICD and ITKE at University of Stuttgart and its application was successfully demonstrated through multiple research projects. The variation of different numbers of anchor points along non-identical frames enabled the fabrication of non-circular, non-uniform fiber bodies in the ICD/ITKE Research Pavilion 2013-14 with different numbers of anchor points (Dörstelmann et al. 2015). The Elytra Filament Pavilion (Koslowski et al. 2017) enabled the production of fiber components with circular apertures, which was achieved with identical frames and identical number of anchor points. The BUGA Fibre Pavilion incorporated a method for the syntax development of an interlaced fiber pattern with adjustable shift values, which allowed for the variation of the component's geometric properties (Dambrosio et al. 2019, Zechmeister et al. 2020), which can be applied to different densities of fiber layups in the fiber composite dome structure (Gil Pérez et al. 2020). Previous research outcome serves as the base for the development of visually permeable yet structural fibrous wall elements as shown in Fig. 1. The geometry of the wall component is a result of reciprocal effects between architecture, structure and fabrication. With coreless wound structures their final shape is determined by the material's behavior during fabrication. Through the sequential process of CFW, fiber interaction determines the material limits that form three-dimensional geometries and ensures structural performance. The more three-dimensional the structure, higher the chances to achieve good fiber interaction.

1.2 Co-Design of Maison Fibre

Co-design of fibrous architecture can be understood as an interdependence between the CFW fabrication process and the resulting design and performance characteristics, and are the result of a feedback-based development of design and engineering methods, fabrication processes and the resulting material and building system (Menges et al., 2022). As part of the continued research in the field of robotically fabricated CFW structures, this modular building system not only performs as a cantilevered canopy or skeletal dome, but extends the application of CWF towards multi-story architectural structures (Dambrosio et al. 2021).

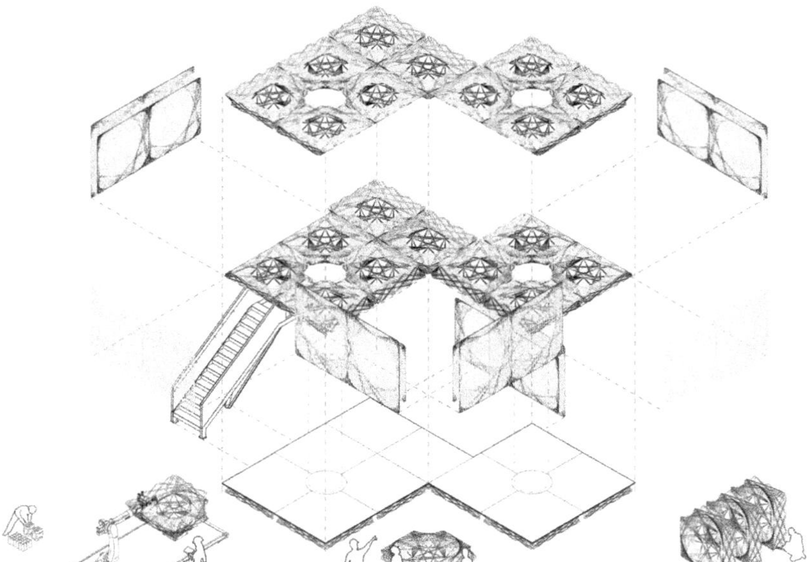

Fig. 2: Modular Building System of the installation © ICD/ITKE/IntCDC, University of Stuttgart.

Here a two-story FRP timber hybrid building was developed for the 17th Venice Architecture Exhibition, which consists of 30 modular arranged components, as shown in Fig. 2, that correspond to the conventional 2.5×2.5 m grid of multi-story structures (Dambrosio et al. 2021). The 20 slab and 10 wall components use a reconfigurable frame to accommodate all component variations and their fiber layup corresponds to specific requirements based on individual tributary loads (Gil Pérez at al. 2022a). This paper focuses on novel computational design exploration strategies of fibrous wall elements. These components are structurally relevant parts of the installation and integrate both building opening, vertical bracing and fall protection within identical boundary conditions that originate from one specific frame configuration.

2 Background

The research paper focuses on the design development of the fiber wall components. Novel criteria to the design of fiber components had to be examined: Instead of using bespoke boundary configurations, as in the BUGA Fibre Pavilion (Bodea et al. 2020) or identical frames for the Elytra Filament Pavilion (Dörstelmann et al. 2017), here a set of individual boundary frames had to be used for the articulation of individual geometries within this boundary condition. The aperture, which is a prominent detail of the component, usually originates from a regular shift of anchor points in the syntax, where two continuous frames with identical numbers of anchor points lead to a uniform-shaped aperture that acts as a fiber body.

For Maison Fibre, both continuous frames and the number of anchor points are not identical. This led to an investigation of scaffolding strategies for uniform apertures in a non-uniform boundary system. A key challenge for the design development of the wall component was the smaller, square-shaped frame, coming from a rigid global configuration, as it allows for the production of both walls and slabs. This configuration in combination with a larger, non-uniform secondary frame, had to be geometrically transformed so that it allows for a high degree of fiber interactions in the context of CFW. The wall components consist of three main elements:

1. The glass fiber screen, which is located on the outside of the component and acts as enclosure and as fall protection
2. The inner fiber body, consisting of a glass fiber scaffold, and a carbon fiber reinforcement layer in distinct locations to provide structural strength and stiffness
3. The outer fiber body which adds to the structural performance of the component and incorporates both glass fiber body and carbon fiber reinforcement layers.

For the structural articulation of the wall components, a distinction was made between ground floor and upper floor components, which is reflected in the amount of material used. Two wall components are always placed next to each other, and the layup is mirrored vertically. The structural requirements of the wall component relate to the design of the carbon fiber layup and the distribution of the connections between slab and wall, and wall to wall. The layup needs to respond to the load transfer between connection points informed by structural FEM. From an architectural design standpoint, it was intended to create elliptical shapes for the apertures, allowing for window-frame-like configurations in the wall components.

In the BUGA Pavilion 2019 (Zechmeister et al. 2020) or the Elytra Filament Pavilion (Dörstelmann et al. 2017), a shifted syntax was considered for the development of a scaffolding layer that is used to add reinforcement in the fiber components. Typically, the reinforcement layup needs to be continuous to transfer loads between components. The more this syntax shift differs from the starting point of the initial list, the smaller the aperture inside the component becomes. This usually leads to a uniform shape of the aperture with a uniform distance to the edge in a component with identical frames. The challenge in the wall components is to establish a strategy for different shapes and numbers of anchor points of the two main frame pieces to create a homogenous aperture shape with a high degree of fiber interactions. Typically, this would rather lead to an irregular aperture. Moreover, if the number of anchor points on one side of the frame are different than on the other, this can cause less fiber interactions in the component that is based on the interlaced syntax method, leading to a structurally undesirable result. In this project, a slab component was tested and used for the calibration of the other component types (Gil Pérez et al. 2022a).

3 Methods

The two winding frames used for the wall components are not identical: Since the floor and ceiling components are connected to the wall components at an angle of 45 degrees, one of the two winding frames had to be smaller. As a result, there is a larger frame with 106 anchor points of 10 cm distance and a smaller frame with only 94 anchor points, sharing the same distance, which is used for a multi-layered configuration of a wall component (Fig. 3).

The structural design of the building system combines multi-level FE modeling with a digital-physical approach that has been used in recent CFW structures (Gil Pérez et al. 2021; Gil Pérez et al. 2022b). This allows to evaluate the structural performance and global load distribution with a full FE model, anddesign the specific fiber layup for the carbon reinforcement layers in a component level (Fig. 4). To complement this approach, calibrate the structural simulations, and prove the integrity of the structure, full-scale prototyping and mechanical testing is performed for a single component type. The co-design of the wall components (Fig. 4) underwent two feedback loops: Firstly, the glass fiber body was investigated with a syntax strategy to obtain the desired geometry. This was then evaluated with a surface FEM giving geometrical feedback

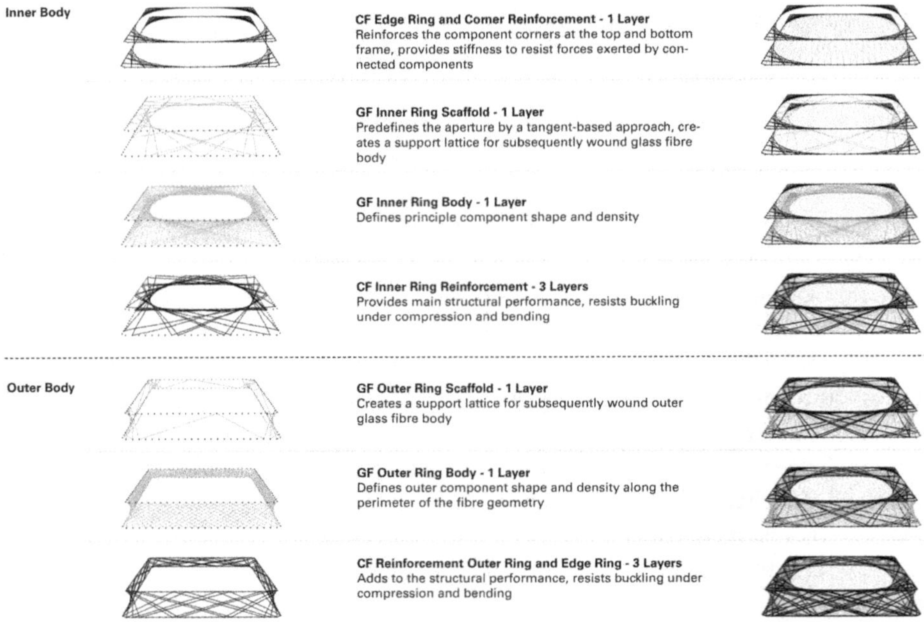

Fig. 3: Layer configuration of the fiber wall bodies.

to the design. Then the carbon fiber reinforcement was designed through a second iterative loop where a beam FEM representing the carbon fibers was used to investigate the fiber layup yielding best structural performance.

3.1 Inner and Outer Glass Fiber Bodies

Due to the required stiffening of the walls within the overall composition of the structure, as well as the specific load bearing capacity that the walls need to fulfill, a uniform fiber body had to be introduced to transfer the loads evenly to the adjacent components. As a consequence, the aperture also had to be uniform, even though the number of anchor points and frame size were different. Here, a new method for the generation of the fiber body syntax was developed, shown in Fig. 5.

For the wall components, a planar contour was defined in the central field of the frames, which is to describe the glass fiber aperture. In the next step, the subdivision of this contour was defined based on the specific distances to the frame edge, which will become an approximation for the aperture of the fiber body. Subsequently, the tangent at the subdivision points along the contour curve can be generated, which provides information about which anchor point can be reached on the corresponding frame at the position of the contour. Using this method, the resulting aperture can be developed using tangents from the central contour, displayed in Fig. 5.

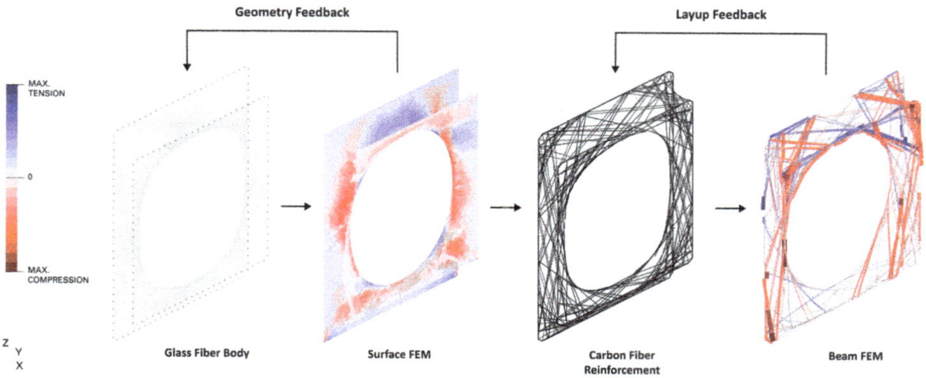

Fig. 4: Co-Design of fibrous walls based on FEM feedback loop.

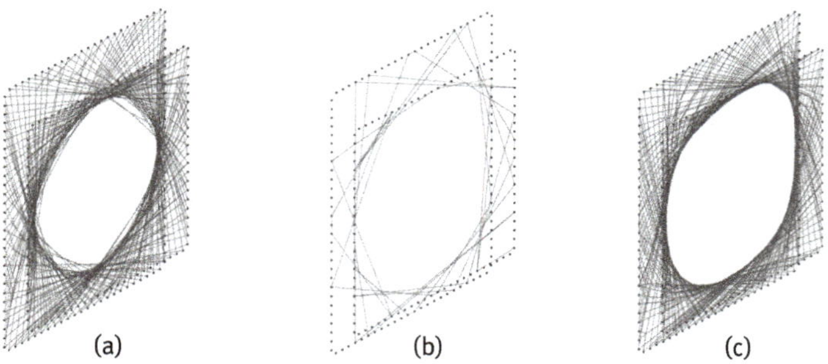

Fig. 5: Inner Ring Body Aperture: Difference between an interlaced shift syntax for asymmetric frames with and without a tangent-based scaffold layer; visualized from fiber net simulations. (a) Inner ring body syntax without a scaffold layer results in a non-uniform aperture with loose fibers (b) Tangent-based scaffold layer (c) Uniform inner ring body syntax based on identical interlaced shifts from (a) but with a precedent tangent-based scaffold layer (b).

As the frames are different and have a different number of anchor points, the aperture does not necessarily correspond to an axisymmetric geometry when all anchor points are used for the fiber layup. To avoid this, the so-called scaffold layer was introduced to the multilayer structure of the fiber body, which creates a syntax with as few tangent fibers as possible to constrain the shape of the subsequent fiber body (Fig. 6). This approach is an approximation of the final shape: since not every tangent along the inner contour curve meets an anchor point along the planarized frame, small deviations must be accepted. The maximum deviation of the component corresponds to the anchor point distance.

Additionally, the strategy to create a syntax from tangent points is to connect consecutive anchor points from the opposite frame. In between these tangent lines, connection lines that don't intersect the planar contour had to be defined. This allows for multiple syntax options. In this case, one of the shortest paths was selected, which was congruent to the fiber layup of the inner ring body, allowing for an almost invisible

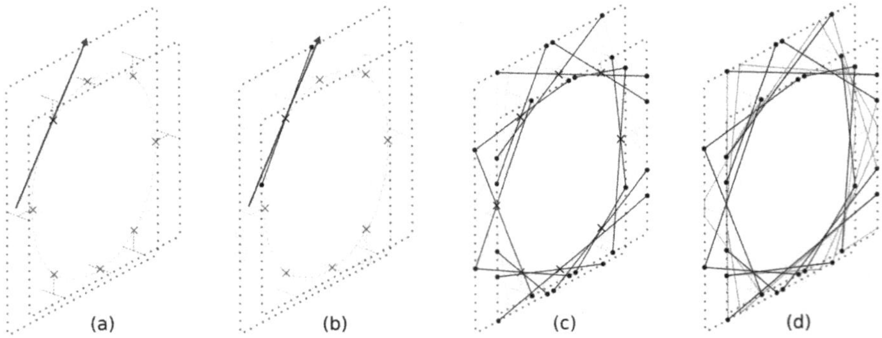

Fig. 6: Inner Ring Scaffold: Basic principles of the tangent-based approximation method for the inner ring glass fiber scaffold layer. (a) Tangent at proposed fiber interaction point, marked as X. (b) Closest anchor points of the tangent, resulting in an adjusted tangent line. (c) Adjusted tangent lines from all proposed glass fiber interaction points. (d) Adjusted tangents implemented in a continuous syntax.

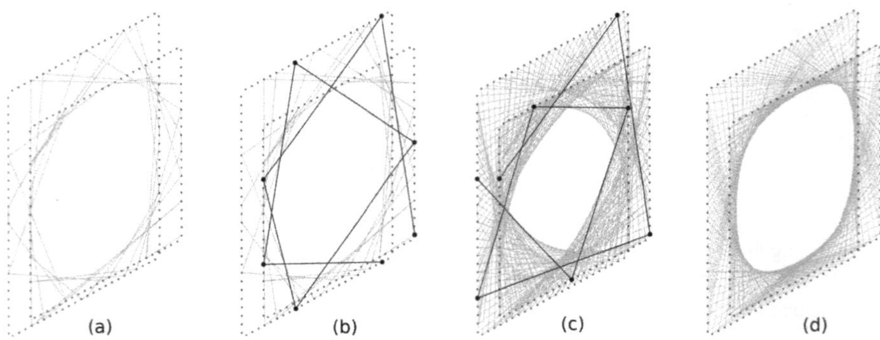

Fig. 7: Inner Ring Body: Interlaced shift syntax of the glass fiber inner ring body (a) Precursor scaffold layer, projected fiber interactions (b) Syntax for the initial winding steps (c) Syntax of the last winding steps (d) Full glass fiber layup.

scaffold layer in the final component. The design of the outline is limited to concave, circular and ellipsoidal basic shapes. A convex curve based on this design principle is not viable for the physical component, as the material behavior of the fibres and their layup in the coreless winding process do not result in positive, but only negative curvatures. However, it is conceivable that polygonised outlines can be realized instead of circle-like shapes. This would require local adjustments along the aperture in the syntax of the scaffold layer. The subsequent fiber body layer (Fig. 7) makes use of a shift value, resulting in an aperture with a smaller radius than the one in the preceding scaffold layer. The combination of a tangent-based scaffold layer and an interlaced syntax (Zechmeister et al. 2020) for the fiber body is a suitable combination for creating the largest possible axisymmetric aperture. If only the tangent-based methods were

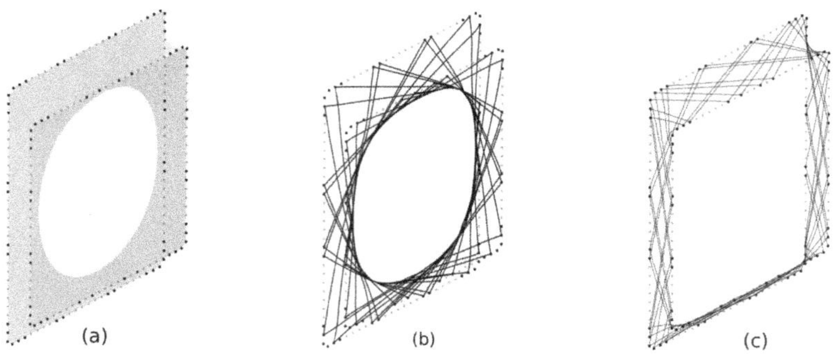

Fig. 8: Location of all connections for carbon fiber reinforcement (a) in the FE shell modell, (b) in the inner body carbon fiber reinforcement layer, (c) in the outer body reinforcement layer.

used for the scaffold layer and the fiber body, there would be a risk that some anchor points are omitted, resulting in visual gaps and irregularities in the fiber body.

The wall components were first investigated geometrically with an FE shell model (Fig. 4) to determine the efficiency and buckling behavior of the resulting inner and outer fiber bodies. Then, the connection locations were established based on the arrangement of adjacent components and the required number of anchors to be connected as analyzed in the global FE model. An additional outer fiber body was introduced, which consists of a glass fiber scaffold ensuring proper fiber interaction along the corners on the subsequently wound glass fiber body. The body uses an interlaced syntax with a low shift.

3.2 Carbon Fiber Reinforcement

Carbon fibers are considered for the load-bearing structural layer and are designed following structural guidance to carry and transfer the loads, while glass fibers are used in a more homogeneous way to provide the required geometry and fiber interaction for the component.

Both slab and wall components needed to provide a double-layered structural reinforcement to transfer the loads and create the depth of the components, based on the results of the FE beam model approach (Fig. 4). The reinforcement for the inner fiber body was structurally required and it connects structurally relevant anchor points with an interlaced shift, indicated in Fig. 9.

To complete the double-layered load-bearing structure of the wall component, an additional carbon fiber reinforcement layer was introduced to the outer body. It uses the same connections as the reinforcement layer of the inner fiber body reinforcement but spans shorter distances. As a last step, the edge reinforcement stiffens the edges of the component ensuring the load induction at the connection points.

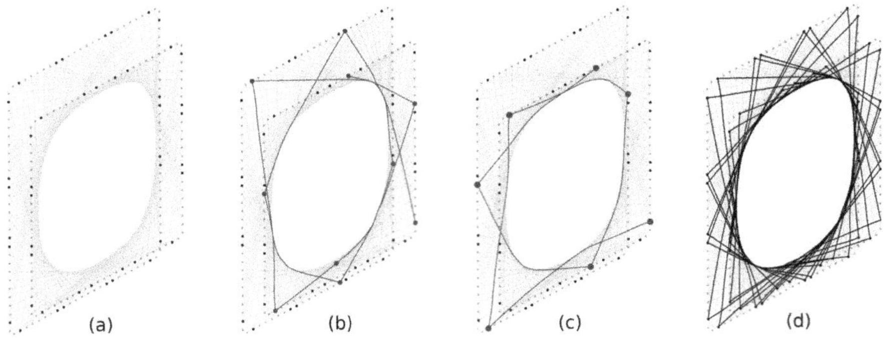

Fig. 9: Precursor layer of the fiber body, projected fiber interactions, indication of structural connections (b) Syntax for the initial winding steps, projected fiber interactions (c) Syntax of the last winding steps, projected fiber interactions (d) Full carbon fiber reinforcement layup.

The final material amount per carbon fiber bundle in this model was calibrated based on the full-scale structural testing performed for the slab component by comparing the internal forces in the fiber layup. In this way, the wall components of the first floor were designed with less fiber layers than in the ground floor. The corner reinforcements were also adjusted based on the FE model. As a result, the wall components on the first floor only weighed 47.6 kg, 22 % less than those on the ground floor (Fig. 10).

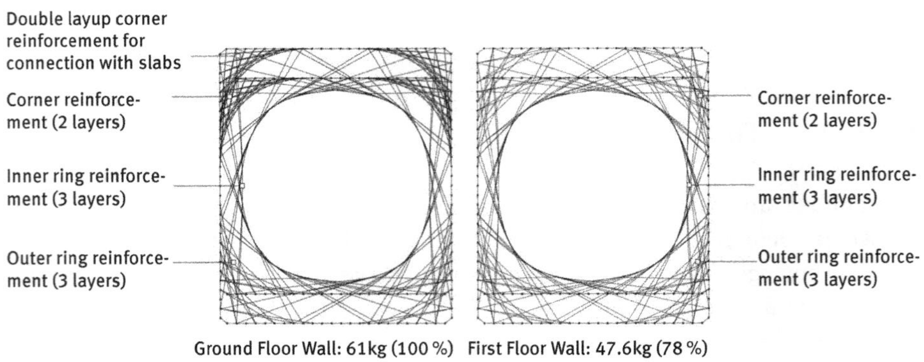

Fig. 10: Optimized carbon fiber material amount in ground and first floor wall components.

4 Results and Reflection

The co-designed modular wall system is based on a continuous knowledge transfer between architectural design, structural engineering and fabrication. The fiber layup optimization results from the structural feedback of digital and physical full-scale testing. The tangent-based approximation method facilitates the creation of a uniform aperture through approximating the circular opening. In addition to the rectangular frame geometries shown here, there's potential to realize freeform frame shapes for aperture-based geometries developed from tangent-based approximation, but this has to be further investigated. The resulting syntax should be as short as possible. Otherwise, fibers in the physical component start to bundle, causing deflections. Another aspect of the examination is the precise integration of the component in the installation. The rigidity of the frame plays an important part to ensure the correct position of the anchor points of each component. Since no frame deflection was observed, the components could be properly installed, shown in Fig. 11.

Fig. 11: Installed wall elements in the upper floor (left) and in the ground floor (right). © ICD/ITKE/IntCDC, University of Stuttgart.

The lightweight material system shown here offers geometric opportunities to reduce the amount of material used for a prototypical building and could potentially lead to a smaller ecological footprint. The LCA of the coreless winding process is currently under investigation, but early research of this and similar material systems has shown its potential in construction (Mindermann et al. 2022). Considering the speed of production – the fabrication, curing and installation of a CFW component can theoretically be achieved within 1–2 days.

5 Conclusion

This article describes the successful implementation of novel design methods of fibrous wall components. The presented methods result in the fabrication of the first coreless filament-wound wall structure built entirely of fiber-reinforced composites as part of a multi-story building. The tangent-based method enables the generation of a syntax that forms a uniform circular opening, which is challenging to accomplish using existing syntax design methods. In addition to the uniform opening, the approach takes into account the positioning of reinforcement layers based on structural feedback and the position of component connections. The presented research has the potential to be utilized in future applications in different component configurations. Further steps can include the benchmarking of fabricated components to their digital models in order to assess the accuracy of the digital design approach by 3D scanning the fabricated structure.

Acknowledgements

The research has been partially supported by the Deutsche Forschungsgemeinschaft (DFG, German Research Foundation) under Germany's Excellence Strategy – EXC 2120/1 – 390831618. The research has also been supported by University of Stuttgart, Cluster of Excellence IntCDC, Ministry of Science and Research and the Arts, Baden-Württemberg. The fiber net simulation was developed by Valentin Noah Hartmann at IPVS-MLR, University of Stuttgart.

References

Bodea, S., Dambrosio, N., Zechmeister, C., Gil Pérez, M., Koslowski, V., Rongen, B., Dörstelmann, M., Kyjanek, O., Knippers, J., Menges, A. 2020. BUGA Fibre Pavilion: Towards Robotically-Fabricated Composite Building Structures. *FABRICATE 2020 – Making Resilient Architecture.* UCL Press. https://www.jstor.org/stable/j.ctv13xpsvw.35

Dambrosio, N., Zechmeister, C., Bodea, S., Koslowski, V., Gil-Pérez, M., Rongen, B., Knippers, J., Menges, A. 2019. Towards an architectural application of novel fiber composite building systems – The BUGA fiber pavilion. In *ACADIA 2019 – Ubiquity and Autonomy,* Proceedings of the ACADIA Conference 2019.

Dambrosio, N., Zechmeister, C., Duque Estrada, R., Kannenberg, F., Gil Pérez, M., Schlopschnat, C., Rinderspacher, K., Knippers, J., and Menges, A. 2021. Design and development of an FRP-timber hybrid building system for multi-story applications in architecture: Maison fibre. In *ACADIA 2021-Realignments: Toward Critical Computation,* Proceedings of the ACADIA Conference 2021.

Dörstelmann, M., Knippers, J., Menges, A., Parascho, S., Prado, M., Schwinn, T. 2015. ICD/ITKE Research Pavilion 2013-14 – Modular CFW Based on Beetle Elytra. *Architectural Design,* vol. 85, no. 5, Wiley, London, pp. 54–59. DOI: 10.1002/ad.1954

Dörstelmann, M., Prado, M., Solly, J., Menges, A., Knippers, J. 2017. Elytra Filament Pavilion: Robotic Filament Winding for Structural Composite Building Systems. *FABRICATE 2017 – Rethinking Design and Construction.* DOI: 10.14324/111

Gil Pérez, M., Früh, N., La Magna, R., and Knippers, J. 2022a. Integrative structural design of a timber-fibre hybrid building system fabricated through CFW: Maison Fibre. *Journal of Building Engineering,* 49. DOI: 10.1016/j.jobe.2022.104114

Gil Pérez, M., Guo, Y., and Knippers, J. 2022b. Integrative material and structural design methods for natural fibres filament-wound composite structures: The LivMatS pavilion. *Material & Design,* 217. DOI: 10.1016/j.matdes.2022.110624

Gil Pérez, M., Rongen, B., Koslowski, V., and Knippers, J. 2021. Structural design assisted by testing for modular coreless filament-wound composites: The BUGA Fibre Pavilion. Construction and Building Materials, 301. DOI: 10.1016/j.conbuildmat.2021.124303

Gil Pérez, M., Rongen, B., Koslowski, V., and Knippers, J. 2020. Structural design, optimization and detailing of the BUGA fibre pavilion. *International Journal of Space Structures,* 35(4), 147–159. DOI: 10.1177/0956059920961778

Knippers, J., Kropp, C., Menges, A., Sawodny, O., and Weiskopf, D. 2021. Integrative computational design and construction: Rethinking architecture digitally. *Civil Engineering Design.* DOI: 10.1002/cend.202100027

Koslowski V., Solly S. and Knippers J. 2017. Structural design methods of component-based lattice composites for the Elytra Pavilion. In *Bögle, A. and Grohmann M. (eds.), Proceedings of the IASS Annual Symposium 2017 Interfaces: architecture engineering science,* September 25-28th, 2017, Hamburg, Germany.

La Magna, R., Waimer, F., and Knippers, J. 2016. Coreless Winding and Assembled Core; Novel Fabrication Approaches for FRP Based Components in Building Construction. *Construction and Building Materials* 127, 1009–16. DOI: 10.1016/j.conbuildmat.2016.01.015

Mindermann, P., Gil Pérez, M., Knippers, J., Gresser, G. T. 2022. Investigation of the Fabrication Suitability, Structural Performance, and Sustainability of Natural Fibres in Coreless Filament Winding. *Materials* 15 (9). DOI: 10.3390/ma15093260

Menges, A., Kannenberg, F., Zechmeister, C. 2022. Computational co-design of fibrous architecture. *Architectural Intelligence,* 1(6). DOI: 10.1007/s44223-022-00004-x

Menges, A. and Knippers, J. 2015. Fibrous Tectonics. *Architectural Design,* Vol. 85 No. 5, Wiley, London, pp. 40–47. DOI: 10.1002/ad.1952

Reichert S., Schwinn, T., La Magna, R., Waimer, F., Knippers, J., and Menges, A. 2014. Fibrous Structures: An integrative approach to design computation, simulation and fabrication for Lightweight, Glass and Carbon Fibre Composite Structures in Architecture based on Biomimetic Design Principles. CAD Journal, 52 (July), pp. 27–39. DOI: 10.1016/j.cad.2014.02.005

Zechmeister, C., Bodea, S., Dambrosio, N. and Menges, A. 2020. Design for Long-Span Core-Less Wound, Structural Composite Building Elements. In Gengnagel, C., Baverel, O., Burry, J., Ramsgaard Thomsen, M. and Weinzierl, S. (eds), *Impact: Design with All Senses,* Cham: Springer International Publishing, DOI: 10.1007/978-3-030-29829-6_32

Shermin Sherkat, Lior Skoury, Andreas Wortmann, Thomas Wortmann

Artificial Intelligence Automated Task Planning for Fabrication

Abstract: Sequencing tasks for robotic fabrication processes of buildings is not automated, contributing to its inefficiency and negative environmental impacts. Automated Task Planning (ATP) is an Artificial Intelligence (AI) method that sequences discrete tasks automatically to achieve a particular goal using a planner (AI planner). However, Construction and Fabrication (CF) has yet to exploit ATP techniques fully, partly because representing 3D geometry in ATP remains challenging. To address this problem, this paper (1) investigates the challenges of using an ATP technique (classical ATP) related to integrating 3D geometry in a fabrication process task planning, (2) models a fabrication domain using the Planning Domain Definition Language (PDDL) – a standard planning language, and (3) generates a plan for a limited version of a fabrication process with a generic AI planner. Thus, fabrication practitioners can get familiar with the advantages and limits of classical ATP technique in fabrication domains.

1 Introduction

Automated Task Planning (ATP) is an AI method that utilizes reasoning to organize sequences of tasks based on their anticipated outcome to achieve a particular goal [1]. This method involves (1) a model of the planning domain and problem created with a planning language, e. g., PDDL, and (2) a planner solver that takes the domain and problem model as input and searches for a task plan.

CF fields require task planning for processes involving labor allocation, equipment and material assignment, and robotic assembly [5]. However, the current methods for task planning in CF primarily rely on manual procedures or process-specific algorithms, resulting in limited efficiency and reusability [7].

Despite the need for automation, current literature on the application of ATP techniques in CF is limited. To our knowledge, only a few papers [8-13] have used ATP techniques in CF, of which only papers [9, 11-12] have leveraged planning languages and AI planners. Thus, ATP benefits in CF are yet to be explored.

ATP has different techniques, namely classical, numerical, and probabilistic planning [5], each defined with a suitable planning language and corresponding AI planners. However, planning languages primarily model abstract tasks and domains, which poses challenges when it comes to representing 3D geometry and its associated details.

https://doi.org/10.1515/9783111162683-019

Hence, the contribution of this paper is threefold:
- Highlighting challenges and potential use-cases of classical ATP techniques for defining a fabrication domain.
- Defining a fabrication planning domain and problem model using classical PDDL,
- Discussing how to model 3D geometry using classical ATP.

We conclude that classical planning techniques can generate task plans for a pick-and-place fabrication problem, enabling the stacking and assembly of individual objects while employing various end-effectors. Nevertheless, we cannot model stacking several objects on top of one because we need to include the number of objects. Additionally, we cannot stack one big object on top of several objects at the same level, as we must consider both the number and height of the objects or the ability to group objects based on their geometric characteristics.

2 Background

ATP techniques create task (action) sequences using a predefined domain and problem model. AI planner is the solver that inputs the domain and problem model, conducts reasoning, and generates a sequence of actions (plan) [2]. The domain model should be generic and reusable for similar problems. Conversely, a problem model varies for each problem and can be solved by various planners that support the modeling language [4].

PDDL is a formal knowledge representation language utilized to express planning models [3]. It is widely recognized as the standard planning language for specific planning problems, including classical planning because it enables the exploration of multiple generic AI planners, algorithms, and optimization techniques. Classical planning, an ATP technique used in discrete and deterministic problem environments such as fabrication [4, 6], will be briefly explained in the following, where we provide a concise overview of the domain and problem models using classical PDDL.

Domain

Main parts of a classical PDDL domain model are *predicates* and *actions* [4].

(1) *Predicates* describe the space and the relation between the objects in the space. For instance, the predicate "At(robot, A)" can indicate the robot's location at position A. Predicates can be true or false, and the ones that are not true are considered false [6].

(2) *Actions* change facts about the world and consist of parameters (variables), preconditions, and effects. For example, action Travel (robot, A, B) in Tab. 3 requires a robot and two locations as parameters to travel from one place to the other. Addi-

tionally, certain preconditions must be true for this action to occur, e. g., the robot being at the first location, denoted by the predicate "At(robot, A)". Then, if all the preconditions are met, the effects will be applied. In this case, the effects would be twofold: the robot is at the second position (B), so the predicate At (robot, B) will be added, and the predicate At (robot, A) will be negated.

Problem

In the problem file, we define an instance of problem using the predefined reusable domain. The main parts of a problem model are the initial and goal states. The initial state includes a conjunction of true predicates before the planning. In the previous example, the initial state describes the robot at position A: "At (robot, A)". The goal state is a conjunction of predicates describing what we want as our goal: "At (robot, B)". Hence, the planner's job is to search for a sequence of actions that would result in all the predicates in the goal state being true. The plan consists of one action in the example provided: Travel (robot, A, B) [6].

Related Works

To our knowledge, only papers [8-13] have utilized AI planning in construction or fabrication. Among these papers, [9, 12] have employed a planning modeling language similar to PDDL that works with a planner they are developing to address long-horizon planning. In the study conducted by Cheng and Hammad [11], a planning language called KQML (Knowledge Query and Manipulation Language) is used to plan the tasks of two cranes working together to lift a heavy object. Huang et al. [8] employed a flowchart for task planning that incorporates structural and geometrical data of a fabrication domain. Therefore, the potential of ATP techniques in CF remains largely unexplored.

Case Study

This paper utilizes a part of the BUGA pavilion fabrication process to test classical ATP. The BUGA Wood Pavilion, constructed on the summer island at the National Horticultural Show 2019 in Heilbronn, is a research demonstrator building. It features a digitally designed structure with a segmented hollow cassette form [17] (Fig. 1). The BUGA pavilion consists of modular parts called cassettes. The fabrication process for building cassettes involves picking objects from their initial position, placing them on a work table, and manipulating them using different end-effectors.

Fig. 1: The BUGA pavilion.

3 Methodology

This section provides an overview of various tasks and setups involved in the fabrication process. Then, it presents a model for the planning domain and formulates a corresponding problem, specifically focusing on describing geometry using classical PDDL.

The Fabrication Setup

The fabrication environment in this example is non-observable, deterministic, sequential, static, discrete, and single-agent [5, 6]. We create a task plan for one cassette, including a lower plate, a beam, and a top plate (Fig. 2).

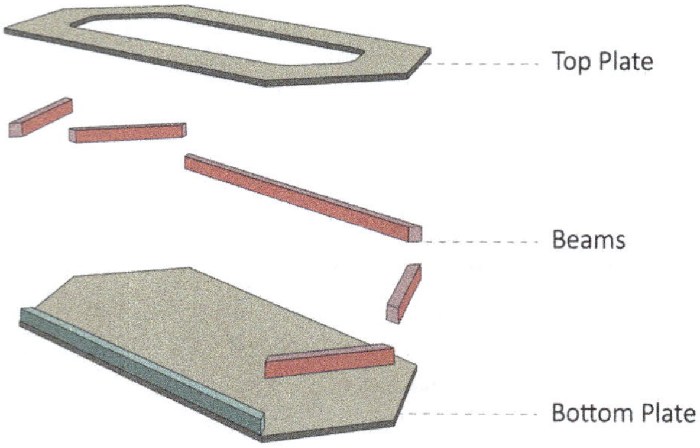

Fig. 2: The components of a cassette, including the lower plate, top plate, and beams.

This example needs three different end-effectors (tools):
- A vacuum gripper that picks the objects
- A glue gun that pours glue
- A nail gripper that puts nails into the pieces

The robot should pick up individual pieces, accurately stack them on a designated worktable, and join them using nails and glue. Once a complete cassette is assembled, the robot picks up the cassette and stores it on a separate table (Fig. 3 and 4).

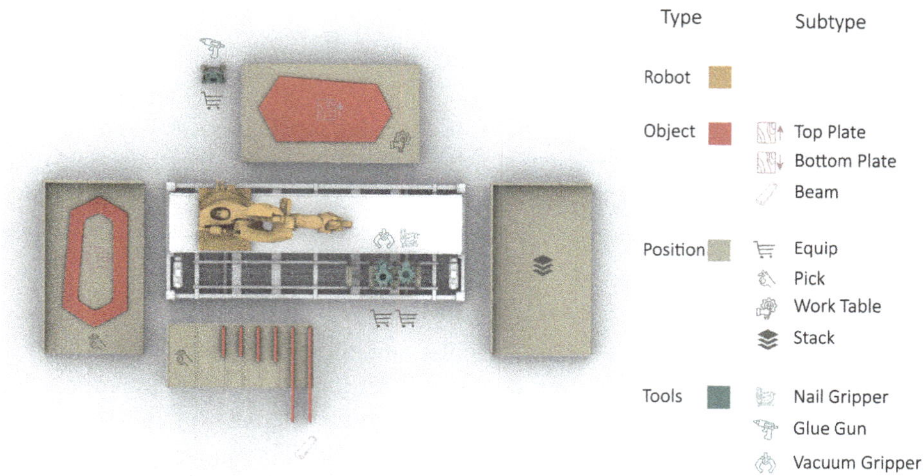

Fig. 3: The various types of objects and the general setup in the pick and place problem.

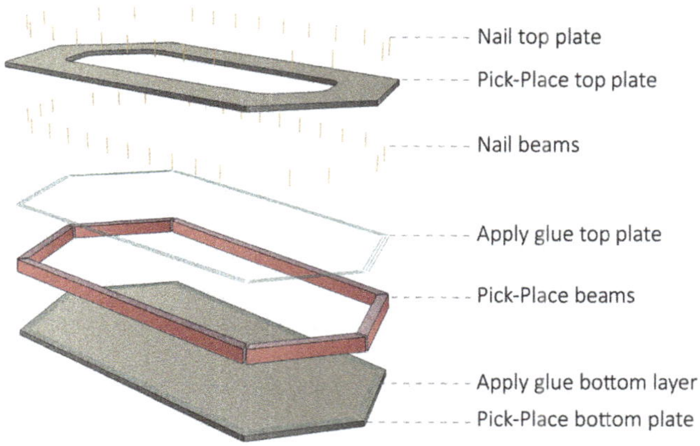

Fig. 4: A complete cassette.

Domain Model

The domain model includes types, predicates, and actions:

Types

Type feature lets us assign types and subtypes to the objects we define. Within the fabrication process domain, four main types are employed: *tools, positions, objects*, and *robots*. Additionally, subtypes are defined for these main types, as outlined below:
- Tools: Vacuum gripper, Nail Gripper, and glue gun
- Positions: EquipPosition, FirstPosition, WorkTablePosition, StackPosition
- Objects: module, LowPlate, TopPlate, Beams

Objects and *Positions* represent geometric information in this process. Notably, these positions and objects are discretized, indicating that they do not store numerical data such as dimensions. Instead, we define discrete positions with known coordinates, and the planner distinguishes between these positions based on their types and subtypes.

Predicates

Predicates are defined with a name, variables, and the type of each variable. For instance, in the predicate "At (?client -Robot ?p -positions)", "?client" is a variable of type Robot. As per the objects defined in the problem model, these variables are subsequently substituted with the corresponding objects. Table 1 presents the 15 predicates we defined.

Tab. 1: Description of predicates.

	Predicate	Description
1	(At ?Client -Robots ?pp -Positions)	a client (robot) is at a location
2	(AtPlace ?Object - Objects ?p -Positions)	an object is at a location
3	(AtTool ?tool -Tools ?ep -EquipPosition)	a tool is at a location
4	(OnTable ?Obj - Objects ?p -Positions)	an object is on a position
5	(TableFree ?t -Positions)	a location is not occupied
6	(Clear ?Obj - Objects)	no object is on top of this object
7	(OnTop ?Obj1 - Objects ?Obj2 - Objects)	an object is on top of another object
8	(Have ?Client -Robots ?Tool -Tools)	a robot is equipped with a tool
9	(Empty ?Client - Robots)	a robot does not have any tool mounted
10	(Active ?tool - Tools)	a tool is activated
11	(Holding ?Client - Robots ?Object - Object)	a robot is holding an object
12	(VG_Empty ?Client - Robots)	vacuum gripper is empty
13	(Finish ?module - Module)	a cassette is built
14	(Glued ?Obj - Objects)	an object is glued
15	(Nailed ?Obj - Objects)	an object is nailed

In Tab. 1, predicates one to four indicate the specific location of each object. These predicates tie a location to a single object. However, they do not give any information regarding other objects. For instance, if object A is not at location P, we do not know if any other object exists in that position. Hence, we need other predicates, such as the number 5, that give us more general information. The same logic goes for predicates six and seven. They are defined to check if one object is on top of another object. However, since classical task planning does not accept numbers as variables, we cannot know how many objects are stacked on top of an object and when the lower object can get the predicate "not (Clear ?obj - Objects)".

Actions are defined with parameters, preconditions, and effects. Table 2 displays the defined actions along with their corresponding parameters. The "Stack (obj1, obj2)" action necessitates the predicate "Clear (obj2)" as precondition and adds the effect "(not(clear obj2))". Hence, the planner cannot stack another object on top of obj2 since the precondition "Clear (obj2)" is negated. Additionally, Obj2 should have the predicate "(not(clear obj2))" only when all the required beams are in place. However, classical PDDL cannot model this information, and the same issue happens with the "unstack (obj1, obj2)" action.

Tab. 2: Description of actions.

	Action	Description
1	Travel (r?, p1?, p2?)	a robot travels from one location to another
2	Equip (r?, t?)	a robot takes a tool
3	DeEquip (r?, t?)	a robot unmounts a tool
4	InitializeTool (r?, t?)	a robot initializes a tool
5	CloseTool (r?, t?)	a tool is turned off
6	Pick (o?, p?, r?)	a robot picks an object (from a table)
7	Unstack (o?, o?, p?, r?)	a robot picks an object (unstack it from the top of another object)
8	Place (o?, p?, r?)	a robot puts an object on a table
9	Stack (o?, o?, p?, r?)	a robot puts an object on top of another object
10	Gluing (o?, p?, r?, gg?)	a robot glues an object with a gluing tool
11	Nailing (o?, p?, r?, ng?)	a robot nails an object with a nailing tool
12	Finish_Off_Cassette (o?, p?, r?, fm?)	Finishes off the assembly of a cassette

Problem model

A problem model consists of the objects, the initial and the goal state. We define the domain for creating one cassette. Thus, the following objects have been introduced (Tab. 4).

Tab. 3: Parameters, preconditions, and effects associated with the actions "Travel" and "Pick".

Action	Parameters	Preconditions	Effects
Travel	?Client - Robots ?from -Positions ?to -Positions	(and (At ?Client ?from) (not (= ?from ?to))	(and (not (At ?Client ?from)) (At ?Client ?to))
Pick	?Obj - Objects ?P - Positions ?Client - Robots ?VG - VacuumGripper	(and (AtPlace ?obj ?P) (Clear ?Obj) (not(TableFree ?P)) (OnTable ?Obj ?p) ...)	(and (Holding ?Client ?Obj) (not (AtPlace ?obj ?P)) (not (Clear ?Obj)) (not(OnTable ?Obj ?p)) ...)

Tab. 4: Objects and their corresponding types in the problem model.

Main Type						
Objects		**Tools**		**Positions**		**Robot**
Name	Subtype	Name	Subtype	Name	Subtype	Name
LP1	LowPlate	VG	Vaccumegripper	FP1,FP2,FP3	FirstPosition	R1
B1	Beams	NG	NailGripper	P2	WorkTablePosition	
Tp1	TopPlate	GG	GlueGun	P3	StackPosition	
M1	module			E1,E2,E3	EquipPosition	

The initial state describes the scene before planning and includes predicates such as "OnTable (Tp1 FP3)" to indicate the object TP1 (top plate) is on FP3 (a first position). The goal state describes the desired outcome and is defined by four predicates (see Fig. 4):

(and
(AtPlace m1 P3)
(AtTool VG E1)
(AtTool NG E2)
(AtTool GG E3))

The predicates in the goal state guarantee that the robot unequips all end-effectors and completes the cassette assembly before relocating it to the final position.

Creating the plan

To create the plan, we utilize a generic planner called "FF-Planner", which leverages the fast-forward algorithm to create the plan. The fast-forward algorithm is a heuristic search algorithm that combines a relaxed planning graph and a goal regression technique to efficiently estimate the number of actions required to reach the goal state [2, 16]. Additionally, we use a PDDL VScode extension to define the domain and debug the plan [15]. The planner then successfully creates a plan for this problem, as described in the next section.

4 Results

This section discusses the results of applying classical ATP to a limited version of the BUGA pavilion fabrication process (Sec. 2). A plan is generated using the FF-Planner, a generic AI planner, based on the defined domain and problem models using classical PDDL. Next, we highlight the advantages and limitations of classical ATP for this fabrication process.

The domain model includes four main types (Tab. 4), 15 predicates (Tab. 1), and 12 actions (Tab. 2). The problem instance models one single cassette that uses one beam, as explained in the Methodology section (Sec. 3).

The planner inputs the domain and the problem models, and successfully creates a plan with 72 steps in 8.313 seconds1. The first 22 steps of the output plan are illustrated in Fig. 5.

```
0.00100: (travel r1 p2 e1)             0.01200: (equipe r1 gg e3)
0.00200: (equipe r1 vg e1)             0.01300: (travel r1 e3 p2)
0.00300: (initialize r1 vg)            0.01400: (initialize r1 gg)
0.00400: (travel r1 e1 fp1)            0.01500: (gluing lp1 p2 r1 gg)
0.00500: (grab lp1 fp1 r1 vg)          0.01600: (closetool r1 gg lp1)
0.00600: (travel r1 fp1 p2)            0.01700: (travel r1 p2 e3)
0.00700: (place lp1 p2 r1 vg)          0.01800: (deequip r1 gg e3)
0.00800: (closetool r1 vg lp1)         0.01900: (travel r1 e3 e1)
0.00900: (travel r1 p2 e1)             0.02000: (equipe r1 vg e1)
0.01000: (deequip r1 vg e1)            0.02100: (initialize r1 vg)
0.01100: (travel r1 e1 e3)             0.02200: (travel r1 e1 fp2)
```

Fig. 5: The first 22 tasks for assembling a cassette with one beam.

In this example, as we are using one cassette with only one beam, there is only one valid sequence of tasks and no optimization is required.

One of the main advantages of ATP is its ability to generate plans for similar planning processes using a reusable domain model. While it takes some time to create the domain model, generating the problem and adjusting the subtypes for other similar problems can be automated because the problem file is the only part that varies.

The domain we created enables planning task sequences for picking and placing single objects on top of each other and utilizes different end-effectors to operate on these objects. It is crucial to define the domain as generally as possible and utilize the main object types for action parameters. This approach eliminates the need to modify the actions for different problems, requiring only adjustments to the subtypes. For instance, we might have other types, such as slabs instead of lower plates.

To create problem instances for different processes, we can categorize the objects based on the main types (objects, tools, and positions) and translate them to PDDL using a Grasshopper component. This component takes the fabrication geometry, assigns them the appropriate type, and generates the initial state and goal state.

However, we only test the domain using one beam because classical planning reaches its limits when stacking multiple objects on top of a larger object. For example, when stacking a beam on a lower plate, the Stack action adds the effect "(not (clear LP1))" and requires the lower plate to have the "(clear LP1)" predicate as a precondition. Therefore, it is not possible to stack the second beam on top of LP1 as the precondition "(not (clear LP1))" already exists. Additionally, when placing the top plate on top of multiple beams, classical planning cannot assign the "(not (clear o?))" precondition as an effect to all the beams underneath.

5 Discussion and Conclusion

Automating task planning in CF is necessary because manual task planning is inefficient and environmentally unfriendly [5]. Therefore, it is important to explore ATP techniques in CF. However, the challenge lies in modeling the fabrication domain, which heavily relies on 3D geometry, using a planning language to leverage ATP potentials in CF. This challenge should be clarified, as few papers have investigated ATP techniques in CF.

This paper (1) highlights classical ATP features and how to integrate 3D geometry for fabrication task planning, along with its limitations, (2) creates a classical PDDL domain for a limited version of the fabrication process of the BUGA pavilion, and (3) tests the domain using an AI planner called FF-planner.

The results show a task plan created to fabricate a single cassette that uses one beam in a static, deterministic, and observable environment. We define the domain using 12 actions, 15 predicates, and four main types. The FF-Planner then creates a plan in 72 steps in 8.313 seconds.

To model geometry using classical PDDL, it is necessary to discretize the values, which means continuous variables such as dimensions or coordinates cannot be included. Additionally, the geometric relations between objects must be represented using predicates that can have a true or false value. We conclude that classical planning using PDDL for a fabrication domain is suitable for (1) stacking single objects on top of each other (e. g., placing one beam on top of one lower plate) and (2) using different tools to assemble and operate on the pieces.

However, classical planning is unsuitable for (1) stacking multiple objects (beams) on top of a single object, as it requires keeping track of the number of top objects, and (2) stacking one large object on top of multiple objects simultaneously as it requires keeping track of the number of lower objects and the ability to group them to assign a predicate to all of them using a single action.

Discretizing geometry is not necessarily a downside, as the environment in fabrication domains is usually static and deterministic, and replanning is unnecessary. Additionally, action models must be generic to ensure reusability across different processes. In this paper, the action "Finish_Off_Cassette" is not generic, as it creates a new object when all the cassette parts are assembled. However, a function can be developed to generate and replace this action for different modules other than cassettes, based on their assembly requirements. Similarly, it is possible to create a domain with as many actions (pick, place, stack, unstack) as the number of beams and add the "Clear(o?)" predicate for the lower plate when the last beam is placed. This method can solve the issue of stacking more than one object on top of one but only works if we generate all these actions for each unique problem.

PDDL 1.2 and ADL extension should be explored in future steps as the former can include numbers, and the latter allows adding conditionals which might solve the grouping issue. Additionally, while many actions can be the same for similar fabrication processes like "nailing" and "gluing", some actions could still be generalized more. For instance, actions pick and place only recognize one vacuum gripper tool. These actions must be adjusted if we require two different types of vacuum grippers for picking different objects.

Acknowledgments

This research was funded by the Ministry of Science, Research and the Arts Baden-Wuerttemberg in the Artificial Intelligence Software Academy (AISA). This research is supported by the Deutsche Forschungsgemeinschaft (DFG, German Research Foundation) under Germany's Excellence Strategy -EXC 2120/1-390831618.

References

[1] Ghallab, M., D. S. Nau, and P. Traverso. *Automated Planning: Theory and Practice.* Amsterdam; Boston: Elsevier/Morgan Kaufmann, 2004.

[2] Geffner, H., and B. Bonet. A Concise Introduction to Models and Methods for Automated Planning. *Synthesis Lectures on Artificial Intelligence and Machine Learning* 22. San Rafael: Morgan & Claypool Publishers, 2013.

[3] Garrett, C. R., R. Chitnis, R. Holladay, B. Kim, T. Silver, L. Pack Kaelbling, and T. Lozano-Pérez. Integrated Task and Motion Planning. *Annual Review of Control, Robotics, and Autonomous Systems,* 2021.

[4] Haslum, P., ed. An Introduction to the Planning Domain Definition Language. *Synthesis Lectures on Artificial Intelligence and Machine Learning* 42. Morgan & Claypool Publishers, 2019.
[5] Sherkat, S., A. Wortmann, T. Wortmann, Potentials of Symbolic AI Planning for Construction.33rd Forum Building Informatics, Technische Universität München, 2022.
[6] Russell, S. J., P. Norvig, and E. Davis. *Artificial Intelligence: A Modern Approach.* 3rd ed. Prentice Hall Series in Artificial Intelligence. Upper Saddle River: Prentice Hall, 2010.
[7] Abioye, S. O., L. O. Oyedele, L. Akanbi, A. Ajayi, J. M. Davila Delgado, M. Bilal, O. O. Akinade, and A. Ahmed. Artificial Intelligence in the Construction Industry: A Review of Present Status, Opportunities and Future Challenges. *Journal of Building Engineering,* 2021
[8] Huang, Y., P. Y. V. Leung, C. Garrett, F. Gramazio, M. Kohler, and C. Mueller. The New Analog: A Protocol for Linking Design and Construction Intent with Algorithmic Planning for Robotic Assembly of Complex Structures. In *Symposium on Computational Fabrication,* 1–17. Virtual Event USA: ACM, 2021.
[9] Hartmann, V. N., O. S. Oguz, D. Driess, M. Toussaint, and A. Menges. Robust Task and Motion Planning for Long-Horizon Architectural Construction Planning. In *IEEE/RSJ International Conference on Intelligent Robots and Systems (IROS),* Las Vegas, USA: IEEE, 2020.
[10] Pradhan, A., and B. Akinci. Planning-Based Approach for Fusing Data from Multiple Sources for Construction Productivity Monitoring. *Journal of Computing in Civil Engineering,* 2012.
[11] Cheng Z., and A. Hammad. Agent-Based Simulation for Collaborative Cranes. In *Winter Simulation Conference,* 2051–56. Washington, DC, USA: IEEE, 2007.
[12] Hartmann, V. N., A. Orthey, D. Driess, O. S. Oguz, and M. Toussaint. Long-Horizon Multi-Robot Rearrangement Planning for Construction Assembly. *IEEE Transactions on Robotics* 39, 2023
[13] Hu, W. Automatic Construction Process of Prefabricated Buildings on Geometric Reasoning. In *Construction Research Congress* 2005, 1–10. San Diego, California, United States: American Society of Civil Engineers, 2005.
[14] Skoury, L., F. Amtsberg, X. Yang, H. J. Wagner, A. Menges, and T. Wortmann. A Framework for Managing Data in Multi-Actor Fabrication Processes. In *Towards Radical Regeneration,* ed. C. Gengnagel, O. Baverel, G. Betti, M. Popescu, M. Ramsgaard Thomsen, and J. Wurm, 2023.
[15] Dolejsi, J., D. Long, M. Fox, G. Besançon, PDDL Authoring and Validation Environment for Building end-to-end Planning Solutions, ICAPS18, 2018.
[16] Hoffmann, J. FF: The Fast-Forward Planning System. *AI Magazine* 22, 2001.
[17] Wagner, H. J., M. Alvarez, A. Groenewolt, and A. Menges. Towards Digital Automation Flexibility in Large-Scale Timber Construction: Integrative Robotic Prefabrication and Co-Design of the BUGA Wood Pavilion. *Construction Robotics* 4, 2020.

Jingwen Wang, Wenjun Liu, Gene Ting-Chun Kao, Ioanna Mitropoulou, Francesco Ranaudo, Philippe Block, Benjamin Dillenburger

Multi-Robotic Assembly of Discrete Shell Structures

Abstract: Discrete shell structures are renowned for their material efficiency and elegance. However, they might require a fair amount of falsework as temporary supports during construction, which could partially offset the gains in material efficiency. This research is an investigation of the use of multi-robotic assembly for the autonomous construction of discrete shells. We present a design-to-fabrication workflow (Fig. 1) that combines structural design, construction sequence assessment, reachability evaluation, and robotic motion planning. We demonstrate the potential of the proposed workflow through a simple and a more complex example. This work contributes to the research field of assembly-aware design and multi-robotic manufacturing to improve assembly efficiency and reduce waste.

Keywords: computational design of stable discrete assemblies, robotic path planning & motion planning, assembly-aware design, design-to-fabrication workflow

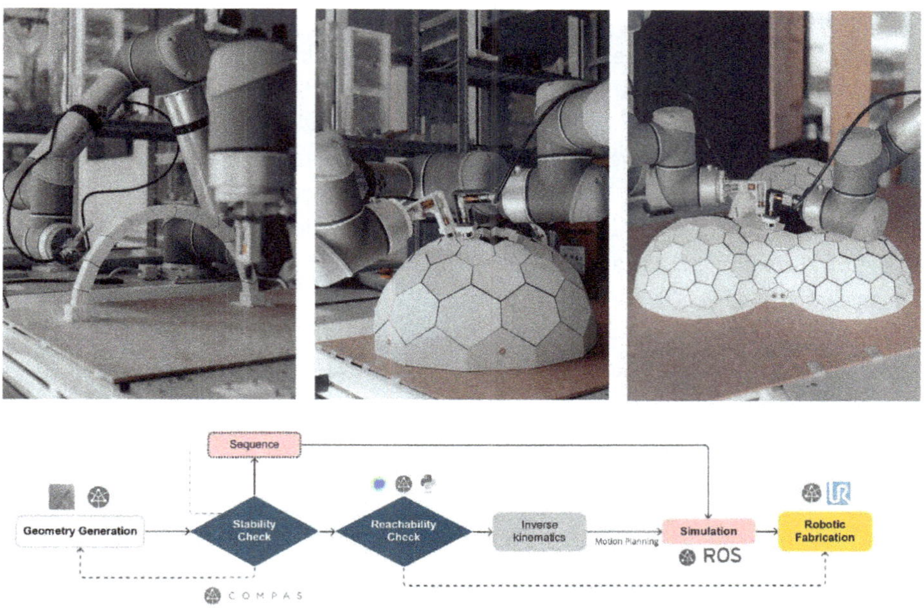

Fig. 1: A fabrication-driven workflow for robotic assembly of discrete shell structures.

https://doi.org/10.1515/9783111162683-020

1 Introduction

Shell structures are common in architecture due to their efficiency and ability to span long distances with minimal material. However, their complex geometry often requires extensive formwork, which can be labor-intensive, wasteful, and costly (Rippmann 2016). Discrete shells consist of distinct elements like blocks or bricks. Compared to continuous shells, discrete shells offer faster construction, and longer lifespan (Rippmann 2016) and can be dismantled and reassembled again. Specifically, this paper focuses on dry-jointed discrete shells, where the elements are not grouted or interlocked together but interact through unilateral (compression) contact. The Armadillo Vault serves as an example (Block et al. 2016).

Although prefabrication of discrete shell components is possible, complex falsework is often needed during assembly. Traditionally, these supports are provided as props and beams that are pre-installed on-site, registered to the correct heights and locations, and removed after decentering the shell. Techniques have been developed to reduce the need for temporary supports, such as (Ponce and Melendez 2015; Deuss et al. 2014). Such strategies, however, impose extensive geometric constraints on the final form. With the increasing use of automation in construction and advancements in computational tools for stability evaluation (see Sec. 2.2), new construction methods are now possible. Specifically, the capacity of robotic arms to move, precisely position, and hold elements in space make them particularly suitable for the fabrication of discrete assemblies.

In this paper, we outline a design-to-fabrication workflow for the autonomous assembly of dry-jointed discrete shell structures, which combines the structural design of the shell with the design of the robotic system. The remainder of this paper is structured as follows: Sec. 2.1 shows the current state-of-the-art regarding the robotic fabrication of discrete assembly; in Sec. 2.2 the main computational tools for the design of discrete shell structures are presented; in Sec. 3, the design-to-fabrication workflow is explained in detail; Design examples and results are presented in Sec. 4, and contribution and discussion are in Sec. 5.

2 State of the art

2.1 Robotic assembly of discrete structures

Fabrication systems with a single robotic arm can easily stack elements to create vertical structures such as walls and columns (Piškorec et al. 2019). However, arching and vaulted structures require more supports to ensure stability during construction. These additional supports can be provided, for example, by temporary scaffolding or by additional robotic arms. This research focuses on the latter strategy in order to remove the need for traditional falsework.

Two robotic arms assembling discrete vaulted structures were previously explored by Wu and Kilian et al. (2020) to erect a compression-only arch, and by Parascho et al. (2020) for the fabrication of a discrete vault with mortar joints. These studies demonstrate multi-robotic system's potential for scaffold-free assembly. Nevertheless, the success of the construction process is highly constrained by the capacity of the robots to access a certain space without colliding with each other, with other obstacles, or with the structure itself. Therefore, an evaluation of the reachability of the robotic system, i. e. the number of Inverse Kinematics (IK) solutions in which a robot can access a given frame in space without collisions, needs to be assessed during the design stage. Specifically, a reachability map (Makhal and Goins 2018) can be constructed to visually show the reachability of a robot for a discrete set of input frames corresponding to possible positions and orientations of the robot's end effector. However, this definition of reachability and of the associated map only provides information regarding the workspace, without specifically targeting the assembly task. This issue is further explained in Sec. 3, Step 3, where we introduce DARM, a novel way of assessing the reachability of discrete assemblies.

2.2 Parameters affecting the stability of discrete shells

Besides reachability, stability is also a crucial factor to consider during fabrication. For constructing a discrete shell assembly, two types of stability should be considered. The global stability of the assembly when finished, and its stability at each construction step. Key factors affecting the stability include:

- *Geometry of the shell:* because of the dry-jointed assumption, the geometry of the shell should follow compression-only shapes. Numerous tools are available for this task, such as RhinoVault (Rippmann 2016).
- *Thickness of the shell:* this depends on the loads acting on the structure and their distribution. Here, we assume a uniformly distributed load. We also refer to the Heyman hypotheses for masonry structures (Heyman 1995) and assume that the strength of the materials is much higher than the stresses imposed by the loads. More details regarding the determination of the correct thickness of the shell are provided in Sec. 3, Step 2.
- *Discretization:* the tessellation, the shape and inclination of the interfaces for dry-jointed discrete shells greatly affect the stability of a discrete shell, mainly during construction. However, searching for the optimal construction sequence is a challenging combinatorial problem to solve (Beyeler et al. 2015) beyond the scope of this paper. Engineering judgment can be used to select a correct assembly strategy, as shown in Kao et al. (2017).

- *Interface friction:* this parameter depends on the material chosen. Higher friction coefficients help the stability of the structure during assembly.

2.3 Stability evaluation of discrete shells

Computational tools are available to evaluate the stability of discrete shell structures both during assembly and after completion. Rigid-block Equilibrium (RBE) Whiting et al. (2009) enables fast and stable 3D discrete element equilibrium analyses, but falls short when dealing with complex interfaces and friction. Kao et al. (2022) addressed this by introducing Coupled Rigid-Block Analysis (CRA) method, which more robustly combines equilibrium and kinematics in a penalty formulation.

3 Proposed workflow

We proposed our own design-to-fabrication workflow for the autonomous assembly of dry-jointed discrete shell structures that integrates both the structural design of the shell and the robotic fabrication (Fig. 2).

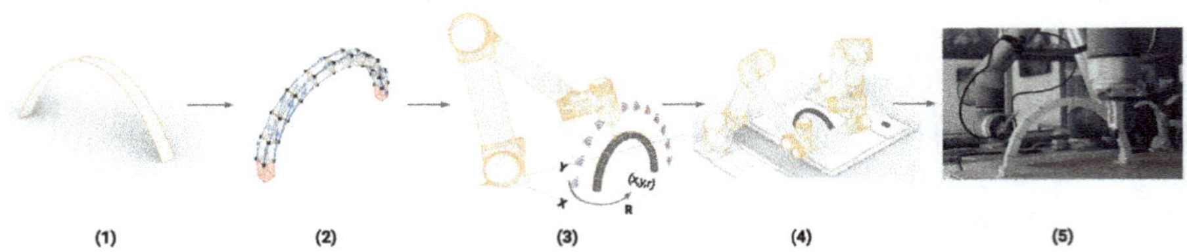

Fig. 2: Proposed workflow: (1) discrete assembly design, (2) stability evaluation, (3) robotic system design, (4) motion planning, (5) fabrication.

Step 1: Discrete assembly design
The first step of the workflow is to parametrically design the shell such that its stability can be evaluated later. The parameterization can be done with any of the tools mentioned in Sec. 2.2, and ultimately depends on the preferences of the designer.

Step 2: Stability evaluation
In this research, we propose the use of CRA and its open-source implementation (Kao 2022) to evaluate the stability.

Step 2a: Global stability. The global stability of the shell is assessed both under vertical and horizontal loads. In particular, horizontal loads are simulated by virtually tilting the structure and measuring the tilting angle: the higher the tilting angle, the more stable the structure is under horizontal actions (wind/earthquakes). The designer can determine the thickness of the shell using the information from the critical tilting angle analysis.

Step 2b: Stability during construction Once the global behavior has been analyzed, a stability analysis is conducted to determine if the assumed construction sequence is valid. For each step of the assembly sequence, CRA is used to assess if the structure is stable and where additional supports are needed. Since our system is composed of two robots, any solution that requires more than one support (the other robotic arm) or to add more than two blocks at the same time must be discarded. If a stable solution cannot be found for the given assembly sequence, the designer can change the sequence, the tessellation or the thickness of the shell. It is important to mention that, for the scope of this research, the assembly sequence is based on engineering intuition and only evaluated a posteriori (see Sec. 2).

Step 3: Robotic system design

Once the structure has been designed and the assembly sequence is validated, the robotic system can be designed, i. e. it is possible to determine the relative positions of the robots and structure. To facilitate the path planning process, we present a variation of the reachability map called *discrete assembly reachability map* (DARM). Unlike the standard reachability map (Section 2), DARM evaluates the reachability of each block of the assembly and provides combined scores that evaluate the feasibility of the entire assembly. It is worth mentioning that the results of DARM depend on the end-effector design and the type of robotic arms used.

Given an assembly in position $P_j(x, y, r)$, with x, y the location on a 2D cartesian plane and r the orientation of the assembly, consisting of n blocks , there are $K_{i,j}$ potential approaching frames for each block i and position j, where i means the index of the block and j means the index of the position. With approaching frames (Fig. 3) we refer to the potential location and orientation of the robot end-effector when placing the block i to its assembled location. The following two scores can be computed to evaluate the reachability for a given position P_j:

- The *ConfigScore* (C_j) is the number of IK solution types that can reach at least one of the approaching frames for each block of the assembly. For a 6-axis robotic arm, the number of IK solution types is 8. C_j is a measure of the capability of the robot to complete the structure for each type of IK solution. If the robot can use the same configuration to construct the structure, the fabrication process will be smooth and faster.
- The *FrameScore* (F_j) is the sum of robot configurations that can reach all the approaching frames (independently from the type).

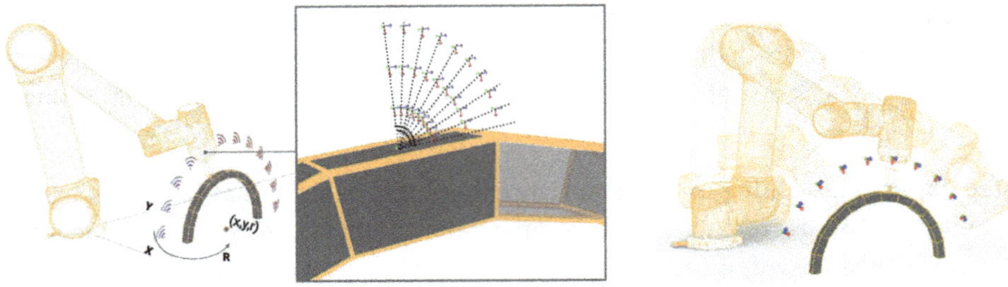

Fig. 3: Left: Position and orientation of the structure $P_j(x, y, r)$. Middle: Potential approaching frames of a block. Right: Smooth fabrication if the robot can finish the assembly with one type of configuration.

Both C_j and F_j are a measure of the redundancy of the system. As we are still not considering the collisions of the robot with the structure and with the other robot, higher scores provide a higher chance that a solution will be found later in the more accurate motion planning process. The DARM algorithm consists of the following steps:

1. For each robot:
   (a) Uniformly sample the workspace in j positions P_j of the assembly.
   (b) Calculate the $K_{i,j}$ potential approaching frames for each block for each sample position.
   (c) Find all valid robot configurations for all approaching frames. In this step, only the robot self-collisions and the collisions with the floor are considered.
   (d) Calculate the ConfigScore (C_j) and FrameScore (F_j) for each position P_j.
   (e) Select the best position as the one with the highest C_j. If there is more than one selected position, then pick the one that has the highest F_j.

   In d), it is possible to only calculate the F_j for the positions with the highest C_j for saving computational power.
2. Fix the relative location of the two robots by moving one of the two such that the best positions (location and rotation) of the structure for both coincide.
3. Use the IK solution types found for each robot (1c) as starting configurations for the motion planning algorithm (further described in Step 4a).

To better illustrate the algorithm, we can consider a simple arch assembly (Fig. 4) to be assembled using one 6-axis UR5 robot with 8 IK solution types. The assembly consists of $n = 10$ movable blocks and two supports (12 elements in total). For each block, we generate $K = 51$ approaching frames. At a sample location $P_0(x_0, y_0, r_0)$, for one of the two robots we have:

- $C_0 = 2$, as the robot can reach at least one approaching frame of each block of $P(x_0, y_0, r_0)$ with two different IK solution types. This was computed using compas_fab analytical solver (Rust et al. 2018) with Pybullet for collision checking (Coumans and Bai 2023).

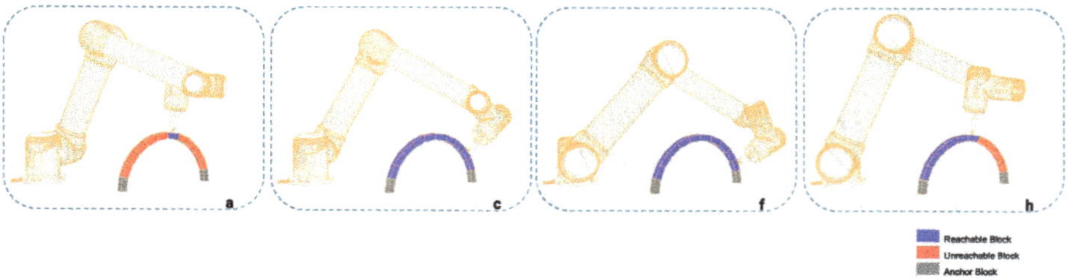

Fig. 4: Reachability of the robot at different IK solution types. Here only IK Types (c) and (f) can reach at least one approaching frame of each block. As a result, the ConfigScore is 2.

– $F_0 = 1377$. This was calculated by removing from the maximum FrameScore $F_0^{\max} = 51 \cdot 10 \cdot 8 = 4080$, the unfeasible solutions due to positions out of reach, self-collision and collisions with the floor.

By calculating the scores for different assembly positions, a score map can be built (Fig. 5) to visualize the preferable location (highest F_j among the ones with highest C_j), highlighted with a red circle in the figure.

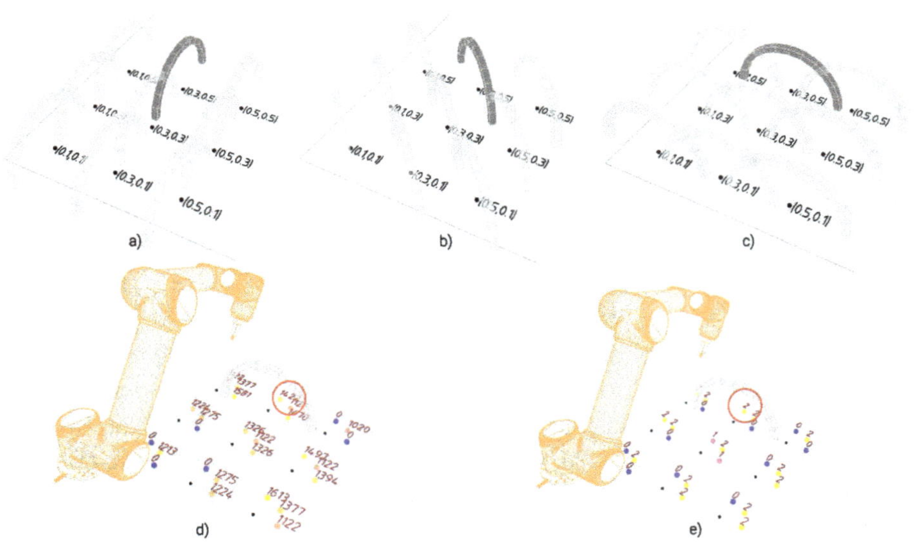

Fig. 5: Use of DARM to design the location and orientation of an arch a)-c) Sample locations and orientations of the arch $P_j(x, y, r)$. Build score maps, d) F_j FrameScore map and e) C_j ConfigScore map. The P_j at red circle is the best location.

Step 4: Motion planning

The motion planning consists of the following parts.

Step 4a: Trajectory design. The robot motion trajectory (Fig. 6) is designed to have (a) a start configuration, (b) a linear pick trajectory to pick up the block from the pick-up station, (c) a free space trajectory to transport the block to approaching frames, (d) a linear place trajectory from approaching frames to the final location, and (h)–(i) a free return trajectory to move back to the start position. The critical trajectory to be planned here is the free space trajectory (c), which covers a large distance and has more possibilities for collisions.

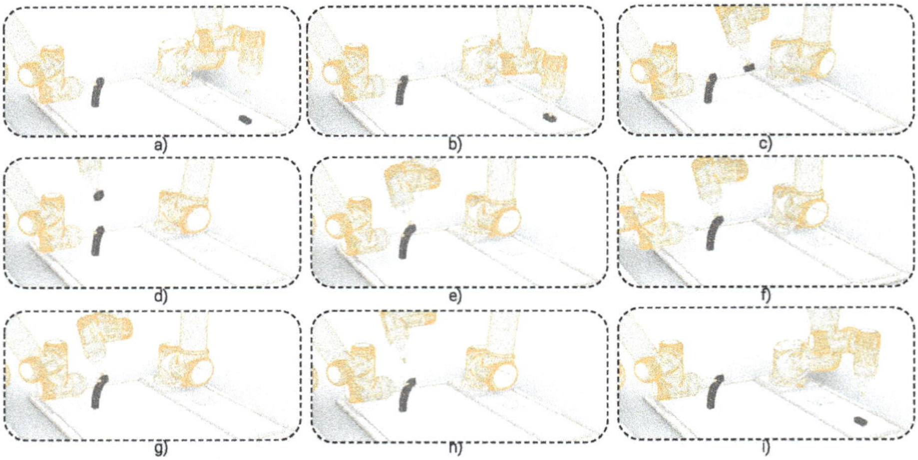

Fig. 6: Robotic motion trajectory: (a) start configuration, (b) linear pick trajectory, (c) free space trajectory, (d) linear place trajectory, (e) place to position, (f)–(g) holder switch holding position, (h)–(i) free return trajectory.

The free space trajectory (c) is planned using the Rapidly-exploring Random Tree (RRT) (LaValle et al. 2001) algorithm, implemented through Open Motion Planning Library, considering all collisions that can occur: self-collisions, collisions between robots, collisions with the table, with the already-built structure. The algorithm is initialized with a starting configuration that significantly affects performance and resulting trajectory; if it is too far from the target configuration, the algorithm may need to explore a large portion of the configuration space before finding a feasible path.

As explained in Step 3, the start configuration for the free space trajectory is derived using DARM. For the selected position P, one or more robot IK solution types are feasible for the entire assembly. We use the configurations from one of these solution types as the starting configuration of the free motion at the pick-up station. Since DARM can be calculated without considering the full range of collisions that can occur, it is possible that in this step a previously feasible configuration is now ruled out. In our experience, this reuse of information from the reachability map largely reduces the computation time of the motion planning and aids in finding good solutions with as few changes in IK solution types as possible.

Step 4b: Collaboration strategy. Next, different roles are assigned to each robot, namely either placer or holder: The placer picks up a block from the pick-up station and places it in the assembly. The holder holds the structure in a certain location so that it remains stable while the assembly process continues.

As a result, the following collaboration strategies are possible. *Asymmetric roles:* one robot is always the *placer* and the other is always the *holder.* Since the two robots have different roles, the process is relatively uncomplicated, but more traveling paths are required, increasing the fabrication time. *Symmetric roles:* the two robots are both *placers* and *holders;* they switch roles after the placement of each block. The planning process is more complicated as the robots cover a larger space with their motion, and their paths intersect more often, but the overall fabrication time is reduced. The impact of the collaboration strategy used will be further discussed in the results (Sec. 4).

Step 5: Fabrication

Once the stability of the structure is assessed and the robotic system and the fabrication planning are defined, the fabrication process can begin.

4 Applications

4.1 Dome

In this section, we apply the proposed workflow to a relatively simple dome assembly (Fig. 7) with a circular base of radius 15 cm.

The geometry of the dome (Step 1) was designed by the form-finding process of Kangaroo2-Grasshopper. The blocks of the assemblies were 3D-printed. To accurately pick blocks of different shapes and sizes, we added a notch connection between the pick-up station and the block (Fig. 9, left (c)). We attach sandpaper (Fig. 9, left (d)) to the contact surface to increase the friction coefficient to 0.8. A 0.5 friction coefficient was conservatively assumed for the analyses in CRA.

The global stability check (Sec. 3, Step 2a) is shown in Fig. 7, right (f). The tilt angle analysis shows limit thickness for several values of the tilt angle. We assumed, for this example, a 40° tilt angle which gave us the thickness of our geometry.

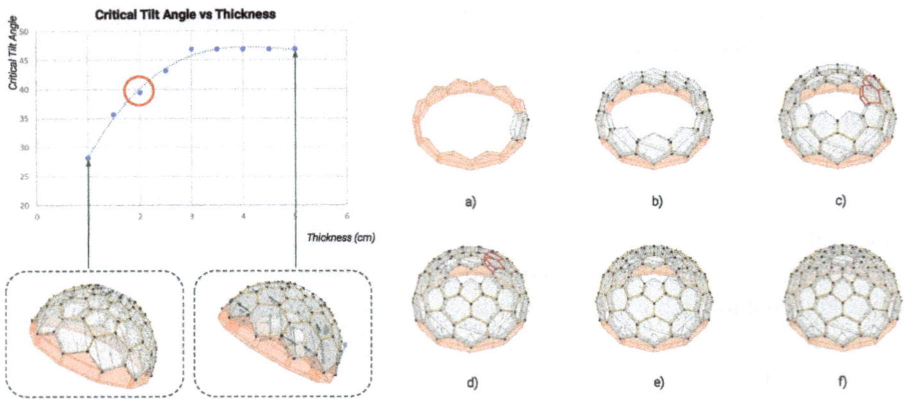

Fig. 7: Left: Critical tilt angle analysis of the dome. Right: Construction sequence analysis of the dome (red base blocks – anchors in CRA analysis, marked red blocks – unstable blocks during construction).

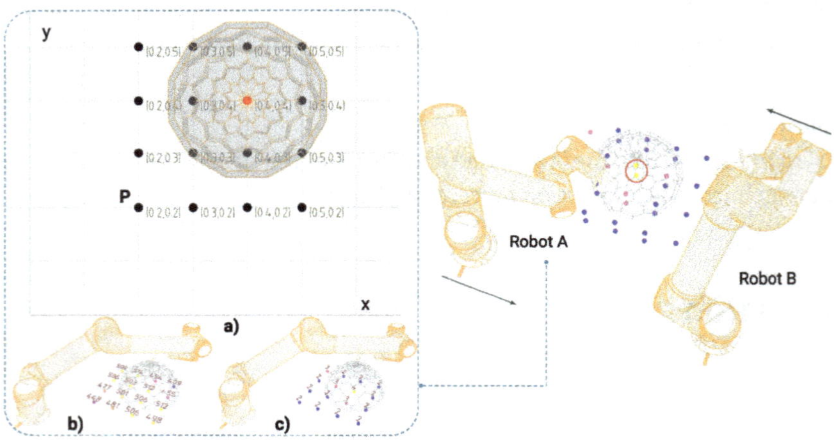

Fig. 8: Left: (a) Sampling of the assembly locations $P(x, y)$ for RobotA, Calculate; (b) frameScore and configScore for each assembly location. Right: DARM for positioning the two robots so that the best positions coincide. The red circle represents the ideal location to have max ConfigSore and FrameScore.

The discretization of the dome was chosen to have a layer-by-layer construction sequence, where each horizontal layer is stable when completed. Figure 7 shows the results obtained using CRA for several steps of the construction sequence (Sec. 3, Step 2b). It is possible to note that in Fig. 7, right (d) two blocks must be placed together with two robots, or one robot needs to hold one of the two while the other robot places the next one. This feedback was used to design the motion plan.

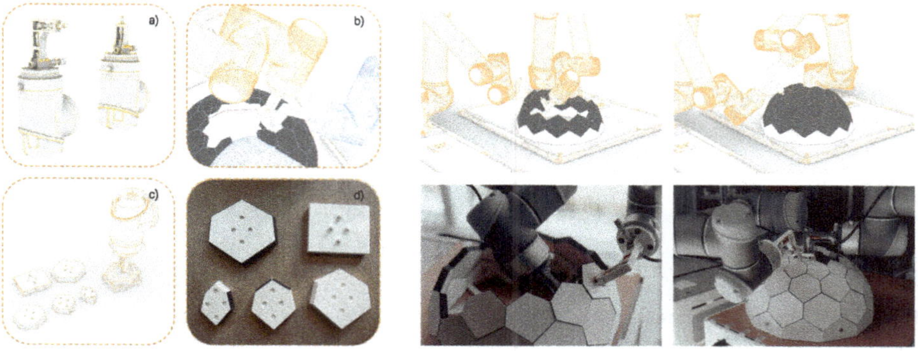

Fig. 9: Left: (a) L-shape and straight end-effectors, (b) end-effectors comparison, (c) notch with pick Station, (d) sandpaper. Right: Asymmetric roles fabrication for the dome and symmetric roles fabrication for the dome.

The setup for this example consisted of two UR5 robots with suction end-effectors; straight and L-shape (Fig. 9, left (a)). The straight end-effector helps reduce the collision between the robot and already constructed structures, while the L-shape end-effector helps reduce the collision of the two robots with each other (Fig. 9, left (b)).

We used DARM (Fig. 8) to locate the position of the structure and robots (Sec. 3, Step 3). Since the dome is axisymmetrical, the rotation is not relevant. Thus we only sampled the workspace for positions in X and Y axis $P_j(x, y)$.

The proposed trajectory design and collaboration strategy from Sec. 3, Step 4 is applied to the motion planning and fabrication of a dome. By using the IK solutions type from DARM as the initial configuration, the motion planning time is significantly reduced. For the given geometry, an asymmetric-roles fabrication is challenging, as the blocks must be held from the intrados side (Fig. 9, right). Thus, we used the asymmetric-roles fabrication only for the lower layers, and symmetric-roles fabrication for the top two layers (Sec. 3, Step 4b).

4.2 Tri-dome

We successfully extended the workflow to a more complex geometry, the tri-dome shown in Fig. 10, left, consisting of three arches (in blue) and three domes (orange). The stability checks are presented in Fig. 10, DARM analysis is shown in Fig. 11 and some steps of motion planning and fabrication process are shown in Fig. 12.

Fig. 10: Left: Geometry of the tri-dome. Right: Construction sequence of a tri-dome.

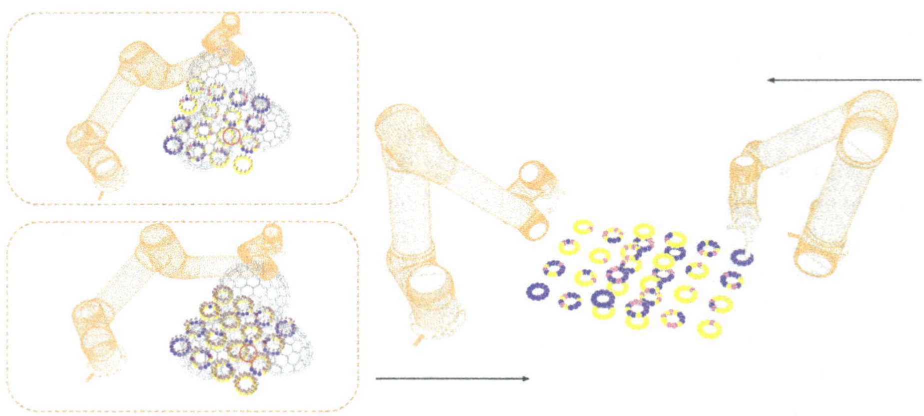

Fig. 11: DARM of the tri-dome. Left: FrameScore map and ConfigScore map. Right: DARM for positioning the two robots so that the best positions coincide.

5 Contribution and discussion

In this paper, we propose a design-to-fabrication workflow for discrete shell structures that, thanks to the use of a dual-robotic system, removes the need for additional falsework. We showed that this workflow can be applied to complex geometries. Our contributions can be summarized in the following points:

- A novel workflow to analyze and fabricate stable discrete assemblies for autonomous construction, with stability check, reachability check, and motion planning.
- A discrete assembly reachability map (DARM) that combines assembly information and robotic reachability information.

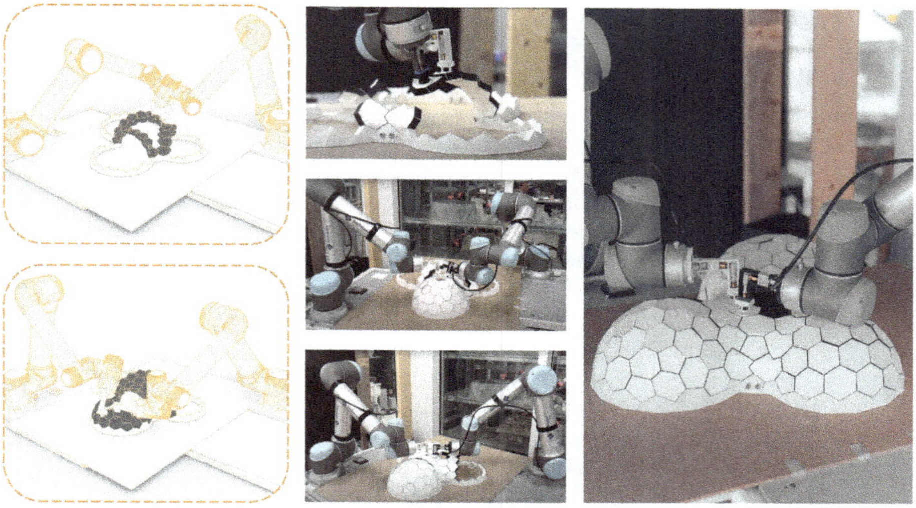

Fig. 12: Motion planning and fabrication of then tri-dome.

The following items are worth to be mentioned:
- *Scaling-up:* although the workflow is generalizable, the current applications are limited by the scale and payload of the chosen robots. For applications in architecture, the payload of the robots should consistently increase.
- *Design space:* The complexity of the structures in this research does not fully represent the possibility of the robotic fabrication strategy. Future works should apply the proposed workflow to more complex geometries and include additional robots.
- *Accuracy:* the inaccuracy of our current workflow comes from imperfections during the 3D printing of the blocks, unevenness of the base, tolerance of the calibration process, and impact forces from the suction end-effectors when initiating or releasing contact with the blocks. This problem could be reduced by implementing force sensors and visual feedback in the workflow.

This research is a contribution to the field of assembly-aware design and robotic path planning. The resulting workflow demonstrates how to embed stability and robotic reachability constraints into the design of spanning structures. Implemented in the early stages of a design, such workflow drastically reduces the need for falsework, thus improving productivity and reducing waste.

References

Beyeler, L., J.-C. Bazin, and E. Whiting. 2015. A Graph-Based Approach for Discovery of Stable Deconstruction Sequences. In *Advances in Architectural Geometry 2014,* Eds. P. Block, J. Knippers, N. J. Mitra, and W. Wang, 145–57. Springer International Publishing.

Block, P., T. Van Mele, M. Rippmann, M. DeJong, J. Ochsendorf, M. Escobedo, and D. Escobedo. 2016. Armadillo Vault – An Extreme Discrete Stone Shell. *Detail* 10 (October): 940–42.

Coumans, E., and Y. Bai. 2023. PyBullet, a Python Module for Physics Simulation for Games, Robotics and Machine Learning. https://github.com/bulletphysics/bullet3

Deuss, M., D. Panozzo, E. Whiting, Y. Liu, P. Block, O. Sorkine-Hornung, and M. Pauly. 2014. Assembling Self-Supporting Structures. In *ACM Transactions on Graphics.* Vol. 33. Association for Computing Machinery. DOI: 10.1145/2661229.2661266

Heyman, J. 1995. *The Stone Skeleton: Structural Engineering of Masonry Architecture.* Cambridge University Press. DOI: 10.1017/CBO9781107050310

Kao, G. T.-C. 2022. COMPAS CRA: Coupled Rigid-Block Analysis (CRA) for the COMPAS Framework. 2022. DOI: 10.5281/zenodo.7043135

Kao, G. T.-C., A. Iannuzzo, B. Thomaszewski, S. Coros, T. Van Mele, and P. Block. 2022. Coupled Rigid-Block Analysis: Stability-Aware Design of Complex Discrete-Element Assemblies. *CAD Computer Aided Design* 146 (May). DOI: 10.1016/j.cad.2022.103216

Kao, G. T.-C., A. Körner, D. Sonntag, L. Nguyen, A. Menges, and J. Knippers. 2017. Assembly-Aware Design of Masonry Shell Structures: A Computational Approach. In *International Association for Shell and Spatial Structures.* https://www.researchgate.net/publication/320188621

LaValle, S. M., and Jr. J. J. Kuffner. 2001. Randomized Kinodynamic Planning. *The International Journal of Robotics Research* 20 (5): 378–400. DOI: 10.1177/02783640122067453

Makhal, A. and A. K. Goins. 2018. Reuleaux: Robot Base Placement by Reachability Analysis. *Second IEEE International Conference on Robotic Computing (IRC),* 137–42. http://arxiv.org/abs/1710.01328.

Parascho, S., I. Xi Han, A. Beghini, M. Miki, S. Walker, E. P. G. Bruun, and S. Adriaenssens. 2020. LightVault: A Design and Robotic Fabrication Method for Complex Masonry Structures. In *Advances in Architectural Geometry,* 350–75.

Piškorec, L., D. Jenny, S. Parascho, H. Mayer, F. Gramazio, and M. Kohler. 2019. The Brick Labyrinth. In *Robotic Fabrication in Architecture, Art and Design 2018,* Eds. J. Willmann, P. Block, M. Hutter, K. Byrne, and T. Schork, 489–500. Springer International Publishing.

Ponce, A. R., and R. R. Melendez. 2015. Curves of Clay: Mexican Brick Vaults and Domes. In *Architecture and Mathematics from Antiquity to the Future: Volume I: Antiquity to the 1500s,* Eds. K. Williams and M. J. Ostwald, 309–24. Springer International Publishing. DOI: 10.1007/978-3-319-00137-1_21

Rippmann, M. 2016. Funicular Shell Design: Geometric Approaches to Form Finding and Fabrication of Discrete Funicular Structures. ETH Zurich, Department of Architecture. DOI: 10.3929/ethz-a-010656780

Rust, R., G. Casas, S. Parascho, D. Jenny, K. Dörfler, M. Helmreich, A. Gandia et al. 2018. COMPAS-FAB: Robotic Fabrication Package for the COMPAS Framework. https://github.com/compas-dev/compas_fab/.

Whiting, E., J. Ochsendorf, and F. Durand. 2009. Procedural Modeling of Structurally-Sound Masonry Buildings. *ACM Transactions on Graphics* 28 (5): 1–9. DOI: 10.1145/1618452.1618458

Wu, K., and A. Kilian. 2020. Designing Compression-Only Arch Structures Using Robotic Equilibrium Assembly. In *Impact: Design With All Senses,* 608–22. Springer International Publishing. https://doi.org/10.1007/978-3-030-29829-6_47.

Mehrzad Esmaeili Charkhab, Yuxi Liu, Boris Belousov, Jan Peters, Oliver Tessmann

Designing for Robotic (Dis-)Assembly

Abstract: Geometrically interlocking building elements offer unique opportunities for the construction industry to minimize waste, maximize reuse and reduce its carbon footprint. Dry-joint elements allow for fast robotic assembly, disassembly, and reassembly of complex structures out of prefabricated modules. The article discusses the robotic assembly of SL blocks, which are modules that interlock with each other. The assembly process is challenging due to the intricate assembly sequences, design hierarchies, numerous potential grasping points, contact-rich assemblies, and instability until the assembly is completed. To implement the robotic assembly of self-interlocking structures, advances in several research scopes are necessary, including geometry, algorithms, and implementation. The study aimed to co-evolve SL block geometry and robotic grippers using a robot-oriented design approach to compensate for tolerances and add self-centering features. The article also presents an assembly environment that includes mechanical fixation cubes to secure the SL blocks during assembly and support cantilevered sections of the structure until the aggregation is stable. The article presents algorithms for robotic task and motion planning and the generation of assembly sequences inspired by recursive algorithms to design 3D interlocking puzzles and directional blocking graphs. The study highlights how the voxel-based representation of complex geometries can be used to prepare directional blocking graphs. The research provides insights into improving the element's geometry and robotic assembly, which could have significant applications in other studies.

Keywords: topologically interlocking, robotic assembly/disassembly, task planning, robotic-material oriented design, directional blocking graph, SL blocks

1 Introduction

Self-interlocking systems connect multiple components, similar in shape, into stable structures. The goals for architectural constructions made from self-interlocking systems are stability and resistance to compression and bending during assembly and in the final structure. The interlocking of elements is meant to replace any form of adhesives or additional mechanical fasteners. The presented system unfolds complex and multi-directional structural capacity by elements that interlock with multiple neighboring elements simultaneously. This creates a network of interlocking connections that transmits and distributes loads and stresses throughout the entire structure. This research explores not only the structural and tectonic aspects of self-interlocking structures but also their assembly, disassembly, and reassembly through robots. Two

https://doi.org/10.1515/9783111162683-021

Fig. 1: SL block structure demonstrator

different approaches with opposing points of departure are explored: Pre-programmed, "blind" robots without sensor feedback assemble "smart" elements designed for robotic assembly with self-calibrating shapes and minimum friction during assembly. In contrast, the second approach uses "smart" robots equipped with tactile and force sensing for feedback during the assembly of "stupid" elements with little to no tolerance and high friction during assembly. The goal of this research is to take the best of both approaches: Elements designed for robotic assembly and responsive robots combined in one system in order to evolve automated assembly into the autonomous assembly. By reaching this goal, we seek to contribute to more circular material, and building element flows within architecture and construction. We expect self-interlocking structures to reduce the amount of composite materials systems that are hard to recycle as we reversibly assemble different mono-material elements. Furthermore, we are convinced that shifting computational resources in design from calculating parametrically unique parts to combinatorial explorations of repetitive elements might help to challenge current notions of permanence in architecture, which are focused on single buildings (Touw 2006). Instead, we advocate for a permanence of building elements that can be assembled, disassembled, and reassembled to be used for dynamic reconfiguration of space or in a series of consecutive buildings.

2 Related Work

Robotic assembly, as part of various initiatives in construction automation, promises to increase efficiency in architecture and construction (Bock 2015). Robotic assembly of reversibly joined building elements is explored in various research projects. Fascetti et al. assemble self-calibrating building blocks called a "Drexel with robot arms" in a controlled laboratory environment (Fascetti et al. 2021). Zhang et al.'s present a design for 3D-printed interlocking blocks and an algorithm that allows these blocks to be assembled by a dual-robot system (Zhang et al. 2021). Jenett et al. presents a material-robot system consisting of mobile robots that can assemble discrete cellular structures. The researchers describe “relative” robots that can locomote on, transport, and place elements and navigate relative to and in coordination with a cellular structure with minimal feedback (Jenett et al. 2019). Petersen and colleagues propose the concept of collective embodied intelligence, which draws inspiration from natural collective behavior. This idea involves using individual agents that lack central control but can perform complex and robust behaviors to construct structures successfully. This approach has various benefits, such as avoiding the need for expensive sensors, global communication difficulties, and complicated mechanical movements. The field of automated construction is gaining popularity and features various examples of robot groups, brick-laying robotic arms, and 3D-printed houses. Although each method has its advantages and limitations, a combination of approaches is likely to pave the way for fully automated construction in the future (Petersen et al. 2017). Latteur and colleagues introduce novel research related to the construction of large-scale buildings using aerial vehicles, such as drones (Latteur et al. 2016). Previous work by the authors (Belousov et al. 2022) explores assembling building blocks with an autonomous robotic arm. A combination of reinforcement learning and planning is used to address complex contact dynamics. The first findings indicate that model-free deep reinforcement learning algorithms, which rely on trial-and-error, can be effectively used for this task, even with minimal prior knowledge of the problem.

3 SL block and hierarchical assemblies

The chosen self-interlocking system for this research is called SL block, developed by Shen-Guan Shih from the National Taiwan University of Science and Technology. An SL block consists of one S-shaped and one L-shaped octa cube. Two SL blocks can be joined into six different conjugate pairs called SL-engagements (Shih 2016) (Fig. 2). Aggregated SL-engagements can form a wide variety of forms called SL-Strands (Shih 2018). The use of design hierarchies that emerge from the complexity and number of aggregated SL blocks is used as a strategy for up-scaling aggregations into building elements (Wibranek et. al., 2021). Thus, large structures can be built from many smallelements, which allows for various future (swarm) robotic assembly systems to be developed (Augugliaro et al. 2014; Fascetti 2021; Goessens 2018).

Our approach consists of three steps: Firstly, take a voxelized form as input and divide it into two classes of voxels. These voxels can be selected and combined into a set of planar parts that promises a global interlocking for the desired structure. The joints-design of planar parts is based on three connection types that we defined in the expanded hierarchy: nesting connection, Borromean connection, and joinery connection (Fig. 3). Secondly, voxelize the planar parts into smaller units when designing with “d” and “y” SL-engagements. For instance, combining “d” SL-engagement to create an SL strand within the dark-grey planar part requires constructing a subpart from smaller

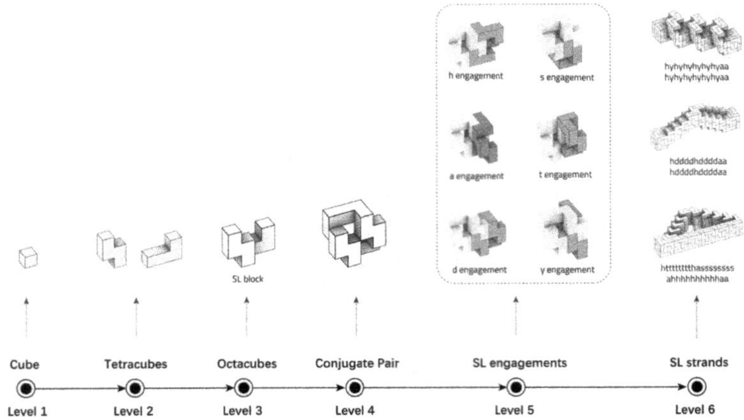

Fig. 2: Hierarchical structure of the SL Block system (based on the hierarchy proposed by Shih).

voxels (see Fig. 4). Thirdly, convert all voxel-based parts into SL block assembly. With the use of the hamiltonian path-finding algorithm, we can compute a closed polyline from non-ordered centroids of voxels. This closed polyline is a sequence of joined vertices, and each vector of vertices can correspond to an SL-engagement. Thus, we can decode the polyline into a string sequence for generating the corresponding SL-strand. Figures 1 and 5 illustrate the interlocking structure we designed and built with SL blocks. In this paper, our focus is solely on one-size voxels, and we are specifically exploring the construction of large-scale structures through hierarchical assemblies of these uniform voxels. Consequently, the design and assembly strategies discussed here are exclusively for this particular size of voxels and the corresponding SL blocks.

SL blocks are self-interlock through topological interlocking but also through friction between the elements (Shih 2020). As friction fit connections are challenging for robots to assemble, we present strategies for improving the SL block geometry for pre-programmed robotic assembly below. The assembly procedures make use of semi-interlocking moments during SL block assembly in which a structure is stable in most directions but still allows for collision-free robotic assembly from one trajectory. These properties are essential for the assembly algorithms described below.

4 SL block geometry

Geometry and robotic grippers were co-adapted with a robot-oriented-design approach (Bock 2015). The pure SL block octa cube geometry was gradually refined. Edges were rounded off to compensate for tolerances of robotic movements as well as imperfections of the aggregated structure. An SL block with fillet edges furthermore self-calibrates when being placed. Additionally, self-centering features and notches in the SL blocks

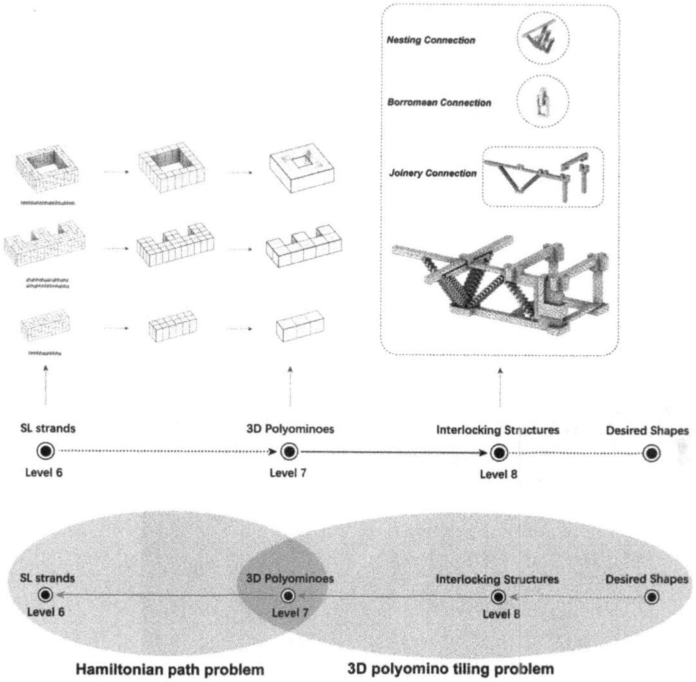

Fig. 3: An expanded hierarchy when using SL blocks to design interlocking structures. Based on a bottom-up analysis of the geometrical hierarchy, we propose two critical Problems that should be solved in the design process according to Top-Down Thinking.

as counterparts of a fork-like robotic gripper were tested (Fig. 6). The conical shape of the notch cross-section calibrates the SL block position during grabbing.

5 SL block fabrication

For the production of the more complex shaped SL block geometry, we tested FDM 3d-printing, laser cutting, and CNC milling (Fig. 7). The production of SL blocks from cross-laminated veneer plywood panels required a decomposition of the SL block into elements that can be CNC milled from sheet material. The geometry of the self-centering notches was adjusted to the tools of the CNC router. To avoid gluing pieces together, we created parts that could be friction-fit to form an SL block using tongue and groove joinery (Fig. 8). These parts can be fabricated from sheet materials with varying thicknesses. Using thinner wooden sheets can create a lighter structure creating porosities and carrying the potential to respond to varying loads and stresses in a structure.

Other SL blocks were cast from concrete in two different ways: collapsible molds and permanent formwork. Doyle et al. (2019) suggest using custom dissolvable form-

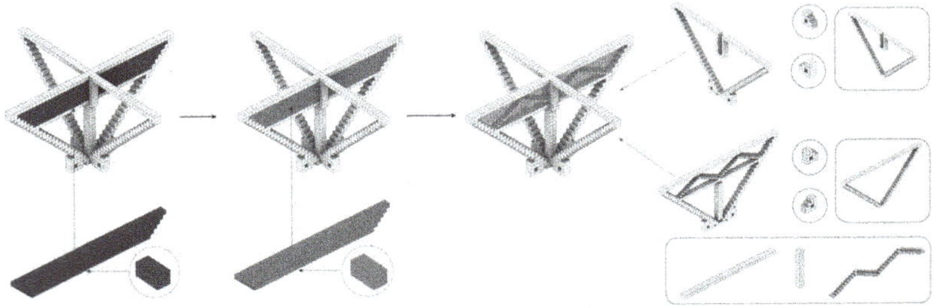

Fig. 4: Redesign of the planar part from smaller voxels. To automate the reconfiguration of the voxel-based part using d SL-engagement, we should further subdivide the dark-grey voxel into four units and then design a subpart from these smaller voxels.

Fig. 5: Left: Rendering of the cantilever structure made from 1030 SL blocks. Right: The physical prototype assembled from concrete, veneer plywood, and 3d-printed SL blocks.

Fig. 6: SL block geometry with self-centering notches and counterpart fork-like gripper.

work and flexible steel reinforcement with traditional concrete construction methods. They propose an experimental approach to integrating these materials and techniques in concrete casting. We tested 3D printing molds using water-soluble filaments for collapsible molds and achieved precise SL blocks in a reliable process (Fig. 9); however, every 3d-printed mold can only be used once, which contradicts the serial character of the SL blocks. Using the 3d-printed hull as permanent formwork creates a thin layer of plastic around the concrete core and between the discrete elements. This makes the assembly of elements easier and protects the brittle concrete during assembly and disassembly. Such a layer might improve durability but leads to undesirable material composites.

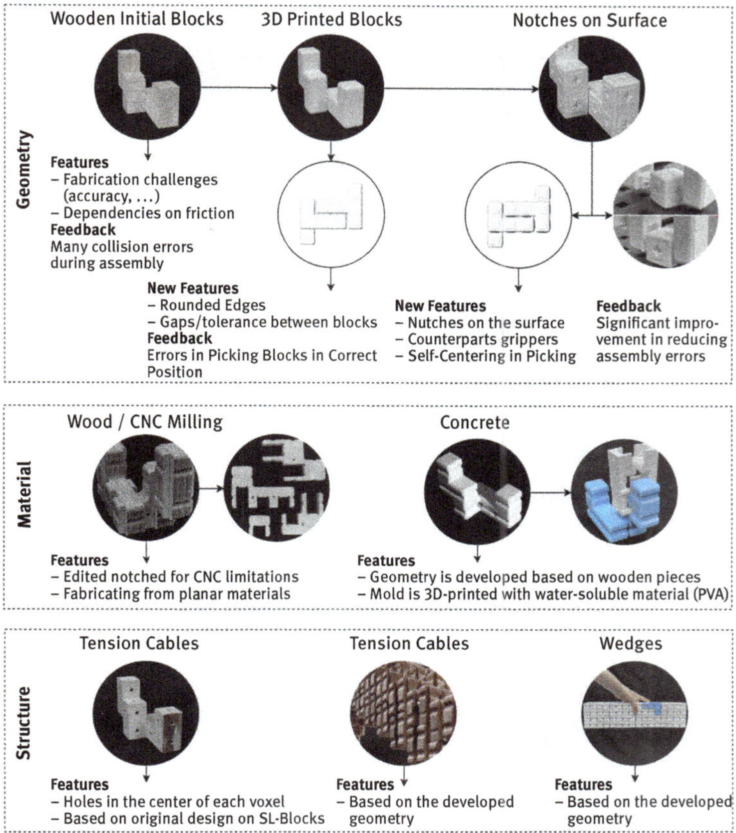

Fig. 7: The diagram depicts how the SL block undergoes a geometric evolution process. The first step involves testing 3D-printed SL blocks to overcome any fabrication difficulties. After achieving the desired shape for robotic assembly, alternative materials were investigated for production purposes. Finally, multiple structural enhancements were experimented with.

6 Robotic Assembly

The assembly process of complex self-interlocking small-scale elements into larger structures is time-consuming, and SL block aggregations are complicated due to the intricate nature of the various assembly sequences, design hierarchies, and the sheer number of elements. SL block assemblies limit movement between blocks in certain directions but allow for it in others (Fig. 10). The degree of freedom depends on the complexity of the aggregation and the interlocking mechanism that could range from mere surface contact to rotational and recursive procedures of connecting elements. For robotic assembly, stability through interlocking and ease of assembly must be balanced during the process. In certain SL engagements, x and y directions are interlocked while

Fig. 8: SL block CNC milled from cross-laminated veneer plywood elements and a demonstrator element made from planar material of varying thicknesses.

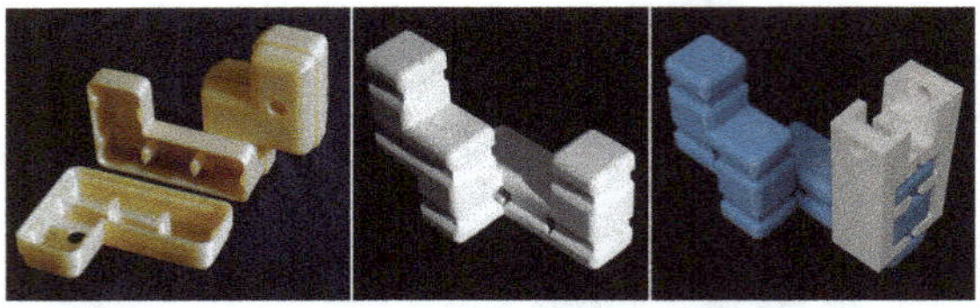

Fig. 9: The concrete SL block is produced using a 3D printed mold made of water-soluble (PVA) material, and the function of the designed fork-like gripper.

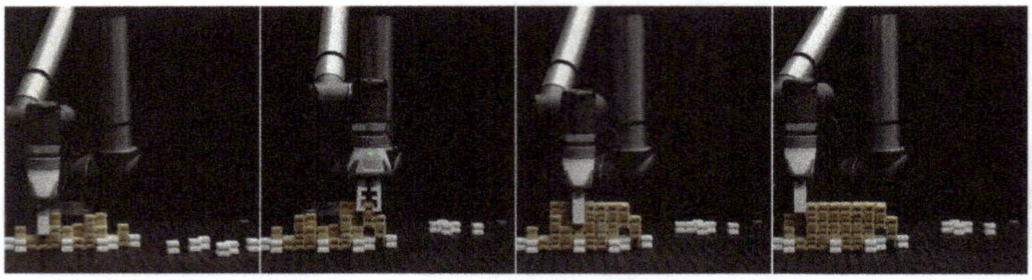

Fig. 10: UR10 robotic arm in the assembly process.

constraining the z-axis exclusively relies on friction. Such contact-rich connections pose challenges for assembly. In response, the above described geometric modifications were made to the SL blocks.

7 Assembly Environment Setup

An additional challenge posed by the strong contact between the elements during assembly is the risk of movement of already aggregated SL blocks during the placement of an additional block. Such an unintended and unregistered movement can hinder the placement of additional blocks. To address this issue, mechanical fixation cubes for the block placement process were added (Fig. 11). They are similar in size to the cubes SL blocks are made of and can be attached to an assembly plate via tongue and groove connections. The blocks aid in the self-calibration of SL block placement, secure the SL blocks during the assembly process, and support cantilevered sections of the structure until the aggregation is stable through placing key-stone-like SL blocks.

Fig. 11: Assembly environment setup parts. From left to right: Assembly Environment Setup, 2SL-Engagement: hddhddh, Support Blocks, Fixation Blocks.

The robot frequently rearranges those blocks during assembly. The below-described algorithm facilitates this process.

8 Robotic Task and Motion Planning

Robotic task and motion planning determine a sequence of actions and movements a robot should perform to efficiently complete a given task. For robotic assembly, task planning involves determining which parts to assemble, in what order, and how to perform each assembly operation. Task planning in robotic assembly can be divided into two main stages: high-level and low-level planning (Fig. 12).

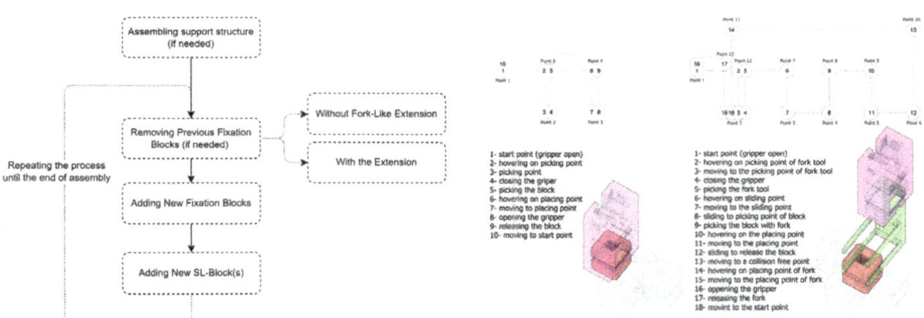

Fig. 12: Left: Robotic task planning in each assembly sequence, Right: General motion planning of the robotic arm with fork-like finger and fork tool extension.

A large number of parts need to be assembled in a specific order and in a tight space without the robots or its end effector colliding with any obstacles or already aggregated parts. Thus, the robot's motion planning algorithm must consider the geometry of the parts and the environment and the constraints of the robot's motion, such as joint limits.

The high-level task planning considers two main scenarios: picking/removing SL blocks, transferring them, and placing them in their intended location. The low-level task planning addresses the proper assembly and disassembly sequences for SL blocks, determining the direction in which they should be assembled or disassembled, finding the gripping points for the SL blocks, and generating arrangements for fixation blocks. Additionally, the algorithm must identify the correct assembly, disassembly, or rearranging sequences for fixation blocks and their optimal gripping points and directions.

Our assembly sequences are inspired by Song's recursive algorithm to design 3D interlocking puzzles (Song 2012) and Wang et al.'s research on identifying disassembly sequences and reversing them in order to find assembly sequences (Wang, 2021). Directional blocking graphs were utilized in several phases, initially to identify the sequence and direction of assembly and subsequently to rearrange fixation blocks and find suitable grasping points that would prevent collisions and enable the blocks to be held in a stable manner, whether it was one SL block, pair of SL blocks or fixation block. We consider an assembly plan as valid if there are no collisions while assembling each part. Assemblies and parts are represented as graphs. We identify removable parts by analysing the directional blocking graph. Each SL block is surrounded by several others limiting its translational degrees of freedom. To numerically represent this situation, every SL block is subdivided into the eight cubes it consists of. These cubes are subsequently converted into lists of x, y, and z-coordinates. These voxel-based geometries interact in a voxelized environment. All data about aggregation geometries and the assembly environment is efficiently stored as point lists. Due to the voxel-based algorithm and the utilization of coordinate comparisons in the x, y, and z coordinates, the computational intensity of the calculations is low, resulting in a relatively fast calculation time.

Every SL block aggregation has two options: either a single SL block can be disassembled, or a pair of adjacent SL blocks can be disassembled together. The algorithm checks these options recursively and prioritizes disassembling single SL blocks.

We name SL blocks that limit the movement of its neighbours "blockees". In order to determine the blockees around each SL block, a list of six possible directions ($\pm z$, $\pm x$, $\pm y$) for assembling the SL block is analysed. The algorithm iterates through each direction and moves the SL block's voxel points by one voxel. If one SL block voxel has the same x, y, and z-coordinates as the voxel of neighbouring obstacles, such as another SL block, fixation, or support blocks, it is considered to not be removable (see Fig. 13). The obstacle coordinates are added to a Blockers list including the index numbers of blockees in each direction. To decrease calculation time, the initial data is calculated and saved initially. Alternatively, voxel points of each SL block can be moved in different directions to create directional checking point lists. Then, each point of each list can be iterated through to check for obstacles in that direction and save the final list. Figure 14 provides visual examples of the Blockees List.

We can determine the sequences and directions for disassembly using a recursive algorithm. In each recursive level, there is a main loop that iterates over each SL block object and another loop in each object that checks directional points. The algorithm compares the X, Y, and Z coordinates of these points to find matches in other blocks or assembly environments. If the coordinates match, it means the block is an obstacle [not movable]; otherwise, it is free to move in that direction. If there is no match, we can

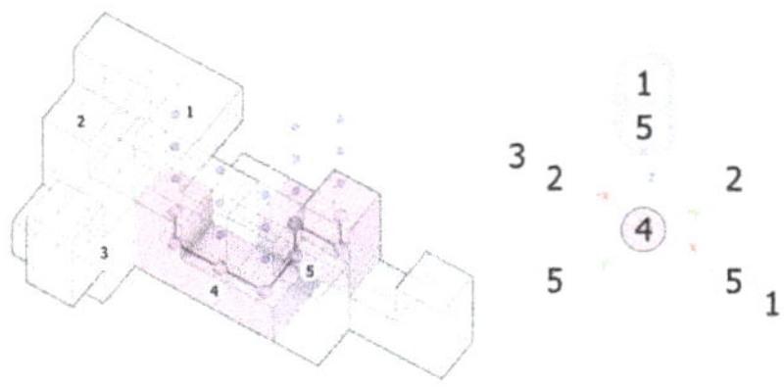

Fig. 13: Creation of part-graphs using voxel representations of SL blocks.

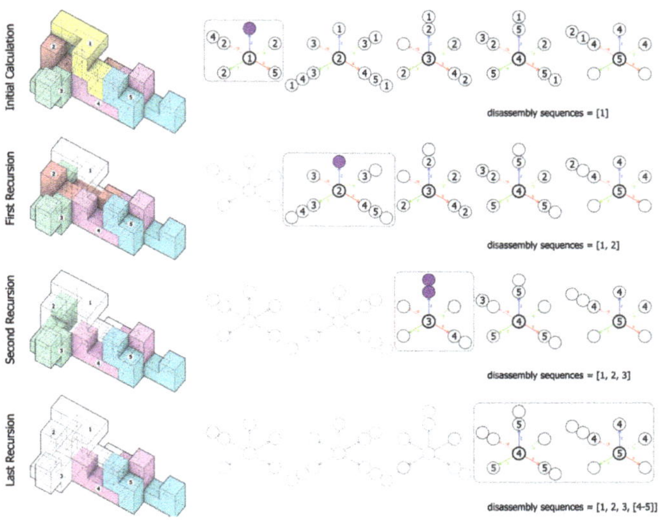

Fig. 14: The recursive disassembly sequence search algorithm relies on directional blocking graphs.

disassemble the block in that direction and add the free block's object and moving direction to the disassembly list while removing it from the main aggregation list. Finally, the list of SL block objects needs updating to remove the index number of the free SL block from all blockees lists, opening up the aggregation space for the next recursion.

When no blocks can be moved individually, it may be necessary to move multiple blocks together. Two pairs of Sl blocks with a specific configuration can be moved together. To identify these pairs, we compare the Sl block index and Blockee indexes in different directions. A pair of movable SL blocks are detected if both SL blocks have a Blockee in the same direction and the index numbers match. These SL blocks are then added to the disassembly list, and the list of SL block objects is updated. This process is repeated recursively until no SL blocks are left in the aggregation list. Finally, by reversing the disassembly list, we obtain the assembly list. This can be seen in the bottom image of Fig. 14, where two blocks can move together in the z-direction.

A procedure has been created to detect the location of fixation blocks for each SL block in assembly sequences to avoid undesired movements during their placement

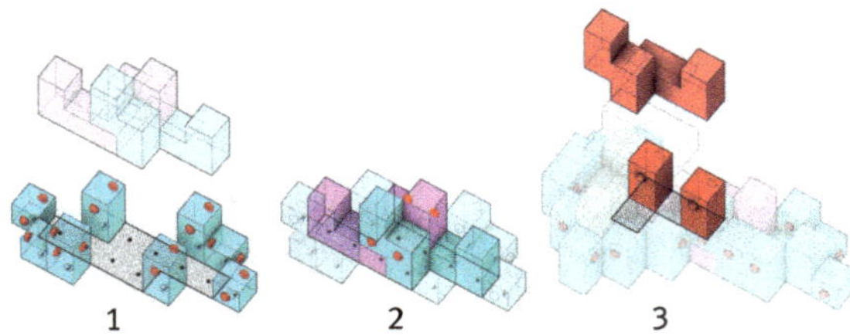

Fig. 15: The diagram illustrates the steps involved in positioning fixation blocks and grasping points, identifying a collision-free grasping point for the SL block and eliminating previous fixation blocks if necessary for the upcoming assembly of the SL block. 1) Finding fixation blocks for each assembly step, Finding gripping points of fixation blocks, 2) Finding collision-free gripping points of SL-Blocks, 3) Finding fixation blocks that need to be removed before the next assembly step (if needed).

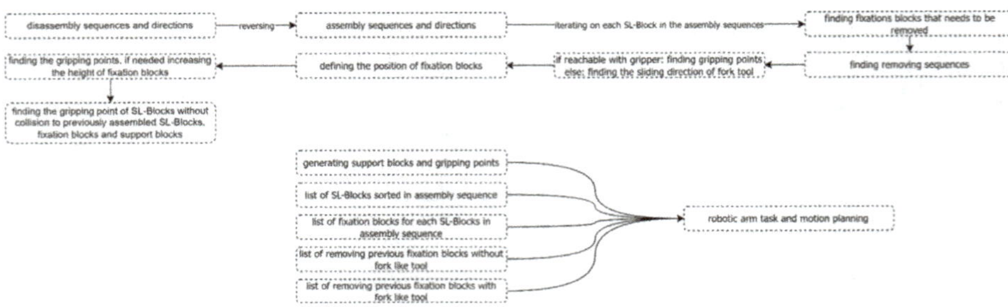

Fig. 16: Algorithmic process following the discovery of assembly sequences.

(Fig. 15, 16). In addition, it's necessary to identify the gripping points of these fixation blocks. The algorithm operates in two stages. Firstly, it recognizes potential locations where support or fixation is needed around the present SL block placement. Then, it checks each location to make sure that there are no collisions between the gripper, the SL block being positioned, and other blocks in the assembly environment. If there is no collision, the gripping point is recorded. However, if there is a collision, the fixation block is extended by including another block on top, and the process repeats. This strategy guarantees that the configuration of fixation blocks with collision-free grasping points can be removable in later steps. Afterward, the algorithm calculates the potential gripping points for the current placement of the SL block, taking into account the position of the fixation blocks. Since there may be various feasible grasping points, it selects the highest one. Finally, a segment of the algorithm confirms if any blocks obstruct the current placement of SL blocks and if so, those blocks are added to the removal list. Next, the algorithm verifies the location of the gripping point for removing the blocks and employs the fork tool extension if necessary (Fig. 12).

The presented research is complemented by the exploration of a system consisting of advanced robotic technology, sensor feedback, and machine learning to control the assembly process (Belousov et al. 2022). To implement a reliable stacking policy, we employ an impedance-based controller (Hogan 1984) in combination with imitation learning (Osa et al. 2018). A human first demonstrates how the SL blocks need to be inserted, and their movement is tracked by an RGB camera Logitech BRIO 4K Pro based on AprilTag markers. Subsequently, the programmatic compliance of the Franka Panda robot arm is leveraged to follow the demonstrated part trajectory. Thanks to the material properties of the SL blocks and the smoothened corners, a reliable insertion policy can thus be learned for each type of connection. Importantly, the approaching movement demonstrated by the human needs to be robust with respect to small perturbations to level out the errors in perception and execution by the robot. However, as long as there is constant contact between the parts, they slide automatically into the desired position. Employing such a robotic compliance controller allows for insertion of SL blocks and SL pairs from top to bottom. Our research aims to merge both approaches and take the most out of robot-oriented elements design and AI-driven robots to contribute to more circular material and building element flows within architecture and construction.

9 Conclusion and Future Work

In our research, we explored design strategies for self-interlocking SL blocks and assembly/disassembly strategies of such aggregated structures. We focussed on the use of pre-programmed robots without sensor feedback that assembled elements designed and optimised for robotic assembly. We regard the combination of dry-joint, reversible structures with automated robotic assembly as necessary for the circular use of building elements. The permanence and durability of elements reused in a series of buildings can contribute to significantly saving material resources and energy. However, economic feasibility might only be achieved through automation. We suggest approaching this challenge with small-scale elements so that large structures can be built from many small elements that will co-evolve with robotic systems to come. Similar to brick dimensions historically evolving in relation to the human agent and a socio-economic context.

We expect self-interlocking structures to reduce the amount of composite materials systems that are hard to recycle as we reversibly assemble different mono-material elements. Furthermore, we are convinced that shifting computational resources in design from calculating parametrically unique parts to combinatorial explorations of repetitive elements might help to challenge current notions of permanence in architecture, which are focused on single buildings (Touw 2006). Instead, we advocate for a permanence of building elements that can be assembled, disassembled, and reassembled to be used for dynamic reconfiguration of space or in a series of consecutive buildings.

Acknowledgments

This research was supported and executed by architecture students at the Digital Design Unit, TU Darmstadt. Among the students involved, Bingqun Liu assisted in the robotic assembly experiments and documentation. We would also like to acknowledge Timm Schneider from IAS TU Darmstadt, who supported our team with the robotic assembly.

Funding for this project was generously provided by Zukunft Bau with project number 10.08.18.7-21.34. The FiF at TU Darmstadt financially supported the initiation of the research.

References

Augugliaro, F., S. Lupashin, M. Hamer, C. Male, M. Hehn, M., M.W. Mueller, J.S. Will, F. Gramazio, M. Kohler, R. D'Andrea. (2014). The flight assembled architecture installation: Cooperative construction with flying machines. *IEEE Control Syst.* 34(4), 46–64.

Bock, T. (2015). The future of construction automation: Technological disruption and the upcoming ubiquity of robotics. *Autom. Constr.* 59, 113–21. DOI: 10.1016/j.autcon.2015.07.022

Fascetti, A., P. Latteur, and S.H. Lim. (2021). Ground-based automated construction of droxel structures: An experimental approach. *Autom. Constr.* 131, 103899.

Goessens, S., C. Mueller, and P. Latteur. (2018). Feasibility study for drone-based masonry construction of real-scale structures. *Autom. Constr.*, 94, 458–80.

Jenett, B., A. Abdel-Rahman, K. Cheung, and N. Gershenfeld. (2019). Material-robot system for assembly of discrete cellular structures. *IEEE Robot. Autom. Lett.* 4(4), 4019–26.

Liu, Y., B. Belousov, N. Funk, G. Chalvatzaki, J. Peters, and O. Tessmann. (2022). Auto (mated) nomous Assembly. In *Trends on Construction in the Digital Era: Proceedings of ISIC 2022* (167–81).

Peng, S., C.-W. Fu, D. Cohen-Or. (2012). Recursive interlocking puzzles. *ACM Trans. Graph.* (31)6, 1-10.

Shih, S.G. (2016). On the Hierarchical Construction of SL blocks – A Generative System That Builds Self-interlocking Structures. In *Advances in Architectural Geometry 2016*, 124–36.

Shih, S.G. (2018). The art and mathematics of self-interlocking SL blocks. *Proceedings of Bridges: Mathematics, Art, Music, Architecture, Education, Culture. 2018.*

Shih, S.G. (2020). Grammars of Interlocking SL blocks. In *Advances in Architectural Geometry 2020.*

Touw, K. (2006). Firmitas re-visited: Permanence in Contemporary Architecture (Master's thesis, University of Waterloo).

Wang, Z., S. Peng, and M. Pauly. (2021). State of the Art on Computational Design of Assemblies with Rigid Parts, *Computer Graphics Forum* 40(2).

Zhang, Y., Y. Koga, and D. Balkcom. (2021). Interlocking Block Assembly With Robots. *IEEE Transactions on Automation Science and Engineering*, 18(3), 902–16.

Hogan, N. (1984). Impedance Control: An Approach to Manipulation. *American Control Conference*, San Diego, CA, USA, 1984, 304–13, DOI: 10.23919/ACC.1984.4788393

Osa, T., J. Pajarinen, G. Neumann, J.A. Bagnell, P. Abbeel, and J. Peters. (2018). An algorithmic perspective on imitation learning. *Foundations and Trends in Robotics*, 7(1-2), 1–179.

Doyle, S.E. and E.L. Hunt. (2019). Dissolvable 3D Printed Formwork, 10.52842/conf.acadia.2019.178

Petersen, K. and R. Nagpal. Complex Design by Simple Robots: A Collective Embodied Intelligence Approach to Construction, 05 July 2017, DOI: 10.1002/ad.2194

Latteur, P., S. Goessens, and C.T. Mueller. Masonry construction with drones, Conference: International Association for Shell and Spatial Structures (IASS) Symposium 2016 At: Tokyo

Wibranek, B. and O. Tessmann O. (2021). Interfacing architecture and artificial intelligence, The Routledge Companion to Artificial Intelligence in Architecture, London, p. 486.

Uday Kusupati, Florin Isvoranu, Seiichi Suzuki, Mark Pauly

RUM: Reconfigurable Umbrella Mesh

Abstract: We propose Reconfigurable Umbrella Meshes (RUMs), a new class of pre-assembled deployable structures. A RUM has a compact rest state that can be easily altered to deploy into different bending-active 3D target shapes. RUMs consist of elastic beams and rigid plates connected by hinge joints and can be assembled in a stress-free fabrication state. The key principle of a RUM is that it encodes the desired deployment shape in the height distribution of its constituent cells. Reconfigurability is achieved by introducing sliding joints to connect the elastic beams with the rigid plates to allow adapting the cell heights. We demonstrate that even for small variations in the cell heights, RUMs can capture a diverse range of shapes. Assembled from identical cells that can be mass-produced, a RUM can deploy into several desired shapes, which makes them well suited for re-usable temporary structures. We provide a computational design tool based on physical simulation, where users can interactively edit cell heights to explore achievable deployed shapes. Numerical optimization enables designers to easily navigate between the reconfigurable design parameters and the deployed shape space. We validate our approach with a physical prototype and demonstrate its various deployed states.

Keywords: geometry, reconfigurable structures, deployable structures, bending-active, optimization

1 Introduction

Shape-morphing structures are physical systems that have the ability to change their spatial configuration. Such systems are of significant interest at multiple scales ranging from micro-scale metamaterials to large-scale architectural designs. Here, we present a new shape-morphing structure that can be easily assembled from identical, mass-produced components in a compact, stress-free state, and that deploys to a wide range of free-form shapes.

Specifically, we propose an extension of *umbrella meshes* recently introduced by Ren et al. (2022). Umbrella meshes are a class of deployable structures that can be optimized to deploy to one pre-defined target surface. We generalize umbrella meshes to be fully reconfigurable, enabling deployment to a continuous range of target shapes for a single physical model.

Umbrella meshes are composed of a regular array of scissor-linkage cells, where each cell's height is optimized to achieve a specific local expansion during deployment. This leads to incompatibilities among neighboring cells causing the elastic structure to deform into the desired curved target surface. As a consequence, each cell is different and needs to be fabricated to measure. Reconfigurable Umbrella Meshes (RUMs) replace

https://doi.org/10.1515/9783111162683-022

the custom cells by identical copies of a single, adaptable cell. This not only allows mass-production of the cells, but also enables programming many different deployed target geometries *after* fabrication and assembly, thus creating a truly reconfigurable and adaptable system.

We use physics-based simulation to handle the complex force coupling of the elastic scissor linkage and provide accurate predictions of the deployed shape of the global bending-active structure. The design handles, i. e., the programmable heights of individual cells, are directly exposed to the user for physical manipulation. We facilitate design exploration both via a forward simulation tool that computes the deployed shape based on direct user manipulation of the design handles, and an inverse algorithm to optimize design handles for specific target shapes.

2 Related Work

In recent years, many efforts have been conducted on the use of rigid-beam scissor mechanisms for deployable structures in architecture and engineering. A review of recent developments can be found in Moy et al. (2022). Maden et al. (2019) presented a study focusing on the terminology and classification of scissor mechanisms used in the research of deployable structures. The predominant categorization of scissor structures in the literature typically relies on three fundamental planar cells, referred to as translational, polar, and angulated, that can be combined to shape 3D scissor cells. Several recent studies have proposed geometric approaches for exploring the design space of scissor structures using polar and angulated cells (Roovers and De Temmerman 2017; Roovers et al. 2013; Dinevari et al. 2021; Arnouts et al. 2018, 2020).

Besides deployable structures, special attention has also been devoted to the topic of reconfigurable structures. Akgün et al. (2007) proposed a new type of planar cell based on adding two revolute joints and beams attached to a basic scissor cell. The principle was further extended in the study by (Akgün et al. 2011) to develop 3D cells. This advancement introduced unique extension and rotation capabilities that enable dividing the behavior of the entire scissor structure into sub-structures. These sub-structures can then transform locally without impacting the other parts permitting to develop a wide variety of geometrical shapes. Rosenberg (2010) introduced the concept of a double scissor-pair cell, which utilizes a unique combination of two polar cells to transform the scissor structure into various shapes. Additional transformation capabilities based on the double scissor-pair cell are enabled through the introduction of notches. Incorporating notches on the beams makes it possible to convert translational cells into polar ones, thereby introducing an extra degree of freedom to the mechanism. More recently, the concept of a universal scissor component, which can be easily reconfigured to achieve all basic scissor cells and manufactured on a large scale, was introduced by Alegria et al. (2016).

Research on active-bending systems proposes a different approach for constructing deployable scissor structures that do not rely on rigid beams. Active bending is defined as a controlled shaping process for inducing curvature through elastic deformations (Lienhard 2014). By incorporating flexible beams into deployable scissor structures, a greater range of shapes can be achieved, as the beams are able to bend and twist. Recent studies on deployable gridshells aim to finely tune geometrical incompatibilities of irregular grids to achieve both deployability and shaping goals. Panetta et al. (2019) presented X-shells as a generalized type of deployable gridshell in which geometrical incompatibilities during deployment push the structure to buckle out-of-plane to achieve a 3D curved shape. An optimization process is proposed to address the design of X-shells based on the use of physical simulations. Related efforts have also been conducted to investigate geometric approaches based on the design space of geodesic grids (Pillwein et al. 2020; Pillwein and Musialski 2021).

3 Reconfigurable Umbrella Meshes

Our work builds upon and extends *umbrella meshes* as proposed by Ren et al. (2022). Umbrella meshes are deployable structures composed of a regular tessellation of scissor linkage cells assembled in a compact rest state. Each cell consists of three pairs of arms joined as vertical scissor mechanisms around X-joints. The arms connect to rigid plates (top and bottom) through T-joints. The cell is deployed by pushing the top and bottom plates towards each other. This causes the arms to rotate and expand the lateral footprint of the cell (see Fig. 1).

Cells can have varying expansion factors of their lateral footprint, indicated by the dashed lines in Fig. 1b. When fully deployed, the expansion factor of a cell depends on the separation between the two plates in the rest state, defined as cell heights. Deploying two connected cells of different heights creates a mismatch in their planar footprints. When several such cells of incompatible expansion factors are combined, the umbrella mesh deforms out of plane during deployment into a curved, surface-like structure.

3.1 Reconfigurable Umbrella Cell

In the physical realizations presented by Ren et al. (2022), cell arms are cut to size to match the heights of one specific optimized deployed shape. Our key observation is that we can easily avoid this irrevocable, shape-specific fabrication of each cell and instead adapt the heights of the cells dynamically (see Fig. 1). Specifically, our new cell design allows adjusting its height at the T-joints. During reconfiguration, the arms of the scissor linkage can slide through the T-joints and be blocked at the desired height

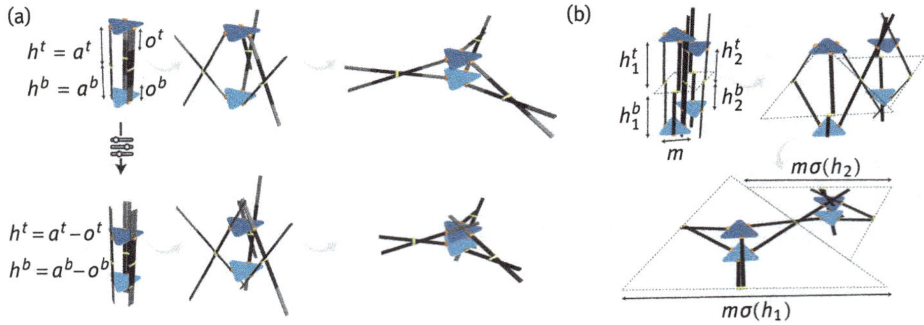

Fig. 1: Reconfigurable Umbrella Cell. Three scissor linkages are connected at the top and bottom triangular plates. Deployment of the cell is achieved by pushing the top and bottom plates towards each other. (a) By sliding these rigid plates downwards in the compact rest state, we can control the length of the scissor arms and correspondingly the lateral expansion during deployment. (b) Two adjacent cells with an identical footprint in the rest state have different expansion (dashed lines) when configured at different cell heights.

value. This reconfiguration becomes possible because the compact state stores no elastic energy, i. e., the scissor arms can easily slide through the T-joints in their unbent rest state. We can thus adjust the corresponding lengths of the active scissor segments h^t and h^b. We call these values the *heights* of a cell. If we denote the fixed length of the top and bottom scissor arms as a^t and a^b, respectively, and the sliding range by o^t (and o^b), then the range of values the height h^t of the top arm of the cell can take is given by $[a^t - o^t, a^t]$, and analogously $[a^b - o^b, a^b]$ for the bottom arm.

Let m be the edge length of the triangular top and bottom plates and $H = h^t + h^b$ be the total height of the cell. H gives the separation of the plates when the cell is in its compact, undeployed state. When deployed, the edge length of the triangular cell footprint increases by $2\sqrt{3}h$, where $h = \min(h^t, h^b)$ as shown in Fig. 1. When the cell is deployed towards a separation s, the length of the edge of the footprint cell increases from m to $m + \sqrt{12h^2 - 3s^2}$. Since we use a regular tessellation of cells with constant triangle size m, the length expansion factor is a function of h and s only:

$$\sigma(h) = \frac{\sqrt{12h^2 - 3s^2}}{m} \tag{1}$$

Figure 1a illustrates the deployment of a cell in two different configurations. Note that the configuration with longer heights expands to a larger footprint as expected. We show one possible physical realization of a reconfigurable umbrella cell in our fabricated prototype in Sec. 7.

A key principle of RUMs is to use identical reconfigurable cells across the entire umbrella mesh, which facilitates mass fabrication and re-use. Figure 1b shows how two identical cells configured to different height values deploy towards planar footprints of different lateral expansion. Identical reconfigurable cells can be assembled to form a RUM which can be configured to different height distributions.

4 Design Exploration

To explore design alternatives, we provide two complementary tools built upon the computational approach presented in Ren et al. (2022): (i) A forward exploration where the designer can directly control cell heights and receives interactive visual feedback on the deployed shapes. This allows building intuition on the range of possible spatial configurations of a given RUM. (ii) An inverse method that computes precise cell heights for several given target shapes in an offline optimization.

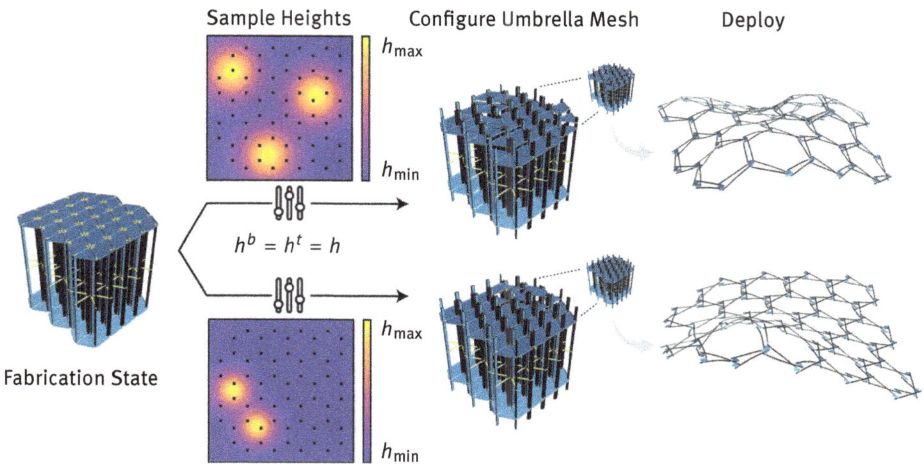

Fig. 2: Direct control of the cell heights allows exploring design variations. A painted density map (center left) is converted to umbrella cell heights (center right). Our algorithm then provides visual feedback of the corresponding deployed shapes (right).

Figure 2 illustrates the forward exploration framework. A user can sample the umbrella cell heights within the reconfigurable range of a pre-determined cell. Here we use a mixture of radial basis functions to generate a continuous height distribution from user input. We overlay the umbrella cell topology to sample the discrete umbrella heights. Alternatively, the user can directly manipulate individual cell heights to generate new configurations of the umbrella mesh. With the chosen height distribution, we use our simulation framework to compute equilibrium state of the deployed structure, actuated by pushing the top and bottom plates of each cell towards each other.

Figure 3 illustrates how inverse design can optimize the cell heights for specific given target shapes. Here we solve for the reconfiguration parameters to achieve a spherical cap (positive Gauss curvature), a saddle (negative Gauss curvature), and a cylinder (zero Gauss curvature). All three of these states can be achieved with the same RUM model. During the optimization we enforce bounds on the design parameters to ensure that the umbrella heights h^t, h^b are within the permissible range ($[a^t - o^t, a^t]$, $[a^b - o^b, a^b]$) defined by the reconfigurable cell. More details on the optimization algorithm can be found in Ren et al. (2022).

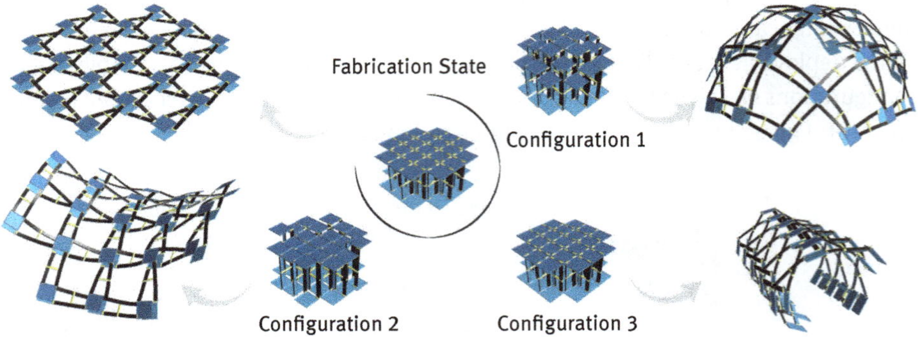

Fig. 3: A single RUM composed of identical cells can be reconfigured to deploy to a variety of target shapes.

5 Shape Space

A critical factor when designing a RUM cell is to select an appropriate range of reconfigurability. For a constituent umbrella cell with reconfigurable ranges $[a^t - o^t, a^t]$, $[a^b - o^b, a^b]$ of the top and bottom heights respectively, we define $h_{min} = \min(a^t - o^t, a^b - o^b)$ and $h_{max} = \min(a^t, a^b)$. These values represent the minimum and maximum possible heights achievable by the chosen reconfigurable cell. Subsequently, we can compute the minimum and maximum expansion factors σ_{min}, σ_{max} from Eq. (1). A larger range requires longer arms and thus more material. If the range is more limited, the space of achievable shapes is reduced. The optimization will then solve for the closest possible surface deployable from a height distribution within the allowed range.

Following Konaković et al. (2016), we can understand the space of all possible deployed shapes of a given RUM using conformal geometry. A conformal map from a design surface to the plane preserves angles and only allows locally isotropic scaling and rotation. This behavior is reflected in the design of our umbrella cells that expand isotropically when deployed, as shown in Fig. 1. Since a conformal map of the surface is invariant under global uniform scaling, the ratio $\frac{\sigma_{max}}{\sigma_{min}}$ defines the maximum conformal scale factor that can be expressed by a specific cell.

The maximum conformal scale factor of an arbitrary surface helps characterize the shape space. Specifically, the Gauss curvature K at a point on a surface is directly related to the local conformal scale factor through the Yamabe equation, $\Delta \log \sigma = -K$, where Δ denotes the Laplace operator. This means that a high *variation* of scale factors, and consequently a high local incompatibility among neighboring cells, leads to high Gauss curvature. Bounds on the maximal scale factor limit the range of obtainable shapes.

For example, the auxetic structures studied in Konaković et al. (2016) have a maximal conformal scale factor of two. The authors show that within this range, one cannot obtain any spherical patch larger than a hemisphere. In contrast, umbrella meshes can in principle have arbitrary maximal conformal scale factor (Eq. (1) shows $\sigma \propto h$) and can thus approximate any surface. Reconfigurable umbrella meshes on the other hand have limits on their maximal conformal scale factor as a function of their reconfigurable range (fixed once the cell is fabricated), consequently affecting the range of shapes they can accurately attain (see Fig. 4). In Fig. 3 we show that a relatively small range of $\frac{h_{max}}{h_{min}} = 1.5$ allows approximating shapes with significantly different curvature profiles.

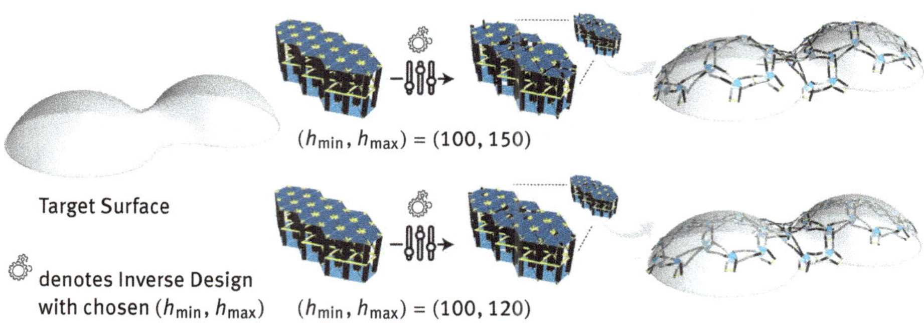

Fig. 4: Influence of reconfiguration range on the approximation of a target surface during inverse design. A reconfigurable cell with a larger height range (top) approximates the surface better than the cells with reduced range (bottom).

6 Sparse Reconfigurability

Reconfigurable cells are more complex than fixed cells due to the required sliding mechanism that allows adjusting the cell heights. Our computational framework also allows building hybrid assemblies of reconfigurable cells and fixed cells of constant height. Having fewer reconfigurable cells constrains the shape space that the structure can approximate, but reduces fabrication cost and the complexity of reconfiguration. This can be particularly useful when the resolution of the umbrella mesh is high, which leads to a large number of reconfigurable parameters, or when the reconfiguration is automated through motorized actuators.

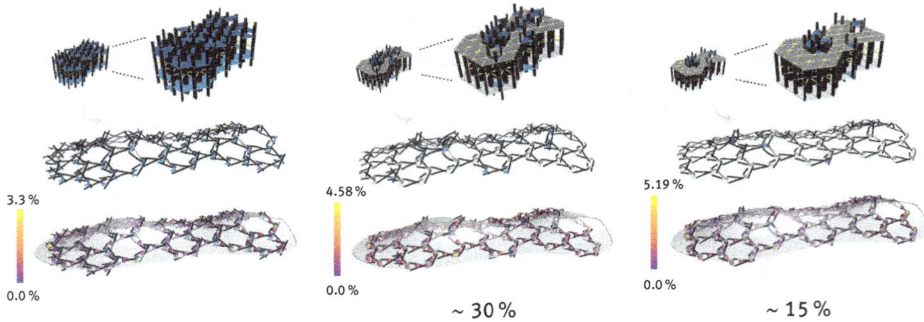

(a) Fully Reconfigurable (b) Sparsely Reconfigurable

Fig. 5: Sparsifying reconfigurability: (Top) demonstrates the optimized deployed structure with the non-reconfigurable cells shown in grey. (Bottom) displays the distance from the target surface relative to the magnitude of the bounding box diagonal. The maximum deviation only increases from 3.3 % to 5.19 % after replacing close to 85 % of the reconfigurable cells with fixed ones.

Figure 5 illustrates reconfigurable umbrella meshes of different extents of sparse reconfigurability approximating a free form surface. Currently, the distribution of reconfigurable and fixed cells is designed manually. An interesting and challenging problem for future work is to automatically find an optimal distribution for a given space of deployable target shapes.

7 Physical Prototype

To observe the physical behavior of RUMs, we designed a prototype cell with sliding joints to allow adjusting the scissor linkage arms. Figure 6 shows the design of the constituent components, reconfiguration by sliding, and deployment of our fabricated cells. The arms are locked in place once the cell's height values are programmed. We discretize the possible values of h for easy control. This simplification did not affect the achievable shape space noticeably. The desired height level is locked using indents on the sides of the arms. Since each cell's h^t and h^b are equal across the three arms, the discrete levels make it easier to ensure this constraint too. Continuous sliding of h values requires a more sophisticated control and locking system but offers higher shape fidelity. The top and bottom plates feature a compliant T-joint that is CNC milled from a 6 mm thick polypropylene sheet. The arms are CNC milled from a 4 mm thick POM-C sheet. The deployment spacing is enforced through 3D printed spacers that are positioned between the plates.

We assembled a RUM consisting of 37 identical cells to demonstrate reconfiguration in practice. In the initial assembly state, all heights are equal and the structure deploys

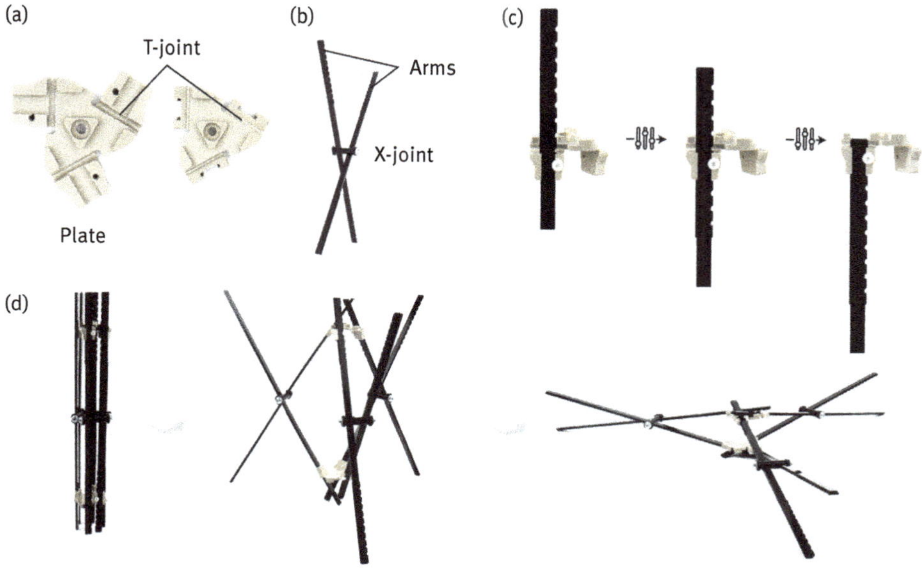

Fig. 6: Components of our Reconfigurable Umbrella Cell: (a) (Left) The top plate in the fabrication state with the T-joints milled and the sleeve for the sliding arm, (Right) The compliant T-joints rotate to form the compact assembly state. (b) The arms assembled with the X-joint (c) The sliding mechanism at the T-joint blocking the heights at three different levels/values. (d) Deployment of a single cell.

into the flat state. Figure 7 illustrates that even such a simple model can be reconfigured to deploy into significantly different shapes of positive and negative Gauss curvature.

8 Applications

Our work focuses on the underlying principle of RUMs and provides an interactive simulation-based approach for designing the deployed structure and exploring potential transformations. RUMs exemplify transformable structures with a high degree of control. We believe the theoretical implications of such transformable structures permeate through several domains and scales. Here we discuss some applications of RUMs in architectural scenarios.

One of our preliminary postulations suggests that RUMs permit adaptability and reuse of architectural structures. Individual cells can be pre-fabricated and assembled off-site to simplify and reduce on-site construction processes. It is also assumed that the compact state of the structure facilitates construction logistics, including transport and storage. The structure is assembled in its compact state and realized through deployment. More fundamental is the capability of RUMs to be taken apart and reconfigured

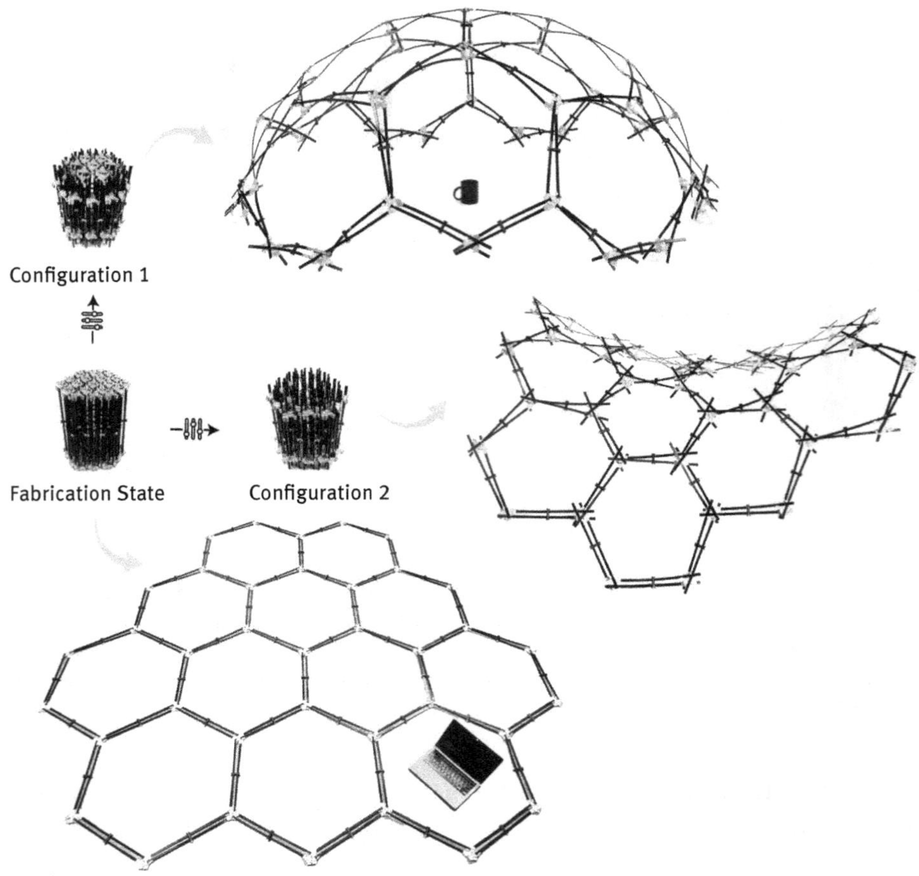

Fig. 7: A fully reconfigurable umbrella mesh composed of 37 identical cells. Sliding the top and bottom plates towards the X-joints effectively shortens the active scissor linkage arms, leading to reduced lateral expansion. In this way, significantly different deployed target shapes can be achieved in a single model. Note how the parts of the arms that are not active, i. e. above the top and below the bottom plates, avoid collisions and form a reciprocal arrangement.

to deploy into a different geometrical shape. This implies that a variety of architectural solutions can be created using the same kit of parts. Between deployments, the structure is returned to its compact state to incorporate a new set of cell heights, or to be disassembled and re-assembled in a new configuration. The inherent modularity of RUMs also allows for topological changes to be easily introduced by altering the number of cells during re-assembly. This type of flexibility permits to introduce major changes in the shape and span of the structure. Under these conditions, it is possible to relocate the structure to a different site and reconfigure its cells to adapt to the needs of that specific location. Eventually, functional, spatial and environmental demands can then drive the transformation of the structure. Inspired by the Great Court at the British

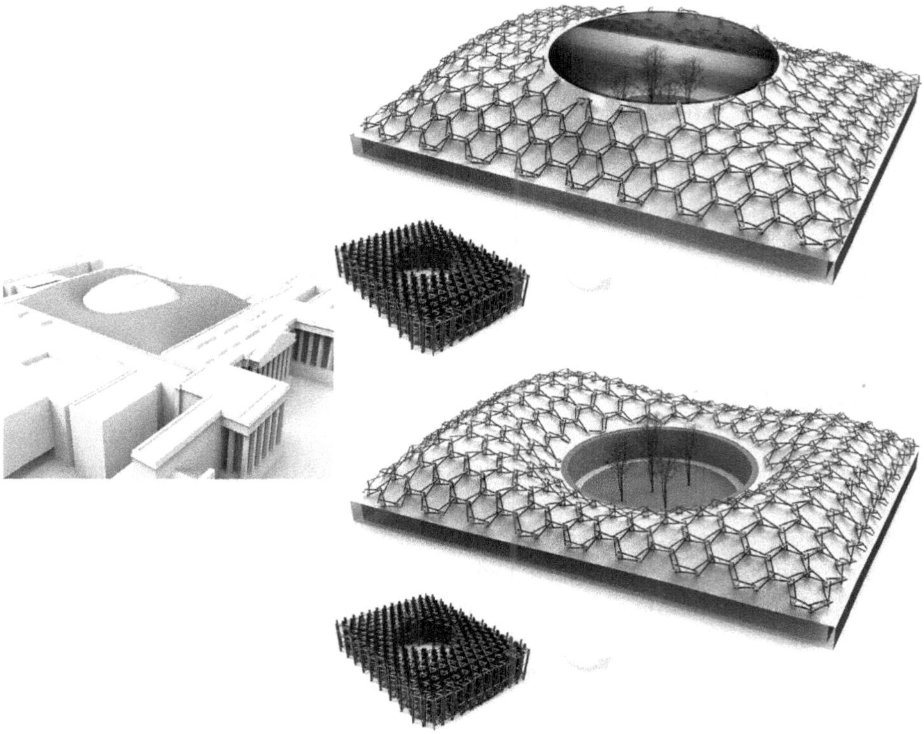

Fig. 8: Concept employing a RUM with two configurations inspired from the geometry of the Great Court at the British Museum.

Museum (Foster and Sudjic 2011), Fig. 8 envisions two possible configurations of a RUM structure. The two configurations provide different spatial layouts for different use cases.

With rapid advances in robotic actuation control, it becomes possible to envision RUMs as a potential type of adaptive structure enabling dynamic transformations. This implies changing the height of the cells without returning to the compact state of the structure. On this subject, we simulated changing the height distribution of the umbrella cells when already deployed to a particular shape. Figure 9 demonstrates a few stages of a transformation of the structure as the initially configured heights corresponding to a saddle shape change to the height configuration for a cylindrical shape. This enables to visualize the concept of adaptive structures controlled by actuators manipulating the cell heights directly in the deployed structure.

In general, efforts are required to explore scalability problems of these types of structures. The materialization of T-joints presents special challenges for the development of cells since there is a need to perform numerous reconfiguration cycles without damage.

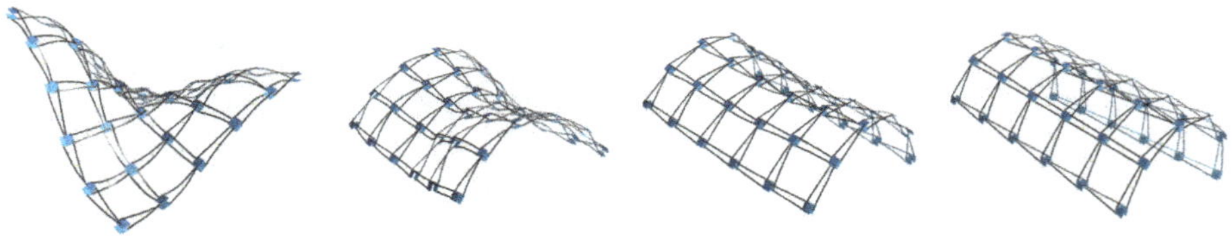

Fig. 9: Active shape morphing: We show a concept where a deployed structure can actively change its shape by reconfiguring the heights when deployed. Here we use a linear interpolation of the height distributions at the ends to generate the intermediate states taking a saddle shape to a cylindrical one.

9 Conclusion

Reconfigurable Umbrella Mesh (RUM) is a new class of deployable structures that features highly adaptable shape configurations, mass-produced, identical, and reusable constituent cells, and a compact-to-surface deployment that supports efficient transport and storage. Future research will be required to define appropriate material systems when scaling the system up/down and analyze the structural behavior of RUMs. Combinations with other transformable materials, such as fabrics, also provide interesting avenues for future research.

Acknowledgements

We are grateful to Michele Vidulis, Filip Goč and Jonathan Chuah for their help in fabricating and assembling the prototype. This research was supported by the Swiss National Science Foundation (Grant FNS 514543 / CF 1156).

References

Akgün, Y., C. J. Gantes, W. Sobek, K. Korkmaz, and K. Kalochairetis (2011). A novel adaptive spatial scissor-hinge structural mechanism for convertible roofs. *Engineering Structures* 33 (4), 1365–76.

Akgün, Y., W. Haase, and W. Sobek (2007, 01). Proposal for a new scissor-hinge structure to create transformable and adaptive roofs. *Proceeding of international association of spatial structures symposium.*

Alegria, Mira, L., A. Thrall, and N. Temmerman (2016, 02). The universal scissor component: Optimization of a reconfigurable component for deployable scissor structures. *Engineering Optimization* 48, 1–17.

Arnouts, L., N. De Temmerman, T. Massart, and P. Berke (2020). Geometric design of triangulated bistable scissor structures taking into account finite hub size. *International Journal of Solids and Structures* 206, 84–100.

Arnouts, L., T. Massart, N. De Temmerman, and P. Berke (2018). Computational modelling of the transformation of bistable scissor structures with geometrical imperfections. *Engineering Structures* 177, 409–20.

Dinevari, N. F., Y. Shahbazi, and F. Maden (2021). Geometric and analytical design of angulated scissor structures. *Mechanism and Machine Theory* 164, 104402.

Foster, N. and D. Sudjic (2011). *The Great Court at the British Museum: Foster + Partners.* Prestel.

Konaković, M., K. Crane, B. Deng, S. Bouaziz, D. Piker, and M. Pauly (2016). Beyond developable: computational design and fabrication with auxetic materials. *ACM Transactions on Graphics (TOG)* 35 (4), 1–11.

Lienhard, J. (2014). Bending-active structures: Form-finding strategies using elastic deformation in static and kinetic systems and the structural potentials therein. *Forschungsberichte aus dem Institut für Tragkonstruktionen und konstruktives Entwerfen der Universität Stuttgart.* Universität Stuttgart Inst. f. Tragkonstr.

Maden, F., Y. Akgün, G. Kiper, E. Gür, M. Yar, and K. Korkmaz (2019). A critical review on classification and terminology of scissor structures. *Journal of the International Association for Shell and Spatial Structures* 60 (1), 47–64.

Moy, J. J., C. S. Tan, S. Mohammad, and A. R. Z. Abidin (2022). State-of-art review on deployable scissor structure in construction. *Structures* 42, 160–80.

Panetta, J., M. Konaković-Luković, F. Isvoranu, E. Bouleau, and M. Pauly (2019). X-shells: A new class of deployable beam structures. *ACM Trans. Graph.* 38 (4), 1–15.

Pillwein, S., K. Leimer, M. Birsak, and P. Musialski (2020). On elastic geodesic grids and their planar to spatial deployment. *ACM Trans. Graph.* 39 (4), 1–12.

Pillwein, S. and P. Musialski (2021, dec). Generalized deployable elastic geodesic grids. *ACM Trans. Graph.* 40 (6).

Ren, Y., U. Kusupati, J. Panetta, F. Isvoranu, D. Pellis, T. Chen, and M. Pauly (2022). Umbrella meshes: elastic mechanisms for freeform shape deployment. *ACM Trans. Graph.* 41, 1–15.

Roovers, K., L. Alegria Mira, and N. De Temmerman (2013, September). From surface to scissor structure. In *Proceedings of the First Conference Transformables 2013,* 275–80. Starbooks. the First Conference Transformables 2013; Conference date: 18-09-2013 Through 20-09-2013.

Roovers, K. and N. De Temmerman (2017). Geometric design of deployable scissor grids consisting of generalized polar units. *Journal of the International Association for Shell and Spatial Structures* 58 (3), 227–38.

Rosenberg, D. (2010). Indeterminate architecture: Scissor-pair transformable structures. Volume 6, pp. 19–39.

Yuta Shimoda, Kai Suto, Sei Hayashi, Tomoyuki Gondo, Tomohiro Tachi

Developable Membrane Tensegrity Structures Based on Origami Tessellations

Abstract: In this paper, we propose a design method of membrane tensegrity structures by solving the inverse problem using freeform origami tessellations by generalizing Resch's patterns. We found that if we generate an origami tessellation and replace the mountain crease of the tessellation with struts and the other parts with a membrane under initial tension, the tessellation can be formed as a tensegrity structure. In addition, this structure may snap through and deform significantly, but we have shown that the mean curvature of a given surface is a valid evaluation index as a condition for a surface that does not snap through. When computed origami tessellation patterns are naively used, the shrinking of membranes caused by initial tension results in a deviation from the target shape. We propose an optimization-based framework that resolves the deviation and ensures that the structures match the given surfaces.

Keywords: tensegrity, membrane tensegrity, inverse problem, origami, origami tessellation

1 Introduction

Developable membrane tensegrity structures are tensegrity structures formed from single planar membranes used as the tensile member and struts used as the compression members. They have advantages in the manufacturing process as they can be constructed in planar states by laying out struts on a single pre-tensioned membrane and erected by releasing the pretensions. Several methods for "forward problems" to find the three-dimensional resulting shape from a two-dimensional layout pattern have been proposed.

One example is the MOOM Pavilion, created by students of the Tokyo University of Science under the supervision of Kazuhiro Kojima and Jun Sato (Ratschke et al. 2017). They developed a staggered arrangement of parallel struts and a tunnel-like structure with arches in the direction parallel to the struts and cables pulling in the transverse direction. The pavilion consists of 131 aluminum tubes and an elastic polyester mesh, and was successfully constructed as an 8×26 m.

Gupta et al. (2020) extended the designable structure by placing compression struts in a reciprocal pattern on the membrane. They also demonstrated a workflow for accurately producing knits using CNC-produced knittings as the main tensile element. The process of creating these structures is form-finding through mechanical simulation

https://doi.org/10.1515/9783111162683-023

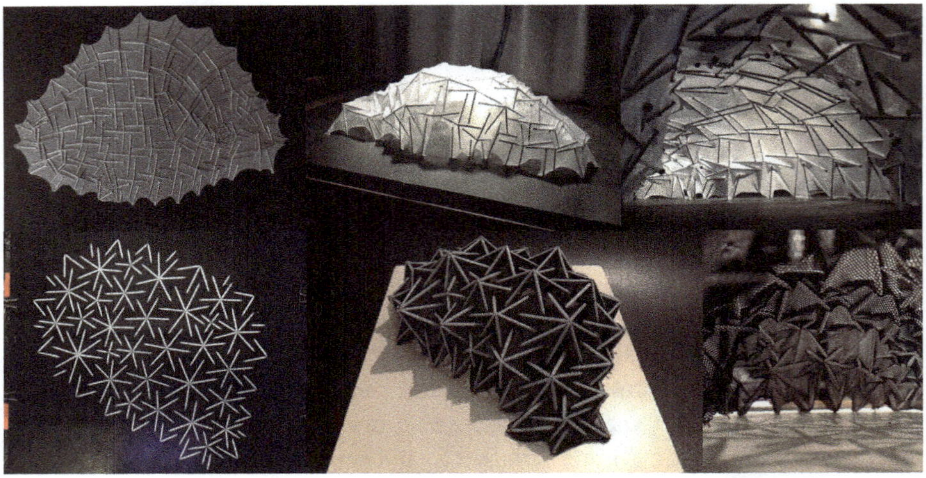

Fig. 1: Process of model making. Top left: Fixing aluminum pipes to stretched stretchable nylon cloth. Top right: When the boundary is cut out and the foot is fixed, the structure stands up in the same shape as the given curved surface.Bottom left: 3D printing PLA directly onto stretched stretchable nylon fabric. Bottom right: When the boundary is cut out and the foot is fixed, a structure with the same shape as the given curved surface rises up.

and model making, solving forward problems to determine what 3D shape can be obtained from a given 2D strut pattern.

However, the inverse problem, i. e., designing the planar pattern when the target shape is given, has not been solved. In this paper, we propose a design method of membrane tensegrity structures by solving the inverse problem using freeform origami tessellations by generalizing Resch's patterns proposed by Tachi (2013). Fig. 1 shows models generated by this method. When the struts are attached to the elastic membrane under tension, or directly 3D printed, and the surrounding area is cut away to release the tension, a structure with the same shape as the given surface rises up. In Sec. 2, we present the basic strategy for converting a polyhedral mesh into a tensegrity structure.

Through experimentation, we have identified two problems and propose the following solution for each problem. (1) Firstly, the obtained structure may buckle into an undesired state depending on the curvature of the input mesh. In Sec. 3, we perform an experiment using constant mean curvature surface to empirically estimate the conditions under which the tensegrity structure can be established without collapse. (2) Secondly, the shrinking of membranes causes a deviation of a surface from the target shape. In Sec. 4, we propose an optimization method to correct deviations due to membrane deformation.

2 Inverse problem of membrane tensegrity structures

Several studies suggest the relationship between origami and tensegrity structures. The analytical model of MOOM pavilion, in which tension members are replaced by cable is very similar to the Yoshimura pattern (Ratschke et al. 2017). The printing pattern on the membrane used in Jourdan et al. (2020) is similar to Resch's pattern. Tachi (2012) points out the theoretical equivalence of an infinitesimal folding mode of a polyhedral surface and the equilibrium of forces to show that the same shape can be used as a shaky polyhedron and as a tensegrity structure.

In this study, we aim to make use of the universality of computational origami design, i. e., origami-based patterns can be computed for given polyhedral meshes Tachi (2009, 2013) for the design of membrane tensegrity structures that can be deployed from flat states.

Specifically, we show that if we generate an origami tessellation by generalizing Resch's patterns proposed by Tachi (2013), and replace the mountain crease of the tessellation with struts and the other parts with a membrane under initial tension, the tessellation can be formed as a tensegrity structure (Fig. 2). The flow of membrane tensegrity structure generation is as follows:

1. Provide a target shape as a polyhedral mesh.
2. *Origamise*, i. e., apply the algorithm of (Tachi 2013) to create a generalized Ron-Resch pattern. We choose star-shaped wrinkle patterns inserted on the edges of the given mesh and let the corrugated surface be developable using an optimization scheme. Here, either *Freeform Origami* (Tachi 2013) or *Crane* (Suto et al. 2022) can be used for optimization calculations.
3. Replace the mountain crease lines of the obtained mesh with compression members and the other creases with tension members under initial tension by setting the target length to its scaled version by some scale factor $s < 1$ (empirically chosen $s \approx 0.90$). Support points are provided around the boundary of the mesh. To obtain the equilibrium position, we use dynamic relaxation. *Kangaroo2* (Piker 2013) was used in our analysis.

After verifying the form under equilibrium, we stretch the membrane in a planar state by some scale factor $1/s$ and attach rigid material along the mountain creases of the crease pattern. Here, for a small-scale model, the attachment of rigid bars can be replaced by 3D printing directly to the membrane as in Tibbits (2016); Jourdan et al. (2020). We release the prestress by cutting out the boundary of the crease pattern; then the membrane's tension produces the curvature and after a small adjustment and the pinning at the support point. We obtain the desired membrane tensegrity structure.

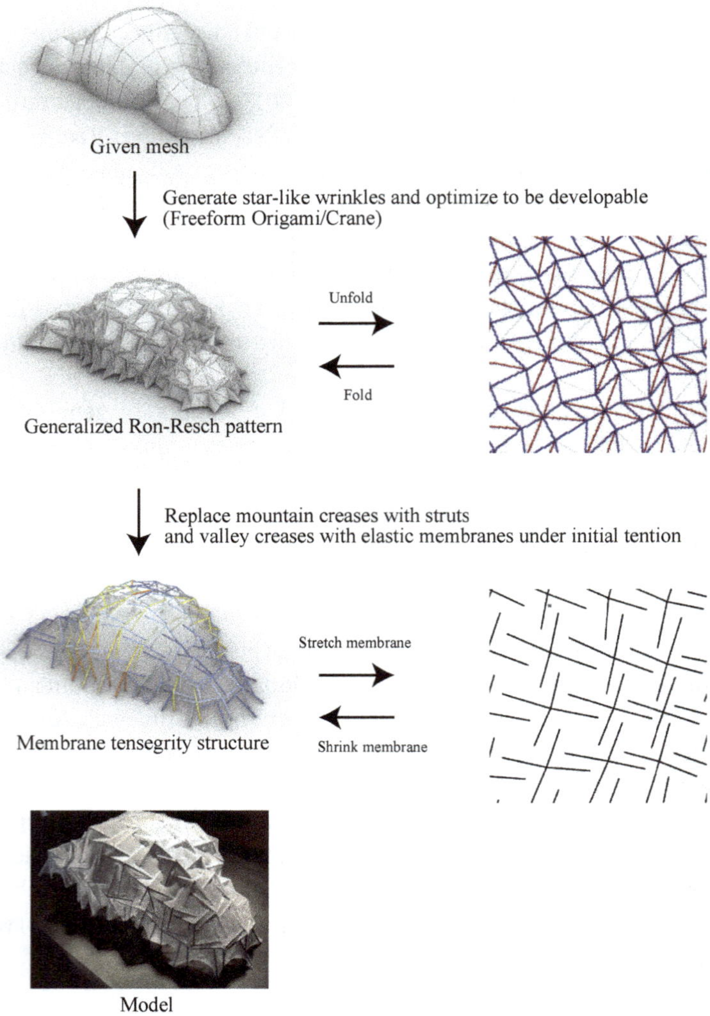

Fig. 2: The process of solving inverse program of a membrane tensegrity structure.

Here, we have obtained two interesting problems. (1) Firstly, the obtained structure may buckle into an undesired state depending on the curvature of the initial mesh and the stretch factor of the membrane. We clarify these conditions in the next section. (2) Secondly, the shrinking of membranes causes a deviation of a surface from the target shape. In Sec. 4, we propose an optimization-based scheme to correct deviations due to membrane deformation.

3 Conditions for membrane tensegrity to maintain its shape

Here, we show the conditions for a surface to which this structure can be adapted. In the simulation phase, i. e., Step 3 in Sec. 2, we apply the equilibrium force by lowering the shrinkage factor s from 1 to a smaller value. Normally, the structure gets stabilized as s lowers; however, we have found that some surfaces with concave parts may lose stability through a snap-through event. Specifically, the star-shaped "spikes" at the concave part of the surface will snap-through when the membrane section is tensioned towards zero length (Fig. 3). When adapting to freeform surfaces, it is important to avoid such snap-through that may lead to the collapse of the entire structure.

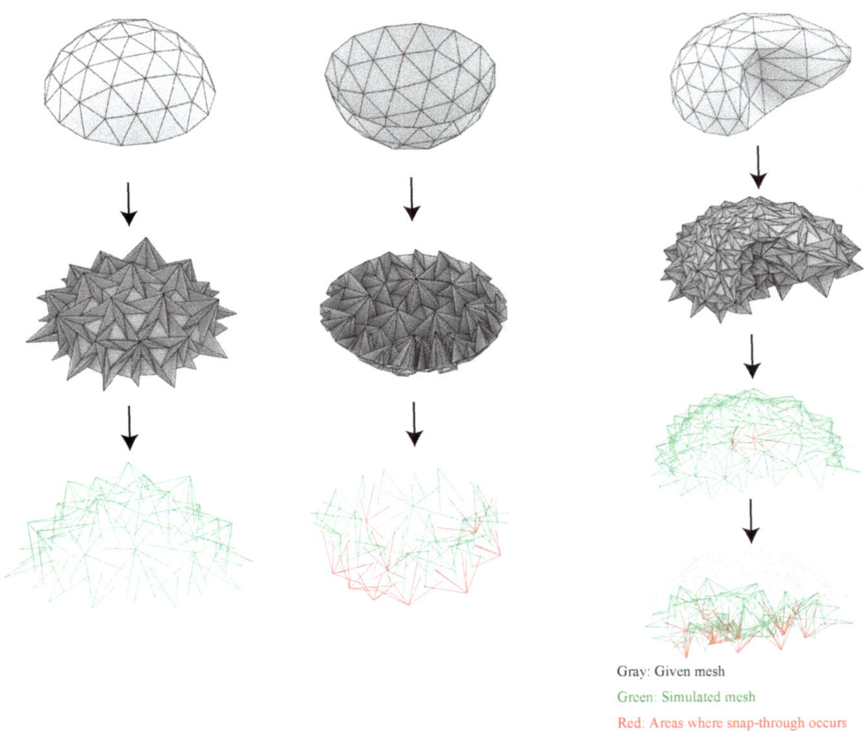

Fig. 3: When the membrane section is tensioned towards zero length, it can be formed as a tensegrity structure if the surface is convex (left), but if it is concave, the vertices will snap through and deviate significantly from the target shape (center). When adapting to freeform surfaces, if concave areas are included, snap-through may occur and the entire structure may collapse (right).

We can observe from the two left figures of Fig. 3 that the system is more stable when the star-shaped spikes orients toward the curvature direction. Also, an example in Fig. 2 shows that even a surface with a concave part can be realized if the surface is “roughly convex”. Therefore, we propose the mean curvature as an empirical criteria for evaluating surfaces that do not snap through. One surface with a constant mean curvature, the unduloid, was discretised and a small mesh consisting of seven degree-6 vertices was taken as the test piece. Generate tensegrity from this small mesh, and then simulated under tension to verify whether snap-through occurs. When the average length of the discretised mesh edges was set to 1 and the mean curvature was 0.05, 0.10, 0.15 and 0.20, it was verified that snap-through occurred at mean curvatures below 0.10 but not at curvatures above 0.15 (Fig. 4).

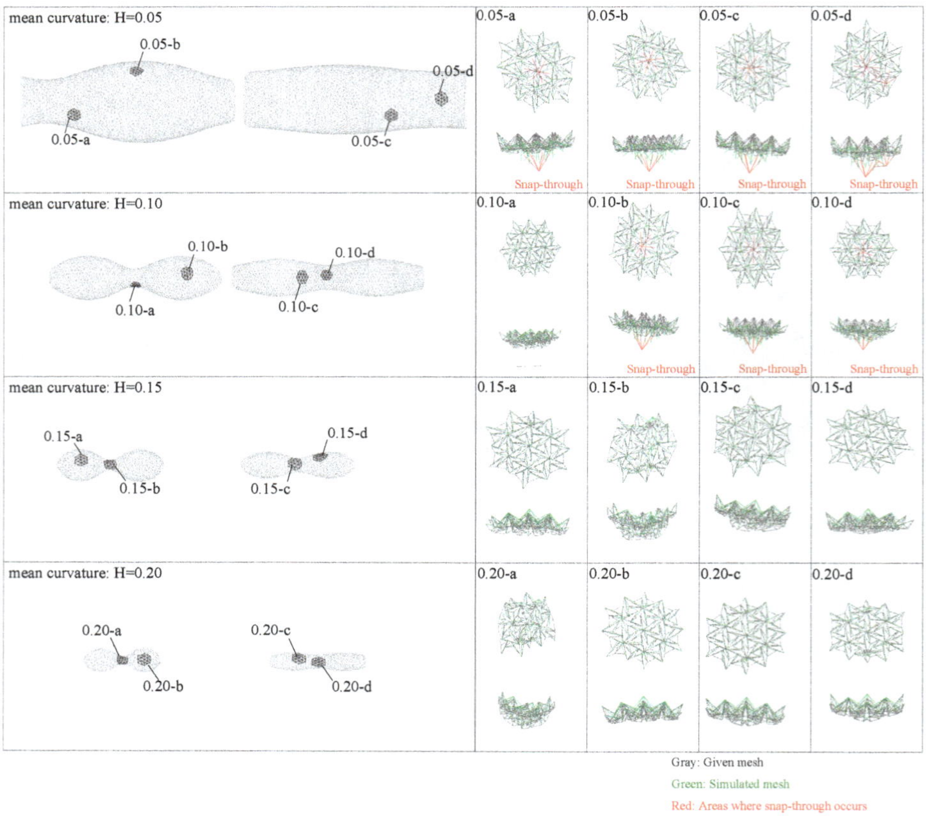

Fig. 4: A small mesh was taken from a surface of constant mean curvature to see if it would snap through under tension on the membrane. Snap-through occurred at mean curvatures below 0.10 but not at curvatures above 0.15.

From the study, it is believed that a stable structure can be obtained by keeping the average curvature above a certain level. In the future, vertices with degrees other than the 6-degree vertex should be considered.

4 Optimization of resolving deviation

We propose an optimization-based framework that resolves the deviation caused by the membrane shrinkage and ensures that the structures match the given surfaces (Fig. 5). The difficulty of the problem is that the actual lengths (relative to original lengths)

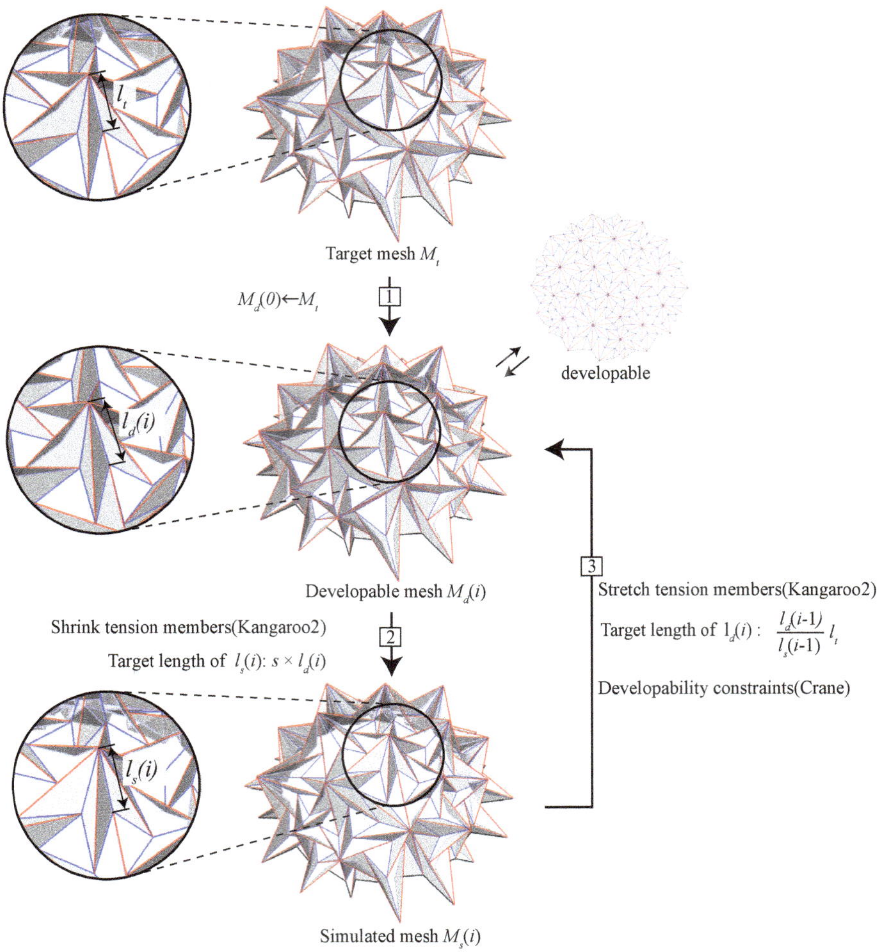

Fig. 5: The process of optimization.

of tension members under equilibrium are chosen between s and 1 as the process of equilibrium and cannot be predefined.

Therefore, our strategy is to iterate the simulation process of the tensegrity structures and correct the scale factor parameters for each step. In each step $i = 0, 1, 2, \ldots,$ we keep three meshes: a target mesh, a deployable mesh, and a simulated mesh.

The target mesh M_t is generalised Ron-Resch pattern of the given surface. The deployable mesh $M_d(i)$ is the membrane stretched under initial tension and is deployable. The simulated mesh $M_s(i)$ is the actual final shape resulting from releasing the tension in the deployable mesh, which should match the target mesh.

In each of the three meshes, the length of the struts (mountain crease lines) should remain unchanged. The lengths of the membrane elements (valley crease lines) are modified during the optimisation process; For each tension edge, we let l_t, $l_d(i)$ and $l_s(i)$ be its lengths in M_t, $M_d(i)$, and $M_s(i)$, respectively. $1/s$ is an initial stretch of the membrane. The optimisation process is as follows,

1. Set $M_d(0) \leftarrow M_t$.
2. Obtain $M_s(i)$ from $M_d(i)$. Specifically, set the target length of $l_s(i)$ to $s \times l_d(i)$ and simulate a dynamic relaxation method.
3. Obtain $M_d(i+1)$ from $M_s(i)$. Specifically, set target length of $l_d(i)$ to $\frac{l_d(i-1)}{l_s(i-1)} l_t$ while applying developability constraints on $M_d(i)$ using *Crane*.
4. Return to step 2 and repeat this until the deviation between the target mesh and the simulated mesh is sufficiently small.

As an example, we tested this process on a tensegrity structure with an average strut length of 1. The average deviation at each vertex was reduced by 54%, from 0.098 to 0.045. The maximum deviation decreased 46% from 0.365 to 0.198 (Fig. 6).

5 Outlook and Future Works

The construction of complex free-form tensegrity can be costly, labor-intensive, and time-consuming due to challenges in reproducing the shape and assembling the structure. Our proposal potentially solves these challenges by enabling various complex three-dimensional forms constructed from a single plane. This method uses a single planar membrane as a guide, and even inexperienced workers can easily construct free-form tensegrity structures. In the future, there is a need to study using full-scale mockups and to clarify the conditions under which tensegrity is established under various loading conditions.

In addition, we tested only the degree-6 vertices case regarding the mean curvature at which snap-through does not occur; we would like to study vertices of other degrees as well. The current algorithm to reduce the deviation converge to a state with non-zero deviation. This may be because we try to equate the intrinsic metric of the equilibrium

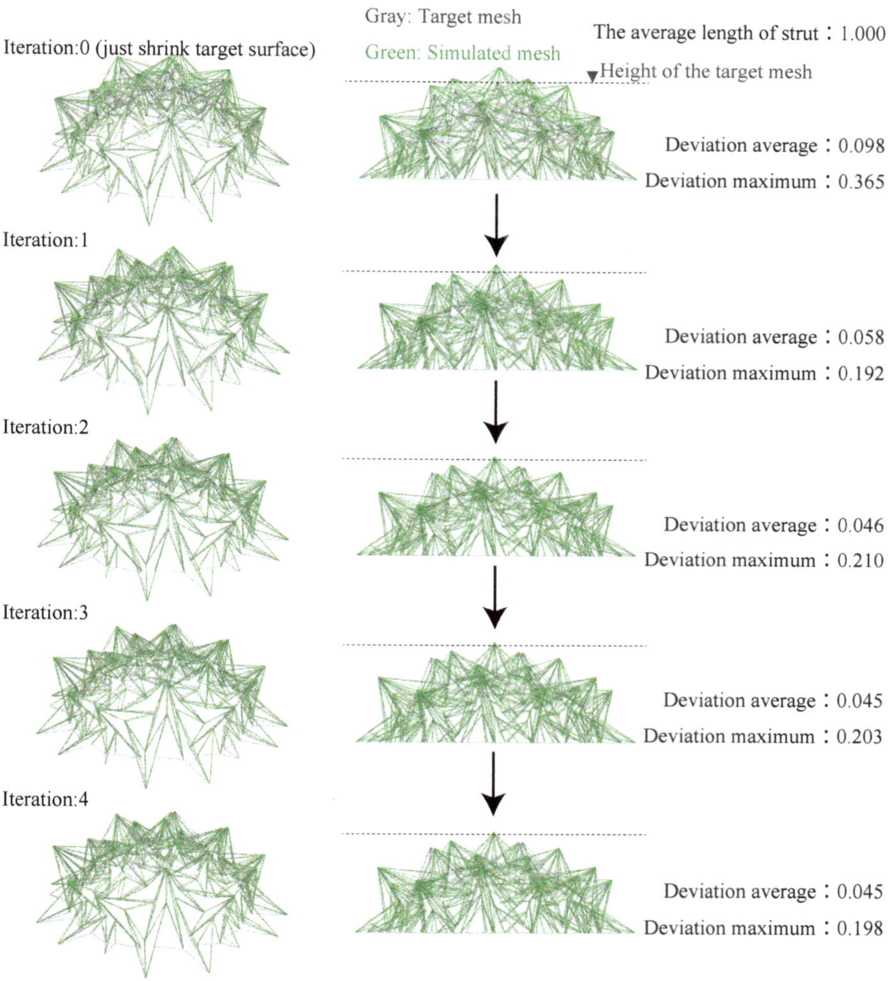

Fig. 6: The result of optimization.

state and the target state already given as a corrugated origami; direct measurement and minimization of the proximity to the given mesh target may improve our result.

References

Gupta, S. S., Y. Y. Tan, P. Z. Chia, C. P. Pambudi, Y. H. Quek, C. Yogiaman, and K. J. Tracy (2020). Prototyping knit tensegrity shells: a design-to-fabrication workflow. *SN Applied Sciences* 2(6), 1–13.

Jourdan, D., M. Skouras, E. Vouga, and A. Bousseau (2020). Printing-on-fabric meta-material for self-shaping architectural models. *Advances in Architectural Geometry 2020.*

Piker, D. (2013). Kangaroo: form finding with computational physics. *Architectural Design* 83(2), 136–37.

Ratschke, N., A. Bögle, and J. Lindenberg (2017). Parametric analysis of tensegrity-membrane-structures. In *Proceedings of IASS Annual Symposia,* Volume 2017, pp. 1–9. International Association for Shell and Spatial Structures (IASS).

Suto, K., Y. Noma, K. Tanimichi, K. Narumi, and T. Tachi (2022). Crane: An integrated computational design platformfor functional, foldable, and fabricable origami products. *ACM Transactions on Computer-Human Interaction.*

Tachi, T. (2009). Origamizing polyhedral surfaces. *IEEE transactions on visualization and computer graphics* 16(2), 298–311.

Tachi, T. (2012). Design of infinitesimally and finitely flexible origami based on reciprocal figures. *J. Geom. Graph* 16(2), 223–34.

Tachi, T. (2013). Designing freeform origami tessellations by generalizing resch's patterns. *Journal of mechanical design* 135(11).

Tibbits, S. (2016). *Self-assembly lab: experiments in programming matter.* Taylor & Francis.

Andrea Micheletti, Kilian Bruckner, Ilaria Giannetti

Parametric Design of Tensegrity-Origami Structures

Abstract: Lightweight, deployable, and reusable structures attracted increasing attention over the years in terms of construction sustainability. In this research area, tensegrity and origami-inspired structures were extensively investigated for different applications. This contribution focuses on the design of two-dimensional geometric patterns of vertices, edges, and faces that can be used for both tensegrity and origami structures, which are here analyzed in connection to each other. The design methodology is based on a semi-analytical form-finding procedure in which two set of equations are simultaneously solved: the rank-deficiency conditions on the equilibrium operator for the tensegrity structure and the developability conditions for the corresponding origami. In addition, the relation between geometry and selfstress of the obtained structures is analyzed by parametric simulations. A 3D modeling workspace – combining Grasshopper™ and Matlab™ codes – supports the design phases from the study of the geometric pattern, satisfying the rank-deficiency and the developability conditions, to the evaluation of the selfstressed tensegrity configuration. The case of a Snelson tensegrity tower functioning as a lightweight, deployable, and reusable pavilion is considered to demonstrate the proposed methodology for the design of tensegrity-origami structures.

1 Introduction

The interest in origami structures for applications in architecture started developing since the beginning of last century with the studies of Joseph Albers at the Bauhaus, and several experiences accumulated over time. Nowadays, detailed analytical models and computer-aided simulations make it possible to design complex origami structures with shape-morphing capabilities (see e. g. Micheletti et al. 2022 and the literature cited therein), and in recent years there was a spike in the interest on origami in other fields such as physics (Evans et al. 2015) and robotics (Treml et al. 2018). In the middle of last century, another peculiar type of structures came to the attention of architects, engineers, and mathematicians: those based on the tensegrity concept, a term coined by Richard Buckminster Fuller, which was popularized by the numerous eye-catching sculptures that the artist Kenneth Snelson realized mainly in US. The numerous diverse studies about and applications of tensegrity systems were recently reviewed in Micheletti and Podio-Guidugli 2022.

The purpose of this work is to investigate the feasibility of tensegrity-origami systems, and possibly to devise a dedicated design methodology. Tensegrity-origami

https://doi.org/10.1515/9783111162683-024

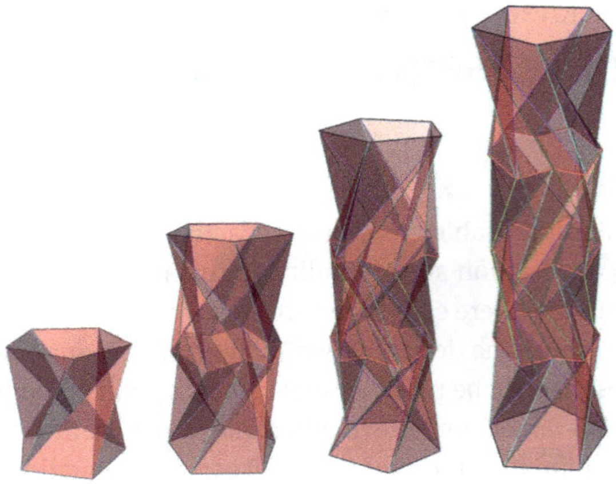

Fig. 1: Representative geometry of tensegrity-origami towers with pentagonal base. From left to right, the number of modules passes from $m = 1$ to $m = 4$.

systems are intended here as planar geometric patterns of nodes, edges, and faces, which can be adopted to realize both origami and tensegrity three-dimensional structures. We focus on the case of Snelson's modular tensegrity towers which can also be seen as tubular triangulated origami (Fig. 1). We combine the typical form-finding conditions of tensegrity systems with the conditions necessary to develop the corresponding origami surface on the plane. We show that tensegrity-origami towers are feasible for any number of modules and present the parametric procedure we utilized to determine their geometry and state of selfstress. A modular tensegrity-origami pavilion is presented as a design example.

2 Semi-analytical form-finding procedure

2.1 Geometry of Snelson's towers

Snelson has experimented on several occasions with the construction of tower-shaped tensegrity structures. Snelson's tensegrity towers are modular. Each module consists of a lattice assembly of struts and cables. The modules have $n > 2$ struts and n-sided polygonal bases. In assembled configuration, modules are juxtaposed along the vertical axis in number m. The assembly of the modules exhibits a cyclic symmetry, with relative rotation angle about the vertical axis equal to $\frac{2\pi}{n}$. The base module of the tower, the tensegrity frustum, is described hereafter by considering the case of triangular bases ($n = 3$). Such module (so-called T3) can be obtained by considering a triangular frustum with cables along the edges, and non-contiguous struts placed along one diagonal of the lateral faces; in addition, one base is rotated with respect to the other one about the vertical axis by a certain twist angle φ, measured as shown in Fig. 2. It is well-known that symmetric stable selfstressed equilibrium configurations of a T3 are required to a have a twist equal to $\varphi = \theta_0 = \frac{\pi}{n} + \frac{\pi}{2}$, while the size of each base and the height of the frustum can be chosen freely (see e. g. Oppenheim and Williams 2000). It is worth noticing that, given the lengths of all the edges, it is possible to realize two mirrored versions of the T3, referred to as modules with opposite handedness.

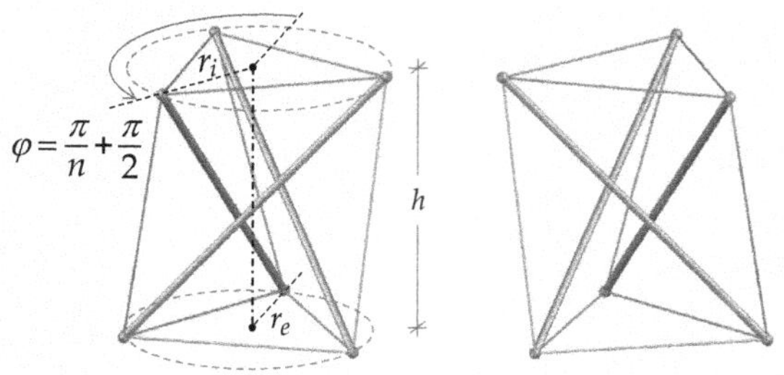

Fig. 2: Two triangular tensegrity frustums, or T3s ($n = 3$), with same geometric parameters and opposite handedness.

The assembly of the modules is obtained by superposing them sequentially along the vertical axis of the tower. The criterion of assembly is based on the superposition of two base modules with opposite handedness. By referring to Fig. 3 (left), a two-lewel tower can be obtained through the following steps: the cables of the lower base of the superimposed module are removed; the lower nodes of the superimposed module are connected with the mid-points of upper cables of the lower module; each of these cables is, thus, divided in two different ones called 'saddle cables'. The number of saddle cables, marked in red, is equal to $2n$. Then, additional n cables, marked in green and called 'diagonal cables', are added to each module. The other types of cables are the 'vertical cables', marked in blue and connecting struts of the same module, and end-base cables, marked in black. The overlap length d is the distance between the centers of the two polygons whose vertices are the lower end nodes of the upper module and the upper end nodes of the lower module.

The assembly of the tower with more than two modules can be obtained by repeating the operations previously described for the superposition of the next module. As Fig. 3 (right) shows, the tower with three levels has two types of modules: the 'end module', placed at the lower and upper extremities, and the 'intermediate module', placed in-between. In multi-level towers, a sequence of intermediate modules between two end modules can be recognized.

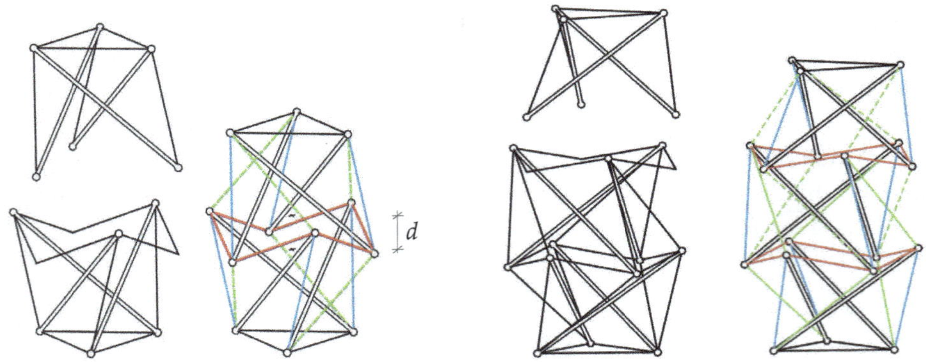

Fig. 3: Sequential assembly of a tensegrity tower.

2.2 Rank-deficiency condition

2.2.1 Equilibrium, kinematic compatibility, and prestress stability

Snelson's towers belong to the class of tensegrities which are simultaneously over- and under-determined structures, which means that they have a rank-deficient equilibrium operator and that they possess both internal mechanisms and selfstress states. Tensegrity towers are also prestress-stable, that is, their selfstress state imparts first-order stiffness to all internal mechanisms.

The equilibrium and kinematic compatibility equations of a tensegrity structure composed of N nodes and E members, or edges, can be written in compact form as

$$\mathbf{A}\mathbf{t} = \mathbf{f}\,, \quad \mathbf{B}\mathbf{v} = \mathbf{e}\,,$$

where $\mathbf{A} = \widetilde{\mathbf{A}}(\mathbf{p})$ is the equilibrium operator, dependent on the $3N$-dimensional nodal position vector $\mathbf{p}$, $\mathbf{t}$ is the E-dimensional vector containing the force coefficients of the members, $\mathbf{f}$ is the $3N$-dimensional vector containing the nodal loads, $\mathbf{B} = \mathbf{A}^T$ is the kinematic-compatibility operator, $\mathbf{v}$ is the $3N$-dimensional vector of nodal velocities, and $\mathbf{e}$ is the E-dimensional vector containing the elongation rates of the members (see Calladine and Pellegrino 1991 for details). Selfstress vectors, $\mathbf{t}_s$, and internal-mechanism vectors, $\mathbf{v}_m$, belong to the nullspace of $\mathbf{A}$ and $\mathbf{B}$, respectively,

$$\mathbf{A}\mathbf{t}_s = \mathbf{0}\,, \quad \mathbf{B}\mathbf{v}_m = \mathbf{0}\,,$$

the former representing the internal forces in equilibrium with null loads, and the latter representing the non-rigid nodal velocities for which the rate of elongation of each member is null.

The number of independent selfstress states and internal mechanisms, denoted respectively by S and M, are related to the number of nodes and edges by the extended

Maxwell's rule (Calladine 1978),

$$3N - 6 - E = M - S\,,$$

in which 6 is the number of rigid-body motions in three dimensions.

For an equilibrium configuration $\mathbf{p}_0$ of a tensegrity system possessing both selfstress states and internal mechanisms, we can express the prestress-stability condition as follows. Let $\mathbf{t}_s$ be the selfstress state, and $\mathbf{g}$ be the $3N$-dimensional vector of geometric loads, that is, the out-of-balance nodal loads induced by a change in configuration along an internal mechanism $\mathbf{v}_m$, which is computed as $\mathbf{g} = \widetilde{\mathbf{A}}(\mathbf{v}_m)\mathbf{t}_s$; then, the tensegrity system is prestress stable if

$$\mathbf{g} \cdot \mathbf{v}_m > 0 \quad \forall \mathbf{v}_m \,:\, \mathbf{A}^T \mathbf{v}_m = \mathbf{0}\,,$$

which correspond to requiring that geometric loads spend positive work on all internal mechanisms.

In summary, when $3N-6-E \geq 0$, which is the case of several instances of tensegrity systems, in order for a configuration with prestress $\mathbf{t}_s$ to be feasible, we require the rank of the equilibrium operator $\mathbf{A}$ to be not maximal (rank-deficiency condition) and that the prestress-stability condition is satisfied.

2.2.2 A particular form-finding solution for Snelson's towers

The equilibrium operator for tensegrity towers with triangular end bases and arbitrary number m of modules is represented by a square matrix, with just one internal mechanism and one selfstress state, $M = S = 1$. For a tower with n-sided poligonal end bases, S is still equal to 1 while the number of internal mechanisms is $2(n-3)+1$. It can be seen that such a tower can be transformed into one with $M = 1$ by adding edges along the cords connecting vertices of end polygons.

Hence, the rank-deficiency condition assumes the simple form

$$\det(\mathbf{A}_r) = 0\,, \tag{1}$$

with $\mathbf{A}_r$ obtained from $\mathbf{A}$ by eliminating 6 dependent rows.

Previous work (Micheletti 2003) showed that the condition (1) for multimodular towers with $m \geq 3$ and polygonal base can be expressed analytically if all intermediate modules are identical, with the two end modules equal to each other, with respect to geometry and selfstress. Such a restriction leads to the pair of scalar conditions,

$$f_e(n, \theta_e, \delta_e, \rho) = 0\,, \qquad f_i(n, \theta_i, \delta_i) = 0\,, \tag{2}$$

which are expressed in terms of the following parameters (see Fig. 4): the number of sides of the bases, n; the overtwists of the end module and inner module, $\theta_e = \varphi_e - \theta_0$

and $\theta_i = \varphi_i - \theta_0$, with $\theta_0 = \frac{\pi}{n} + \frac{\pi}{2}$; the ratios between the overlap distance d and the height of the end module and inner module, $\delta_e = d/h_e$ and $\delta_i = d/h_i$; the ratio between the radii of the circles circumscribed to the bases of the end modules, $\rho = r_e/r_i$ (cf. Fig. 4). For $m = 2$, condition (1) correspond to just the first of (2). The two functions $f_e(n, \theta_e, \delta_e, \rho)$ and $f_i(n, \theta_i, \delta_i)$ are written explicitly as

$$f_e(n, \theta_e, \delta_e, \rho) = \delta_e^2(\rho \sin\theta_0 + \sin\theta_e) + \delta_e(1 - \sin\theta_0)(\rho - 2\sin\theta_e) + 2(1 - \sin\theta_0)\sin\theta_e,$$

$$f_i(n, \theta_i, \delta_i) = \delta_i^2(\sin\theta_0 + \sin\theta_i) + \delta_i(1 - \sin\theta_0)(1 - \sin\theta_i) + (1 - \sin\theta_0)\sin\theta_i\,.$$

For later use, we introduce the two functions, $\delta_e = \widetilde{\delta}_e(n, \theta_e, \rho)$ and $\delta_i = \widetilde{\delta}_i(n, \theta_i)$, implicitly defined by the conditions (2) for $\theta_e, \theta_i \in [-\frac{\pi}{2}, 0]$ and $\delta_e, \delta_i \in [0, 1]$.

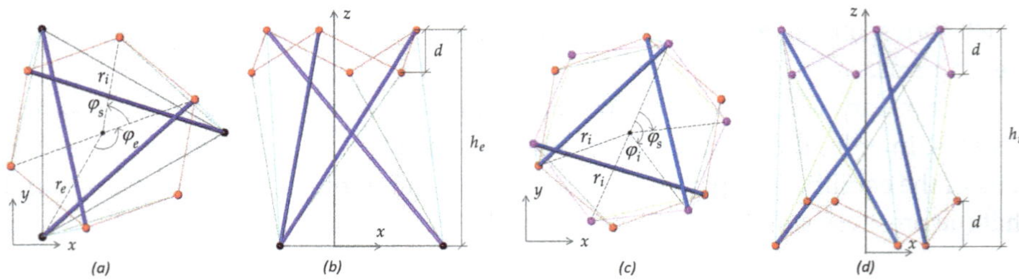

Fig. 4: Top and side view of end module (a,b) and intermediate module (c,d).

2.3 Developability conditions

Given a cyclic symmetric tensegrity tower, we consider a corresponding spatial origami surface obtained by replacing each triangle of edges by a triangular face (except for the end bases if $n = 3$). Then we cut such origami surface along a path of edges connecting the end bases, and we consider a target developed configuration in which all triangles are brought to lie in the same plane, maintaining the same incidence relation between nodes, edges, and faces. The developability condition are imposed by requiring the angles at an internal node to sum up to 2π, and the angles at a boundary node to sum up to π.

In the following, we treat the cases $m = 2$, $m = 3$, and $m \geq 4$ separately. It is worth observing that, for $m = 1$ the tower is a tensegrity frustum, and the corresponding origami is trivially developable (cf. Fig. 5): a cut along one of the lateral edges permit to develop it on a plane.

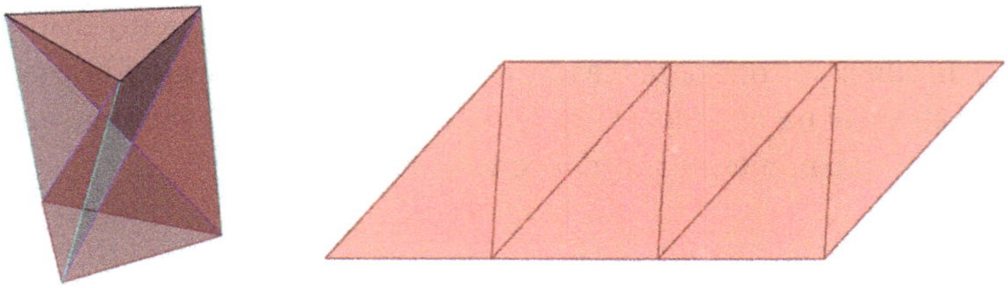

Fig. 5: The particular case of a tensegrity prism and the corresponding developed geometry.

2.3.1 Two-level towers

Cyclic-symmetric two-level towers have one type of internal nodes and one type of boundary node. We number the triangular faces and denote the interior angles of face k by $\alpha_k, \beta_k, \gamma_k$, as shown in Fig. 6.

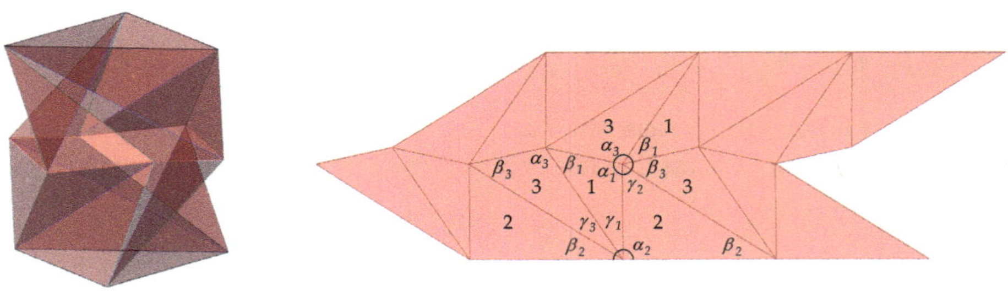

Fig. 6: A two-level tower and the corresponding developed geometry.

The developability conditions on boundary nodes and internal node are expressed respectively as

$$\phi_e = 0\,, \quad \phi_i = 0\,, \tag{3}$$

where the corresponding angle summations are functions of the geometric parameters already introduced,

$$\phi_e = \widetilde{\phi}_e(n, \theta_e, r_e, r_i, h_e, d) = \alpha_2 + \beta_2 + \gamma_1 + \gamma_3 - \pi\,,$$
$$\phi_i = \widetilde{\phi}_i(n, \theta_e, r_e, r_i, h_e, d) = \alpha_1 + \alpha_3 + \beta_1 + \beta_3 + \gamma_2 - 2\pi\,.$$

It is worth noticing that, since $\alpha_k + \beta_k + \gamma_k = \pi$, $k = 1, 2, 3$, we have $\phi_e + \phi_i = 0$, which implies that it is enough to impose the developability condition on one type of node to ensure that also the condition on the other type is satisfied. In order to satisfy conditions (3) together with the first of (2), we may use the function $\widetilde{\delta}_e$ defined in Sec. 2.2.2 to eliminate the dependence of $\widetilde{\phi}_e$ on d, obtaining

$$\phi_e = \widehat{\phi}_e(n, \theta_e, r_e, r_i, h_e) = \widetilde{\phi}_e(n, \theta_e, r_e, r_i, h_e, \widetilde{d}(n, \theta_e, \rho))\,,$$

with $\widetilde{d} = h_e\widetilde{\delta}_e$. Now, the condition $\widehat{\phi}_e(n, \theta_e, r_e, r_i, h_e) = 0$ permit to implicitly define $r_i = \widetilde{r}_i(n, \theta_e, r_e, h_e)$. In summary, one possible way to find a feasible tensegrity-origami pattern is to take the string of parameters $\{n, \theta_e, r_e, h_e\}$ as independent variable, and to determine the remaining parameters r_i and d consequently.

2.3.2 Three-level towers

Three-level cyclic-symmetric towers have three types of nodes, that is, one type of boundary nodes and two types of internal nodes. We express the developability conditions as

$$\phi_e = 0\,, \quad \phi_{i1} = 0\,, \quad \phi_{i2} = 0, \tag{4}$$

where the angle summations are (Fig. 7)

$$\begin{aligned}
\phi_e &= \widetilde{\phi}_e(n, \theta_e, r_e, r_i, h_e, d) = \alpha_2 + \beta_2 + \gamma_1 + \gamma_3 - \pi\,,\\
\phi_{i1} &= \widetilde{\phi}_{i1}(n, \theta_e, \theta_i, r_e, r_i, h_e, h_i, d) = \alpha_1 + \alpha_4 + \beta_3 + \beta_5 + \gamma_2 - 2\pi\,,\\
\phi_{i2} &= \widetilde{\phi}_{i2}(n, \theta_e, \theta_i, r_e, r_i, h_e, h_i, d) = \alpha_3 + \alpha_5 + \beta_1 + \beta_4 + \gamma_4 + \gamma_5 - 2\pi\,.
\end{aligned}$$

Analogously to the previus case, since $\alpha_k + \beta_k + \gamma_k = \pi$, $k = 1, \ldots, 5$, we have $\phi_e + \phi_{i1} + d\phi_{i2} = 0$, which means that it is only necessary to impose two out of the three conditions in (4).

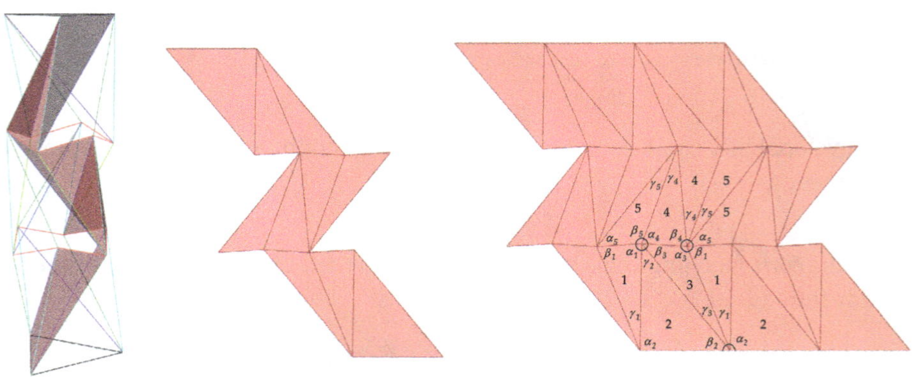

Fig. 7: A three-level tower and the corresponding developed geometry.

In order to satisfy the developability conditions (4) together with the rank-deficiency conditions (1) we may first proceed by expressing r_i and d in terms of $\{n, \theta_e, r_e, h_e\}$, in the same way as in the case of a two-level tower, then we can use the function $\widetilde{\delta}_i$ introduced in Sect. 2.2.2 to obtain $\widetilde{h}_i = \widetilde{d}/\widetilde{\delta}_i$ and then to define

$$\begin{aligned}\widehat{\phi}_{i1}(n, \theta_e, \theta_i, r_e, h_e) \\ = \widetilde{\phi}_{i1}\left(n, \theta_e, \theta_i, r_e, \widetilde{r}_i(n, \theta_e, r_e, h_e), h_e, \widetilde{h}_i(n, \theta_e, \theta_i, \rho), \widetilde{d}(n, \theta_e, \rho)\right) .\end{aligned}$$

At this point, the developability condition $\phi_{i1} = \widehat{\phi}_{i1}(n, \theta_e, \theta_i, r_e, h_e) = 0$ permit to solve for

$$\theta_i = \widetilde{\theta}_i(n, \theta_e, r_e, h_e) = 0 .$$

In summary, for $m = 3$, a possible way for obtaining a tensegrity-origami pattern is to choose the string of parameters $\{n, \theta_e, r_e, h_e, \}$ as independent variable to determine r_i, d, θ_i, and h_i.

2.3.3 Multilevel towers

When passing from a tower with $m = 3$ to one with $m \geq 4$, the difference consists in just one additional type of internal nodes. However, it is easy to see that the sum of the angles at that node is identically equal to 2π. Therefore, there is no further condition to impose, and the case $m \geq 4$ can be reconducted to the case $m = 3$.

3 Spatial modeling workspace

A modeling workspace supports the visualization of the 3D geometry of the structure, the simulation of the developed configuration, and the tensegrity selfstress design. The spatial modeling workspace features the use of the parametric and algorithmic modeling tool Grasshopper^{TM} (Rutten 2013), while the use of a MatLab^{TM} code supports the tensegrity self-stress design, exploiting the geometric data generated by the Grasshopper^{TM} code.

3.1 Spatial and developed configurations

The development of a robust Grasshopper^{TM} modeling tool permits the visualization of the tower and the related developed configuration by implementing the analytic conditions introduced in Sec. 2. The modeling steps are described next considering the case of a three-levels tower (cf. Fig. 8). The geometric parameters involved in this process are recalled in Tab. 1.

Tab. 1: Geometric parameters.

$m = 3$	number of modules
n	number of struts per module
$\theta_0 = \pi/n + \pi/2$	twist of a tensegrity frustum
$\varphi_e = \theta_0 + \theta_e$	twist of the end module
$\varphi_i = \theta_0 + \theta_i$	twist of the intermediate module
r_e	radius of the end bases
r_i	radius of the intermediate modules
h_e	height of the end module
h_i	height of the intermediate module
d	overlap between adjacent modules
$\varphi_s = \pi/n$	rotation between modules
$\rho = r_e/r_i$	ratio between r_e and r_i
$\delta_e = d/h_e$	overlap ratio of the end module
$\delta_i = d/h_i$	overlap ratio of the intermediate module

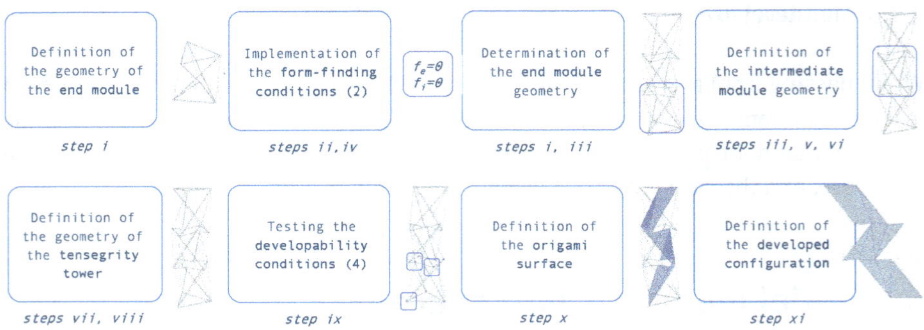

Fig. 8: Steps of the modeling algorithm.

Modeling algorithm of tensegrity-origami towers

1. Construction of the end module (tensegrity frustum): definition of n, r_e, and h_e; definition of the vertices of the base polygon; a copy the latter vertices is then translated along the z axis by h_e and rotated about the vertical axis by φ_e to obtain the upper vertices of the module; definition of the lines connecting vertices corresponding to cables and struts.
2. The first condition in (2), $f_e = 0$, is implemented in the code and solved for δ_e, determining also $d = \delta_e h_e$.
3. A copy of the upper vertices of the end module is translated downward along the vertical axis by d, and then rotated about that axis by φ_s to obtain the lower vertices of the intermediate module.
4. The second condition in (2), $f_i = 0$, is implemented in the code and solved for δ_i, determining also $h_i = d/\delta_i$.

5. A copy of the vertices obtained in step 3) is translated along the vertical axis by h_i and then rotated by φ_i about that axis to obtain the upper vertices of the intermediate module.
6. A copy of the latter vertices is translated downward along the vertical axis by d and then rotated about that axis by φ_s to obtain the lower vertices of the other end module.
7. A copy of the latter vertices is translated along the vertical axis by h_e and then rotated by φ_e. Then, their the nodal coordinates x and y are rescaled by $\rho = r_e/r_i$ to obtain the vertices of the other end base.
8. Definition a of the lines connecting vertices corresponding to cables and struts. A color code is defined to identify the different type of cables: vertical cables (blue), diagonal cables (green), saddle cables (red), base cables (black).
9. Test of the developability conditions (4) by adjusting the value of r_i.
10. Definition of the tensegrity-origami surface.
11. Definition of the developed configuration.

3.2 Tensegrity self-stress design

The geometric data of the tower, generated by the Grasshopper™ algorithm, are imported in Matlab™ as nodes coordinates and connectivity lists of struts and cables. The selfstress design of a given tower is then performed by exploiting a custom-written Matlab™ code.

The single selfstress state of a tensegrity tower can be computed by looking for a non-null solution of the equation $\mathbf{A}\,\mathbf{t}_s = \mathbf{0}$ to determine the nullspace of the equilibrium operator. It is worth recalling that tensegrity towers possess just a single selfstress state, which determines the ratios between the axial forces in the edges. It is crucial to perform a parametric analysis of selfstress to avoid having cables with an axial force that is too low with respect to that in the other cables and struts, given that external loads may make low-stress cables to go slack, or high-stress struts to buckle. Once a selfstress state with these features has been selected, the structural design can be carried out following standard codes on prestressed structures (see Cadoni and Micheletti 2012 for details).

As an example, Fig. 9 (left) shows the ratios between the tensile forces in cables and the magnitude of the compressive force in struts, $|t_p|$, as a function of θ_e for $m = 2$, $n = 5, \rho = 1.1215$, and $r_e/h_e = 1.2$ ($\delta_e = \widetilde{\delta}_e(n, \theta_e, \rho)$). It can be observed that for θ_e tending to zero the tension in diagonal cables tends to be null, while when θ_e decreases the tension in vertical and base cables decreases too. Fig. 9 (right) shows the colormap of the axial forces at the configuration corresponding to $\theta_e = 18^{\text{o}}$.

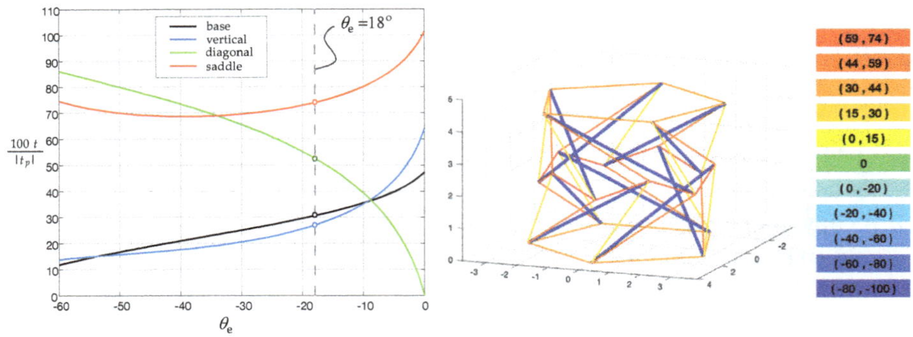

Fig. 9: Left: ratios between the tensile forces in cables and the magnitude of the compressive force in struts as a function of θ_e for a two-level tower with $n = 5$, $\rho = 1.1215$, and $r_e/h_e = 1.2$ ($\delta_e = \tilde{\delta}_e(n, \theta_e, \rho)$). Right: colormap of the axial forces at the configuration corresponding to $\theta_e = 18°$.

4 Case study: modular pavilion

An architectural exploration of the tensegrity-origami structure is developed for the design of a modular pavilion, employing the spatial modeling workspace. The conceptual definition of the geometry is homeomorphic to a hemicylinder, obtained by sectioning the tubular geometry of the tower by a vertical plane. The procedure to define the geometry features three main steps: first, a two-levels pentagonal-base tensegrity tower ($m = 2$, $n = 5$) was generated, with $h_e = 5\,\text{m}$, $r_e = 6\,\text{m}$, and $\theta_e = \frac{2\pi}{n} + \frac{\pi}{2} = 18°$; second, the tower configuration is sectioned by a vertical plane; third, the corresponding origami surface was generated and then developed on the plane. Fig. 10 (top) shows the three-step sequence of the generation of the pavilion geometry and the developed configuration of the origami, while Fig. 10 (bottom left and center) shows the corresponding tensegrity and the origami structures.

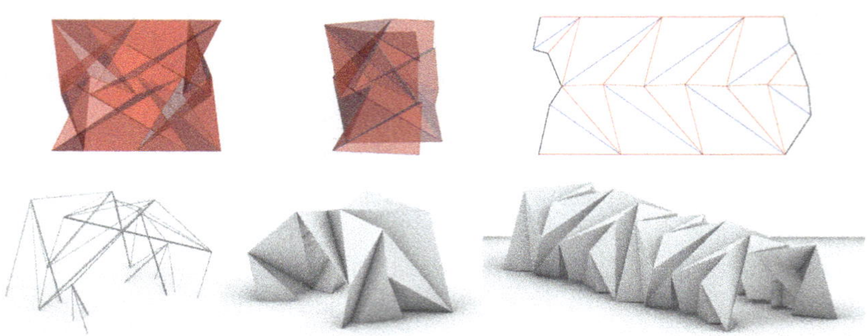

Fig. 10: Top: three-step pavilion generation. Bottom left and center: tensegrity and origami pavilion versions. Bottom right: a three-unit modular structure.

The geometry of the pavilion can be considered as a base unit of modular configurations, obtained by serial assembly of units along the longitudinal axis of the hemicylinder (Fig. 10, bottom right).

A scale physical model is used to verify the realizability of a two-levels pentagonal-base tensegrity tower and the developability feature. The simulation of the developed configuration is used to define the pattern and the length of clustered cables and struts. Aluminum bars are used for the struts while nylon tie is used for the cables. Fig. 11 shows the pattern of clustered cables, together with the physical model, assembled and developed.

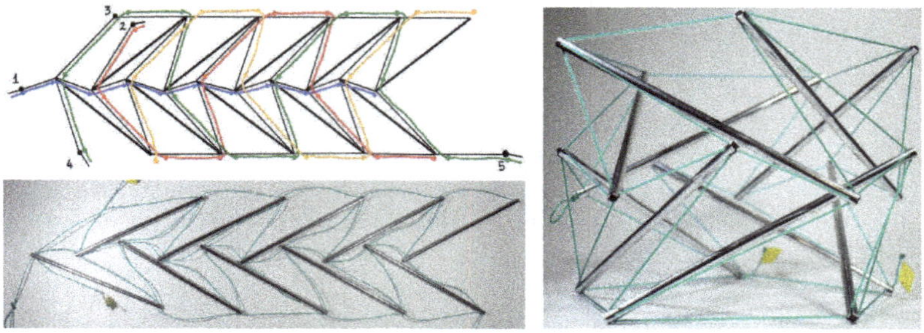

Fig. 11: Desktop model of the two-level tower shown in Fig. 9 (right).

5 Concluding remarks

The main advantage of the proposed tensegrity-origami design is that of an easy assembly of the developed structure on the ground, which can then be erected and prestressed at once. In addition, a membrane cladding could be easily installed during the assembly phase because of the flat geometry. Future work may regard the interaction between the membrane and the tensegrity system. It is worth remarking that one of the requirements of tensegrity systems, as compared to other solutions for reusable structures, is that of the specialized labor force necessary for prestress implementation and maintenance.

The proposed methodology can be extended to design origami with certain properties by exploiting the known features of tensegrity systems. For example, it would be possible to design multistable origami structures by following a procedure analogous to that presented in Intrigila et al. 2022. Such an extension could impact, other than architectural applications, all those fields in which tensegrity and origami are currently investigated, e. g. robotics, material science, or aerospace engineering.

References

Cadoni, D. and A. Micheletti. (2012). Structural performances of single-layer tensegrity domes. *Int. J. Space Struct.*, 2-3:167–78.

Calladine, C. R. (1978). Buckminster Fuller's 'tensegrity' structures and Clerk Maxwell's rules for the construction of stiff frames. *International Journal of Solids and Structures*, 14:161–72.

Calladine, C. R. and S. Pellegrino. (1991). First-order infinitesimal mechanisms. *International Journal of Solids and Structures*, 27:505–15.

Evans, A. A., J. L. Silverberg, and C. D. Santangelo. (2015). Lattice mechanics of origami tessellations. *Physical Review E*, 92:013205.

Intrigila, C., A. Micheletti, N. A. Nodargi, E. Artioli, and P. Bisegna. (2022). Fabrication and experimental characterization of a bistable tensegrity-like unit for lattice metamaterials. *Addit. Manuf.*, 57:102946.

Micheletti, A. (2003). The indeterminacy condition for tensegrity towers, a kinematic approach. *Revue Française de Génie Civil*, 7:329–42.

Micheletti, A., I. Giannetti, G. Mattei, and A. Tiero. (2022). Kinematic and static design of rigid origami structures: Application to modular yoshimura patterns. *Journal of Architectural Engineering*, 28(2):04022009.

Micheletti, A. and P. Podio-Guidugli. (2022). Seventy years of tensegrities (and counting). *Arch. Appl. Mech.*, 92(9):2525–48.

Oppenheim, I. J. and W. O. Williams. (2000). Geometric effects in an elastic tensegrity structure. *Journal of Elasticity*, 59(1-3):51–65.

Rutten, D. (2013). Grasshopper 2007.

Treml, B., A. Gillman, P. Buskohl, and R. Vaia. (2018). Origami mechanologic. *PNAS*, 115(27):6916–21.

Robby Kraft, Rupert Maleczek, Klara Mundilova, Tomohiro Tachi

From Quad Filling to Wrinkled Surfaces

Abstract: Fabricating complex geometries from flat sheets has many practical advantages such as cost-efficient fabrication and space-efficient transportation. In this paper, we explore a family of shapes that consist of two types of curved-crease molecules, which can be composed as a modular design. Alternatively, we explore how to optimize the target shape towards a global origami development, that is, a development that does not require additional slits or holes.

1 Introduction

Developable surfaces, or developables, are surfaces that can be flattened into the plane without stretching or tearing. These single-curved surfaces are often times easier and thus more cost-efficient to fabricate than doubly-curved surfaces. Therefore they have various applications in many disciplines, such as architecture, engineering and design [6, 8].

Inspired by straight crease origami, bend surfaces can be folded along curves. This enables a new space of achievable shapes that can be approximated in the straight crease origami world by only infinitely many tiny creases. Shapes obtained by folding along curves share many benefits of straight crease origami: creases can add structural stiffness, folded shapes decrease material off-cut in production, and the designs are easier to transport in their original flat state [4, 13].

Although physical fabrication of curved creases with paper may seem straightforward, the digital design of curved folded shapes is not trivial. In particular, approximating a target shape with a specific curved-crease design is challenging. Only recently, two approaches were studied. Jiang et al. [7] study principal pleatedstructures and propose a method for the reconstruction of shapes as curve pleated structures. Maleczek et al. [10] construct an edge rounded version of a polyhedral surface by first rounding each edge with a right circular cylinder and then folding the cylinders incident at a vertex into (generalized) cones.

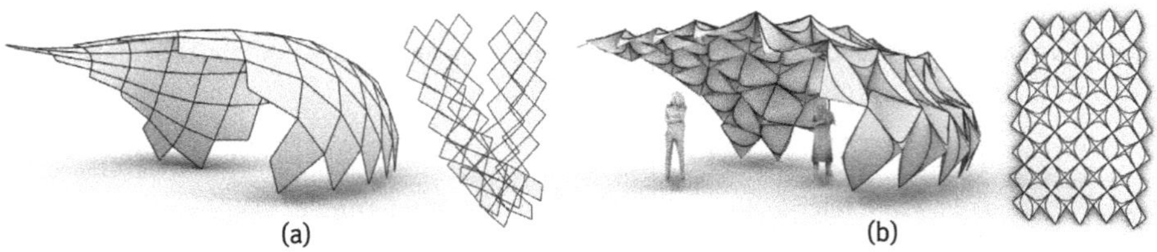

Fig. 1: A pavilion structure using the lens molecules (b), and its non-devlopable target geometry (a).

https://doi.org/10.1515/9783111162683-025

In this paper, we propose a new approach to approximate a user-specified polyhedral surface with developable quadrangular curved crease modules; modules which are able to be customized in design across a few different parameters. We demonstrate how the developable state is generated, how to target alternative developed shapes, and using this, generate a globally developable surface without cuts. Our method is implemented as a grasshopper plug-in for Rhino6/7 [1, 2].

Our paper is organized as follows. In Sec. 2, we show how to fill a quadrangular face with two curved crease designs: the cone-cone, and the lens. We then propose a subdivision scheme to prepare an arbitrary mesh for our surface-filling algorithm in Sec. 3. Finally, we explain how to target a developable shape from a range of configurations and build a global development without holes in Sec. 4.

2 Quad Filling

In this section, we show how to fill a single non-planar quad $Q = \{\mathbf{V}_1, \mathbf{V}_2, \mathbf{V}_3, \mathbf{V}_4\}$ with a curved crease design, consisting of cylinders and cones[1]. In particular, we consider two developable surface layouts (see Fig. 2):

- *Cone-cone:* A cone with apex $\mathbf{V}_1$ and rulings $\mathbf{V}_1\mathbf{V}_2$ and $\mathbf{V}_1\mathbf{V}_4$ is folded into a cone with apex $\mathbf{V}_3$ such that the crease passes through the points $\mathbf{V}_2$ and $\mathbf{V}_4$.
- *Lens:* A central cylinder is folded into two cones with apices $\mathbf{V}_1$ and $\mathbf{V}_3$ such that both creases between the cylinder and the cone pass through points $\mathbf{V}_2$ and $\mathbf{V}_4$.

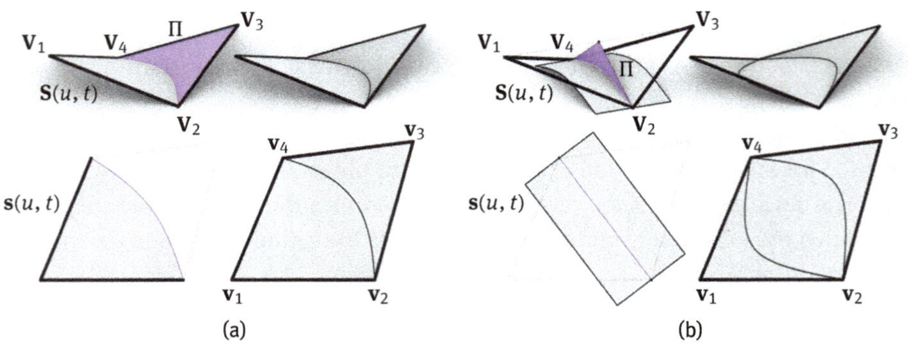

Fig. 2: Two steps of the quad filling method for the cone-cone molecule (a) and the lens molecule (b).

1 In this paper, when we refer to cones and cylinders, we are referring to the generalized versions unless specified otherwise. Cones, or conical surfaces, are surfaces composed of lines that all intersect at a single point known as the apex of the cone. Cylinders, or cylindrical surfaces, on the other hand, are surfaces composed of lines that are all parallel to each other.

In both cases, our method consists of two steps:

1. *Construct first surface:* First, we construct a cylindrical or conical surface $\mathbf{S}(u, t)$ inside the quad. A curve which defines the shape of this surface and further parameters can be specified by the user. Details are provided in Sec. 2.1.
2. *Construct remaining surface(s):* Then, we construct the fold between $\mathbf{S}(u, t)$ and a cone with apex $\mathbf{V}_4$ (cone-cone), or two cones with apices $\mathbf{V}_1$ and $\mathbf{V}_4$ (lens). Details are provided in Sec. 2.2.

2.1 Construction of the first surface

Intuitively, we obtain the first surface $\mathbf{S}(u, t)$ by extruding a not self-intersecting planar curve $\mathbf{P}(u)$ with start point $\mathbf{V}_2$ and end point $\mathbf{V}_4$. In the following, we will call $\mathbf{P}(u)$ the *base curve*. In particular, when constructing a cone, we extrude $\mathbf{P}(u)$ w.r.t. the center $\mathbf{V}_1$ by connecting every point of $\mathbf{P}(u)$ with $\mathbf{V}_1$ using a line. When constructing a cylinder, $\mathbf{P}(u)$ is extruded in the direction that is perpendicular to the curve's incident plane. The resulting surface can therefore be parametrized as $\mathbf{S}(u, t) = \mathbf{C}(u) + t\mathbf{R}(u)$, where in case of a cone we set $\mathbf{C}(u) = \mathbf{V}_1$ and $\mathbf{R}(u) = \frac{\mathbf{P}(u)-\mathbf{V}_1}{|\mathbf{P}(u)-\mathbf{V}_1|}$, and in case of a cylinder we set $\mathbf{C}(u) = \mathbf{P}(u)$ and $\mathbf{R}$ to be the normalized vector perpendicular to the curve's base plane.

The choice of $\mathbf{P}(u)$ influences not only the shape of the first surface $\mathbf{S}(u, t)$, but also the shape of the second, constructed cone $\mathbf{S}_2(u, t)$ that connects to $\mathbf{S}(u, t)$ with a curved crease. When choosing the planar curve $\mathbf{P}(u)$ arbitrarily, we might observe some undesired artifacts, such as local self-intersections of $\mathbf{S}_2(u, t)$ and "complete" or "no" folds between $\mathbf{S}(u, t)$ and $\mathbf{S}_2(u, t)$, see Fig. 3 (left).

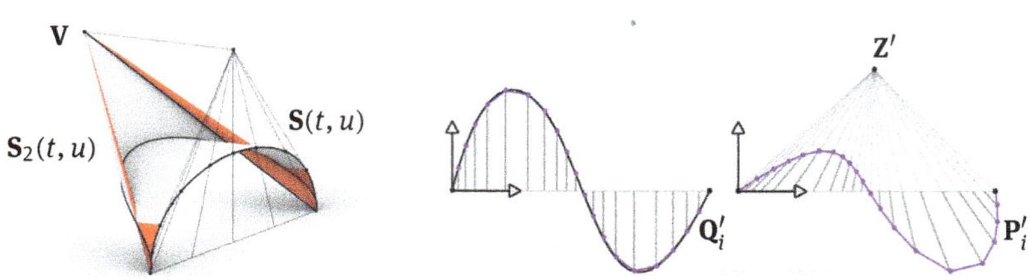

Fig. 3: Left: Intersecting surfaces. Right: Illustration of a projective mapping used for the construction of a $\mathbf{Z}'$-central base curve of the first surface.

In the following, we discuss (1) a constraint on the rulings of the constructed surface $\mathbf{S}(u, t)$ such the curved-crease connection with $\mathbf{S}_2(u, t)$ does not result in self-intersections, and (2) how to modify a user-specified input curve $\mathbf{Q}(u)$ using projective transformations to satisfy this constraint.

2.1.1 Non-self-intersecting cones and central functions

Suppose we want to determine the fold between a developable surface $\mathbf{S}(u, t)$ and a cone with apex $\mathbf{V}$. The developable surface $\mathbf{S}(u, t)$ contains a family of lines, the so-called *rule lines*. Let us now consider the family of *ruling planes* T, that is, planes that consist of two consecutive rulings joined by a point on the curved crease, in our case a ruling of $\mathbf{S}(u, t)$ and the cone apex $\mathbf{V}$, see Fig. 4. If $\mathbf{S}(u, t)$ is a conical surface with apex $\mathbf{V}_1$, the family of planes is a bundle of planes with axis $\mathbf{V}_1\mathbf{V}$. If $\mathbf{S}(u, t)$ is a cylindrical surface with ruling direction $\mathbf{R}$, the family of planes is a bundle of planes with its axis passing through $\mathbf{V}$ and having direction $\mathbf{R}$.

As the rulings continuously vary along $\mathbf{S}(u, t)$, so do the planes in T. The second surface does not have any self-intersections if all planes in T are distinct, that is, don't "double back", see Fig, 3 (left). We argue that surfaces are not repeating by considering a planar section of the bundle of planes with a plane Π that passes through $\mathbf{V}$ and intersects all planes at least once. If all intersecting lines of $\Pi \cap T$ are unique, so are the planes. The case-specific location of plane Π is discussed in Sec. 2.1.2, construction step (a).

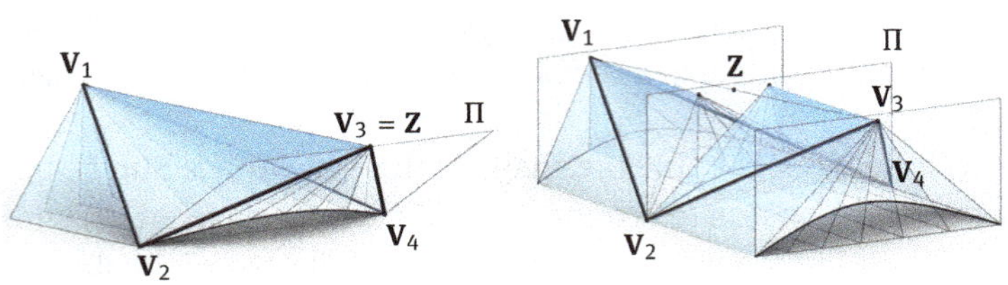

Fig. 4: Illustration of the bundle of ruling planes used in the argumentation in Sec. 2.1.1. Left: cone-cone molecule. Right: Lens molecule.

Consequently, when locating the base curve $\mathbf{P}(u)$, we want it to induce non-repeating planes, resulting in a curve that has unique connecting lines with the apex. In the following, we call such a curve a *central function*[2]. We can obtain a central function with center $\mathbf{Z}$ from the graph of a function by applying a projective transformation that maps the ideal point of the y-axis to the center $\mathbf{Z}$, see Fig. 3 (right).

In case of the cone-cone design, we construct the first surface $\mathbf{S}(u, t)$ such that its base curve is a $\mathbf{V}_3$-central function. In case of a lens design, we determine the fold between the cylinder and two cones on either side. Thus we would need a central function w.r.t. two centers (the orthogonal projections of $\mathbf{V}_1$ and $\mathbf{V}_3$ on Π). If the orthogonal projections are not the same, we observed in our experiments that it is usually sufficient to approximate this "doubly-central" function by a $\mathbf{Z}$-central function, where $\mathbf{Z}$ is the intersection of the line spanned by $\mathbf{V}_1$ and $\mathbf{V}_3$ with Π.

2 This terminology is based on the "common" 2D functions, which are central functions w.r.t. the point at infinity of the z-axis.

2.1.2 Surface fitting

On a highlevel, the construction of the first conical or cylindrical surface amounts in locating plane Π and constructing an appropriate central base curve $\mathbf{P}(u) \in \Pi$ which reassembles the shape of a given input curve.
In addition to the coordinates of the 3D quad, the inputs for the surface filling are a *shape-defining curve*, that is a curve $\mathbf{Q}(u) = (u, f(u), 0)$ where $f(u)$ is a function with $f(0) = f(1) = 0$, and a *scale parameter* s. By changing the shape of this curve, the user can influence the initial surface, which affects the location and shape of the computed crease curves, offering more design freedom. Moreover, if we are constructing the lens design, we have an additional parameter ϕ that influences the orientation of Π.

Although the following operations can be performed analytically, we explain the following in terms of our implementation using a sampling of the input curve. We adopt this discretization approach because the computation of the crease curve relies on establishing an accurate correspondence between the 3D surface and its developed representation. Furthermore, the subsequent section requires information about the opening angle of the developed conical surface or the length of the base curve of the cylinder. Both of these quantities can be readily approximated from a finely sampled mesh.

Our proposed algorithm works as follows, see Fig. 5:

(a) *Locate plane* Π*:* For a given 3D quad, we define the design-dependent base plane Π. In case of a cone, let Π be the plane containing the points $\mathbf{V}_2$, $\mathbf{V}_3$ and $\mathbf{V}_4$. In case of a cylinder, let Π be a plane containing $\mathbf{V}_2$ and $\mathbf{V}_4$, whose orientation can be influenced by a parameter ϕ (the rotation about $\mathbf{V}_2\mathbf{V}_4$). In both cases, let $\mathbf{V}$ be the intersection of Π with the line spanned by $\mathbf{V}_1$ and $\mathbf{V}_3$ (in the cone design, $\mathbf{Z} = \mathbf{V}_3$).

(b) *Scale and orient* $\mathbf{Q}(u)$*:* We scale the user defined curve so that the distance between its endpoints equals $|\mathbf{V}_2 - \mathbf{V}_4|$. In addition, we scale the curve in y-direction by the user-specified parameter s. Finally, we move the curve from the xy-plane to Π, such that $\mathbf{Q}(0) = \mathbf{V}_2$ and $\mathbf{Q}(1) = \mathbf{V}_4$, such that $\mathbf{Z}$ lies in the $y > 0$ half-plane. Note that the resulting polyline might not be $\mathbf{Z}$-central and thus not suitable for the fold construction without self-intersections.

(c) *Projective transformation:* We apply a projective mapping to transform $\mathbf{Q}(u)$ into a $\mathbf{Z}$-central function. In the following, we utilize a local 2D coordinate system where $\mathbf{V}_2$ corresponds to the origin, $\mathbf{V}_4 - \mathbf{V}_2$ to the x-axis, and $\mathbf{Z}$ lies on the half-plane with

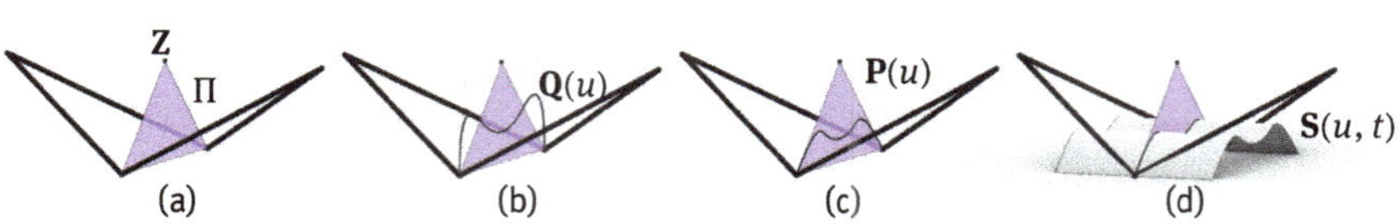

Fig. 5: Surface fitting. Illustration of the construction steps 1(a) – 1(d).

$y > 0$. Let $\mathbf{Z}'$ denote the coordinates of $\mathbf{Z}$ in this local coordinate system. Moreover, let $\mathcal{Q}' = (\mathbf{Q}'_1, \mathbf{Q}'_2, \dots, \mathbf{Q}'_n)$ be a sampling of curve $\mathbf{Q}(u)$ in this local coordinate system. We use the following projective transformation to turn $\mathcal{Q}'$ in a $\mathbf{Z}'$-central polyline $\mathcal{P}' = (\mathbf{P}'_1, \mathbf{P}'_2, \dots, \mathbf{P}'_n)$, that is,

$$(\mathbf{Q}'_x, \mathbf{Q}'_y) \mapsto \mathbf{P}' = \left(\frac{k\mathbf{Q}'_x + \mathbf{Z}'_x\mathbf{Q}'_x}{k + \mathbf{Q}'_y}, \frac{\mathbf{Z}'_y\mathbf{Q}'_y}{k + \mathbf{Q}'_y} \right) \quad \text{where} \quad k = \mathbf{Z}'_y - \min_i \mathbf{Q}'_{i,y}.$$

Note that because of the choice of k, the points are bounded. In particular, the lowest y-coordinate of $\mathbf{P}'_i$ is the lowest y-coordinate of $\mathbf{Q}'_i$.

(d) Finally, depending on the application, we can either smoothly or linearly interpolate the points of $\mathcal{P}$ to obtain $\mathbf{P}(u)$ (in world-coordinates) and construct the design-depended initial surface.

2.2 Construction of the remaining surface(s)

In the following, we briefly review the patch-to-cone construction presented by Mundilova [11], to construct a crease between a given developable surface patch and a cone. For more details on the implementation, see Mundilova et al. [12].

2.2.1 Preparing the development

To construct the crease that connects the given developable surface $\mathbf{S}(u, t)$ with a cone, we first need to find the unrolled surface $\mathbf{s}(u, t)$. In our implementation, we utilize the above approximation of $\mathbf{S}(u, t)$ by a discretized cone (family of triangles) or discretized cylinder (family of planar quads with parallel edges). We place the faces successively into the xy-plane and denote the 2D counterparts of $\mathbf{P}_i$ as $\mathbf{p}_i$, the 2D counterpart of the ruling direction $\mathbf{R}_i$ incident to $\mathbf{P}_i$ as $\mathbf{r}_i$ (in case of the cylinder, all $\mathbf{r}_i$ are the same). Furthermore, let $\mathbf{v}_2 = \mathbf{p}_1$ and $\mathbf{v}_4 = \mathbf{p}_n$. Note that in case of a lens design, because $\mathbf{P}(u)$ is a curve that lies in a plane orthogonal to the ruling direction, its development lies on a straight line. Finally, we find the points $\mathbf{v}_1$ and $\mathbf{v}_3$ corresponding to $\mathbf{V}_1$ and $\mathbf{V}_3$ on the appropriate side such that

$$|\mathbf{v}_1 - \mathbf{v}_2| = |\mathbf{V}_1 - \mathbf{V}_2| \quad \text{and} \quad |\mathbf{v}_1 - \mathbf{v}_4| = |\mathbf{V}_1 - \mathbf{V}_4|,$$
$$|\mathbf{v}_3 - \mathbf{v}_2| = |\mathbf{V}_3 - \mathbf{V}_2| \quad \text{and} \quad |\mathbf{v}_3 - \mathbf{v}_4| = |\mathbf{V}_3 - \mathbf{V}_4|.$$

2.2.2 Patch-to-cone construction

We construct the crease between the developable surface and a cone by considering distances between points on the same ruling. As developing a developable surface to the plane preserves the distances between points on the same ruling, we derive a constraint from comparing two corresponding distances.

To utilize a unified expression for both cases, let $\mathbf{C}_i = \mathbf{V}_1$ and $\mathbf{c}_i = \mathbf{v}_1$ in case of a cone, and let $\mathbf{C}_i = \mathbf{P}_i$ and $\mathbf{c}_i = \mathbf{p}_i$ in case of a cylinder. Furthermore, $\mathbf{V} = \mathbf{V}_3$ and $\mathbf{v} = \mathbf{v}_3$.

Note that for a crease between a developable surface and a cone, every point on the 3D or 2D crease curve is connected to the corresponding vertex $\mathbf{V}$ or $\mathbf{v}$ with a ruling of the constructed cone. We parametrize the points on corresponding rulings in 3D and 2D by $\mathbf{L}_i(t) = \mathbf{C}_i + t\mathbf{R}_i$ and $\mathbf{l}_i(t) = \mathbf{c}_i + t\mathbf{r}_i$. For every pair of corresponding 3D and 2D rulings, we find the crease points $\mathbf{F}_i = \mathbf{L}_i(t^*)$ and $\mathbf{f}_i = \mathbf{l}_i(t^*)$ as points that satisfy

$$|\mathbf{L}_i(t_i^*) - \mathbf{V}|^2 = |\mathbf{l}_i(t_i^*) - \mathbf{v}|^2 \qquad \Longrightarrow \qquad |\mathbf{P}_i - \mathbf{V} + t_i^*\,\mathbf{R}_i|^2 = |\mathbf{p}_i - \mathbf{v} + t_i^*\,\mathbf{r}_i|^2$$

that is, the pair of corresponding points such that the distances between the corresponding points and the 3D and 2D apices are the same.

Solving the above equation for t_i^* results in

$$t_i^* = \frac{1}{2}\frac{|\mathbf{v} - \mathbf{c}_i|^2 - |\mathbf{V} - \mathbf{C}_i|^2}{(\mathbf{v} - \mathbf{c}_i)\cdot\mathbf{r}_i - (\mathbf{V} - \mathbf{C}_i)\cdot\mathbf{R}_i}. \tag{1}$$

Note, that the crease points $\mathbf{F}_i$ and $\mathbf{f}_i$ only provide the location of the crease curve as the intersection of two developable surfaces, and do not always produce a clean and useful result. As discussed in [12] in more detail, a “good" curved crease point $\mathbf{F}_i$ connecting a developable surface patch and a cone has the following characteristics:

- *Valid patch combination:* The computed crease separates a the given developable surface and the constructed cone into four surface patches. Out of the four possible combinations, only two are developable. As discussed in [12], we require that the denominator is greater zero for the values of each ruling, resulting in $(\mathbf{v} - \mathbf{c}_i)\cdot\mathbf{r}_i > (\mathbf{V} - \mathbf{C}_i)\cdot\mathbf{R}_i$.
- *Valid range:* We want to make sure that the crease exists in a suitable range of the developable surface (e. g., does not pass through the apex to the other part of the cone). This property translates to the numerator of t_i^* being greater than zero, that is, $|\mathbf{v} - \mathbf{c}_i|^2 > |\mathbf{V} - \mathbf{C}_i|^2$.

When the above inequalities are satisfied for all sampled points, we have found a valid curved crease. In case of a cone-cone design, we then only interpolate the crease and construct extrusions to $\mathbf{V}_1$ and $\mathbf{V}_2$. In case of the lens design, we compute the crease for the other cone with apex $\mathbf{V} = \mathbf{V}_1$. We then extrude both creases to the corresponding cone apices and construct the cylinder as a loft with parallel rulings between the two curves.

3 Modular Curved-Crease Designs

Equipped with the theory of the previous section, we apply the quad-filling method to each face of a non-planar quad mesh. If the faces of the mesh are planar or not all quads, we can apply the following algorithm to construct a mesh M' with non-planar faces from the given mesh M.

1. Initialize M' with the set of vertices of M and an empty face set.
2. For every face, add a new vertex at a user-specified amount in normal direction from the face's center. The corresponding parameter establishes the degree of non-planarity.
3. For every interior edge, create a face containing the edge's endpoints and the two vertices corresponding to its adjacent faces.

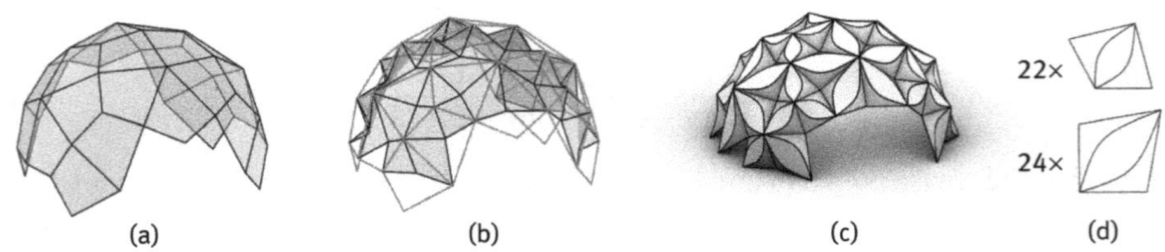

Fig. 6: Illustration of the modular curved-crease design workflow: (a) Initial mesh M. (b) Subdivided mesh M'. (c) Quads filled with curved-crease molecules. (d) Development consisting of two different types of developed molecules.

Development. Applying the quad-filling method from Sec. 2 to a non-planar quad mesh results in a shape comprised of curved-crease modules. Each curved-crease module can be unrolled, resulting in a family of curved-crease patterns with quadrangular boundaries. However there is no guarantee that the sum of developed angles incident to a common vertex is exactly 2π, and thus the pattern might not be globally developable. For fabrication purposes, the decomposition into smaller molecules can be beneficial. Alternatively, knowing the dimensions of the unrolled quads allows the use of other positioning heuristics of the quads, such as polygon nesting [3] or Origamizer-based kirigami patterns [5].

Design variations. Tiling entire surfaces with foldable modules is a design approach used in many fields. Some design variations of foldable lenses are shown in Maleczek et al. [10]. In our setup, each non-planar quad of a mesh can be filled with one of six curved crease molecules (see Fig. 7). This design freedom can be used to explore patterns generated by the composition of curved creases in quads.

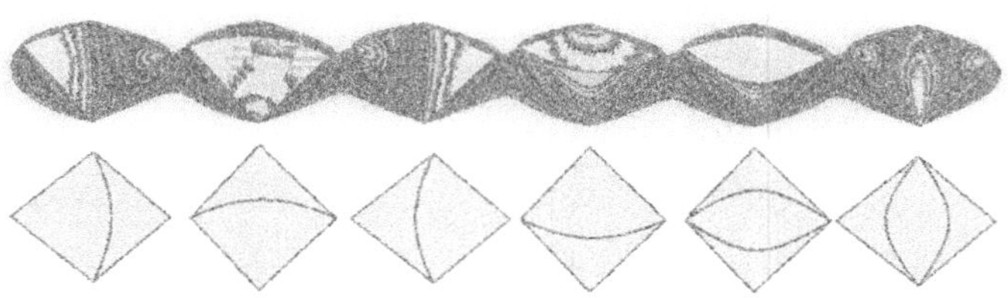

Fig. 7: For every non-planar quad, we have six different choices of the curved crease molecules induced by the same curve (four cone-cone molecules, two lens molecules).

4 Global Origami Development

In the previous sections, we considered the non-planar faces of a quad mesh as individual folded patches, the final mesh being a joining of these parts. However in special cases, the mesh faces can be treated as a connected entity which can be globally developed and fabricated from a single large sheet of material without slits. To achieve this, one must be flexible about the design parameters used to generate each quad, as neighboring quads affect each other, altering each other's interior structures. The global development is discovered through an optimization process whereby the variables are adjusted slightly until all constraints are satisfied or not feasible. A successful development, if found, often times results in a shift of the vertices away from the surface, effectively adding material, and creating a "wrinkling" of the surface.

To begin, we must generate a topologically identical mesh to our given 3D mesh to serve as our crease pattern. This mesh must be planar and contain no overlapping faces or edges. For simple geometries, this can be manually generated, or it can be obtained through a Tutte embedding [15] or the ARAP method [9, 14]. If we choose so, the vertices can be allowed to move in the xy-plane and we make their positions a set of variables in our optimization problem. If we intend for the 2D vertices to be fixed, only the coordinates of the 3D vertices are the variables in our optimization problem.

Solving a fixed-vertex crease pattern, although more heavily constrained, is a simpler case through which we can understand the solver's additional constraints. Because we are dealing with quad meshes, often times a uniform square grid makes for a good initial crease pattern, as shown in Fig. 8.

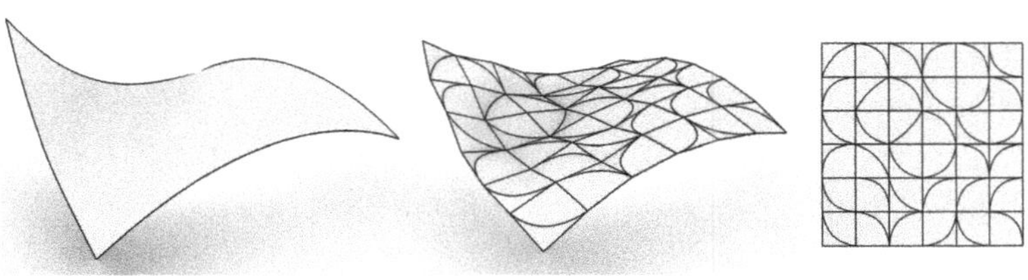

Fig. 8: A wrinkled surface based on a square grid that is globally developable.

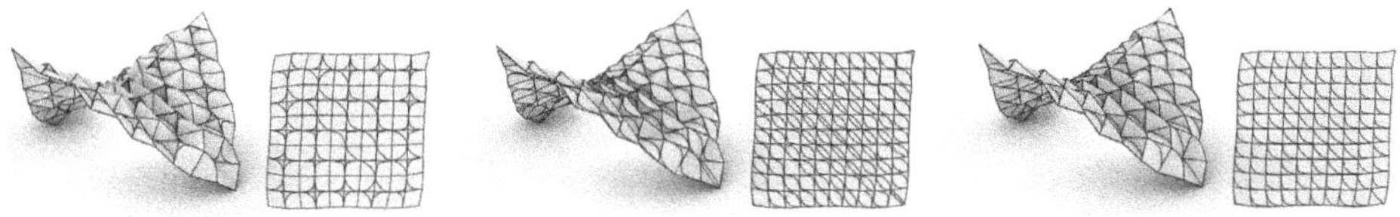

Fig. 9: Three globally developable curved-crease designs that are based on the same mesh and their crease patterns.

4.1 Constraints

To obtain suitable geometries, we impose the following constraints on the 3D (and optionally 2D) vertices of the meshes:

1. Every edge must have identical length in 3D and 2D.
2. Corresponding quad diagonals must be shorter in 3D than in 2D.
3. Every 2D quad's diagonal length should be bounded by above to ensure that the constructed first surface can be realized from a suitable **Z**-central function.

As the solver executes, vertices will be moved around, in some cases dramatically. Therefore adding regularization constraints, such as anchoring points to their original position, might be necessary. Alternatively, it might be helpful to constrain a sparse selection of vertices to the design surface, to keep closeness to the original shape, or constrain some boundary vertices to the boundary edges, to prevent the vertices from scaling down to a point.

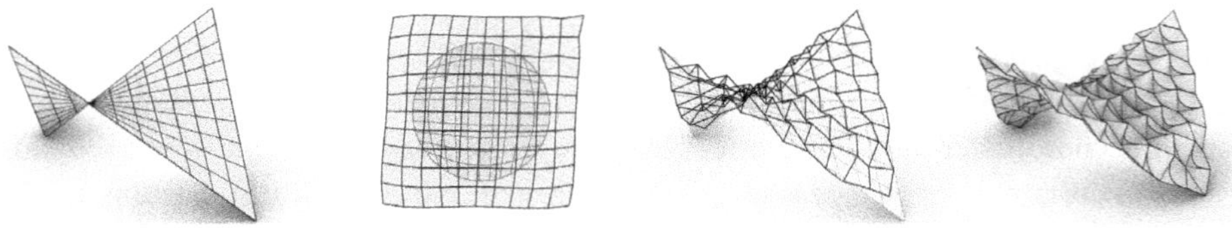

(a) (b) (c) (d)

Fig. 10: Illustration of the steps for in Section 4: (a) Initial 2D mesh. (b) Constrained and scaled initial 2D mesh. (c) Wrinkled 3D mesh based on the lengths of the 2D mesh. (d) Mesh with quad faces filled with curved-crease molecules with target diagonals.

4.2 Implementation and practical considerations

We have implemented an optimization process in Kangaroo, a dynamic relaxation engine plugin for Rhino/Grasshopper. The solver uses “goals” and moves the mesh’s vertices in real time during calculation, until it finds an equilibrium configuration.

We have found that it is more often the case that the solver desires more material to satisfy constraints, consequently, it is beneficial to scale the developed constrained mesh in relation to the 3D. The authors found that multiplying the edge lengths by 2 to 10 percent of their original lengths led to good results.

Due to the edge-length constraint, this scaling operation will cause the 3D vertices to move away from the surface, resulting in larger angle differences and potential challenges such as surface intersections. To maintain better control over the scaling process, pre-wrinkling the 3D mesh can be beneficial. Pre-wrinkling is a process by which the designer can displace the vertices away from the surface in a manner of their choosing.

During the optimization, vertices will start to move on the target geometry and will therefore scale down to a point if not anchored or limited in motion with respect to the original boundaries.

4.3 Filling a 3D quad with prescribed development

Upon success of the above described optimization process of the 3D and 2D mesh, we aim to fill each 3D non-planar quad with a curved crease molecule whose development perfectly fits the corresponding 2D face. The scale parameter s, as discussed in Sec. 2.1.2, provides a degree of design flexibility. Adjusting this scale parameter affects the intrinsic opening angle of the cone in the case of the cone-cone design, or the intrinsic distance between two points on a cylinder in the case of a lens design.

Determining an appropriate scale parameter that corresponds to the desired opening angle or diagonal distance of the target 2D quad can be framed as finding the root of a scalar function. We have observed that in many valid configurations, this scalar function demonstrates (almost) monotonic behavior.

5 Conclusions and Outlook

We have presented a method that enables the generation of foldable patches or globally developable shapes by filling non-planar quads with curved-crease molecules. In particular, we discussed how to fill a single non-planar quad with two types of curved-crease molecules, and how to optimize for global developability of the resulting shape. In addition, we highlighted design specific considerations, such as the generation of patterns.

As discussed above, further adjustments and investigations are necessary to improve the results when dealing with complex surfaces. Currently, achieving global developability for such surfaces remains challenging. Additionally, exploring the degrees of freedom and tangent continuity between different patches will be subject to future research. The authors are confident, that with appropriate adaptations, the presented strategy allows for different design and fabrication applications.

References

[1] Grasshopper. https://www.grasshopper3d.com/. Accessed: 2023-05-31.

[2] Rhino6/7. https://www.rhino3d.com/. Accessed: 2023-05-30.

[3] Bennell, J. A. and J. F. Oliveira. The geometry of nesting problems: A tutorial. *European Journal of Operational Research,* 184(2):397–415, 2008.

[4] Demaine, E., M. Demaine, D. Koschitz, and T. Tachi. A review on curved creases in art, design and mathematics. *Symmetry: Culture and Science,* 26(2):145–61, 2015.

[5] Demaine, E. and T. Tachi. Origamizer: A practical algorithm for folding any polyhedron. In *33rd International Symposium on Computational Geometry (SoCG 2017),* 2017.

[6] Glaeser G. and F. Gruber. Developable surfaces in contemporary architecture. *Journal of Mathematics and The Arts,* 1:59–71, 03 2007.

[7] Jiang, C., K. Mundilova, F. Rist, J. Wallner, and H. Pottmann. Curve-pleated structures. *ACM Transactions on Graphics,* 38:1–13, 11 2019.

[8] Lawrence, S. Developable surfaces: Their history and application. *Nexus Network Journal,* 13: 701–14, 2011.

[9] Liu, L., L. Zhang, Y. Xu, C. Gotsman, and S. J. Gortler. A local/global approach to mesh parameterization. In *Computer Graphics Forum,* vol. 27, pp. 1495–1504. Wiley Online Library, 2008.

[10] Maleczek, R., K. Mundilova, and T. Tachi. Curved crease edge rounding of polyhedral surfaces. *Advances in Architectural Geometry,* 2020.

[11] Mundilova, K. On mathematical folding of curved crease origami: Sliding developables and parametrizations of folds into cylinders and cones. *Computer-Aided Design,* 115:34–41, 2019.

[12] Mundilova, K., E. Demaine, R. Foschi, R. Kraft, R. Maleczek, and T. Tachi. Lotus: A curved folding design tool for Grasshopper. In *Proceedings of the 41st Annual Conference of the Association of Computer Aided Design in Architecture (ACADIA),* ACADIA 2021, pages 194–203, 2021.

[13] T. G. Nelson. Art to engineering: Curved folding and developable surfaces in mechanism and deployable structure design. 2018. PhD thesis.

[14] Sorkine, O. and M. Alexa. As-rigid-as-possible surface modeling. In *Symposium on Geometry processing,* vol. 4, pp. 109–16. Citeseer, 2007.

[15] Tutte, W. T. How to draw a graph. *Proceedings of the London Mathematical Society,* 3(1): 743–67, 1963.

Iman Fayyad

Bending Cylinders: A Geometric Syntax for Zero-Waste Architecture

Abstract: This research explores the generation of a novel syntax for architectural space driven by a geometric construction strategy that aims to eliminate material waste. The primary goal of the investigations is to avoid off-cuts by creating a vocabulary of varied three-dimensional forms using the full dimensions of standard rectangular sheet material (4 ft × 8 ft) as a building block. By engaging the isometric relationship between flatness and three-dimensional form, the research investigates how waste-conscious approaches to geometric innovation can have typological, topological, and tectonic effects on architectural language. Specifically, the work deploys a subset of curved crease folding that uses planar reflections (sectional mirror operations) to create unique spatial and structural forms through composite cylindrical and conical surfaces. Whereas known studies in curved crease folding typically explore singular (one-off, figural) compositions, this process develops the operation into an aggregation strategy to suggest large inhabitable structures as both figural and field-like conditions. Through geometric literacy, this process attempts to achieve three primary advancements: fusing the design process with constructability, designing the part and whole simultaneously, and creating a workflow whereby complex three-dimensional forms are designed, fabricated, stored, and transported in their compact two-dimensional state. This proposed methodology allows us to consider material economy and labor early in the design process, before construction, and eliminates the additional step of post-rationalizing complex curvature. Finally, the research illustrates a built full-scale prototype that tests the effectiveness of the process in the form of a low-cost shade structure in a small neighborhood park.

1 Introduction: Geometric Aesthetics of the Anthropocene

This research considers methodologies that use 4 ft × 8 ft rectangular sheet material – the most readily available, standardized building material – as the primary tectonic, proportioning, and structural device to create new vocabularies of space. These new geometries serve structural, spatial, ergonomic, and aesthetic roles; their logic of assembly informs a range of constraints, from ideal seating heights to creating novel typological forms or modifying familiar ones.

In this paper, I identify two independent materially driven methodologies as reciprocal venues for exploring the typological, topological, and tectonic potentials of minimizing material waste through an isometric relationship between flatness and

https://doi.org/10.1515/9783111162683-026

three-dimensional form: first, curved crease folding, a geometric technique for designing structural surfaces; and second, the aggregation of standard sized rectangular sheet material, a construction strategy for cladding large complex curved surfaces. By pairing these two operations, we create a process that attempts to fuse design with constructability and a strategy to design the part and whole simultaneously. This pairing allows us to consider material economy and labor as part of the design process prior to construction and eliminates the additional step of post-rationalizing complex curvature.

Both methodologies have a long history of productively leveraging their relationship with flatness in their own ways. Curved crease folding is a technique that is inherently materially efficient due to its dependency on developable surfaces to create rigid three-dimensional forms entirely out of flat, two-dimensional sheet material. Developable surfaces, or surfaces that can unroll to flatness without distortion, are uniquely suited to processes that simultaneously design the part and whole as they isometrically map two dimensions onto three. This property means we can immediately establish a workflow between a standard given 2D part (both a material type and dimension) and its three-dimensional implications.

Mapping standardized, mass-produced sheet material onto complex curvilinear forms has become a ubiquitous practice. Most complex curvilinear surface geometries undergo some process of rationalization by being constructed out of planar quadrilaterals (for instance, glass grid shells). The research here seeks to deploy the rectangular sheet not to rationalize complex geometry but rather to generate it. The grammar of flexible rectangular flat sheet material is preserved and translated to map rectilinear Cartesian space onto a curvilinear three-dimensional form, setting new standards of spatial proportions based on the dimensions of the part rather than the whole.

This research studies the spatial, geometric, aesthetic, and typological potentials of using whole rectangular sheets of standard flexible sheet material – specifically, 4 ft × 8 ft – to create composite assemblies of geometric surfaces through the piecewise stitching of individual parts so that they may function at an inhabitable architectural scale. Economical use of the material is paramount, emphasizing the importance of the rectangular geometry of the folded sheets as this tectonic consideration allows for the minimization or elimination of material waste.

2 Curved Crease Folding as a Tectonic Language

Paper folding techniques have been documented and studied exhaustively for centuries. Curved crease folding was first seen through the work of Josef Albers and Irene Shawinsky in the Bauhaus in the 1920s and 1930s. The work of later pioneers in the field (David Huffman, Ronald Resch, and Martin and Erik Demaine) focused mostly on recreating doubly ruled and doubly curved surfaces, such as the hyperbolic paraboloid, out of concentric circular or rectangular pleats on a sheet of paper (Davis et al. 2013).

However, given the ease and economy with which complex calligraphic objects are produced, much of the research, even that which has undergone rigorous mathematical calibration ("computational origami"), has remained purely sculptural (ErikDemaine.org, 2013). Or, in the case of more scientific research, curved crease mechanisms are studied with regards mostly to their relevance to cellular structures, such as intestinal villi and pollen grains (Bende et al. 2015). The studies furnished in this analysis differ from the computational process described by Demaine, in which the three-dimensional outcome is approximated by preferencing the two-dimensional patterning. Instead, this process guarantees a desired three-dimensional outcome by reverse engineering it through the relationship between conic sections in three dimensions and their corresponding conic sections in the transformed flat state. In other words, knowledge of the relationship between an ellipse in three-dimensional space and a hyperbola on a flat plane allows us to precisely derive the type of developable surface (cone or cylinder), its radius, orientation relative to the x-y plane (the ground), and the angle at which a plane that intersects it would need to be to extract an ellipse of a desired eccentricity.

In his seminal work *Structural Systems*, the German architect Heino Engel suggested aggregations of 'bent' or mirrored cylinders for long-span roofs, described simply as "linear structural systems composed of cylindrical surfaces" (Engel 1997a). Structurally, these configurations act as barrel vaults in what Engel refers to as "surface-active" structural systems. However, the execution of these theoretical studies has seldom materialized in the built environment beyond the scale of a tectonic façade detail.

This research looks specifically at a subset of curved crease folding that will be referred to in this paper as "sectional mirrors", also commonly referred to as "reflecting creases" (Demaine et al. 2018) or "curved folding from mirroring operations" (Lebee 2015). While studied extensively, there is not much advancement in understanding the consequential tectonic effects at the architectural scale in the fields of origami and curved crease folding, specifically, on the tectonic translation on modes of habitability and type (materiality, proportion, and structure). Additionally, the difference between *idealized mathematical* curved creasing and *physical paper* creasing does not acknowledge the incorporation of material properties and thicknesses beyond the infinitesimally thin sheet of paper into the computational process (Mundilova 2019).

The application of this research is demonstrated in the design and construction of a full-scale prototype which tests five primary variables of the study: 1) the limitations of a construction sequence of on-site flat-to-form creasing; 2) the structural capacity of the curve-creased membrane geometry; and 3) the time and cost-effectiveness, and ease of installation of a pavilion that deploys this technique.

3 Conic Section Mirroring: Cylindrical Mapping of the Rectangular Sheet Building Block

Any developable surface, such as a cylinder, cone, or tangent developable, can be split in two by an intersecting plane tangent to a curve on the surface, known as an osculating plane.[1] If one part of the original curved surface is mirrored across that plane, the contiguity of the two parts is preserved along the planar curve of the intersection, allowing them to rejoin into a single poly-surface that can unroll coherently into its original shape. The curvature of the surface (direction of concavity) flips about the planar mirroring curve (in this case, a semi-ellipse), creating a single poly-surface of opposing curvatures that share an edge. The result adds a new curve to the surface in the flat state that translates to that planar crease in the 3D state (Fig. 1).

A few aspects of this technique are compelling for considerations of material and structural economy. First, creasing flat sheets in this way adds structural rigidity to singly ruled surfaces which otherwise are not inherently structural unless oriented a certain way (like a barrel vault). So, on the one hand, it becomes a productive way to introduce an extrinsic curve to a mathematical surface (a curve identified as not belonging mathematically to the u- or v- curve fields of a surface). Second, a mathematical relationship can be easily calibrated between the 2D sheet and the 3D form by relating conic sections. In this case, an ellipse of a specific proportion describes a section through a circular cylinder cut at a certain angle, corresponding to a hyperbola of that same angle in the flat state. This approach means a 3D object is designed purely in two dimensions, guaranteeing a pre-determined three-dimensional outcome (the inverse process of Huffman, Resch, and Demaine's work). Lastly, the system allows aggregating and stitching of multiple parts in the flat state which fold into a coherent three-dimensional whole (Fig. 2), introducing a unique and atypical strategy where three-dimensional modules combine in a remote space as 2D objects, using two-dimensional operations on a single plane. The processes necessarily mean we design the part and the whole, and the 2D and 3D – hence the cut-sheet (design) and outcome (construction) – simultaneously.

This line of inquiry attempts to expand on the vocabulary and use of sectional mirrors in two ways: first, to establish descriptive (not analytical or algebraic) mathematical relationships between conic sections derived out of planar intersections with developable surfaces (e. g., the translation of ellipses on a cylinder to corresponding hyperbolas when the cylinder unrolls to a flat plane), which allows for a more intuitive and accessible way of modeling and drawing these geometries without the

---

1 An osculating plane is defined as a plane that contains the tangent vector and normal vector to a curve at any point along its length, becoming the x-axis and y-axis of that plane, respectively. The normal vector of that plane (z-axis) is defined by the binormal vector of the curve. For a planar curve, the osculating plane, by definition, contains the entirety of the curve.

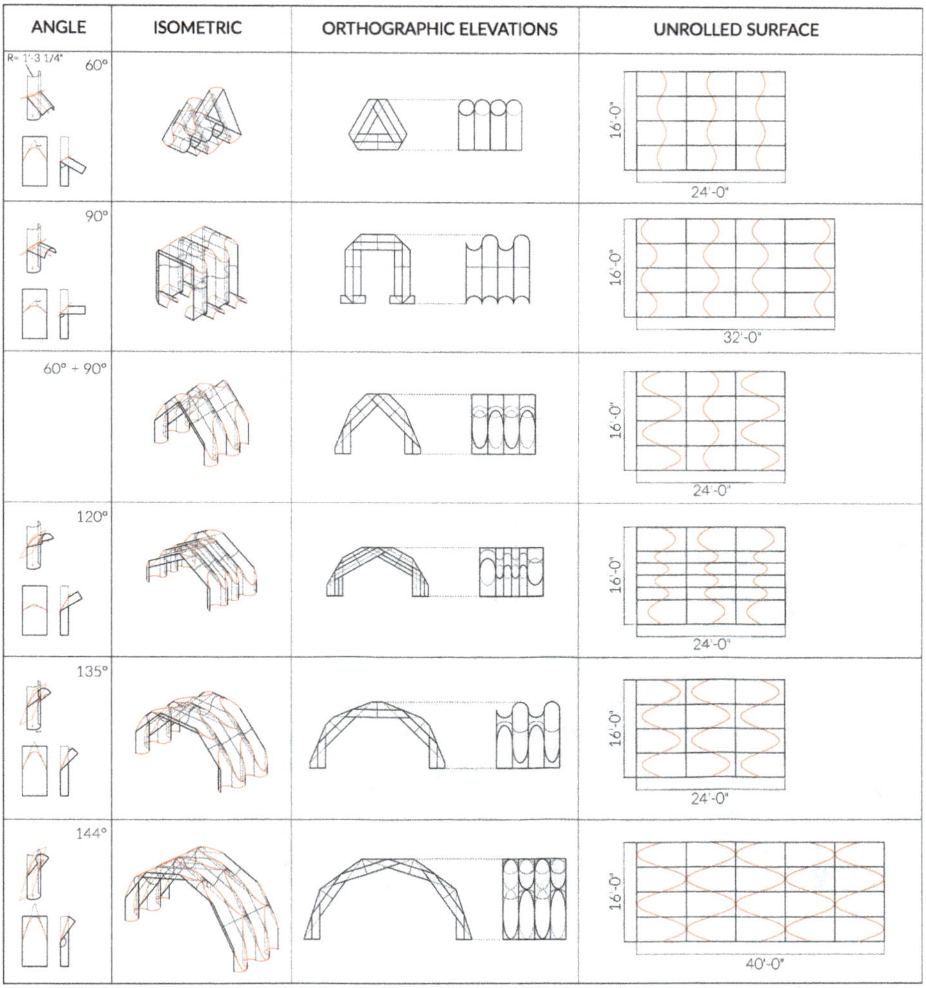

Fig. 1: Assembles of vaulted and gable geometry hybrids using osculating mirrors on cylinders at angles varying from 60° to 144°. Individual cylinders are stitched tangentially with adjacent ones in 2D to produce infinite fields of undulating forms.

need for specialized computational processes. And second, it preserves and maps the original dimensions and rectilinearity of readily available sheet material onto three-dimensional conical or cylindrical surfaces as potential strategies for reducing or eliminating material waste in digitally driven design and fabrication processes. Most existing experiments in curved creasing do not rely on or need to respect the rectilinearity of the original sheet.

The primary interest in these studies investigates formal distinctions between enclosures of different kinds by leveraging the geometric characteristics unique to

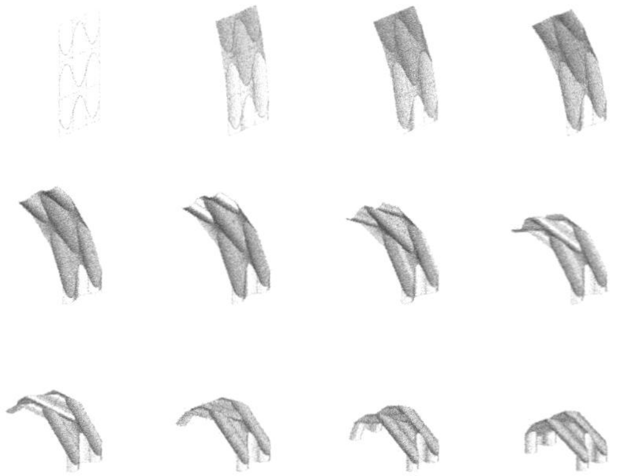

Fig. 2: Assembly of nine individual sheets in the flat state inflated to a three-dimensional gable/barrel vault form.

Fig. 3: Assemblies of vaulted and gable geometry hybrids at four different angles and scales.

these surfaces (Fig. 3). Forexample, a cylinder's ability to combine circular geometries and linear edges allows us to ask: can a barrel vault and a gabled roof share the same compositional surface?

4 Bending Cylinders: Shading Structure Prototype for Climate Resilience

A full-scale prototype in the form of a shading structure was constructed in July 2021 in collaboration with the City of Cambridge's Public Space Lab and Community Development Department (Fig. 4) as part of the "Resilient Cambridge" program (CambridgeMA.gov, 2021), which addresses climate inequity in the city. It was imperative

Fig. 4: Left: Aerial photograph of the shading pavilion. Right: The constructed prototype assumes the form of a gable in profile while containing a membrane surface that resembles a barrel vault.

to use construction methods that use recyclable material, produce little to zero waste, and require limited skill and cost in assembly, given a very low budget and the structure's temporary nature. The project addresses three key issues that are important in advancing architectural design in public discourse: social equity, climate equity – two undeniably intertwined issues – and innovation in construction and material technology, in this case, an application of the research on sectional mirror operations to introduce unfamiliar vocabularies of form.

This project uses a cylindrical module of 120° as the primary unit for the "roof-wall". 90° modules transition the vertical wall into the horizontal ground in the form of seating units that visually and structurally anchor the pavilion onto the site. The membrane comprises 1/8th-inch thick sheets of HDPE (high-density polyethylene), a lightweight, translucent, and recyclable plastic that is also relatively cheap. The dimension of each module is calibrated around the most readily available stock size of HDPE to use each in its entirety. No part of any sheet was cut and thrown out. Instead, the sheets are merely scored with hyperbolic curves on a CNC bed and then bent to shape. The same six units create the undulating walls, roof, and seating (Fig. 5). Once dismantled, the plastic can be reused (stored flat or installed at a different location as loose outdoor furniture) or recycled.

While the studies of angular compositions shown in Fig. 1 and 3 show various spatial configurations, the constructed pavilion uses a module of a 60° mirrored cylinder (yielding 120° roof and eave angles) as it proved optimal in terms of four interrelated (dependent) parameters that account for cost, material waste, and dimensional concerns. Up to three 4 ft × 8 ft modules (an entire bay) were stitched together flat and folded into their three-dimensional undulating cylindrical form (Fig. 6). The modules are rotationally aggregated to produce a sine-like wave in the flat state (Fig. 7). This undulation translates visibly in every projection of the three-dimensional object in plan and elevation.

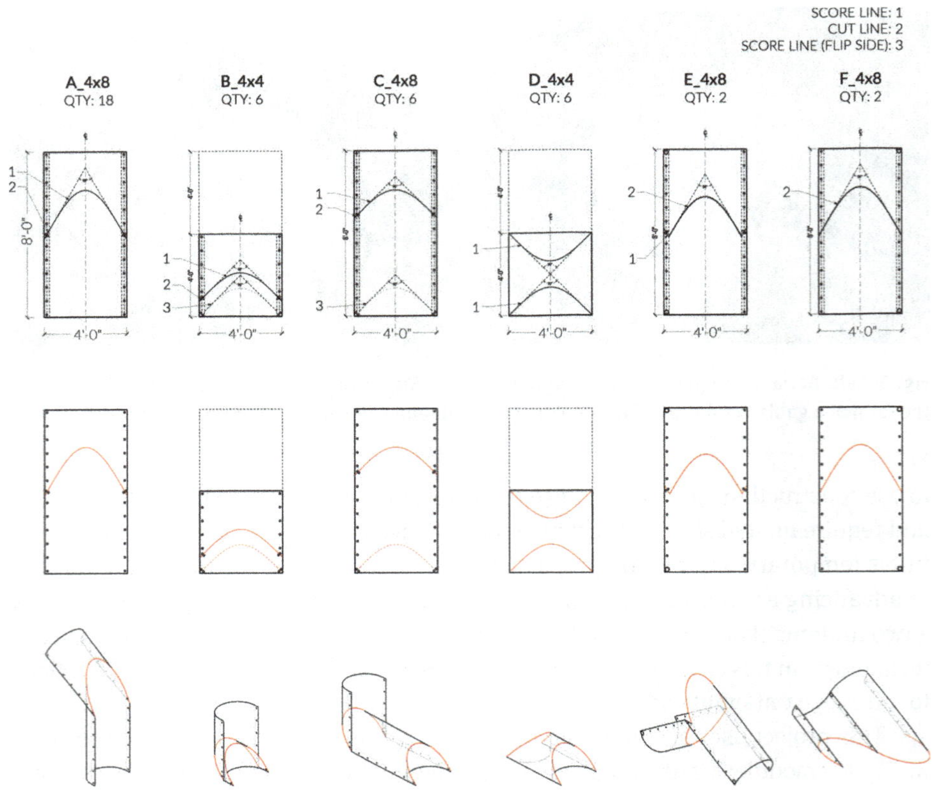

Fig. 5: The six units of construction and their corresponding cutsheets (unrolled surfaces).

5 Observations and Continued Research

The simplicity and ease with which these forms translate from the flat template to the three-dimensional state is beneficial for design and assembly but has limitations. The one-way three-dimensional form emulates an accordion, making it dynamic and indeterminate in one axis. Due to the material's stiffness and the curved crease's inability to hold its shape as effectively as a straight crease, the entire composition retains its memory of flatness and wants to unroll naturally. On the one hand, the curve adds structural stiffness to the planar surface but only if it is held in that curved position by external members. As a result, extrinsic structural systems are required to preserve the integrity of the surface membrane: first, an external structural lumber moment frame that operates as a clamp or bookend to resist the lateral thrust of the curved HDPE sheets; and second, steel tie rods operating in both compression and tension at the roof ridge lines and eaves, and cross bracing in the central bay of the composition, to resist shear.

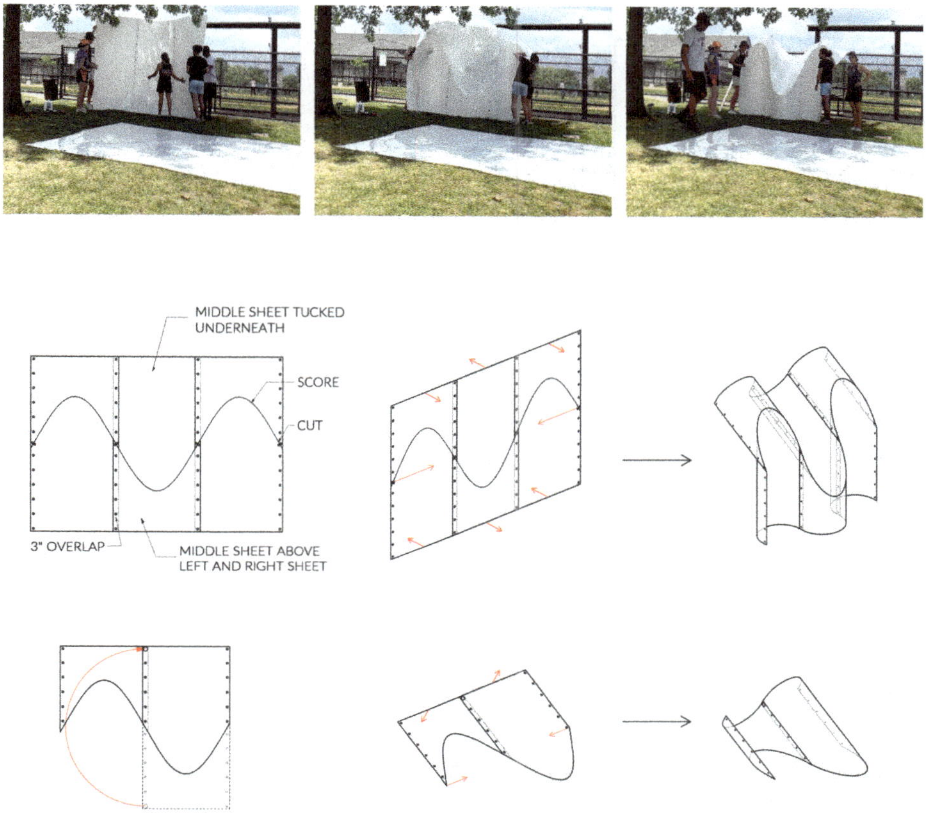

Fig. 6: Flat to 3D transformation process. Three identical Type A modules stitched together with mechanical fasteners along a tangent planar overlap while flat, then curled up into the 3D state collectively.

There are various ways of addressing the structural limitations of the one-way system. The studies in Fig. 8 and Fig. 9 introduce curvature along the perpendicular axis to turn the one-way accordion-like system into a two-way self-supporting system that can turn the corner in two axes. By introducing curvature in a perpendicular axis, the curved sheets assume the dual responsibility of both membrane and structure. This outcome is achieved most effectively by intersecting cylinders in perpendicular planes (via sectional mirroring planes) and recreating the familiar form of a groin vault (Fig. 10). However, the difference here is that the cylinders are not trimmed at the intersection but cut and folded to provide adequate surface area for the adjacent participating cylinders to adhere to one another. Extraneous material left behind belongs precisely to this zero-waste curved crease folding system, elevating the capabilities of surface membranes to behave like reciprocal structures (where individual parts rely on each other in equilibrium) while assuming a sculptural and spatial form of enclosure. Here,

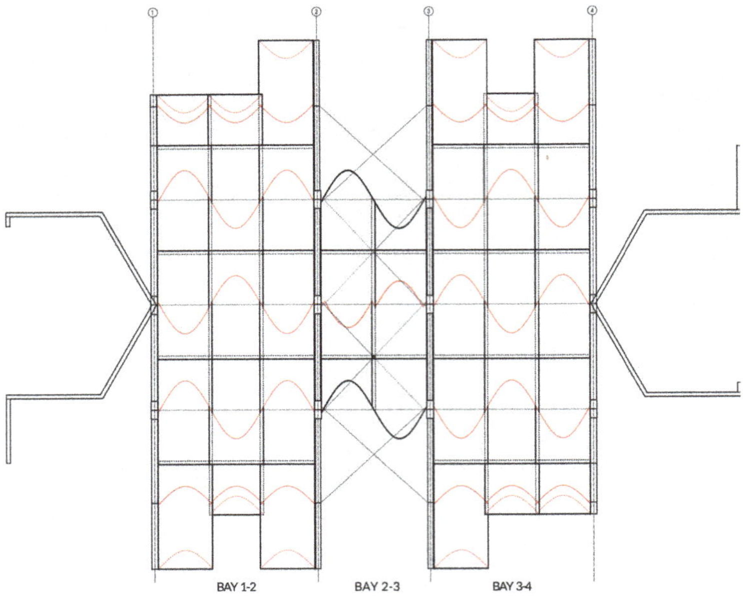

Fig. 7: Unrolled composition of membrane shown with structural members and the 3″ material overlap for fastening sheets with each other (in three-dimensions, this overlap is a planar surface tangent to the cylindrical portion).

we are considering a scenario where membrane, structure, and tectonic assembly sequence are fused into a single vocabulary.

The next phase of the research investigates how the surface can become self-supporting, creating opportunities to attach to itself on either end (as a structural 'loop') or by self-intersecting to introduce curves in the perpendicular plane to counter the torsional effects of the one-way accordion system. An example would be to incorporate structural members within the patterning of the planar mirroring curves. Figure 11 demonstrates a specific proportion of creasing whereby the elliptical curves of adjacent units are almost tangent, thereby acting structurally as arches leaning on one another in mutual compression. If these arches are manifest as structural strips of wood or metal (these would be standard linear members when flat), they can be installed while the sheets are in the flat state and then anchored to one another through tension cables once the membrane takes its 3-dimensional form.

Additionally, eliminating mechanical fasteners to stitch multiple sheets together would be desirable for a more efficient assembly process. Figure 12 demonstrates preliminary studies of more literal "stitching" techniques between two sheets using rope, strips of HDPE, or other similar linear material that can be easily woven and tied without using mechanical fasteners or tools. Further explorations are being developed through a system of additional cuts along the edge that allow the sheets to interlock. Axel Kilian explored a similar endeavor assembling flat sheets to produce double-

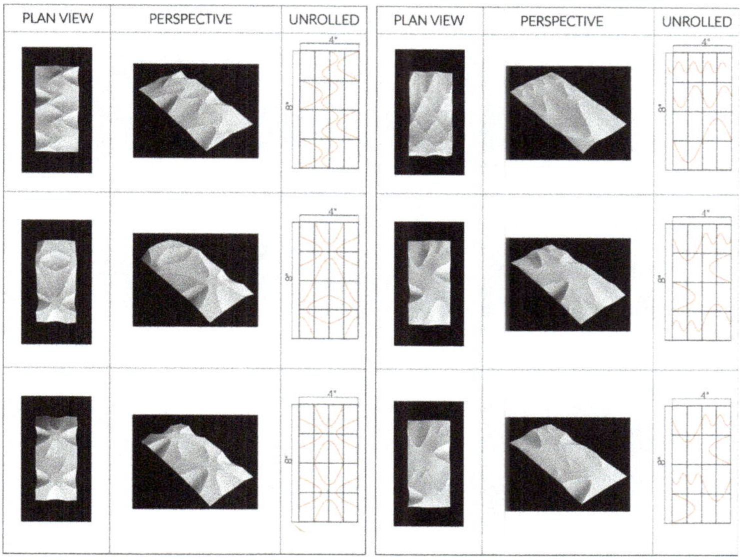

Fig. 8: Inscribing hyperbolic curves on a sheet perpendicular to one another to produce a surface of double curvature to increase the structural rigidity of the surface. These compositions are designed purely in 2D, akin to Huffman and Demaine's work.

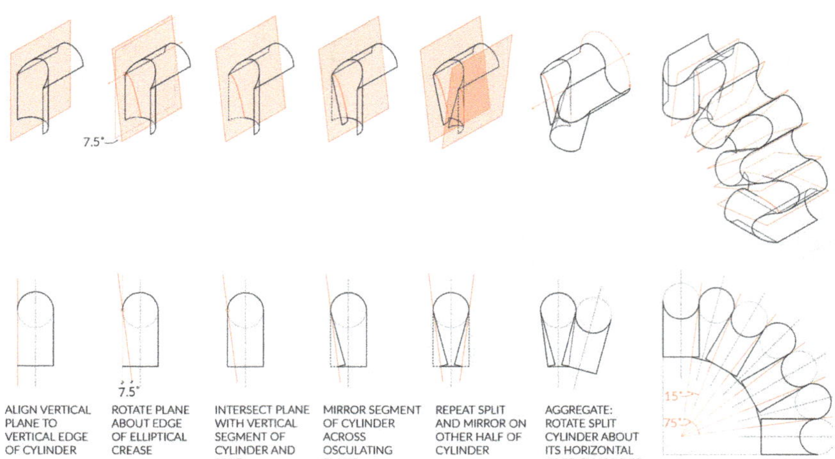

Fig. 9: Introducing sectional mirrors in the perpendicular axis to force the system to rotate in the other direction. This allows the one-way system to evolve into a two-way system by turning the corner in another plane and making the membrane serve structural purposes.

curved surfaces through what he calls a "puzzle joint". This "zipper" system allows adjacent sheets to interlock via friction-fit using the geometry of the surface itself rather than fasteners (Kilian 2003).

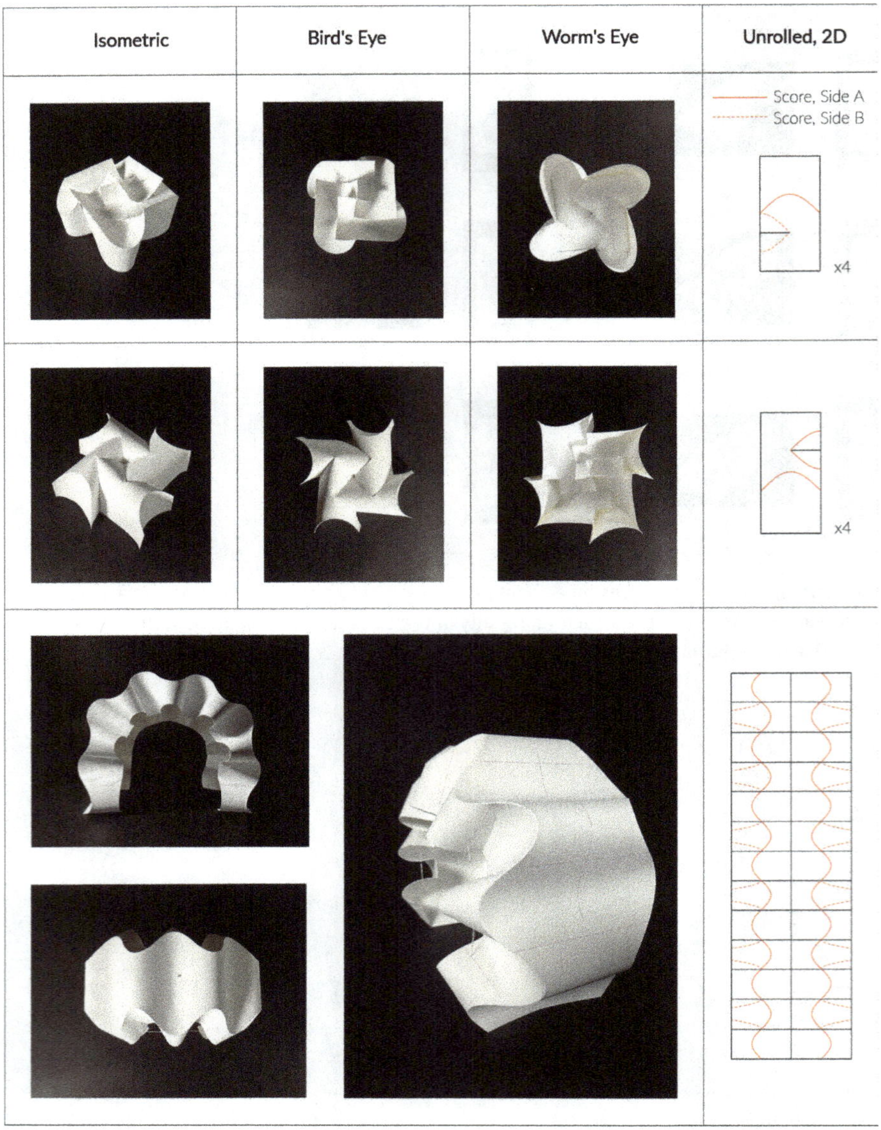

Fig. 10: Drawings and models of groin vault geometries, and a surface using sectional mirrors in the perpendicular axis to allow the system to turn the corner in the perpendicular plane.

Another area of improvement could test the limits of the number of sheets that can be stitched flat and folded up to inflate an entire structure in a single operation. Lastly, this prototype used HDPEsheets but the technique should be tested using various other flexible sheet materials such as metal, wood, fiberglass, other plastics, or a hybrid composite of methods and materials (e.g., sandwiching fabric between two layers of thin plywood).

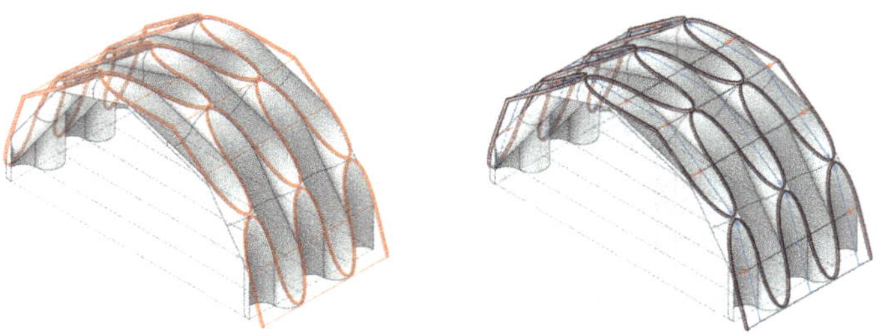

Fig. 11: Strips of wood or metal incorporate structure into the patterning of the creases, acting as arches. The arches lean on each other in mutual compression (blue arrows) while cables would pull them together in tension laterally (red arrows).

This research is continually evolving to advance building practices, develop new geometric vocabularies, and create a system that combines design and construction processes. It requires shifting attention to the geometric intelligence of the individual material part, rather than the complexity of a digital immaterial whole, allowing the tectonic definition to lead the aesthetic, structural, and topological characteristics of the surface. The primary ambition to elevate the tectonic assembly to the design process (rather than remain in the construction process, where it is most typically considered) is to limit or eliminate material waste in the process without sacrificing formal invention.

References

Bende, N. P., A. Evans, S. Innes-Gold, L. A. Marin, I. Cohen, R. Hayward, and C. Santangelo. 2015. Geometrically controlled snapping transitions in shells with curved creases. *Proceedings of the National Academy of Sciences (PNAS).*

City of Cambridge, MA. Resilient Cambridge: Climate change preparedness and resiliency plan. www.cambridgema.gov/-/media/Files/CDD/Climate/resilientcambridge/resilientcambridgeplan.pdf.

Davis, E., E. D. Demaine, M. L. Demaine, and J. Ramseyer. 2013. Reconstructing David Huffman's origami tessellations 1. *Journal of Mechanical Design* 135, 11.

Demaine, E., and M. Demaine. 2008. Computational Origami. Last modified December 7, 2013. https://erikdemaine.org/curved/MoMA/.

Demaine, E. D., M. L. Demaine, D. A. Huffman, D. Koschitz, and T. Tachi. 2018. Conic crease patterns with reflecting rule lines. https://arxiv.org/abs/1812.01167.

Engel, H. 1997. *Structure Systems.* Stuttgart: Gerd Hatje Publishers.

Kilian, A. 2003. Fabrication of partially double-curved surfaces out of flat sheet material through a 3D puzzle approach. In *Connecting: Crossroads of Digital Discourse,* Proceedings of the 2003 Annual Conference ACADIA, Indianapolis, USA, October, 23–26. 75–83.

Lebee, A. 2015. From Folds to Structures, a Review, *International Journal of Space Structures,* 30 (2).

Mundilova, K. 2019. On Mathematical Folding of Curved Crease Origami: Sliding Developables and Parametrizations of Folds into Cylinders and Cones, *Computer-Aided Design,* Vol. 115. 34–41.

Tahouni, Y., T. Cheng, D. Wood, R. Sachse, R. Thierer, M. Bischoff, and A. Menges. 2020. Self-shaping curved folding: a 4D-printing method for fabricating self-folding curved crease structures, Symposium on Computational Fabrication (SCF '20). Association for Computing Machinery, 1–11.

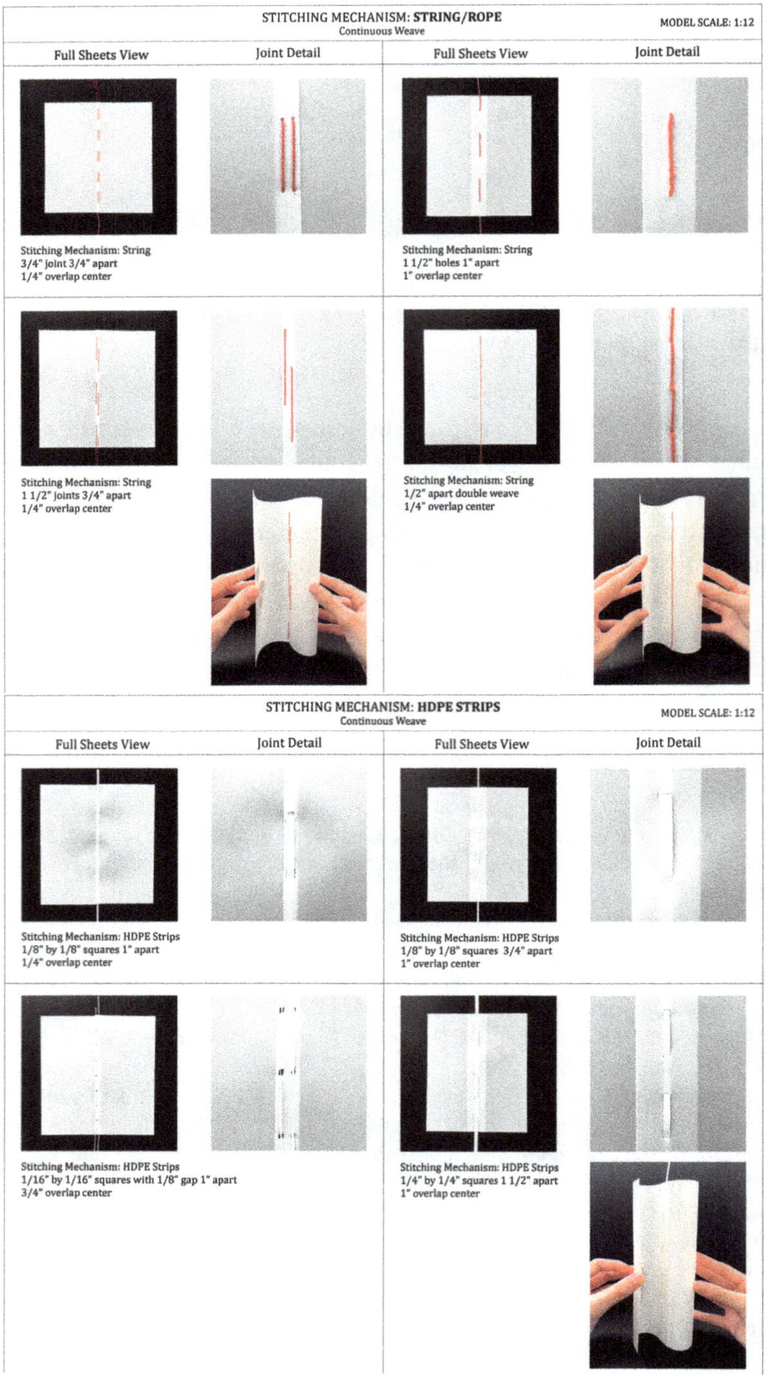

Fig. 12: Scalar studies of alternate sheet fastening mechanisms (*continued on next page*).

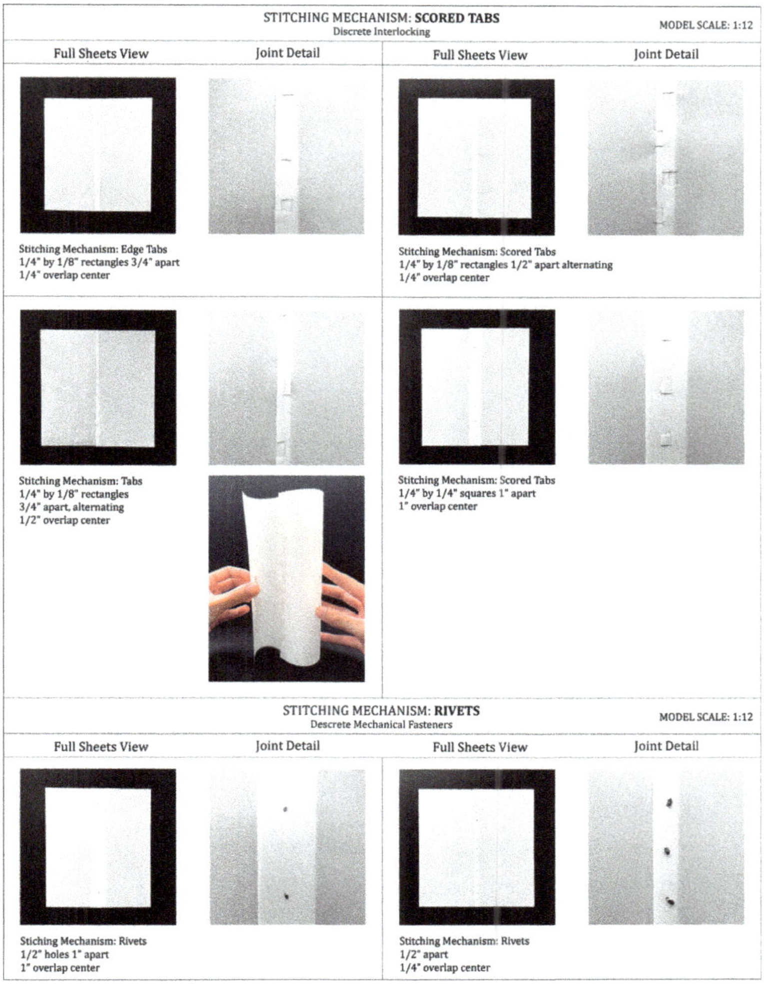

Fig. 12: (*continued*) Scalar studies of alternate sheet fastening mechanisms; weaving techniques through pre-drilled holes using string or HDPE strips (same material as the sheet itself), interlocking mechanism via scored tabs within the sheet itself, requiring no additional or extrinsic material for attachment, and the default metal fastening technique used in the prototype shade pavilion, represented here using rivets.

Lotte Scheder-Bieschin, Tom Van Mele, Philippe Block

Curved-Crease Flat-Foldable Bending-Active Plate Structures

Abstract: Curved-crease folded (CCF) bending-active plates efficiently form complex curvilinear geometries with structural applications. Instead, this research proposes joining stacked plates along common curved creases into flat-foldable configurations. These unfold with an accordion-like one-degree-of-freedom mechanism into corrugated spatial structures. The proposed system, termed curved-crease unfolding (CCU), allows for simple 2D prefabrication, flat-packed transport, and rapid on-site deployment. Its globally double-curved and articulated structural geometry extends the design space of CCF and finds application as structure and formwork. This research translates the fundamental geometric design principles of CCF to CCU for planar creases and demonstrates the design space for a multi-crease corrugated structure with non-zero thickness. A parametric model is implemented for the geometric construction and kinematic deployment in the COMPAS framework. Its deployment is validated by capturing the mechanical behavior with finite element simulation in the software SOFiSTiK. The paper demonstrates the non-developability conditions for convex synclastic and concave anticlastic creases. For the special cases of planar creases, angle correlations are formulated with direct inversion from CCF using discrete differential geometry. The trigonometric correlation for the kinematic deployment is applied to the discrete mesh representation. Inclining subsequent osculating planes reveal restricted geometric applicability regarding crease planarity. The non-zero thickness is modeled with an axis-shift approach. Finally, a rule catalog for global shape control is derived based on crease profiles and plane layouts with inclinations resulting in synclastic and anticlastic multi-crease designs. These would be challenging to construct otherwise and are enabled solely based on its formation principles.

Keywords: curved-crease folding, flat-foldable unfolding, bending-active plates, non-zero thickness, corrugated structures, discrete geometric modeling, deployment simulation, FE simulation, COMPAS.

1 Introduction

Active bending of plates allows for the efficient formation of lightweight, curvilinear structures without formwork. However, single bending-active plates under external loading deform dramatically and risk stability failure. Their performance improves significantly when mutually restrained along curved creases (Scheder-Bieschin et al. 2022). Thus, curved-crease folded bending-active plates find structural applications

https://doi.org/10.1515/9783111162683-027

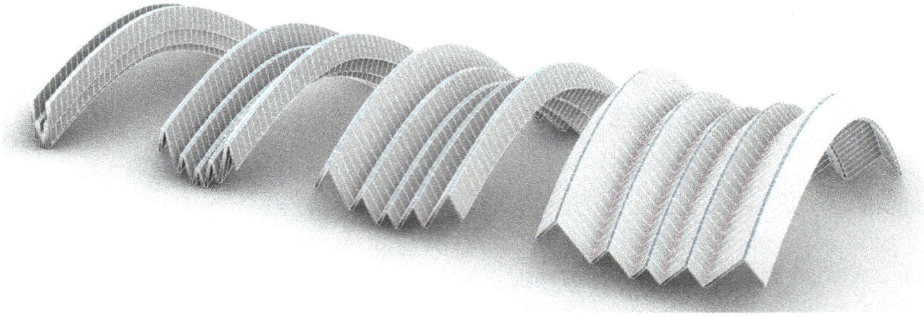

Fig. 1: Deployment of the CCU system from its closed flat-folded configuration to spatial unfolded configuration.

and, moreover, possess complex geometries (Maleczek et al. 2020). Due to plate thicknesses beyond foldability and limited panel size, separate plates are assembled with curved hinges and folded from a flat configuration.

When plate joining is necessary anyhow, stacked plates can be joined along common curved creases into a flat-folded configuration which is then unfolded. In accordance with curved-crease folding (CCF), this research introduces the terminology of curved-crease unfolding (CCU). A non-developable CCU system of multiple flat-folded developable plates unfolds with an accordion-like one-degree-of-freedom mechanism into a corrugated spatial structure of global double curvature (Fig. 1). CCU is compactly flat-packable for transport and can be rapidly deployed on-site into a material-efficient system of structural geometry, which expands the constrained design possibilities of CCF.

The objective of this research is to translate the fundamental geometric design principles of CCF to CCU for a single planar crease system, considering non-zero plate thickness. It aims to demonstrate the design space and associated rules of CCU for a multi-crease corrugated structure validated through finite element (FE) simulation.

2 Background

The feasibility of CCU was demonstrated as a proof-of-concept as formwork for corrugated concrete shells (Scheder-Bieschin et al. 2022). Furthermore, CCU can be identified combined with CCF in closed cross-sections as a single module in a timber bridge (Rihaczek et al. 2022) or array in computational vault explorations (Tachi 2013). The proposed system essentially resembles the vault's horizontally-cut half, avoiding joining more than two creases for fabricability.

Tachi (2013) explicitly introduced the concept of closed-folding based on four quadrants, however, without thickness. Shimoda et al. (2020) demonstrated an axis-shift approach with uniform thickness panels on a flat-foldable rigid-origami system with a one-directional mechanism termed 'bellow-type'. This essentially equals the proposed bending-active CCU system with the difference that the polygonal creases are more densely discretized for the curved creases and that the discrete transversal hinges of rigid panels are replaced by dense ruling lines of the bending-active strips. Consequently, the ruling direction can readjust increasing design freedom (Sec. 5.1), and the thickness approach considers continuous strain (Sec. 5.3).

3 Methods

This research approaches the geometric design of CCU in comparison to CCF using discrete differential geometry and plate thickness. It focuses on translating the geometric formation principles explicitly, hence, without employing dynamic relaxation or optimization approaches. Angle conditions are formulated for the non-developable CCU versus their developable CCF counterpart. As the paper focuses on planar creases, the design rules for the ruling layout are derived for the special case. For the kinematic modeling of the deployment mechanism, we apply trigonometric correlations in the discretized mesh representation and show implications for the behavior of subsequent creases for a multi-crease corrugated structure. The non-zero thickness is handled with an axis-shift approach. By exploring variations in crease profile and plane layouts with inclinations, a rule catalog is derived for global shape control into doubly-curved designs, revealing the design space.

The computational modeling is implemented in COMPAS, an open-source Python-based framework for research in AEC (Van Mele et al. 2022). As curved-crease unfolded structures are composed of developable strips – cylindrical, conical, or tangent surfaces – which are isometric to a planar configuration, they are discretely represented in a half-edge mesh data structure with quadrilateral planar faces. Beyond the geometric construction in a parametric design model for CCF and CCU, the simulation comprises their geometric deployment and an interface for structural verification.

The simulation with the FE software SOFiSTiK (SOFiSTiK AG 2022) captures the mechanical behavior with bending resistance and potential instabilities. The plates are modeled as QUAD shell elements with plywood of 8 mm thickness on a scale of 2.5 m and with a hinge condition along the curved creases. The plates are actuated by initial pre-bending to avoid buckling and then by crease-alignedinwards forces and perpendicular outwards forces to simultaneously cause plate bending and unfolding. The forces are applied in an iterative procedure with third-order analysis.

4 Fundamental design rules for a single curved crease

4.1 Geometric principle of CCF and CCU

The geometric principle of CCF and CCU is demonstrated using the concept of four quadrants (Fig. 2 top) introduced by Tachi (2013) for the special case of a planar crease that lies within a single osculating plane, allowing for mirror reflection (Sec. 4.2). Reflecting a single plate and its continuation results in four plates encompassing four quadrants. Quadrants Q1 and Q3 are generated by CCF by folding a continuous developable plate, with Q1 folding leftwards and Q3 folding rightwards. Their crease curve (in green) is convex for one plate and concave for the other. Quadrants Q2 and Q4 are generated by CCU by unfolding two mirror-reflected sheets joined along their shared crease. Q2's crease curve is convex for both plates (in blue), while Q4's crease curve is concave for both plates (in pink). Thus, CCU introduces two new types of creases. In a single curved crease with inflection points, the types transition from Q2 to Q4 from its convex to its concave curvature (Fig. 4 middle-left).

T, **N**, and **B** denote the tangent, normal, and binormal vectors in the orthonormal Frenet-Serret frame of the crease curve, while **B** is also the normal of the osculating plane **O**. Let α denote the reflection angle between the crease normal vector **N** and the surface tangent vector **S**, which lies in the **NB**-plane. $\alpha > 0$ for the actuated folded (CCF) and unfolded (CCU) 3D configurations, and $\alpha = 0$ for the unfolded (CCF) and closed-folded flat (CCU) 2D crease patterns. For CCF, the folding angle is $\pi - 2\alpha$, while for CCU, the folding angle is directly 2α (Fig. 2).

The developability conditions of the four quadrants are demonstrated on a single vertex in Fig. 3 with the sector angles denoted by $\beta_{0...3}$, counterclockwise, between the crease curve edges **C** and ruling vectors **R**. The sum of sector angles $\sum_{i=0}^{3} \beta_i = 2\pi$ for Q1 and Q3, indicating their assemblies are developable; $\sum_{i=0}^{3} \beta_i < 2\pi$ for Q2, indicating its assembly is synclastic; and $\sum_{i=0}^{3} \beta_i > 2\pi$ for Q4, indicating its assembly is anticlastic, despite the plates being monoclastic.

For a planar crease, in Q2 and Q4, the neighboring sector angles on the adjacent plate are equal, $\beta_0 = \beta_3$ and $\beta_1 = \beta_2$. Consequently, the sums of both sector angles on each side are equal, $\sum_{i=0}^{1} \beta_i = \sum_{j=2}^{3} \beta_j$ and $< \pi$ for the convex crease in the synclastic Q2 and $> \pi$ for the concave crease in the anticlastic Q4. However, this sector angles correlation cannot apply for the developable Q1 and Q3, as $\sum_{i=0}^{1} \beta_i < \pi \neq \sum_{j=2}^{3} \beta_j > \pi$ for the convex – and concave crease side, respectively. However, a crossover correlation exists for the ruling angle γ between the ruling **R** and tangent **T**, quantified depending on the special cases (Sec. 4.2).

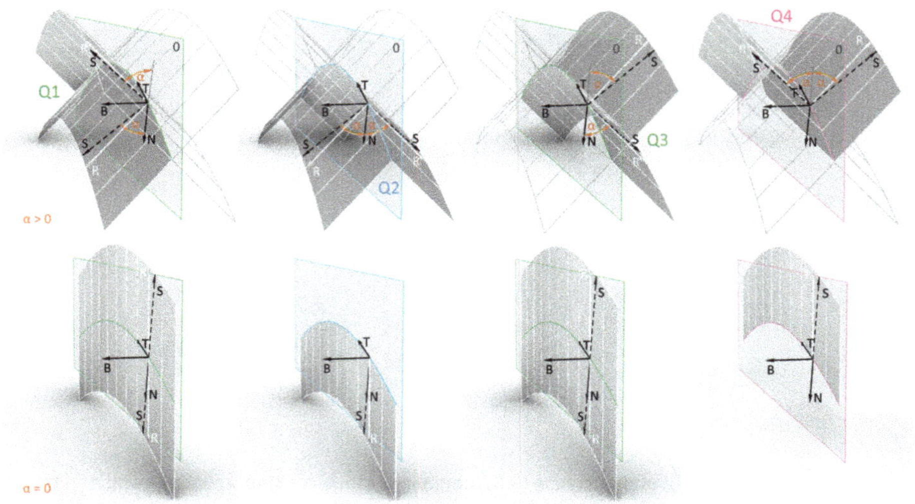

Fig. 2: Geometric principle of CCF and CCU with the concept of four quadrants. Top: actuated 3D state. Bottom: 2D crease pattern.

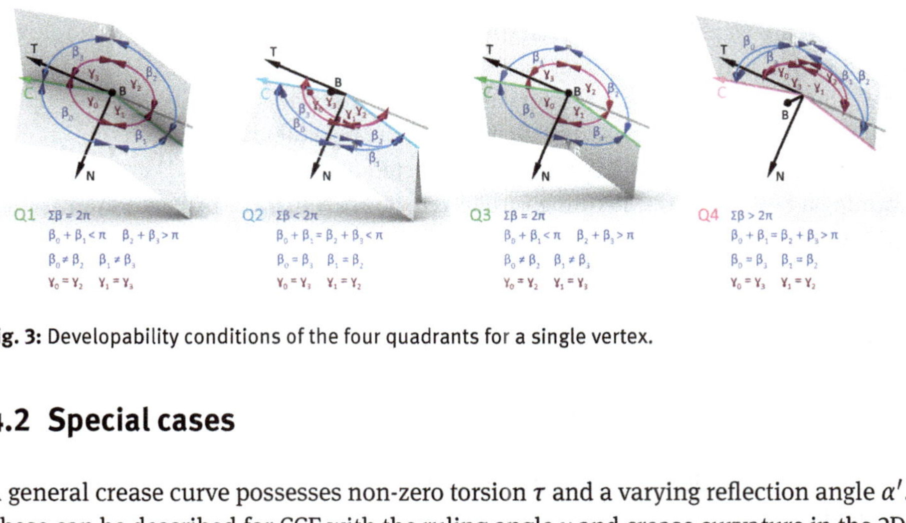

Fig. 3: Developability conditions of the four quadrants for a single vertex.

4.2 Special cases

A general crease curve possesses non-zero torsion τ and a varying reflection angle α'. These can be described for CCF with the ruling angle γ and crease curvature in the 2D pattern κ_{2D} with Eq. (1) and Eq. (3) by Fuchs and Tabachnikov (1999). The equations are adapted for CCU with Eq. (2) and Eq. (4):

$$\alpha'_{CCF} = (\cot\gamma_0 - \cot\gamma_3)\, 0.5\, \kappa_{2D} \tan\alpha \tag{1}$$

$$\alpha'_{CCU} = (\cot\gamma_0 + \cot\gamma_3)\, 0.5\, \kappa_{2D} \tan\alpha \tag{2}$$

$$\tau_{CCF} = -(\cot\gamma_0 + \cot\gamma_3)\, 0.5\, \kappa_{2D} \tan\alpha \tag{3}$$

$$\tau_{CCU} = -(\cot\gamma_0 - \cot\gamma_3)\, 0.5\, \kappa_{2D} \tan\alpha \tag{4}$$

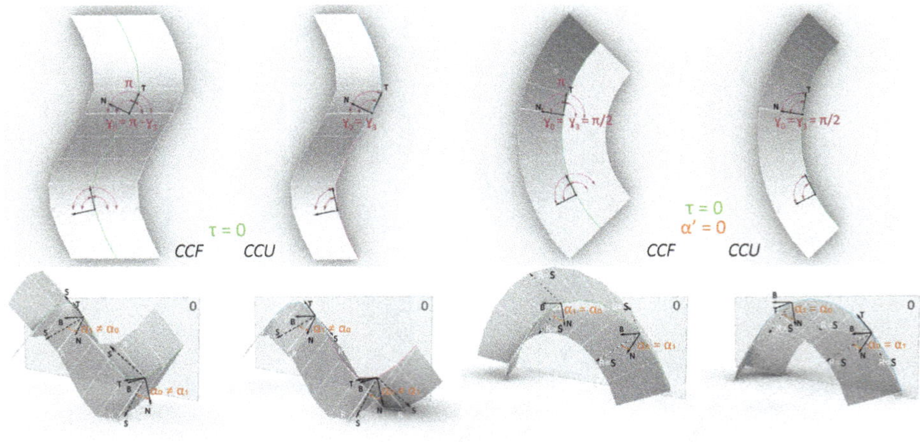

Fig. 4: The special cases of a planar crease (left), combined with a constant reflection angle (right) of CCF translated to CCU.

Special cases exist regarding crease planarity, reflection angle consistency, and a combination of both for the classic CCF (Fuchs and Tabachnikov 1999; Mundilova 2017) and are translated to CCU. In the special case of constant reflection angle $\alpha =$ const. with a non-planar crease, its change must equal $\alpha' = 0$. Hence Eq. (1) implies that, unless the crease is a straight line $\kappa_{2D} = 0$ or the system is flat $\alpha = 0$, the neighboring ruling angles across the crease must be equal in CCF, such that $\gamma_0 = \gamma_3$. This indicates that the rulings must be reflected across **T** in the 2D crease pattern, forming a kink. In CCU, the ruling is reflected across **N** in the 2D crease pattern, such that $\gamma_0 = \pi - \gamma_3$.

In the special case of a crease lying in a single osculating plane with a non-constant reflection angle, its torsion must equal $\tau = 0$. Hence Eq. (3) implies that the ruling must be reflected across **T** and **N** in CCF, such that $\gamma_0 = \pi - \gamma_3$. Consequently, the rulings are collinear in the 2D crease pattern and reflected across **O** in the 3D configuration (Fig. 4 left). In CCU, both rulings are identical, such that $\gamma_0 = \gamma_3$ or $\gamma_1 = \gamma_2$. Any developable surface type reflected at a single plane satisfies this special case. The ability of 3D reflection along a single plane permits the applicability of the reflection method.

In the ultra-special case of the combined special cases of crease planarity and constant reflection angle, both ruling conditions for CCF of their reflection across **T** $\gamma_0 = \gamma_3$ and their collinearity $\gamma_0 = \pi - \gamma_3$ must be satisfied. This implies that all ruling angles must be identical to $\gamma_0 = \gamma_3 = \pi/2$, indicating that the rulings **R** are orthogonal to **T** and coincide with the surface tangent **S** (Fig. 4 right). With the inversed equations and angle conditions, the same applies to CCU.

4.3 Deployment

The deployment mechanism is based on the equation

$$\cos\alpha = \frac{\kappa_{2\mathrm{D}}}{\kappa_{3\mathrm{D}}} \tag{5}$$

by Fuchs and Tabachnikov (1999). It indicates that the 3D crease curvature denoted by $\kappa_{3\mathrm{D}}$ increases proportionally to the reflection angle α during actuation. Eq. (5) holds true regardless of whether the crease is folded as CCF or CCU, as the kinematics is constrained by the planarity of a curved edge of a single developable.

In the discrete case of a mesh, the curvature $\kappa_{2\mathrm{D}}$ can be computed using the curvature radius $\mathbf{r}_{2\mathrm{D}}$ in the direction of $\mathbf{S}$. The magnitude of $\mathbf{r}_{2\mathrm{D}}$ can be determined with Eq. (6) using the trigonometrical relationship of the right-angled triangle formed by $\mathbf{r}_{2\mathrm{D}}$, half of the adjacent crease edge $\mathbf{C}/2$, and the angle $\mu_{2\mathrm{D}}$ between $\mathbf{S}$ and $\mathbf{C}$. This angle equals the average of the adjacent sector angles $\sum_{i=0}^{1}\beta_i/2$ (Fig. 5a). Similarly, the curvature $\kappa_{3\mathrm{D}}$ can be computed using the curvature radius $\mathbf{r}_{3\mathrm{D}}$ in the direction of $\mathbf{N}$. The magnitude of $\mathbf{r}_{3\mathrm{D}}$ can be determined with Eq. (7) using a triangle in the osculating plane $\mathbf{O}$ formed by $\mathbf{r}_{3\mathrm{D}}$, $\mathbf{C}/2$, and the angle $\mu_{3\mathrm{D}}$ between $\mathbf{N}$ and $\mathbf{C}$ (Fig. 5b).

$$\kappa_{2\mathrm{D}} = \frac{1}{|\mathbf{r}_{2\mathrm{D}}|} = \cos\mu_{2\mathrm{D}}/(|\mathbf{C}|/2) \tag{6}$$

$$\kappa_{3\mathrm{D}} = \frac{1}{|\mathbf{r}_{3\mathrm{D}}|} = \cos\mu_{3\mathrm{D}}/(|\mathbf{C}|/2) \tag{7}$$

As the crease length $|\mathbf{C}|$ remains constant during isometric transformation, Eq. (5) can be simplified for the discretized case to

$$\cos\alpha = \frac{|\mathbf{r}_{3\mathrm{D}}|}{|\mathbf{r}_{2\mathrm{D}}|} = \frac{\cos\mu_{2\mathrm{D}}}{\cos\mu_{3\mathrm{D}}} \tag{8}$$

a trigonometrical relationship of the triangle generated by $\mathbf{r}_{2\mathrm{D}}$, $\mathbf{r}_{3\mathrm{D}}$, which is the projection of $\mathbf{r}_{2\mathrm{D}}$ onto $\mathbf{O}$, and the reflection angle α (Fig. 5c). Shimoda et al. (2020) derive the same cosine relationship from the length preservation specific to the ultra-special case.

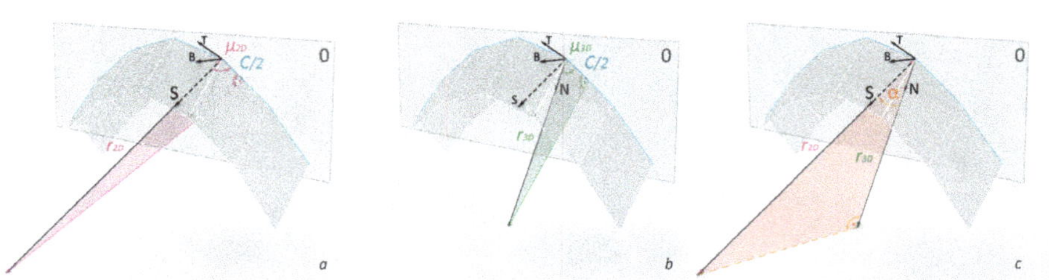

Fig. 5: Trigonometrical relationship of (a) $\mathbf{r}_{2\mathrm{D}}$, (b) $\mathbf{r}_{3\mathrm{D}}$, and (c) together with α.

With Eq. (8), the deployment of a crease κ_{3D} for a prescribed κ_{2D} can be simulated seamlessly for any α. Based on the trigonometric relations with γ and α, we obtain the ruling vectors $\mathbf{R}$ for the leading plate $\mathbf{R} = \cos\gamma_0\,\mathbf{T} + \sin\gamma_0 \cos\alpha\,\mathbf{N} + \sin\gamma_0 \sin\alpha\,\mathbf{B}$, and the CCF plate $\mathbf{R}_{CCF} = \cos\gamma_3\,\mathbf{T} - \sin\gamma_3 \cos\alpha\,\mathbf{N} + \sin\gamma_3 \sin\alpha\,\mathbf{B}$, or CCU plate $\mathbf{R}_{CCU} = \cos\gamma_3\,\mathbf{T} + \sin\gamma_3 \cos\alpha\,\mathbf{N} - \sin\gamma_3 \sin\alpha\,\mathbf{B}$, again with inversed signs. Finally, as the isometric transformation dictates constant lengths and sector angles, the planar quad meshes are reconstructed.

5 Design rules for a multi-crease CCU system

5.1 Behavior of subsequent creases

Further plates can be constructed by trimming and reflecting the rulings at subsequent reflection planes. During deployment, the subsequent plates are reconstructed based on the leading actuated crease and the isometry constraint. However, not any subsequent plane orientation results in planar creases during the intermediate states of actuation. If both planes are parallel, they rotate with identical progression around their parallel actuation rotation axes (Fig. 6a). If one plane is inclined around the other's actuation rotation axis, both planes rotate with different progression around parallel rotation axes (Fig. 6b). In both cases, the crease curvature changes proportionally; hence the subsequent crease remains planar. However, when one plane is inclined perpendicular to the other's actuation rotation axis, the rotation axes of the planes are non-parallel (Fig. 6c). Consequently, the crease curvatures change disproportionally, and the subsequent crease does not remain on a plane (Fig. 6d). Unlike rigid origami where the rulings are fixed, the rulings can reorient during actuation in a continuously bend structures such that the CCU system can unfold nevertheless as demonstrated with FE-simulation (Fig. 6e). The system undergoes a bistable behavior as described by Rihaczek et al. (2022), with minor crease warping until the system reaches its target with planar creases.

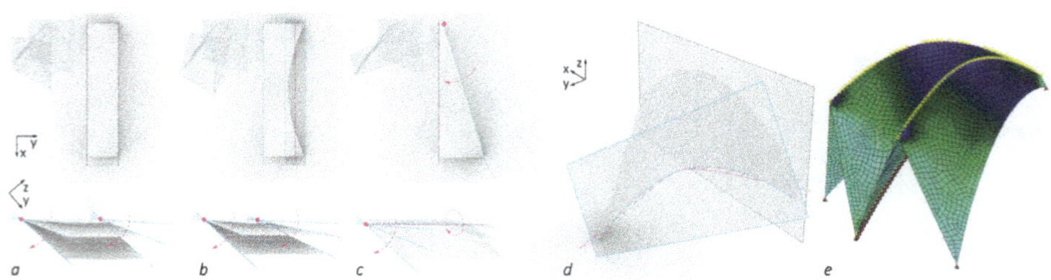

Fig. 6: Planarity of subsequent creases in relation to the osculating plane actuation rotation axes.

5.2 Translation into multi-crease corrugated structures and validation with FEA simulation

An array of parallel osculating planes generates alternating synclastic and anticlastic creases that result in a corrugated multi-crease vault structure (Fig. 1). Its geometric unfolding mechanism is verified with FEA in Fig. 7 with good congruence with deviations of 1 % of the span.

Fig. 7: FE simulation of sample from Fig. 1 (tension and compression from active bending in blue and red, respectively).

5.3 Non-zero thickness of bending-active plates

Bending-active plates require thickness for structural strength, but the induced bending stresses must respect material strength limits. This is ensured by limiting the allowable plate curvature to $\kappa_{\text{plate}} \leq f_y/(E\,0.5t)$ where f_y, E, and t denote yield strength, Young's modulus, and plate thickness (Lienhard 2014). The stresses induce positive strain $\varepsilon_+ > 0$ (red) at their extrados, negative strain $\varepsilon_- < 0$ (blue) at their intrados, while the central fiber remains neutral $\varepsilon_0 = 0$ (yellow dashed) (Fig. 8).

The bending-active plates of uniform thickness are connected using a rotation axis-shift approach (Shimoda et al. 2020). For CCF, this allows unfolding completely flat (Fig. 8a), while for the CCU, this allows folding completely closed (Fig. 8c). In CCF, an intrados connects to an extrados; thus, the strain contradicts as $\varepsilon_+ \neq \varepsilon_-$. In contrast, in the mirror-symmetric CCU, intrados connects intrados at synclastic creases and extrados to extrados at anticlastic creases; thus, the strain is compatible with $\varepsilon_+ = \varepsilon_+$ and $\varepsilon_- = \varepsilon_-$ respectively. Yet, the connection must accommodate strain as there are no one-sided-only discrete hinge zones as in rigid origami. Unlike CCF, where the rotation axes can align in one plane, either to all plates' extra- or intrados (Fig. 8b) or to the neutral fiber layer with tapering to avoid self-collision, the CCU does not allow complete closure with the volume trim method.

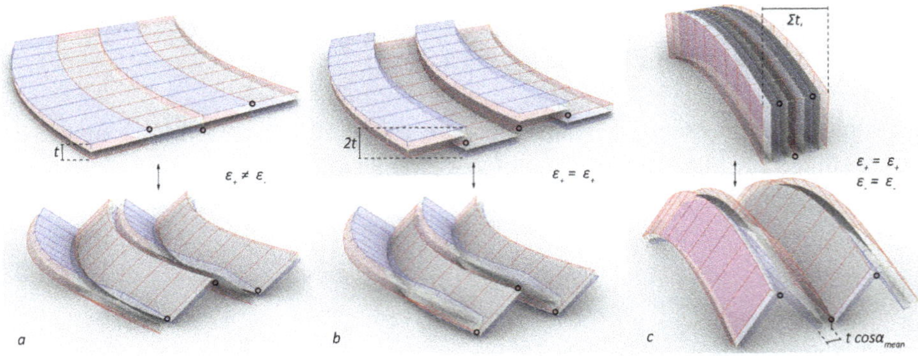

Fig. 8: Non-zero thickness (a) in CCF with axis-shift or (b) axis alignment, (c) in CCU with axis-shift.

The geometric modeling of the thickness requires offset in the vertex's normal directions by t. In the flat-folded state of CCU, the thickness equals the sum of all panels $t_{\mathrm{CCU}} = \sum_{i=0}^{n} t$. In the actuated state, only with rulings perpendicular to the crease as in the ultra-specific case, the offset crease curve is planar (as for Shimoda et al. (2020)). Otherwise, the offset curve is corrected by intersecting the rulings with an offset plane by $\cos_{\mathrm{mean}}$ and reversely offset to adjust the original curve.

5.4 Global shape control by variations in crease curvature

Besides the developable surface type and osculating plane layout, the global shape of a CCU structure is controlled by the profile of the crease curve, such as an arc, Elastica, catenary, funicular under asymmetric loads, and freeform curve with inflection points (Fig. 9). However, the freeform risks reaching unfolding angle limitations; ruling directions from a combination of two conical surfaces would be more suitable. The 2D crease pattern (closed-folded behind) must be computed with Eq. (8) such that it reaches its 3D target under a prescribed α (unfolded front).

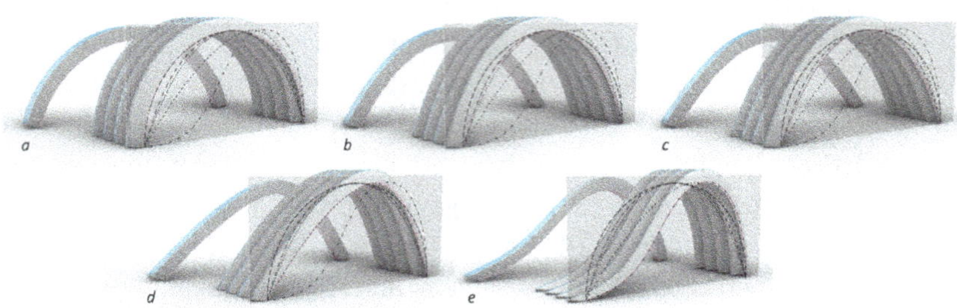

Fig. 9: Corrugated vaults with profiles of (a) arc, (b) Elastica, (c) catenary, (d) funicular with asymmetric loads, (e) freeform curve with inflection points.

5.5 Global shape control by variations in plane distance and inclinations

Even though the plates remain monoclastic throughout the isometric actuation, the multi-crease corrugated structure can be manipulated into various target global shapes through plane distance and inclination. A constant distance of parallel planes results in monoclastic vaults; decreased distance between the plates inclined towards the center generates synclastic vaults, while decreased distance between the plates inclined away from the center generates anticlastic vaults (Fig. 10). The shape is generated to fit target curves identical along the apex and supports.

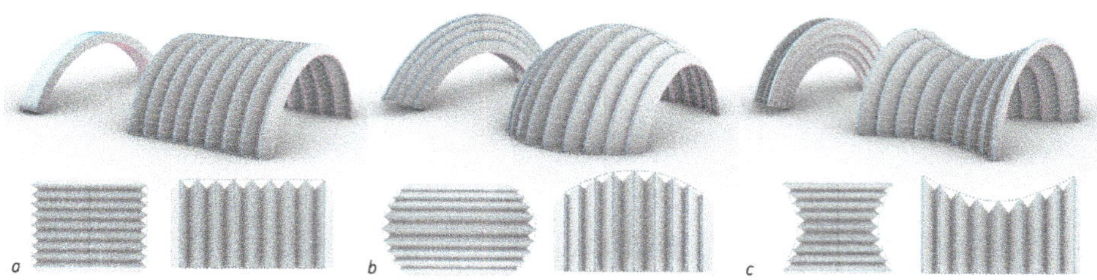

Fig. 10: Corrugated CCU vault with (a) monoclastic, (b) synclastic, (c) anticlastic curvature by plane distance solely. Top: closed and unfolded. Bottom: plan and side view.

By inclining a plane around its actuation rotation axis, the plane distance at the apex and the supports increases and decreases simultaneously. Consequently, the apex can maintain constant height while the support distance varies, and vice versa depending on plane distances, or the ridge and valley creases are articulated alternatingly at the apex and supports (Fig. 11). Furthermore, the plane inclination changes the reflected rulings' directions, resulting in variations in the corrugation depth.

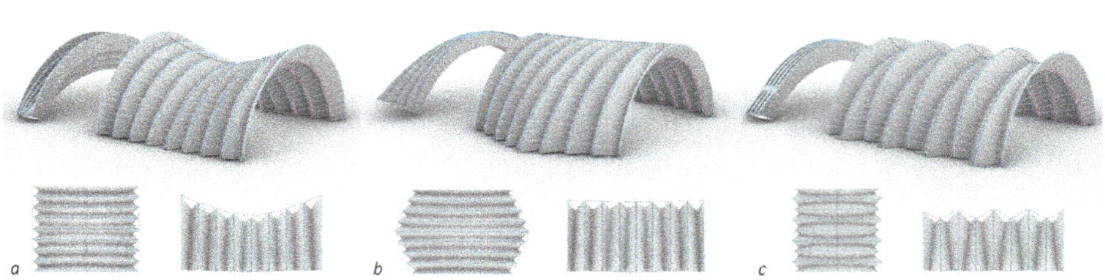

Fig. 11: Corrugated CCU vault by plane inclination around the actuation rotation axis.

By inclining a plane perpendicular to its actuation rotation axis (consider implications from Sec. 5.1), the plane distance decreases towards one end of the crease. This results in alternating non-axisymmetric articulations to the sides (Fig. 12a). Alternatively, with consecutive inclinations in radial layouts, the structure unfolds into a corrugated dome or funnel column (Fig. 12b,c).

Fig. 12: Corrugated CCU (a) vault, (b) dome, and (c) funnel by plane inclination perpendicular to the actuation rotation axis.

6 Reflections and outlook

This research translated the geometric principles of CCF to the flat-foldable CCU system, demonstrating direct inversion in various angle correlations. It demonstrated applicability of the discrete deployment formulation and non-zero thickness with axis shift, including shape adjustment beyond the ultra-special case as in Shimoda et al. (2020). For a multi-crease system, it revealed geometric applicability of plane inclinations and demonstrated possible ruling readjustment during a bistable behavior in flexible bending-active in contrast to rigid-origami structures. This is to be further and more thoroughly investigated, but it potentially increases the design space significantly. Lastly, it demonstrated global shape control with positive and negative Gaussian curvature from developable plates and introduced a novel design language that extends the design space of CCF.

Future work will set out on design explorations of more intricate geometries and their associated formation rules by combining the control variables of crease curvature, plane distance and inclinations in both axes, by combining the CCU with CCF while avoiding self-collision, and by introducing bifurcations and non-planar creases, which will enable transitions between different types of developable surfaces and significantly expand the design space. Through feedback from funicular form-finding and FEA, the overarching objective is to obtain structurally informed structures. The numerical application developed will be distributed open-source and via an interactive Rhinoceros3D Plugin developed within the COMPAS ecosystem. Structural considerations will investigate the mechanical effect on plates and hinges and the structures' scalability potential.

Ultimately the corrugated curved plate system is to be applied on an architectural scale, materialized as a structure itself (Rihaczek et al. 2022) or as formwork for corrugated concrete shells or vaulted floors (Scheder-Bieschin et al. 2022). The bending-active plates can be materialized with plywood connected using various textile hinge strategies (Rihaczek et al. 2022; Scheder-Bieschin et al. 2022) and with GFRP connected using hinge areas of lower stiffness through fiber layout (Körner et al. 2016) or non-impregnation (Choma 2021). The crease assembly of CCU is more complex due to self-obstruction (Scheder-Bieschin et al. 2022) than the flat-assembly of CCF (Maleczek et al. 2020); however, the flat-foldable system folds more compactly than CCF. The curved-crease unfoldable bending-active system offers the advantage of a high degree of simple 2D prefabrication, flat-packed transport, and rapid on-site deployment. It results in complex, structurally-articulated 3D geometries, which would be challenging to construct otherwise and are enabled solely based on its formation principle.

References

Choma, J. 2021. Foldable composite structures. United States US10994468B2.

Fuchs, D., and S. Tabachnikov. 1999. More on Paperfolding. *The American Mathematical Monthly* 106(1): 27–35.

Körner, A., A. Mader, S. Saffarian, and J. Knippers. 2016. Bio-Inspired Kinetic Curved-Line Folding for Architectural Applications. In *ACADIA 2016,* 270–79.

Lienhard, J. 2014. Bending-Active Structures: Form-finding strategies using elastic deformation in static and kinematic systems and the structural potentials therein. Thesis, Stuttgart.

Maleczek, R., Gabriel S., A. Metzler, and C. Preisinger. 2020. Large Scale Curved Folding Mechanisms. In *DMSB 2019,* 539–53.

Mundilova, K. 2017. Geometry and Interactive Design of Curved Creases. Thesis, Wien.

Rihaczek, G., M. Klammer, O. Basnak, A. Körner, R. La Magna, and J. Knippers. 2022. Timbr Foldr: A Design Framework and Material System for Closed Cross-Section Curved Folded Structures. *JIASS* 63(4), 272–88.

Scheder-Bieschin, L., T. Van Mele, and P. Block. 2022. Curved-Crease Folding of Bending-Active Plates as Formwork: A Reusable System for Shaping Corrugated Concrete Shell Structures. In *Acadia* 2022.

Shimoda, Y., T. Tachi, and J. Sato. 2020. Flat-Foldable Rigid Origami with Uniform-Thickness Panels. In *AAG 2020.*

SOFiSTiK AG. 2022. SOFiSTiK. www.sofistik.com.

Tachi, T. 2013. Composite Rigid-Foldable Curved Origami Structure. In *First Conference Transformables 2013.*

Van Mele, T. et al. 2022. COMPAS: A Computational Framework for Collaboration and Research in Architecture, Engineering, Fabrication, and Construction. www.compas.dev

Thorsten Pofahl, Alex Seiter, Martin Trautz, Lisa-Marie Reitmaier, David Baill, Gerhard Hirt

Form Finding of a Sheet Metal Shell by Generative Design and Pareto Optimization

Abstract: The study focuses on the design and optimization of a shell structure made from a 1 mm thick aluminum sheet. The objective of the research is to create an optimized structure that meets the requirements of structural performance, material usage, and production efficiency. We employed a generative definition for the structure and implemented a multi-criteria optimization approach to balance conflicting objectives. The structural morphology of the shell is considered on both macro and micro scales. At the macro scale, various shell shapes, ranging from simple spherical domes to undulated shells with different cross-sections, are explored. The shape of the shell is optimized to achieve both structural performance and production efficiency. Undulating the shell edge is found to positively impact the structural behaviour, although it increases the complexity of the components. On the microscale, structural embossments in the form of beads are introduced to enhance the moment of inertia, buckling resistance and natural frequencies. A parametrically defined bead arrangement is employed resulting in various patterns. The optimization process follows a multi-objective approach, aiming to simultaneously optimize structural performance, tool material usage, and production time. The objectives are derived from the geometry itself or calculated through a linear buckling analysis. The optimization is not weighted initially, allowing all objectives to carry equal priority. The solutions are then represented in non-normalized result domains and Pareto surfaces to evaluate their performance and compatibility. Through the analysis of 22 generations and 4400 parameter sets, the research provides insights into load-bearing capacity, serviceability, manufacturing effort, and manufacturing costs. The results highlight the trade-offs between different objectives and the interplay between shell geometry, stability, and production parameters.

1 From sheet metal to shells

The basic design for the demonstrator is a circular shell on 5-point supports with a diameter of 4.2 m, made from a 1 mm thick aluminum sheet. See Fig. 1 for design and Fig. 2 for production process. We adapted stretch forming and incremental sheet forming processes to produce the complex sheet metal components [1, 2]. Both the overall shape and structural detailing adapt to these primary constraints and are optimized in terms of structural performance, material usage and production efforts. We implemented a generative definition for the structure with a broad spectrum of possible forms and as few input parameters as possible. The analysis procedures are

https://doi.org/10.1515/9783111162683-028

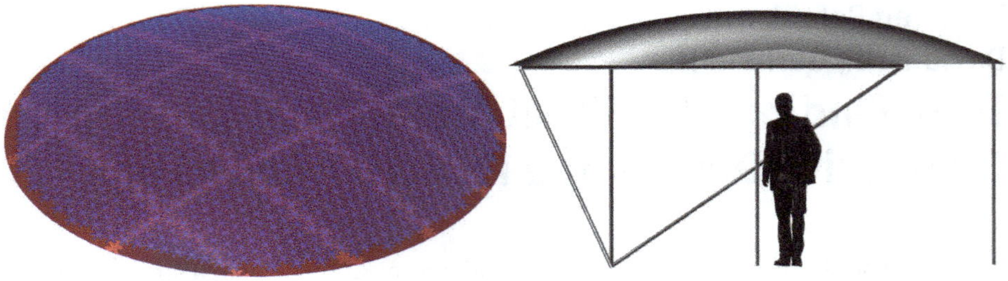

Fig. 1: Demonstrator design in view, perspective, and detail.

Fig. 2: MDF tool block adapts geometry, IBU forming tool traces beads, finished component, Panel joint with pattern overlay.

fully integrated and allow for the usage of an evolutionary algorithm to find reasonable conditions for our design, each balancing opposing goals differently. Based on this set of solutions, we derived general observations on the overall behavior and principles of our system.

2 Structural morphology

Structural morphology is the overall consideration of form, structure and forces. It is a non-hierarchical design principle on different scales [3]. In our case, we are considering the macro-structure of the shell at the global scale and the individual structural embossments at the local scale. The overall optimization on all scales results in the optimal "layout", which includes information about the topology, shape, size of the structural components, and materiality. Optimization processes that address several objective functions simultaneously are used to search for the optimal "layout", following the abstraction of natural processes [4]. The principles of structural morphology guide the adjustment of the basic topology of the shell shape and bead pattern, using a global optimization process that considers both structural behaviour and production

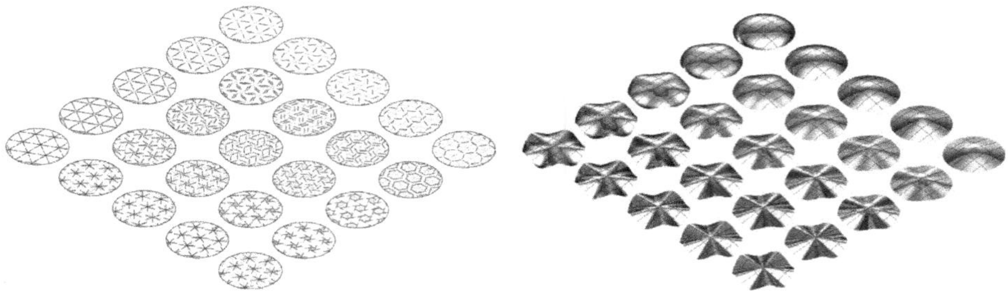

Fig. 3: Microstructures (left), macrostructure (right).

effort. Rather than being combined in an objective function with a priori weights, these partially conflicting objectives are examined in a multi-criteria process, resulting in a solution set of structural variants that fulfill different objective criteria with varying weightings. This approach allows for the individual components of the problem definition to be derived from the composition of the solution set and for the optimization goals to be weighted based on a multitude of conceivable solutions.

2.1 Macroscale – shell structure

The spectrum of possible shell shapes ranges from a simple spherical dome to a multi-curved undulated shell shape with linear or parabolic cross-section. A spherical dome allows all components to undergo stretch-forming on identical or nearly similar molds. The only differences between components are in their cut and pattern formation. Adapting the mold to the next shape to be produced by milling results in a minimum of mold material and production time. Although this shell geometry is not necessarily the optimum shape from a structural point of view, creating the shape based on a parabola or catenary can lead to improved structural behaviour. Undulating the shell edge results in a rotationally symmetrical alternation of synclastic and anticlastic curvature changes, which can also positively impact the structural behaviour (Fig. 3, right). However, both measures increase the degree of component complexity. The complexity is measured by the volume of the required tool block into which the part geometries are oriented and sorted (Fig. 4, left). To represent as wide a spectrum of shapes as possible with a minimum number of parameters, a parametric definition of the shell was developed in this investigation. The parametric definition of the form is based on radial longitudes, which are defined as a basic spline curve and adjusted through variation of three control points. In the radial coordinate system of the sphere, the x-component of the center point influences the shell cross-section and is defined as a variable (Fig. 4, right, in green). The undulation is achieved by adjusting the z-component of the curve end point at every second curve (Fig. 4, right, in red).

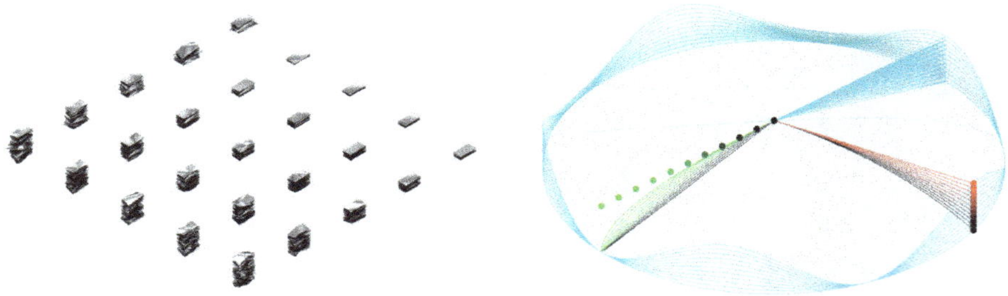

Fig. 4: Parameters of the local and microscale.

2.2 Microscale – structural embossments

Beads are predominantly utilized to fulfill utility functions such as enhancing the moment of inertia, buckling resistance and natural frequencies (ωk), as well as modifying reaction forces and moments [5,6]. In the context of structural morphology, beads serve as the additional structural elements at the local scale. The arrangement of individual beads creates basic shapes, which are then patterned in relation to each other. The degree of additional global bending stiffness induced by the beads relies on the base shape, pattern, and pattern density. A parametrically defined pattern is employed in this investigation. The definition allows a seamlessly transformation of the patterns, resulting in a kaleidoscopic effect in the superposit. The pattern definition is based on a triangle mesh, where truncated edges are rotated around a movable point. One of the parameters of this definition is the rotation angle, which varies between 0 and $0.5\,\pi$. Another parameter determines the point on the length domain, about which the edge is rotated. Based on these two parameters, a spectrum of patterns representing triangular and hexagonal meshes, with linear, star-shaped and elemental properties and different degrees of entanglement can be generated (Fig. 3, left). Entanglement of the patterns is equivalent to the elimination of axes without significant moments of inertia. It is important that the patterns are interlocked in at least two perpendicular directions, preferably in three axes intersecting at 60°. The greater the entanglement, the more continuous the bending stiffness and the more homogeneous the component. A third parameter scales the mesh within a domain of 100–200 %. The production time of these patterns is directly proportional to the total length and complexity of the forming paths of the milling and forming tool.

3 Multi-objective optimization

A multi-objective optimization is an optimization problem with multiple objectives [7]. It can be written as:

$$\min_{x \in X} = (f_1(x), f_2(x), \ldots, f_k(x))$$

With $k \geq 2$, $k \in N$ as the number of objective functionals, and the set X as the admissible set of the explorable set, which is usually $X \in R^n$ but depends on the n-dimensional input domain. The admissible set is usually defined over bounded functions. The vector objective function:

$$f : X \to R^k, \ x \mapsto (f_1(x), f_2(x), \ldots, f_k(x))$$

If an objective function is to be maximized, the negative or the inverse is minimized. We set: $Y \in R^k$ as a mapping of X, $x^* \in X$ as an admissible solution, and $z^* = f(x^*) \in R^k$ as a result. In Multi-Objective Optimization, there is usually no admissible solution that minimizes all objective functions simultaneously. The solution is done by considering a Pareto-optimal solution. This means none of the objectives can be improved without worsening at least one objective. An admissible solution $x_1 \in X$ is Pareto-dominant over a solution, $x_2 \in X$ if:

$$\forall i \in \{1, \ldots, k\}, \ f_i(x_1) \leq f_i(x_2) \ \text{ and } \ \exists\, i \in \{1, \ldots, k\}, \ f_i(x_1) < f_i(x_2)$$

The solution $x^* \in X$ (and the corresponding result $z^* = f(x^*)$) is called Pareto-optimal if there is no more dominant solution. The set of Pareto-optimal solutions written as X^* is the Pareto front (3D Pareto surface).

3.1 Implemented objective functions

The optimization problem sets high structural performance, minimum tool material usage, and short production time as objectives. The values of these objectives are derived from the geometry itself or calculated through a linear buckling analysis.

$$\begin{aligned}
z_1 &= \text{value of deformation} &&= \text{stiffness of the system} \\
z_2 &= \text{value of instability} &&= -1/\text{load factor} \\
z_3 &= \text{value of the IBU pathlength} &&= \text{forming time} \\
z_4 &= \text{value of tool material} &&= \text{milling} + \text{material effort}
\end{aligned}$$

For each generation of the optimization, the following result domains are created.

$$\min z_i = \min\{z_i\} \text{ und } \max z_i = \max\{z_i\}, \ D_i(z_i) = [\min z_i; \max z_i]$$

Normalized to domain [0; 1] for result representation and interpretation:

$$z_i^n = \frac{z_i - \min z_i}{\max z_i - \min z_i}, \quad \max z_i = \max\{z_i\}, \quad D_i^n(z_i) = [\min z_i^n; \max z_i^n] = [0; 1]$$

The optimization is not weighted, indicating that all objectives carry the same priority. Subsequently, the results can be manually weighted later. Moreover, it is feasible to eliminate solutions in case the result values surpass or fail to meet specific threshold values.

3.2 Geometric evaluation (Grasshopper)

We used the Rhinoceros Grasshopper environment (GH) for the overall process chain, including geometrical evaluation. To evaluate the material requirements for forming tools, we first fit the panels into the stretch forming machine's installation space in a production-friendly manner and sort them into a compact volume, which requires optimization based on a traveling salesman problem (TSP). Then, we compute the tooling and material efforts for successively milling individual forming tools from this volume: $h(V_{\min}) = z_4$. To determine the forming time, we sum up the individual lines of the patterns (= milling paths) to obtain a total length. The forming time equivalent to the path length: $\sum_{i=1}^{n} L_i = z_3$.

3.3 FEM (Karamba)

Thin shell structures, akin to thin compression members, are typically evaluated based on their stability, rather than the stress in their cross sections. In the state of membrane stress, significant strain energy can be stored without much deflection. However, if this energy is converted into bending energy, it can lead to instabilities and result in a loss of load-bearing capacity [8, 9]. Consequently, the maximum deflection of the structure with Th. II. O. under a constant area load, as well as the load factor of the structure, which is determined by a linear buckling analysis (LBA), are used to assess the structural performance of the shell structure. Nonlinear analyses (GNA, GNIA, or GNMIA) that provide more accurate results are not considered since the ranking of the results within the solution domain is decisive, rather than the absolute values themselves. Additional information on model building can be found in references [10, 11]. The beads are represented by bar elements joined with hinges. The calculations are carried out within GH, using the finite element (FE) method with the Karamba plug-in. The automatically obtained maximum deflection is: $\max u_z = z_1$. Linear buckling analysis is the solution to an eigenvalue problem based on: $(K + \lambda_i K_g)\Psi_i = 0$, where K is the (linear elastic) stiffness matrix, K_g is the geometric stiffness matrix calculated for a reference load, λ_i is an eigenvalue (load factor), and Ψ_i is a corresponding eigenvector (buckling shape). The LBA assumes negligible deflections before the bifurcation of the load path (not the case for nonlinear consideration). The lowest eigenvalue is referred to as the critical load factor λ_{kr} [8]. For the load factor (negative because the maximum of the objective function is searched for), the following applies consequently: $\lambda_{kr} = -z_2$.

3.4 Optimization (Octopus)

The optimization problem is solved with the Octopus plug-in within GH. The plug-in is based on the evolutionary algorithm HypE. This algorithm contains a fitness assignment scheme based on the Lebesgue measure and can both compute this measure exactly or estimate it using Monte Carlo sampling. The estimation allows a compromise between fitness accuracy and computation time, which makes hypervolume-based search feasible even for problems with many objective functions [12].

4 Results

A total of 4400 parameter sets were calculated over 22 generations, with each generation consisting of 200 sets. Subsequently, various result representations are analyzed to draw general conclusions and identify phenomena.

4.1 Non-normalized result domains

The non-normalized result domains indicate the potential range of solutions for each objective function. They cannot be directly combined or weighted. Nevertheless, they provide insight into loadbearing capacity, serviceability, manufacturing effort, and manufacturing costs. Moreover, the results can be roughly weighted based on these estimates. The non-normalized result domains of the individual objective functions are as follows:

$$z_1 = \text{value deformation} = [0.01\ldots3.20]\,\text{cm}$$
$$z_2 = \text{value stability} = [0\ldots90]$$
$$z_3 = \text{value IBU pathlength} = [243\ldots498]\,\text{m}$$
$$z_4 = \text{value tool} + \text{material} = [0.11\ldots2.08]\,\text{m}$$

4.2 Pareto surface(s)

In the present study, there are four objective functions. To represent a Pareto surface in Cartesian space, the results of three of the objective functions can be plotted on the three axes (x, y, z). Figure 6 (left) shows the solutions depending on the objective functions, tool material, stability, and IBU from the first (red dots) to the last generation (white dots). The result ranges are normalized. The Pareto surface is also shown for the last generation. Based on the plot, the following can be determined:

- The higher the generation, the more two-dimensional the solution space becomes due to the IBU path ($z_3 \rightarrow 0$).
  $\Rightarrow$ Many Pareto-optimal solutions achievable with less IBU.
- Still, the most stable systems ($z_2 \rightarrow 0$) require the longest IBU paths ($z_3 = 1$).
  $\Rightarrow$ A high density of indentations leads to stable systems.
- The most stable systems ($z_2 \rightarrow 0$) can have low demand for tooling material ($z_4 = 0$).
  $\Rightarrow$ Simple shell geometries can be stable.
- In the Pareto-optimal solutions, there are many results with a low IBU path length ($z_3 \rightarrow 0$). Many of these solutions have a low tool material requirement ($z_4 \rightarrow 0$) too. These solutions are not very stable ($z_2 \rightarrow 1$).
  $\Rightarrow$ Simple shell geometries with low density of indentations lead to unstable systems.

One can identify the extent to which the objectives can be achieved and the strength of their compatibility. Additionally, it is possible to determine whether the optimization converges. Choosing the best candidate is proven to be difficult, and no conclusions can be drawn regarding the system's geometry dependencies.

4.3 Multi-axis representation

In this representation, we define each objective function one-dimensionally using its respective y-axis, onto which the solution results are plotted. The results of a given solution are then connected by a line, yielding a polyline for each solution (Figure 5). This creates a representation that is one-and-a-half-dimensional in nature, effectively summarizing the individual objective functions much like a stock market curve. It's worth noting that the optimal solution would be $f(x) = 0$.

A. This graph is equivalent to Fig. 4 but plotted in one-and-a-half-dimensional space. The scales have been normalized, with 0 representing optimal performance and 1 indicating suboptimal performance. Result variance is observed to decrease at higher generations, except for the objective stability (which remains unchanged). This trend is particularly pronounced in the case of Pareto-optimal solutions, which are dominated by minimal deflection and IBU values (with a few outliers for IBU). Additionally, the variance of used tool material is approximately half that of stability.
   $\Rightarrow$ IBU and deflection are dominant.

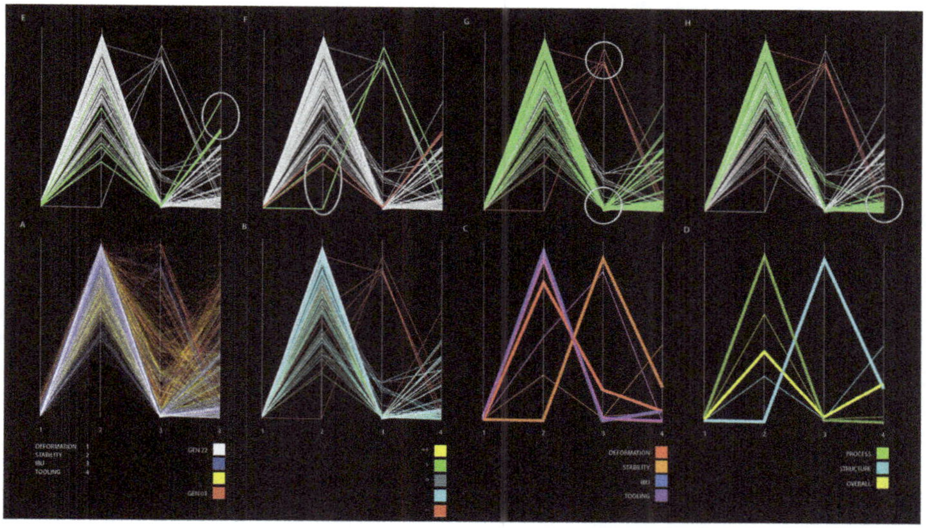

Fig. 5: Multi-Axis representation.

B. In this graph, the Pareto-optimal solutions of the last generations are plotted. However, an evaluation of the solutions is done here: All results $f(x_i) \in [0; 1]$ of a solution are summed up: $\sum_{i=1}^{4} f(x_i)$. The solution with the minimum sum value is the best (++, yellow), with the maximum sum value the worst (– –, red). The solutions with high IBU are the worst within the Pareto-optimal solutions. All the best results have minimum IBU.
   $\Rightarrow$ IBU is dominant.

C. The graph shows the first- and second-place optima of the last generation with the respective minimum results of the individual objectives. We observe that the Pareto-optimal solution with the best stability also possesses the poorest IBU and tool material values of all such solutions. Furthermore, the remaining Pareto-optimal solutions, each of which has a single minimum value, exhibit very poor stability values. Intriguingly, the solution characterized by maximum stability proves to be among the worst performers. However, the second most stable solution performs quite well across all metrics. We conclude that the best solutions are those that minimize deflection and IBU, exhibit good-to-moderate stability, and employ tooling material in a sparing-to-very-sparing fashion.
   $\Rightarrow$ Stability and IBU, as well as stability and tool material, behave contrary to each other. $\Rightarrow$ IBU and tool material behave equivalently.

D. This diagram combines IBU and material cost into a single "process" objective, while deflection and instability are grouped together as the "structure" function. A third curve provides a comparison to the overall view displayed in Graph B by showing the optimum solution.

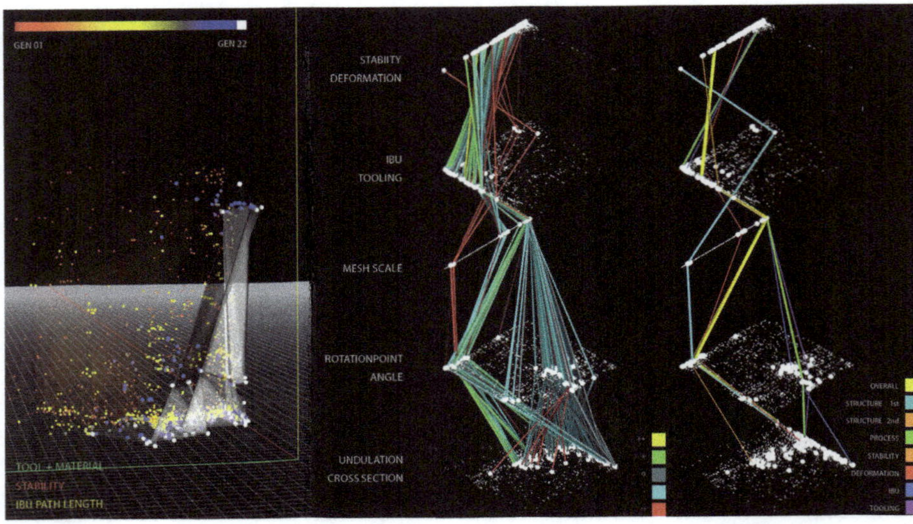

Fig. 6: 3D solution space pareto surfaces of all generations (left), Parameter-Referenced Multi-Level representation (middle and right).

- Structure solution corresponds to the best performing stability solution, which also exhibits very good deflection properties.
- Process function behaves according to the partial parameter curves.
- The generalized function achieves very good IBU and deflection properties, as well as average stability and tool characteristics.

Graphs E-H depict the curves from the most recent generation, with each graph highlighting a different subset of the Pareto set. These subsets were selected based on specific criteria, which are marked by circles and ovals.

E. Stability exhibits a wide spectrum; stability is achieved by much IBU (green) or high use of tooling material (red).
F. Much IBU is used when usage of tooling material is low (red), and the algorithm prefers less IBU, despite fluctuating results of the remaining objective functions (green).
G. High usage of tool material leads to stable systems, even with low IBU.
H. Low usage of tool material leads to low stability (green), except combined with much IBU (red).

4.4 Parameter-referenced multi-level representation

This form of representation aims to gain insights into the interdependencies of geometrical parameters and objectives by combining input parameters and target values (Fig. 6). It is divided into five diagram levels that summarize related values and display them as white dots. The domains of the representation levels are normalized over the results of the total solution space, and all values set and achieved in the optimization process are displayed as smaller dots. Each solution of the last optimization generation is represented as a polyline that links the corresponding solution and objectives. The lower three levels of the diagram depict the parameters for the shell geometry, pattern, and scaling and can be directly linked to the graphical representation in Fig. 3 and Fig. 4 while the upper two levels display the objectives. The upper diagram combines the structural objectives of deflection and stability, while the lower diagram shows the process objectives of tool material consumption and IBU process time. The zero value, or optimal solution, is located at the leftmost point of each diagram. Figure 6 displays the input and result values from Figure 5B, with the 131 solutions from the last optimization generation weighted by the sum of their target values and shown in color. The best solution is represented by yellow, while green represents good results, turquoise represents average results, and red represents poor results. This representation allows for conclusions to be drawn about the structural forms of the final solution space and their overall structural and process performance. The input parameters exhibit a strong containment pattern.

- Many shell shapes have an arbitrary cross-section and lack undulation. However, shells based on circular cross-sections with medium undulation perform better.
- Discontinuous triangular meshes are the predominant pattern and tend to score better. Strongly entangled three-rayed stars in various forms also appear frequently.
- Six-sided polyhedra play a less significant role.
- Pattern scaling shows a strong tendency towards low values (big stars), with only a few outlier solutions having high scaling (small stars).
- The result space for production targets is strongly limited, particularly with regards to the use of IBU, where most solutions use the minimum amount. However, some outliers rely heavily on IBU, including the best-performing solution structurally.
- In the solution space for structural parameters, the one-dimensional confinement is easily explained by the form-finding of shells. Through optimization, the total deflection was significantly reduced, but the spectrum of achieved stabilities is almost completely covered by the initial range of solutions.

In Fig. 6 (right), a subset of the variants that correspond to Figures 9C and 9D is displayed. The input parameters exhibit high variance in the partial selection, making it advisable to interpret the structural shapes depicted in Fig. 7, which shows the shell geometries bead pattern and buckling shapes.

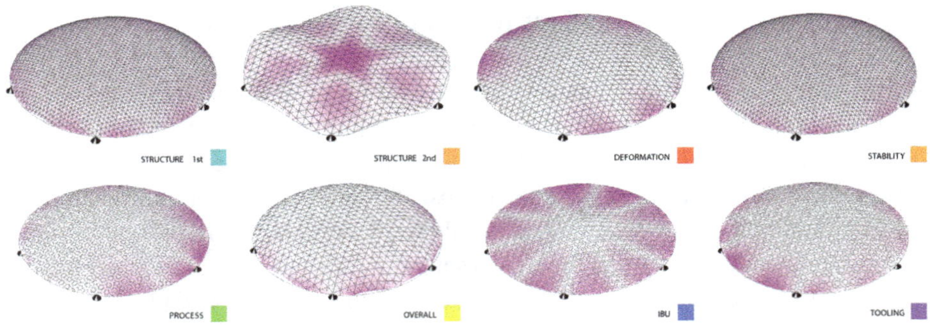

Fig. 7: Geometries of best solutions (buckling spapes).

- All variants except for a few outliers have a pattern of six-sided stars with non-intersecting rays aligned with the base network.
- The elliptical profile (shown in structure 1st in Fig. 7) provides a classic stiffening effect at the free edge, resulting in high stability. Changing from an elliptical profile to a linear one leads to a significant loss of stability but only a slight improvement in deflection performance.
- Dense patterns (shown in stability 1st in Fig. 7) contribute to stability.
- Shell geometries without undulation are dominant, except for the second-best result in static performance. The best solution for deflection has a parabolic to linear profile, which is equivalent to a suspension shape for point-supported shells with attachable bending stiffness. In this case, aclassic stiffening effect at the free edge is not necessary.
- Figure 7 indicates that the deflections of the best solutions are small relative to the total domain, suggesting that in the overall analysis, deflection is not the most crucial factor. This aligns with the assumption that for thin shells, stability issues related to static performance dominate.

5 Summary

The study presents a form-finding process based on the principles of structural morphology that optimizes the macro- and microstructure of the shell without hierarchy. Interpreting the results is fairly complex, due to the number of input parameters and objectives. Despite this, we observe clear tendencies in both shell shape and pattern generation, as well as deviating structural variants that highlighted trade-offs and reciprocal relationships: The solution space for this study includes both undulated and non-undulated options. Upon evaluation, many solutions with slight to medium undu-

Fig. 8: Assembled Demonstrator.

lation and a specific cross-section (between circular and parabolic) fall within the good range. Interestingly, the best solutions are not undulated and cover the entire range of cross-sections, except for the best solution for structural performance. Regarding patterns, clustering can be observed in the solution space. Discontinuous patterns that map the axes of the grid tend to perform better than geometrically interlacing ones, representing the best results. However, there are still many solutions with interlacing triangular stars and hexagonal patterns with closed contours that perform moderately well. Coarse scaling appears to produce clear improvements. Good solutions are achieved with low use of IBU. In terms of structural performance, all results exhibit low deflection. However, the range of solutions for stability could not be narrowed down. Interestingly, the best performers in process and structure tend to perform poorly in the other area, leaving the overall solution space quite large. Nonetheless, many variants perform well in both process and structure metrics, showing that solutions that meet both criteria can be found through trade-offs. Surprising results from this study include how little the solution spaces could be narrowed down. Additionally, entangled patterns do not necessarily lead to the best performers in stability. Fig. 8 shows the assembled demonstrator.

References

[1] Trautz, M., T. Pofahl, A. Seiter, G. Hirt, L.-M. Reitmaier, and D. Bailly. (2022). Karosseriebautechnik für die Architektur. Detail, H. 7./8. 2022, 22–24.

[2] Trautz, M., T. Pofahl, A. Seiter, G. Hirt, L.-M. Reitmaier, and D. Bailly. (2022). Leichtbaukonstruktionen aus Feinblech. Stahlbau 91, H. 6, 375–84.

[3] Motro, R. M. (2009). *Structural Morphology and Configuration Processing of Space Structures,* Multi-Science Publishing Co Ltd.

[4] Stach, E. (2010). Structural morphology and self-organization, Design and Nature, Pisa Italien, 29–40.

[5] Widmann, M. (1984). *Herstellung und Versteifungswirkung von geschlossenen Halbrundsicken,* Springer Berlin.

[6] Emmrich, D. (2005). Entwicklung einer FEM basierten Methode zur Gestaltung von Sicken für biegebeanspruchte Leitstützstrukturen im Konstruktionsprozess, F.-bericht Nr. 13 IPEK.

[7] Miettinen K. (1999). *Nonlinear Multiobjective Optimization.* Springer

[8] Aksel'rad, Ė. L. (1983). *Schalentheorie.* Stuttgart: Teubner.

[9] Kollár, L., and E. Dulácska. (1975). *Schalenbeulung.* Düsseldorf, Werner Verlag,

[10] Seiter, A., T. Pofahl, M. Trautz, L.-M. Reitmaier, D. Bailly, and G. Hirt. (2020). Metal Shells, 14. *Baustatik – Baupraxis.* 23.–24. März 2020, Universität Stuttgart, 439–46.

[11] Seiter, A., T. Pofahl, M. Trautz, L.-M. Reitmaier, D. Bailly, and G. Hirt. (2019). Design and Analysis of Freeform Shell Structures Composed of Doubly Curved Sheet Metal Panels. *Form and Force: IASS Symposium 2019,* 60–67.

[12] J. Bader and E. Zitzler. HypE: (2011) An Algorithm for Fast Hypervolume-Based Many-Objective Optimization. *Evolutionary Computation,* 19(1):45–76.

Jamie Queisser and Saqib Aziz

Searchfield: Navigating n-Dimensional Design Spaces

Abstract: Algorithmic design systems enable access to a variety of performance simulation tools, allowing to evaluate and explore design variations. Design tasks are often defined through multiple input parameters, thus spanning large high-dimensional solution spaces. The designers are faced with challenges when navigating and comprehending this solution space while striving to meet the requirements given by the design brief and aiming to achieve certification guidelines (BREEAM or DGNB). The described method projects the high-dimensional solution space to two dimensions while preserving similarities between individual solutions, making it intuitive for designers to navigate. The solution space is overlaid with performance estimates to assist the designer during the decision-making process and features methods for selecting and filtering according to design and performance criteria. A design project currently in development by the corresponding authoring firm, has been used to serve as a case study probing the design principles.

Keywords: design space exploration, multivariate visualization, dimensionality reduction, machine learning, sustainable design

1 Introduction

The AEC industry has experienced a rapid digital transformation, replacing outdated design and fabrication processes with advanced digital automation and collaborative platforms (Carpo 2017). Building Information Modelling (BIM) has successfully facilitated real-time information exchange among designers, engineers, and contractors within a unified design framework. Research suggests that decision-making in planning decreases exponentially over time towards later design development stages, emphasizing the importance of early-stage performative design optimization for improved economic and ecological performance of buildings (Kohler and Moffatt 2003). While BIM software has limitations in early design stages, design firms often rely on dynamic Computer-Aided-Design (CAD) modeling software like Rhinoceros3D and Grasshopper3D for form exploration and creation (McNeel 2010).

This research examines how technology can facilitate collaborative exploration of design spaces by designers and engineers, considering interdependencies among different domains. The study also investigates how to communicate decision-making and performative optimization processes in the early design stage in a way that architects can easily relate to. To address these issues, an experimental multi-domain and

https://doi.org/10.1515/9783111162683-029

collaborative design interface is introduced that allows users to collectively navigate through decision-making and solution spaces efficiently.

Unsupervised machine learning (ML) identifies patterns and structures in data without predefined labels or categories (Wang and Biljecki 2022). It differs from supervised learning, which relies on labeled examples for learning, and is used when there is no pre-established target or output variable to be predicted. This paper uses Principal Component Analysis (PCA, unsupervised ML) to project multi-dimensional vectors to comprehensible and user-friendly 2D interactive graphs and Linear Regression (supervised ML) for performance estimation, as described in Sec. 3.3.2 and 3.4.3.

2 Background

2.1 Design Process in Architectural Practice

Characterizing the diverse design processes in architectural practice is a challenging and generalized endeavor. Architectural design tasks are often considered wicked problems, as defined by Rittel and Webber (1973). Each design task, whether it involves the entire building or sub-tasks, presents numerous decisions and perspectives that reflect the unique design culture of an architectural office. The decision-making process is highly iterative, driven by changing criteria and requirements encompassing aesthetics, performance, technical considerations, regulations, and personal convictions. The pursuit of a perfect or final design outcome is subjective, with the focus being on achieving an optimal design instance within the given design requirements and project deadlines. In practice, design is typically divided into manageable tasks among team members. Solutions for these subdomains are iterated upon to generate design variants, resembling a decision tree where some branches remain unexplored while others are pursued or revisited.

2.2 Performance Evaluation of Architectural Design

The AEC industry continuously advances environmentally driven design criteria, including standards like LEED, BREEAM, and DGNB, which promote guidelines for natural resources such as natural daylight. The European EN 17037 guideline serves as a unified and updated European standard, replacing country-specific regulations, and provides standardized strategies for assessing natural daylight qualities in interior spaces. Adoption of this standard improves the evaluation of natural daylight quality by utilizing consistent Lux parameter-based calculations instead of less meaningful glass-to-floor area ratios. This promotes human health, well-being, and performance while also leading to energy savings. In this research, EN 17037 performance-based evaluations are integrated into the design workflow using custom programming and simulation software such as Honeybee & Ladybug (Ladybug 2022) for Grasshopper3D. The inclusion of validation tools in generative programming platforms has made performance-based evaluations more accessible to non-experts, although the presence of multiple and sometimes conflicting performance criteria can pose challenges in evaluations.

2.3 Current Research

The design process seeks to find the best trade-off among often conflicting requirements, whether based on client's brief or architect's convictions. The range of possible solutions is often referred to as "morphospace" (Raup 2006) or "phase space". In this paper, we use the term "solution space" to alignwith the concept of "parametric models" (Woodbury 2010) familiar to architects. Parametric models enable rapid generation of design variations, leading to large solution spaces. Efficient parsing of solution spaces for optimal design solutions is needed. This research presents a workflow combining existing methods to assist in complex design tasks in architectural practice.

Octopus, Opossum, and Wallacei are significant tools for interactive evolutionary computation in architecture. These frameworks, developed by Vierlinger (2013), Wortmann (2017), and Makki et al. (2022) respectively, enhance Grasshopper by offering accessible optimization parameters and interactive features. These user-friendly applications enable multi-objective searches and empower designers with greater control over the numeric optimization process.

Thomas Wortmann addresses the issue of visualizing high-dimensional solution spaces and performance value estimation of unexplored solutions. He surveys various methods of data-visualization and argues that the use of star-coordinates is well suited to display high-dimensional vectors in two dimensions and more intuitive than others, even though it has its own limitations. The two-dimensional representation is overlaid with the estimated performance values to create a performance map (Wortmann 2017).

In a previous work a novel method aimed at increasing this efficacy of solution space navigation through self-organizing maps was explored by Marschall, Aziz and Gengnagel (2016). The approach of self-organized fitness landscapes (SOFL) facilitates the convergence towards optimized design trade-offs between numeric and subjective criteria.

The workflow proposed in this paper builds on these concepts aiming to foster greater accessibility to a larger range of architects by offering an intuitive way of handling high-dimensional solution spaces and performance criteria. While black box optimization techniques are driven by maximizing numerical values and generating only a subsample of the solution space, they tend to make it difficult to integrate design intent and exploration.

3 Methodology

3.1 Introduction

The proposed methodology aims to represent entire solution spaces while incorporating performance estimates for each design option. This is achieved through the development of a Rhino Grasshopper suite called “SearchField”. This plug-in enables the generation of discrete solution spaces by systematically exploring all combinations and permutations of input parameters. It also provides dimensionality reduction using an adapted Principal Component Analysis (PCA) to visualize high-dimensional spaces in 2D. By subsampling the solution space, performance simulations can be conducted, and a Linear Regression model estimates performance values for all design options. The methodology includes features for selecting design solutions and filtering the solution space based on design and performance criteria.

“SearchField” is seamlessly integrated with native Grasshopper components and plugins, facilitating a streamlined design workflow that enables designers to generate and evaluate options quickly while considering key performance metrics (Fig. 1).

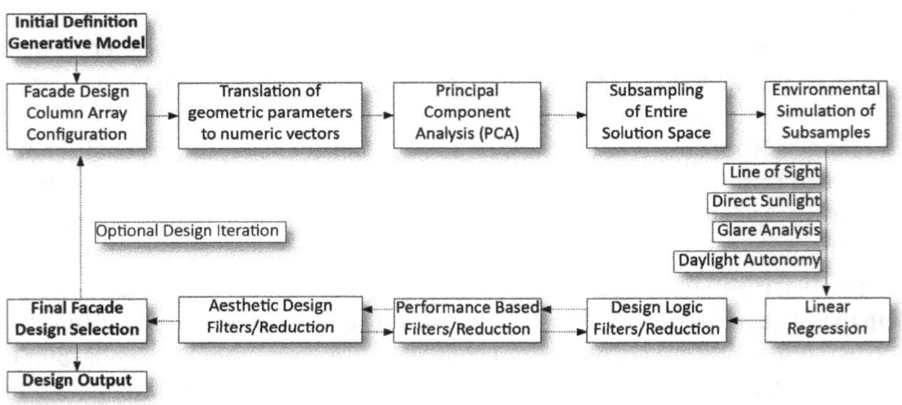

Fig. 1: Process map illustrating workflow using SearchField methodology.

3.2 Case Study

The methodology and custom Grasshopper components were tested on the facade design of an office building currently in planning at the partnering architectural office. The design team aimed to evaluate and refine the basic principle of a glazed facade with a second exoskeleton layer. Irregularly spaced columns supporting the facade would emphasize individual stories and incorporate horizontal overhangs. The investigation focused on the placement of columns for design intent and performance.

The facade consists of 8.1-meter-wide segments, spanning one story in height. Each segment has 16 evenly spaced grid points for potential column placement. To investigate column placement parametrically, this logic is translated into a solution vector. Each dimension or integer in the vector represents the number of grid spacings between columns (Fig. 2b). The vector dimensions range from 1 to 16, with the requirement that the L1 Norm sums to 16. Vectors with lower dimensions are padded with zeros to ensure a consistent dimension of 16 (Fig. 2a). The overall facade appearance is determined by the designer's selection and mapping of segments based on design and performance criteria (Fig. 2c).

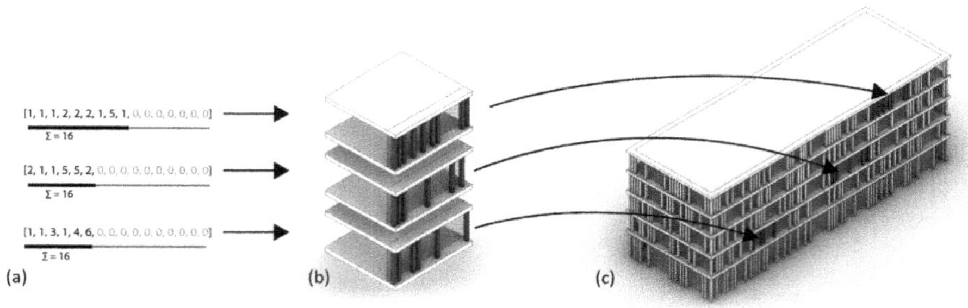

Fig. 2: (a) Solution vectors describing column placement. (b) Geometric interpretation of the vectors. (c) Population of the entire facade.

3.3 Solution space and Dimensionality Reduction

3.3.1 Solution Space

The SearchField components generated all solution vectors, encompassing the entire solution space of 32,768 solutions. While this approach has computational limitations due to combinatorial explosion, it is suitable for architectural design tasks within the capabilities of a standard desktop computer. The generation of the entire solution space took 50 ms, for larger solution spaces, such as 279,936, the time required increases to 500 ms.

This approach offers an advantage over optimization algorithms as it provides the designer with a comprehensive understanding of the solution space's structure. It allows for exploration of the entire range of possibilities without relying on algorithmic guidance.

3.3.2 Dimensionality Reduction

Visualizing high-dimensional data is challenging, but dimensionality reduction techniques, such as those discussed by John Harding (2016), can help by reducing it to two dimensions. Harding's approach, utilizing self-organizing maps, preserves similarity by ensuring that similar solution vectors in high-dimensional space remain close in a lower-dimensional embedding.

To efficiently project high-dimensional vectors to two dimensions, Randomized Principal Component Analysis (PCA) (Martinsson, 2016) using ML.Net was employed (Fig. 3a). Computation time for a dataset of 32,768 objects was 700 ms, and for a dataset of 279,936, it was 2.5 s. The resulting vectors were then transformed by rearranging them on orbits based on their high-dimensional L2 norm while preserving their radial position (Fig. 3b). Vectors with identical L2 norms represent the same input parameter set, albeit in different permutations, making the L2 norm a useful similarity metric in the design process.

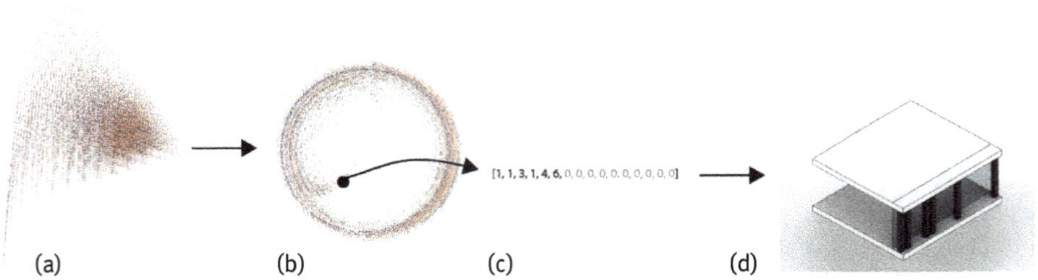

Fig. 3: (a) Projection of solution space to 2D using PCA. (b) Rearrangement of 2D points on orbits using the high-dimensional L2 norm without changing their radial position. (c) Selection of one solution vector. (d) Geometric interpretation.

In the context of star coordinates, Wortmann (2017) addresses the issue of overlapping solutions when vectors with the same L2 norm are projected to the same low-dimensional location. Resolving this, spreading out the two-dimensional vectors on the orbits ensures symmetry and reflects the weightings of the vectors, with larger values at the beginning or end.

Finally, Rhino and Volvox (Zwierzycki et al. 2016) are used to display the two-dimensional solution space as a point cloud, where each two-dimensional vector refers to its high-dimensional counterpart. This representation allows exploration of the high-dimensional space in two dimensions, providing insights into the original data's structure.

3.4 Subsampling, Performance Evaluation and Estimation

3.4.1 Subsampling

The solution space is subsampled to 328 samples, representing 1 % of the total space (Fig.4). Environmental performance evaluation of each sample follows criteria outlined in Sec. 3.4.2, according to DIN EN 17037.

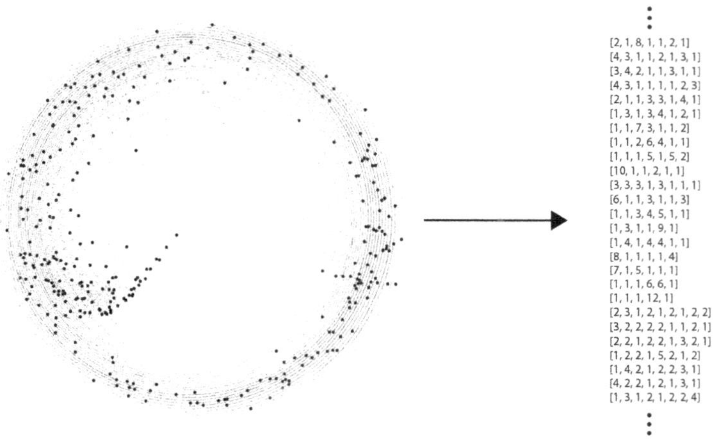

Fig. 4: Subsampled solution space with 328 samples and their vector representation.

To detach performance simulation from the design process, simulations are precomputed for each sample. This allows integration of various performance criteria and simulation tools, enabling any team member (internal or external) to run the simulation. Results are saved in a structured format and linked to the corresponding solution vector.

3.4.2 Performance Evaluation

Our study, based on DIN EN 17037 "Tageslicht in Gebäuden", aimed to conduct four evaluations to develop an optimum facade design selection strategy: spatial daylight autonomy (sDA), line of sight to the outside (LOS), direct sunlight (DS) on the facade, and glare analysis (GA). Ladybug and Honeybee plug-ins accessed through Grasshopper were used for sDA and glare analysis, while custom programming in C# within Grasshopper was developed for LOS and DS analysis.

It's worth noting that each analysis requires varying levels of detail (LoD), regarding the geometric input and environmental meta information (e. g., material attributes for the sDA analysis), and simulation runtimes, ranging from seconds to hours or days.

Each of the 328 subsampled solutions underwent performance evaluation using an $8.1 \times 6\text{m}^2$ shoebox geometry for all four cardinal directions.

Line of sight

In the initial design iteration, a subset of column configurations was assessed for visibility to the building's surroundings. This assessment considered three factors: horizontal viewing angle,distance to the outside perimeter, and visibility of at least 75 % of the room surface. A custom script was developed to generate a multidimensional score for each criterion, ranging from 3 (lowest score) to 9 (highest score), see Fig. 5. Using this script, all 328 design instances were quickly evaluated.

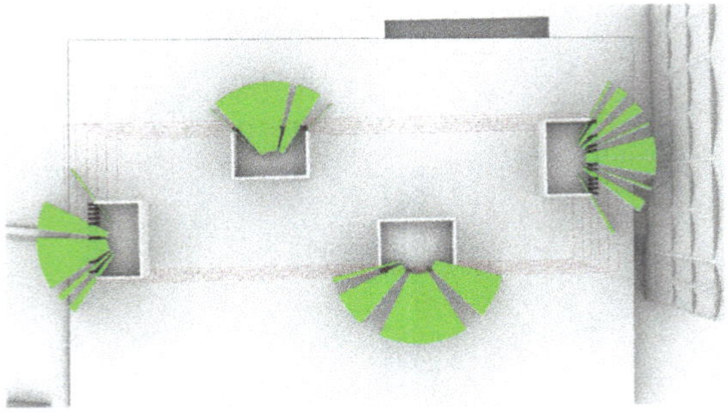

Fig. 5: LOS analysis, exemplary simulation sequence.

Direct Sunlight

Direct Sunlight (DS) analysis evaluates natural sunlight quality in buildings. It commonly uses the equinox date of March 21st as a reference. The duration of direct sunlight exposure is determined by the allowed sun elevation angle, e. g., 11° for Germany (Fig. 6). Boundary conditions such as climate zone, facade geometry, building construction, and obstructions affect sunlight accessibility. Results below or equal to 1.5 hours are considered poor, while 4 hours or more indicate good and well-lit spaces.

Glare Analysis

The glare analysis utilized the EnergyPlus Weather (EPW) format for Berlin and the Ladybug and Honeybee plug-in suite in Grasshopper. The Evalglare method calculated the Daylight Glare Probability (DGP) based on user assessments, categorizing discomfort levels. A high dynamic range imagery (HDRI) technique was employed using a 180° fish-eye camera lens to capture source-target images and generate color-coded False Color representations (Fig. 7). The analysis involved cycling through subsamples for each cardinal direction, exporting resulting images and DGP values.

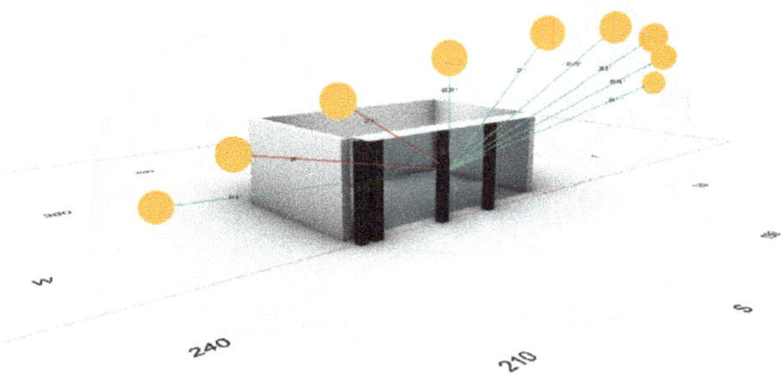

Fig. 6: DS Analysis, exemplary simulation sequence.

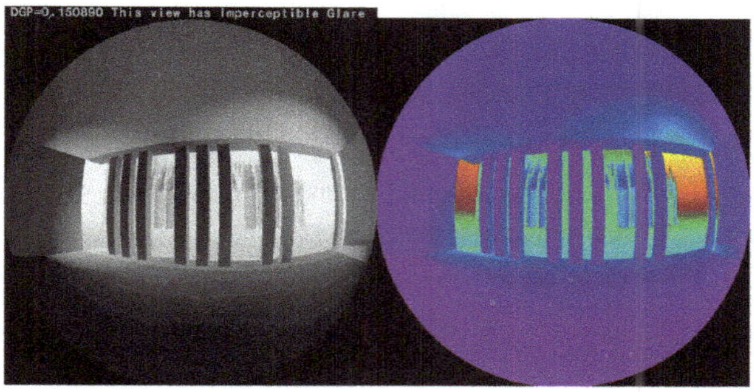

Fig. 7: Glare Analysis, exemplary simulation sequence.

Spatial Daylight Autonomy (sDA)

Spatial Daylight Autonomy (sDA) was used to assess daylight provision in this study, a widely used performance metric in building design and analysis (Nabil and Mardaljevic 2005). sDA quantifies the spatial and temporal distribution of daylight in interior spaces by measuring the percentage of occupied hours per year where a space exceeds a predefined target illuminance level. Factors such as window positioning, building orientation, shading devices, and surface reflectivity are considered in calculating the sDA value. Higher sDA values indicate better illumination throughout the year, improving occupant comfort, health, and productivity. The simulation process automated performance evaluation for each variant's cardinal directions, streamlining the analysis. Specific target values, such as a minimum of 300 lux for 50 % of daylight hours in 50 % of the area for a "poorly lit space", are used to determine adequate daylight levels. A high score is achieved when spaces achieve equal to or greater than 750 lux for 50 % of the area and daylight hours (Fig. 10).

Tab. 1: Recommendations for the supply of daylight through dayligth openings in vertical and inclined surfaces.

Level	E_T	$F_{plane,\%}$	min E_T	$F_{plane,\%}$	$F_{time,\%}$
low (score 1)	300 lx	50 %	100 lx	95 %	50 %
middle (score 2)	500 lx	50 %	300 lx	95 %	50 %
high (score 3)	750 lx	50 %	500 lx	95 %	50 %

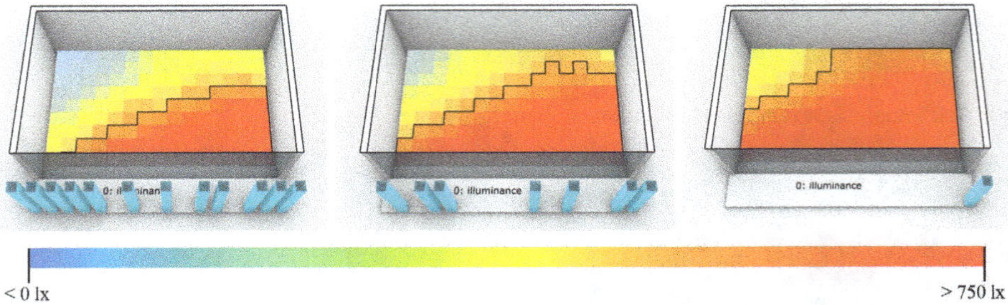

Fig. 8: Exemplary visualization of achievable scores applying DIN EN 17037 guidelines. Left: Sample 20 / West, Score 1; middle: Sample 200 / West, Score 2; right: Sample 332 / West, Score 3.

3.4.3 Performance Estimation

After performing the evaluation of the samples, Linear Regression Models were trained for each performance criteria to estimate the performance of the remaining 32,440 solutions. The chosen Machine Learning model was Linear Regression with a trainer based on the Stochastic Dual Coordinate Ascent (SDCA) method, known for optimizing convex objective functions (Yu et al. 2011). The trained Linear Regression Model was applied to the high-dimensional vectors, and Fig. 9 demonstrates its reliable performance prediction.

Using color-coding, a heat-map can generate a visual representation of performance values in the 2D solution space. This enables an intuitive identification of well-performing directions, aiding in the quick identification of the most favorable solutions.

Tab. 2: R-Squared (R2): Represents predictive power of model, the closer to 1.0 the better Mean absolute error (MAE): measures how close the predictions are to the actual outcomes, the closer to 0.0 the better Root Mean Squared Error (RMSE): measure of accuracy, the closer to 0.0 the better.

	LOS	DSs	DSw	DSn	DSe	GAs	GAw	GAn	GAe	sDAs	sDAw	sDAn	sDAe
R2	0.81	0.83	0.83	0.83	0.83	0.82	0.76	0.69	0.74	0.55	0.61	0.60	0.59
MAE	0.04	0.01	0.01	0.01	0.01	0.01	0.01	0.01	0.02	0.08	0.09	0.08	0.08
RMSE	0.06	0.02	0.02	0.02	0.02	0.01	0.01	0.02	0.02	0.11	0.12	0.11	0.11

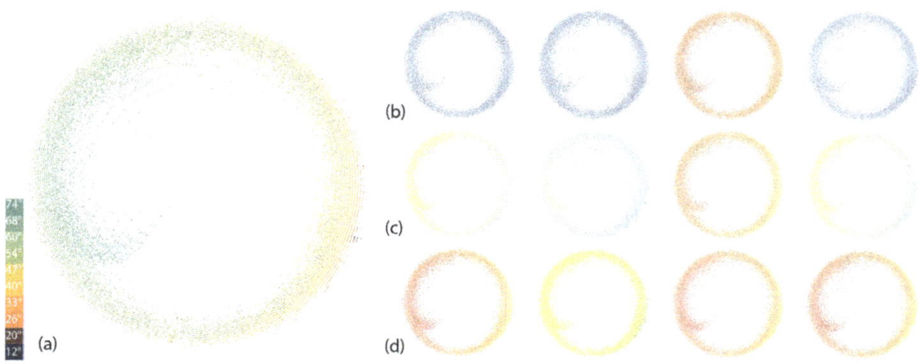

Fig. 9: Performance estimates: (a) Horizontal view angle; (b) sDA east, north, south, west; (c) Glare east, north, south, west; (d) DS east, north, south, west.

3.5 Design Process

3.5.1 Interaction, Selection and Filtering

To efficiently display the solution space and select options, two data structures are utilized: a Rhino point cloud and a KD-Tree (Panigrahy 2008). By specifying numerical ranges for performance criteria, the solution space can be filtered and reduced, allowing for a quick evaluation of design directions that meet performance requirements. Design filters, defined using regular expressions (.NET Regular Expressions 2022), are applied to the high-dimensional solution vectors, which describe design criteria through their properties. The actual design geometry is generated and displayed upon selecting solution vectors. The SearchField components currently offer two selection methods: around a point with a radius and a slice (Fig. 10).

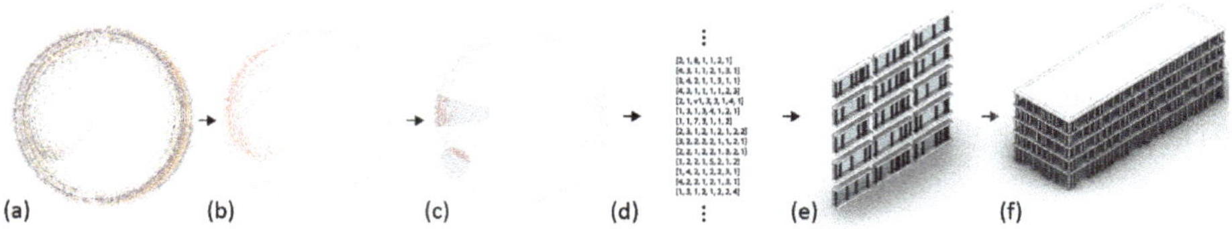

Fig. 10: (a) Projected solution space; (b) Filtered solution space; (c) Selection of design variants; (d) High-dimensional solution vectors of selection; (e) Design specific interpretation of vectors; (f) Population of facade using segments.

3.5.2 Iterative Design Process

The designer can explore and map selected solutions to the building (Fig. 10a), gaining insights into design variations and their performance properties. Filters can be defined to exclude solutions that don't meet specific criteria. For example, to filter out all facade segments that have two columns directly nextto each other, all vectors containing the integer 1 would be excluded, resulting in 610 possible design options for the facade segments. Alternatively, filters can be applied to include solutions within a desired performance range. The designer can freely browse options and gradually filter out undesired solutions, resulting in a refined and traceable design.

In the initial design process, criteria were defined to allow a maximum of 6 adjacent columns with a spacing of up to 10 grid units. Performance filters were then applied to include only options in the upper third of south-facing performance values, resulting in 1906 design options (Fig. 11). In the final step, stricter design criteria were implemented, limiting adjacent columns to a maximum of 3 and maintaining a maximum spacing of 10 units. Performance demands were adjusted accordingly, with different criteria for each cardinal direction. For the south-facing direction, the mid-third of performance values was used, resulting in 441 options. The east-facing and west-facing directions considered the upper third for Glare Analysis and the mid-third for Direct Sunlight, Line of Sight, and spatial Daylight Autonomy, yielding 474 options. The north-facing direction allowed for all upper thirds, resulting in 75 options. This final design outcome represents an optimal solution achieved by incorporating design and multiple performance criteria within these boundary conditions.

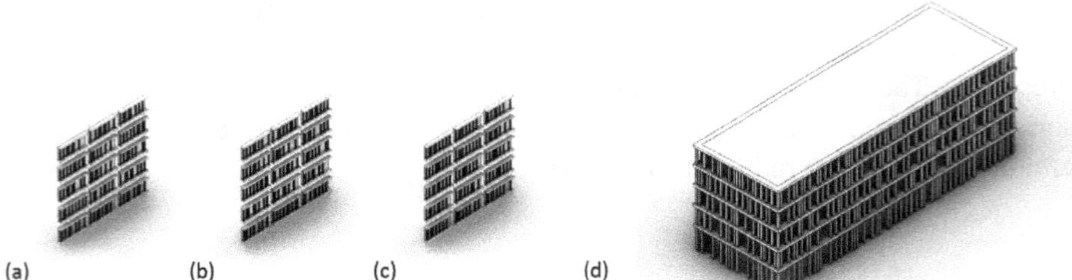

Fig. 11: (a) South-facing facade elements 15/441. (b) East- and West-facing facade elements 15/474. (c) North-facing facade elements 15/75. (d) Final design of column placement.

4 Conclusion and Future Work

Our approach enables intuitive navigation of high-dimensional solution spaces, integrating performance criteria and domain expertise without requiring deep tool understanding. It improves architectural design outcomes by incorporating performance evaluation in the early stages. Currently, we focus on spaces up to 5 million solutions, using dimensionality reduction and machine learning for fast computation. Future steps include expanding to larger spaces, employing advanced techniques like Manifold Learning for projection and neural networks for value estimation. We aim for a step-by-step design process that narrows the solution space and refines performance metrics.

Future research should focus on improving collaboration between architectural designers and domain experts, refining and exploring the validation process and analysis environment for performative evaluations. A standardized and robust digital evaluation environment is crucial to reduce human errors in geometry and numeric input declaration for simulations.

We developed an interactive interface for designers to navigate and preselect design variations based on different criteria. To understand the high-dimensional solution space, we implemented a visualizationassistance using arrayed concentric 2D point-clouds. Using 3D surrogate geometries, designers can allocate cluster characteristics on the 2D map. Further research and empirical testing are needed to create a user-friendly interface and compare it with other computational design strategies. Our goal is to demonstrate that SearchField enables informed decision-making and exploration of a significantly larger solution space.

References

Carpo, M. 2017. *The Second Digital Turn: Design Beyond Intelligence.* The MIT Press. DOI: https://doi.org/10.2307/j.ctt1w0db6f.

Harding, Jo. 2016. Dimensionality Reduction for Parametric Design Exploration. *Advances in Architectural Geometry.* Eds. M. Kohler et al. Vdf Hochschulverlag ETH Zurich.

Hamedani, A. and F. Huber. 2012. A comparative study of "DGNB", "LEED" and "BREEAM" certificate system in urban sustainability. DOI: 0.13140/2.1.3177.0568.

Kohler, N. and S. Moffatt. 2003. Life-cycle analysis of the built environment. *Industry and environment* 26.2. 17–21.

Makki, M., M. Showkatbakhsh, Y. Song. 2019. Wallacei Primer 2.0. Available online: https://www.wallacei.com/. Accessed on 12 December 2022.

Marschall, M., S. Aziz, and C. Gengnagel. (2016). Alternative means of navigating parameter spaces.

McNeel, R. 2010. Rhinoceros 3D, Version 6.0. Robert McNeel & Associates. Seattle. WA.

Martinsson, P.-G.. 2016. Randomized methods for matrix computations and analysis of high dimensional data. DOI: 10.48550/arXiv.1607.01649.

Nabil, A. and J. Mardaljevic. 2006. Useful daylight illuminances: A replacement for daylight factors. *Energy and Buildings* 38, 905–13. DOI: 10.1016/j.enbuild.2006.03.013.

.NET Regular Expressions. Microsoft Docs, 15 September 2021, https://docs.microsoft.com/en-us/dotnet/standard/base-types/regular-expressions. Accessed 5 August 2022

Panigrahy, R. 2008. An Improved Algorithm Finding Nearest Neighbor Using Kd-trees. In: Laber, E. S., Bornstein, C., Nogueira, L. T., Faria, L. (eds) LATIN 2008: Theoretical Informatics. LATIN 2008. Lecture Notes in Computer Science, vol 4957. Springer, Berlin, Heidelberg.

Raup, D. 2006. The Geometry of Evolution Adaptive Landscapes and Theoretical Morphospaces. pp. 57–70. Cambridge University Press. DOI: 10.1017/CBO9780511618369.005.

Rittel, H. W. J., and M. M. Webber. 1973. Dilemmas in a General Theory of Planning. *Policy Sciences*, 4(2), 155–69.

Software Addon Ladybug and Honeybee https://www.ladybug.tools/

Suk, J. Y. and M. Schiler. 2012. Investigation of Evalglare Software, Daylight Glare Probability and High Dynamic Range Imaging for Daylight Glare Analysis. *Lighting Research and Technology*. DOI: 10.1177/1477153512458671.

Sarah Mokhtar, Renaud Danhaive, Caitlin Mueller

Neural Implicit Fields for Performance-Informed Geometries in Building Design

Abstract: A high-performing built environment is essential for a livable future in cities that will face increased pressures from densification and climate change. Designing such buildings requires not only access to complex simulation workflows, but also advanced means of geometry representation, so that concepts can be easily synthesized, developed, and improved in early-stage design. Fundamental to this challenge is the ability to generate and semantically manipulate structured representations of buildings in iterative evaluation workflows through real-time performance feedback. The introduction of continuous implicit fields has emerged as an alternative approach to explicit 3D representations that is topology-, complexity- and resolution-agnostic, capable of representing high-level details in a compact, efficient, and structured manner as well as promotes implicit-explicit and 3D-2D cross-modality without compromising differentiability that is essential for design interactivity and optimization. The opportunities that such representations present for design and simulation in building applications are presented through a novel framework in this paper. The focus is on evaluating their usability for downstream design applications. The findings of this study highlight the potentials for universal representations that are interchangeable across simulation disciplines, spanning over a high-fidelity latent space that enables geometric interpolation and optimization across its performance landscape.

1 Introduction

1.1 Motivation

1.1.1 Open challenge for high-performance design in architecture: representing the built environment

A high-performing built environment is essential for a livable future in cities that will face increased pressures from densification and climate change. Designing such environments requires not only access to complex simulation workflows to capture their perceptual and performative qualities, but also advanced means of geometry representation, so that concepts can be easily synthesized, developed, and improved in early-stage design. The built environment is inherently complex as it is composed of a wide range of morphologically distinct structures, dominated by buildings that vary substantially in scale, detail, and complexity. One key challenge lies in the ability of a computational representation to capture high-fidelity features in an efficient and

https://doi.org/10.1515/9783111162683-030

structured manner for buildings of arbitrary complexity, while concurrently balancing between the diversity potential and realism of the learned representation space. To support performance-driven design interaction and targeted optimization, the representation should additionally enable generation and meaningful manipulation of building geometries in iterative evaluation workflows that are coupled with real-time performance feedback.

1.1.2 Learning 3D shapes for design and simulation

The integration of deep learning models for simulation workflows has attracted substantial attention in design and engineering fields due to their ability to learn highly dimensional, non-linear, and complex phenomena. These workflows, however, require effective methods of capturing the complexities and irregularities of 3D shape data. Explicit representations have been widely used in 3D learning including voxels and point clouds, which are restricted to low resolutions and lack explicit topology; octrees, which lack differentiability; meshes, which in learning approaches rely on template deformations restricting topology variations; and low-order representations, such as depth-maps and hand-crafted feature descriptors, which cannot accurately capture the 3D structure (Ahmed et al. 2018). The introduction of continuous implicit fields has emerged as an alternative approach to 3D representation that is topology-, complexity- and resolution-agnostic, capable of representing high-level details in a compact and efficient manner and promotes implicit-explicit and 3D-2D cross-modality without compromising differentiability that is essential for design iteration and optimization (Remelli et al. 2020). It additionally provides the means to work with data in the wild, and through its continuous latent space enables the expansion of geometrical design explorations beyond hand parametrizations.

1.1.3 Neural implicit fields for building design

The capacity to develop high-fidelity compact representations for the built environment, enabled by recent advances in the field, motivates further explorations into the opportunities that such representations present for the design and simulation of building geometries. In contrast to other design and engineering applications that typically deal with objects of pre-defined scale, architectural applications necessitate an approach that can scale up to entire buildings. The dearth of openly accessible structured building geometry-to-performance datasets and the computational expense of creating synthetic ones additionally motivates the need for universal building representations that can interchangeably be used across simulation disciplines and interpolate geometrically across performance landscapes.

1.2 Related Work

1.2.1 Learning continuous shape parametrizations using neural implicit fields

Neural implicit fields have been developed to learn continuous shape parametrizations that map 3D coordinates to a signed distance field or occupancy function. Global shape-conditioned formulations (Park et al. 2019), in addition to providing high-quality single-shape reconstructions, enable latent-space interpolation within entire classes of shapes. This is achieved by conditioning an auto-decoder fully-connected neural network through latent code concatenation. The network learns to predict signed distance values for associated input coordinates, through encoding individual shapes in the latent variables and the entire class of shapes in the network weights. The training is achieved through backpropagation of a loss function accounting for the deviation between predicted and actual signed distance values, a sparsity loss for latent codes and often associated with weight regularization terms. The implicit representation can be discretized to explicit representations at arbitrary resolutions (Xie et al. 2022). The potentials of neural implicit fields have been highlighted with a focus on reconstruction of objects across scales, but their suitability for and integration in building design applications has not yet been explored.

1.2.2 Exploring and optimizing latent spaces

The continuous latent space embedded in the learning of neural implicit fields expands their applicability beyond just efficient and compact encodings. Learning latent space representations have been employed more generally to compress higher-dimensional spaces for semantic synthesis and organization through capturing key features and structural similarities (Grossmann et al. 2022) as well as for generative modeling through sampling a learned underlying probability distribution or performing latent vector arithmetic operations (Asperti and Tonelli, 2022). Exploring learned latent spaces more broadly and systematically has attracted less attention, but a few approaches have been explored including random or guided latent space walks (Li et al. 2021), interpolation and extrapolation (Park et al. 2019), as well as sampling dimensionality-reduced latent spaces using linear and non-linear embeddings (Jahan et al. 2021). Shape performance optimization operating on object-level learned implicit parametrizations has also been proposed to promote topological diversity while maintaining semantically correct generation of representations (Remelli et al. 2020). The wide array of latent space exploration approaches, particularly for neural implicit fields, offers a novel, yet largely unexplored, territory for smoother performance-informed design integration.

1.2.3 Design space exploration and latent space learning in architecture

Performance-driven design has been commonly approached in the architectural practice through coupling parametric modeling with performance simulation as a strategy to balance between quantitative and qualitative design goals. This process is associated with the challenge of generating a design space that is morphologically diverse and semantically meaningful. Fit-for-purpose handcrafted parametric models, however, are often not extendable, not scalable, biased, and constrained by the mode and resolution they were originally defined in.

Recent experimental work challenged some of these limitations through latent space learning including floor plan generation and semantic characterization (Chaillou 2021), performance-constrained generative modeling for structural design (Danhaive and Mueller 2021), structural morphology ideation using sketching (Ong et al. 2021), among others. While these have shown substantial promise, they are still constrained by their pre-defined dataset generation modes of production. Neural implicit fields learning expands such latent approaches by enabling text-image-3D cross-modality (Poole et al. 2022; Sanghi et al. 2022) through differentiable rendering and explicit-to-implicit pipelines (Remelli et al. 2020; Guillard et al. 2021), enabling the learning to act directly on expandable datasets of geometries rather than handcrafted parametrizations, while providing the flexibility to sample geometries at varying resolutions that match application-specific or simulation tool constraints.

1.3 Contribution

The opportunities that neural implicit field representations present for design and simulation in building applications are presented here through a novel design parametrization-to-exploration end-to-end framework. This is achieved through: (1) demonstrating the diversity and morphological qualities of a building latent space directly conditioned on geometry with no explicit parametrization, (2) developing a structured approach to latent space exploration for design that concurrently captures semantic characteristics of the design and performative landscape and provides control over shape generation, and finally (3) integrating performance-driven feedback through local gradient descent of differentiable analytical equations for building shape refinement and optimization.

2 Methodology

2.1 Design framework

The developed design parametrization-to-exploration framework consists of four main stages, shown in Fig. 1: (1) acquisition and sampling of a large and morphologically diverse dataset of building geometries, (2) training of continuous shape parametrizations, (3) definition of differentiable performance evaluation functions, and (4) latent space exploration. The framework provides the flexibility of continuously augmenting the trained neural implicit model with more building geometry data as it becomes available, and which can be sourced from datasets that are wild (e. g., from 3D scans or individually modeled building datasets), synthetic (e. g., sampled from hand-coded parametric models), or a combination of both. Defining differentiable objective functions for common building performance goals is challenging and, in many cases, not possible due to the black-box nature of most simulation tools. These can, however, be replaced with proxy models through direct abstractions or deep learning pipelines, which are becoming increasingly available.

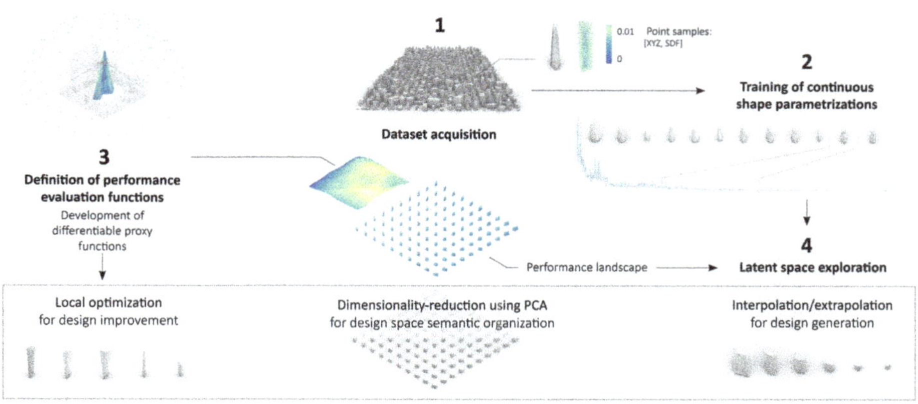

Fig. 1: Design framework for performance-driven design with learned implicit representations. A differentiable performance metric is evaluated over a human-legible latent space to support comparing, exploring, and generating highly diverse geometries for buildings massings without the need for hand-coded parametrizations.

2.2 Synthetic dataset generation

Large open-source morphologically diverse structured datasets of building geometries are scarce. While neural implicit fields learning does not rely on explicit parametrization, in the absence of suitable and accessible geometric datasets for training, a synthetic dataset was developed for this paper. The latter can interchangeably be replaced by and/or augmented with wild data. Thousands of individual building shapes were generated synthetically through sampling a parametric model with variables including size, height, shape, corner type, massing carving, lift-up conditions as well as vertical and horizontal variations. Figure 2 shows the 2,000 building geometries used for training and validation and highlights a sub-selection for detailed inspection of some of the generated forms. Considerations were made to ensure a balance between diversity and representation potentials of the dataset with respect to real urban contexts. Building geometries were centered in 400×400 m bounding boxes and the smallest detail across all ranged from 1 to 4 m. For each building mesh, point samples were selected based on a sub-set rejection sampling approach defined as a function of the distance d to the mesh (Davies et al. 2020): where the probabilistic acceptance criterion is defined as $e^{-\beta|d|}$, where β is 80. Starting with 100M Latin Hypercube samples with the bounding box, a smaller 1M points were re-sampled and augmented with 500k near-surface and 500k random additional samples for which corresponding signed distance values were computed.

Fig. 2: Synthetic building dataset.

2.3 Training the continuous signed distance field

Learning shape-conditioned representations was achieved through training an auto-decoder network comprising of 5 fully-connected layers of 512 hidden neurons and ReLU activation functions. A latent code length of 256, initialized based on a normal distribution with standard deviation of 0.01, is used to condition the network. The loss was defined in two terms: a L1 loss accounting for the sum of absolute deviation between predicted and actual signed distance values, and a latent sparsity term defined as the mean-square of latent code values. The parameters selected for training were selected based on a trade-off between model performance as identified by informal experimentations and computational feasibility. To assess the learned latent space's ability to represent out-of-training samples, for each geometry in the validation set, the network weights are fixed, and their latent codes are optimized. Meshes are reconstructed using Marching Cubes (MC), and the reconstruction quality is primarily evaluated through visual inspection.

2.4 Differentiable performance evaluation – view factor

Physics-based simulations are essential to performance-informed building design, and while they often rely on tools and methods for which differentiability is not straightforward, the development of differentiable proxy models offer an alternative. A differentiable view factor approximation function was developed based on a naïve implementation of Moller Trumbore ray-triangle intersection algorithm coupled with differentiable relaxation of non-differentiable operations. This function is adapted to capture three distinct proxy performance evaluations: (1) sky view factor, a metric that describes the geometrical relationship between surfaces and the sky, and typically provides insights into exposure to daylight and urban microclimate performance, (2) cumulative incident solar radiation, and (3) access to views. The first calculates intersections between rays cast from 145 points sampled on the sky hemisphere based on the Tregenza subdivision scheme to a uniformly sampled offset of the building mesh surface. The second, additionally, associates the 145 points with weighing factors that correspond to cumulative radiation values extracted for the summer season in Boston using Radiance. The third replaces the sky points with points sampled uniformly on a horizontal surface representing a park view.

Figure 3 shows the performance distribution of all geometries in the dataset for the three evaluation metrics and identifies best, worst and average performing shapes for each. A faster and non-differentiable version of the performance objective function was used for evaluating building shapes outside of optimization loops for computational efficiency.

2.5 Latent space exploration

Latent space explorations were structured into three main strategies: interpolation for design generation, sampling dimensionality reduced embeddings using principal component analysis (PCA) for design space semantic organization and local optimization for design improvements. Pairwise interpolations between building shapes in the training sets are achieved through linear interpolations across latent vectors. To explore the latent space in its PCA dimensions, uniform samples are defined across

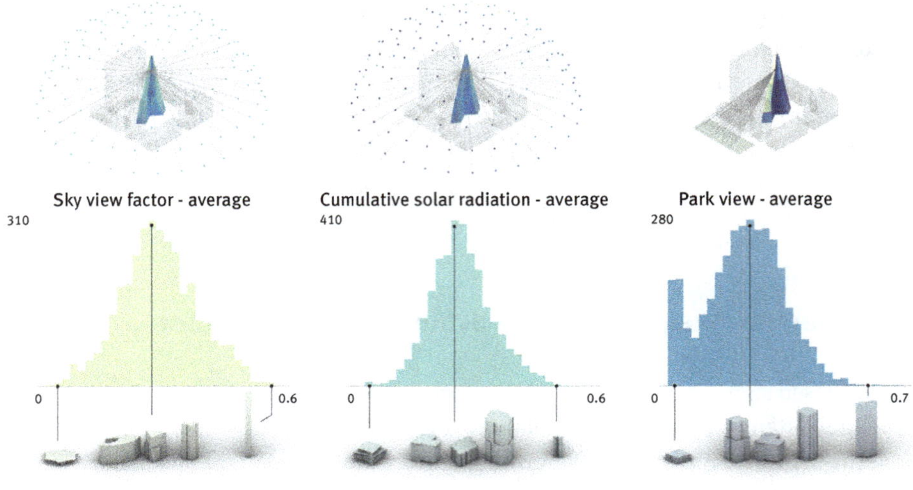

Fig. 3: Performance distribution of dataset across evaluation metrics.

two select principal components (PC) while keeping the mean value for others, followed by inverse mapping to recover their corresponding latent parameters. For local optimization, a starting latent code corresponding to a building of choice is initialized and for each gradient descent step: the mesh is reconstructed using MC through sampling the decoder, surface sample points are extracted and evaluated against the differentiable view factor evaluation, loss terms are computed, and the derivative of the explicit-implicit representation is derived based on the formulation by (Remelli et al. 2020) in terms of surface normals. Two soft constraints are integrated to promote feasible designs: a volume constraint term that accounts for the deviation between the volume of the original and optimized design, and a conservative regularization term that encourages designs that are close to the training dataset. The latter is defined as a l_2 norm of the difference between the latent vector and its closest k latent vectors from the training set. The combined loss term is defined by the following expression:

$$L_{\text{comb}} = \frac{\sum_{s=1}^{n} VF(s)}{n} + \alpha \cdot \sum_{z^i \in Z_k} \frac{\|z - z^i\|_2^2}{|Z_k|} + \beta \cdot |V_{\text{orig}} - V_{\text{cur}}|$$

It combines three terms: a performance term that captures the mean performance loss across n point samples, a deviation from k latent vectors z trained latent space term and a volume constraint term, for which their contribution to the total loss is tuned using α and β.

3 Results

3.1 Model training and reconstruction performance

The neural implicit field was trained on 1,565 geometries from the synthetic dataset based on a 90/10 split, and for each batch, 16,384 points are randomly selected out of the originally sampled 2M. The progression of learning is shown in Fig. 4, highlighting the decrease in loss as the model learns and the reconstruction quality improvement through a building sample. The ability of the network to reconstruct building forms from their associated latent codes is shown in Fig. 5 which shows pairs of building ground truth meshes from the training set and validation sets with their corresponding reconstructed shapes at MC resolution of 128.

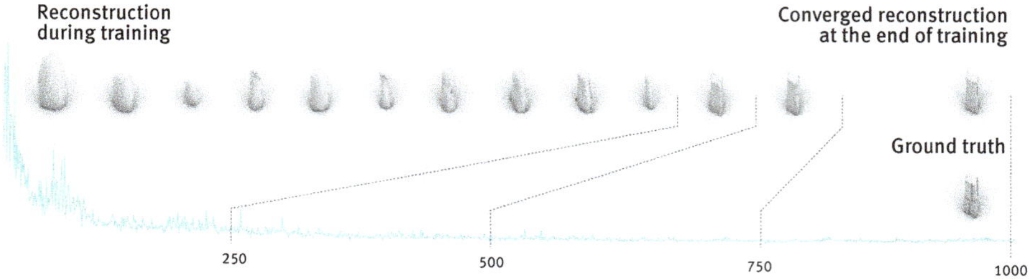

Fig. 4: Loss progression during training and corresponding reconstruction.

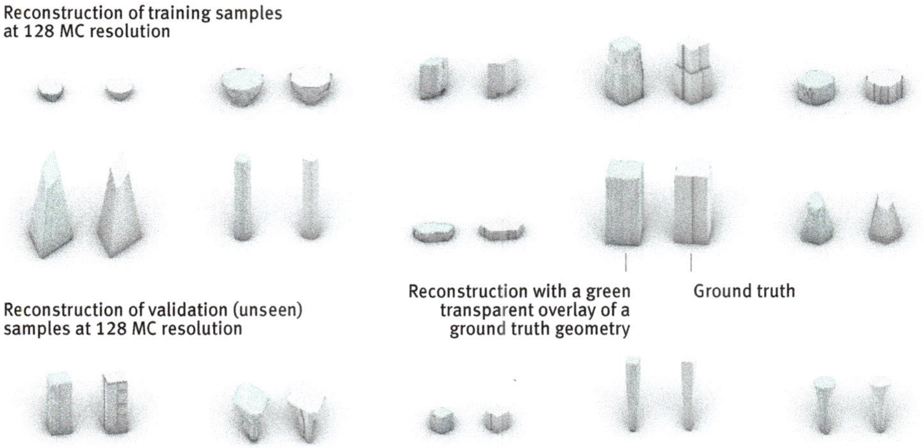

Fig. 5: Reconstruction quality of training and validation samples.

To highlight the deviation between pairs, a green transparent overlay of the ground truth geometry is added on top of reconstruction for clarity. By visual inspection of reconstructions, boundaries of building shapes are consistently identified, with erroneous artifacts associated with surface smoothness and smaller details, particularly boundary shape protrusion variations, that are challenging for the model to learn. These observations apply to both training and out-of-training samples, which is an indication that the learnt latent space is continuous and has a representation ability that extends beyond the data it was trained on. While reconstruction quality is an initial indicator of the model's learning success, for performance-informed design applications, the smoothness and representation of the learnt latent space is the primary goal.

3.2 Exploring and evaluating the latent space

3.2.1 Interpolation for design generation

Pairwise latent vector interpolation between diverse building shapes, shown in Fig. 6, was performed to inspect the representation fidelity and space continuity of the learned latent space. For each pair of buildings from the training set, shown at the two ends of A-R sub-figures, four intermediate buildings are generated from the model. Smooth interpolations are achieved across diverse building features that vary in scale, complexity, and detail. These include global features such as shape, vertical variation, but also smaller-scale and more challenging features such as the lift-up conditions in E, K and L as well as massing carvings in A and O. In addition to demonstrating the quality of the latent space, interpolation can be used as a design technique in itself; for example, a design team could input two candidate design geometries and use latent space interpolation to generate intermediate designs that capture features of both.

3.2.2 PCA for design space semantic organization

To capture the key semantic building features learned by the model, uniform sampling across 18 PC axes (dimensionally reduced from a latent space size of 256) was performed at a resolution of 10×10 and visualized in Fig. 7. Zoom-ins identifying the corner building shape conditions at the range extremities help identify the semantic meaning of each set of PCs.

It is unsurprising that the first set of components represent height, size, proportion, and rotation, which are key global building design features that vary considerably across the built environment. The shape and vertical variation are captured in subsequent components, and fine and more local details such as massing carvings, lift-up conditions and horizontal variations only identified in later components. The performance landscape can be captured by evaluating these samples with respect to the

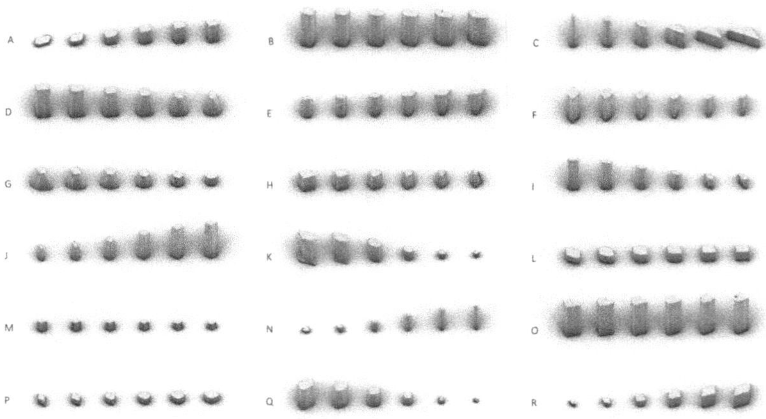

Fig. 6: Design generation using latent vector interpolations.

three metrics, and as shown in Fig. 8. It provides a smooth representation of performance variation across key semantic parameters of building design as identified by the trained model. This shows a compelling and direct application of neural implicit fields for performance-informed design, in which meaningful interpretation of design parameter-to-performance associations can help make informed decisions without any explicit design space geometrical parametrization.

3.2.3 Local optimization for design improvements

By integrating differentiability to the performance evaluation function, the continuous latent space and associated shape embeddings offer the ability to improve the shapes of buildings by operating on object-level low-dimensional and semantically meaningful implicit parametrizations. These learned parametrizations are essential for exploring large continuous shape possibility spaces to ensure feasible solutions to an optimization process coupled with task-specific volumetric and/or spatial constraints. Figure 9 shows the shape evolution of a select building at 50 iteration intervals during an optimization for two metrics and three soft constraints conditions. As the gradient flow is largely dependent on the performance evaluation of mesh reconstructions using MC at every iteration, any reconstruction failure due to the incomplete definition of surface boundary or volume collapse leads to interruptions in the optimization process. In the absence of volumetric constraints, while this leads to higher performing geometries at the onset, this shape collapse is observed early in the process. The two penalty terms: volume penalty and distance from training set were tested to overcome this. More consistent building shape propositions are observed, but with the now tougher challenge of concurrently improving performance values. The sensitivity of the shape

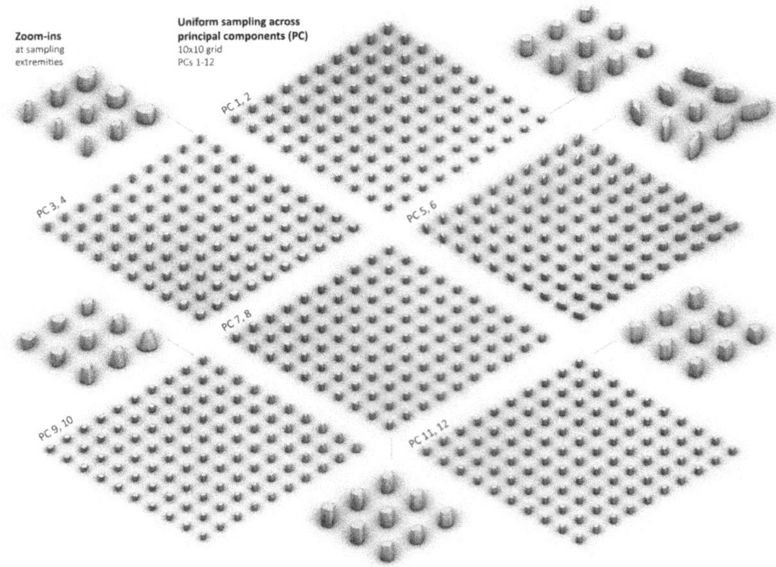

Fig. 7: Uniform sampling across PCA.

optimization to the performance evaluation function is shown in the formal variations observed for the park view and sky view factor objective functions. While the first promotes taller slender buildings oriented to maximize unobstructed park views, the second maximizes exposure to the sky by orienting the vertical slope of building façade up and rotating the shape to minimize surfaces obstructed by the high-rise building in the surrounding context.

4 Conclusion

Motivated by the representation challenges associated with performance integration in architectural design, this paper presented a novel design parametrization-to-exploration end-to-end framework using neural implicit fields. The findings indicate that morphologically diverse design spaces and semantic associations between parameters and performance metrics can be made without explicit parametrizations. Results show that smooth and meaningful latent design spaces are possible but that a balance between reconstruction quality, computational efficiency and latent space quality is challenging. For design and simulation applications, research into neural implicit fields should expand the focus from just reconstruction quality and scalability to latent space optimization and performance integration. This work expands the possibilities for performance-informed building design through: (1) offering a continuous design

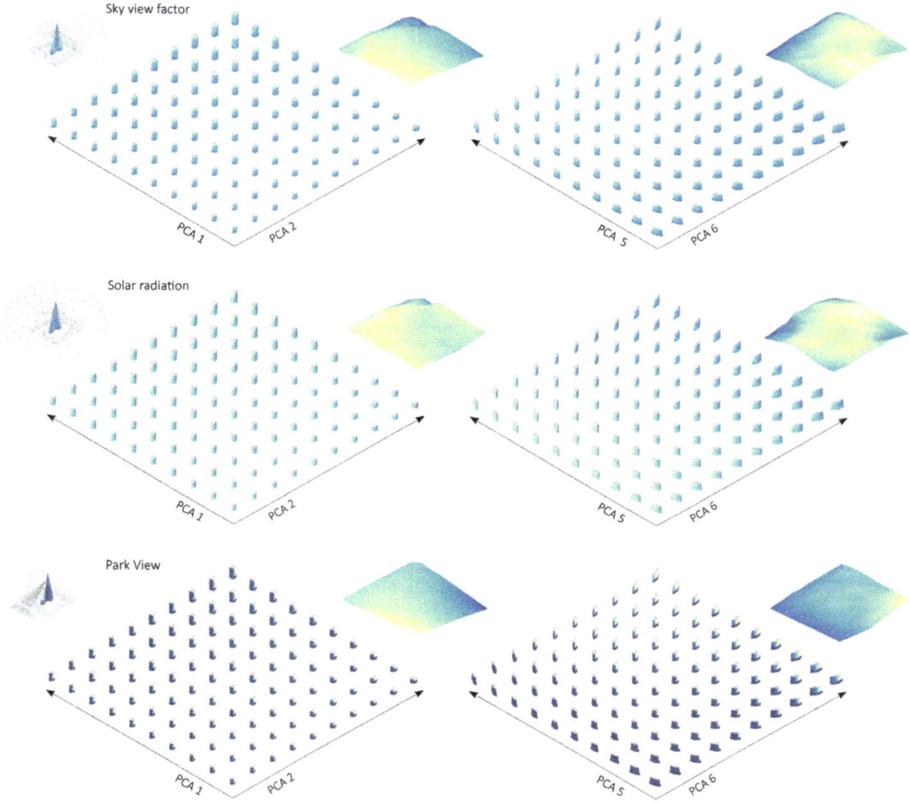

Fig. 8: Performance landscape across PCA 1, 2, 5 and 6 for three metrics.

space parametrization of building shapes without the need for handcrafting parameters making them expandable to wild and synthetic datasets alike, (2) expanding the integration of optimization workflows to complex geometries through coupling efficient fixed-length and semantically-meaningful representations with differentiable performance evaluation, and (3) enabling representation learning of buildings shapes across scales and resolutions, across performance assessment frameworks and potentially across modes (sketches, 3D geometries, etc.) through differentiable rendering. The development of universal topology-, complexity- and resolution-agnostic building representations that can interchangeably be used across creative design and simulation disciplines have the potential to expand design space explorations to the global performance landscape of design possibilities.

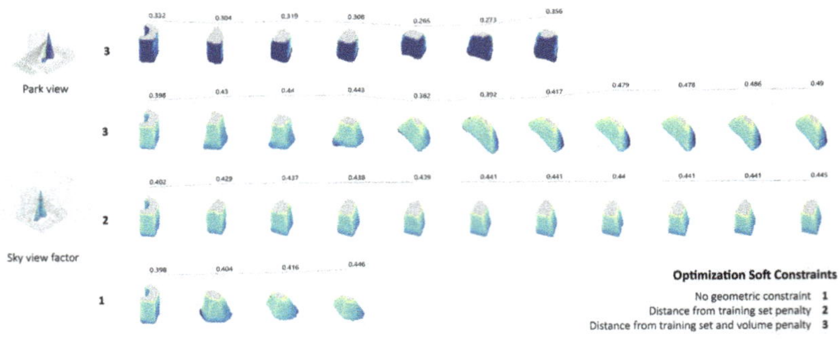

Fig. 9: Shape optimization progression. Reported values give the performance component of the loss terms.

References

Ahmed, E. et al. (2018). A survey on deep learning advances on different 3D data representations, https://arxiv.org/abs/1808.01462 [Preprint].

Asperti, A. and Tonelli, V. (2022). Comparing the latent space of generative models, *Neural Computing and Applications,* pp. 1–18.

Chaillou, S. (2021). AI and architecture: An experimental perspective, In *The Routledge Companion to Artificial Intelligence in Architecture,* pp. 420–41.

Danhaive, R. and Mueller, C. (2021). Design subspace learning: Structural design space exploration using performance-conditioned generative modeling, *Automation in Construction,* 127(103664).

Davies, T., Nowrouzezahrai, D. and Jacobson, A. (2020). On the effectiveness of weight-encoded neural implicit 3d shapes, https://arxiv.org/abs/2009.09808 [Preprint].

Grossmann, N., Gröller, E. and Waldner, M. (2022). Concept splatters: Exploration of latent spaces based on human interpretable concepts, *Computer & Graphics,* 105, p. 7384.

Guillard, B. et al. (2021). Sketch2Mesh: Reconstructing and Editing 3D Shapes from Sketches, In *Proceedings of the IEEE/CVF International Conference on Computer Vision,* pp. 13023–32.

Jahan, T., Guan, Y. and Van Kaick, O. (2021). Semantics-Guided Latent Space Exploration for Shape Generation, *Computer Graphics Forum,* 40(2), pp. 115–26.

Li, G. et al. (2021). Discovering density-preserving latent space walks in GANs for semantic image transformations, In *Proceedings of the 29th ACM International conference on multimedia,* pp. 1562–70.

Ong, B. W. X., Danhaive, R. and Mueller, C. (2021). Machine learning for human design: Sketch interface for structural morphology ideation using neural networks, In *Proceedings of the 7th International Conference on Spatial Structures,* pp. 1–12.

Park, J. J. et al. (2019). Deepsdf: Learning continuous signed distance functions for shape representation, In *Proceedings of the IEEE/CVF conference on computer vision and pattern recognition,* pp. 165–74.

Poole, B. et al. (2022). *Dreamfusion: Text-to-3d using 2d diffusion,* https://arxiv.org/abs/2209.14988

Remelli, E. et al. (2020). Meshsdf: Differentiable iso-surface extraction, *Advances in Neural Information Processing Systems,* 33, pp. 22468–78.

Sanghi, A. et al. (2022). Clip-forge: Towards zero-shot text-to-shape generation, In *Proceedings of the IEEE/CVF Conference on Computer Vision and Pattern Recognition,* pp. 18603–13.

Xie, Y. et al. (2022). Neural fields in visual computing and beyond, *Computer Graphics Forum,* 41(2), pp. 641–76.

Demi Fang, Sophia V. Kuhn, Walter Kaufmann, Michael A. Kraus, Caitlin Mueller

Quantifying the Influence of Continuous and Discrete Design Decisions Using Sensitivities

Abstract: Early-stage design decisions significantly impact the performance of architectural geometries and structures, especially in the context of high-performance buildings. However, most design problems involve a combination of continuous and discrete variables as decisions, making it challenging to understand their individual contributions to performance. For instance, tradeoffs between geometry (continuous) and material type (discrete) often lead to non-obvious performance behaviors, complicating the inclination to establish "rules of thumb" for low-carbon structural design. In response, we propose a new approach for assessing the influence of mixed variables on performance based on variable sensitivities, with a focus on architectural applications. A conditional variational autoencoder is trained to learn a continuous representation of variables and their relationship to performance, and gradients of the learned model are used to compute the relative influences of variables on performance. Multiple scales of insights are attainable: 1) locally at a design instance of interest, and 2) globally across the whole design space. The method is demonstrated on a case study of designing a structurally feasible gridshell for low embodied carbon.

1 Introduction

In the face of the climate crisis, designers are more motivated to produce high-performing, low-carbon designs. Computational approaches offer a promising way to do so, combining quantitative performance evaluation with generative techniques. Parametric modeling and optimization techniques currently available for performance-driven design are valuable, but two ongoing challenges persist: 1) the technical challenge of exploring or optimizing mixed-variable (i. e. both discrete and continuous design variables) design spaces, which are common in architectural design but understudied in computational methods; and 2) the tendency for computational tools to prioritize providing designers with high-performing designs, rather than on revealing generalizable knowledge and enriching human intuition about the design problem.

Recently emerging computational methods can be leveraged to overcome these challenges and support new paradigms of performance-driven design. This paper proposes and demonstrates one such method: conditional variational autoencoders (cVAE) are used to build a learned representation of a mixed variable design space and performance as a continuous knowledge landscape. The resulting sensitivities can be computed automatically, and subsequently leveraged to provide both local and

https://doi.org/10.1515/9783111162683-031

global insights that inform design decision-making. This approach addresses both aforementioned challenges: 1) it offers a smooth way to handle discrete variables in combination with continuous ones, and 2) it offers a technically rigorous, data-driven way to produce insights on the design space. The method is demonstrated on a gridshell design case study.

2 Background

2.1 Navigating mixed variable spaces in design

Discrete variables may pose a technical challenge to design space exploration. Existing methods for design space exploration of mixed variable design spaces, in order from least to most rigorous, include:

1. Relying on "rules of thumb" (e. g. "always use timber where possible"), which lack specificity, carry cognitive bias, and are often too general.
2. Comparing just a few solutions across discrete choices, which risks omitting high-performing regions in the design space.
3. Evolutionary methods, which iterate through the design space but lack guarantees of convergence and a comprehensive understanding of influential decisions in addition to a significant computational effort.
4. Mixed integer programming (MIP), a well-established optimization method that provides a single optimal solution but lacks decision-making insights and typically requires commercial tools for implementation.

The proposed cVAE approach combines the benefits and overcomes the limitations of these existing strategies, providing a learned, continuous representation of the design space to reveal cognitive insights about design decision-making in the design problem.

2.2 Extracting design insights from sensitivity analysis

Various methods are available to quantify the influence of design decisions on performance. Classically, designers might attempt to understand design variable importance through Exploratory Data Analysis (EDA), for example by identifying steep slopes in objective-variable plots. This becomes impractical for high-dimensional design spaces.

Instead, we examine the opportunities offered by computing sensitivities, which can be offered locally or globally. Local sensitivity analysis examines the impact of input variable changes on the model output at a specific point and is widely used in various fields such as ecology and environmental modeling (Cho and Jung 2003). In contrast, global sensitivity analysis assesses these impacts (of inputs on outputs) across the entire design space and has been used in many areas, including engineering, physics, and economics (Saltelli et al. 2008).

Sensitivities are seldom used to provide insight in design decision-making, likely because they are challenging or expensive to compute directly for parametric design problems in which performance is computed via black box simulations or via hard-to-differentiate operations such as matrix inversions (e. g. FEM). Existing methods for approximating sensitivities include:

1. Finite differencing (e. g. Brown and Mueller 2018) which is limited to local recommendations (as opposed to global rules of thumb) in the design space due to linear approximations of performance function evaluations and are therefore both computationally expensive and inaccurate.
2. Automatic differentiation (AD) on a physics-based simulation (e. g. Hu et al. 2020). This is a promising future direction as exact analytical sensitivities are directly available. However, AD is not yet available for most performance function solvers, and sensitivities in the sense of derivatives are not defined for discrete variables.

This paper's cVAE approach employs AD to a learned continuous representation of the mixed variable design space to instantaneously obtain reasonably accurate sensitivity estimates once the model is trained.

2.3 Using generative models to produce design insights

Generative deep learning models are typically used in design for either optimization or "generative design" purposes (Regenwetter et al., 2022). Here we use a generative model for a different purpose: extracting design insights from a mixed variable design space.

Newer classes of generative models – e. g. conditional generative adversarial networks (cGAN) and conditional VAEs (cVAE) – improve upon conventional generative models by not only reducing dimensionality of high-dimensional design spaces but also incorporating performance as a condition; this enables the generation of high-quality designs that satisfy performance criteria (Danhaive and Mueller 2021). These developments are promising for performance-driven design applications, ensuring generated designs meet desired performance criteria during design space exploration. This work employs a specific cVAE architecture for co-learning a forward map between features and performances as well as a conditional distribution over the design space.

Despite cVAEs' potential to handle both challenges of mixed variable spaces and extracting sensitivities, few have demonstrated its usefulness in design decision-making. Balmer et al. (2022) recently present local and global sensitivities together with data visualizations for decision-making. We extend it to summarize decision-making insights as "influence metrics", a data-driven variation of the typical and manual "rules of thumb" familiar to designers.

3 Methods

The proposed method is summarized in Fig. 1 and in the following sections.

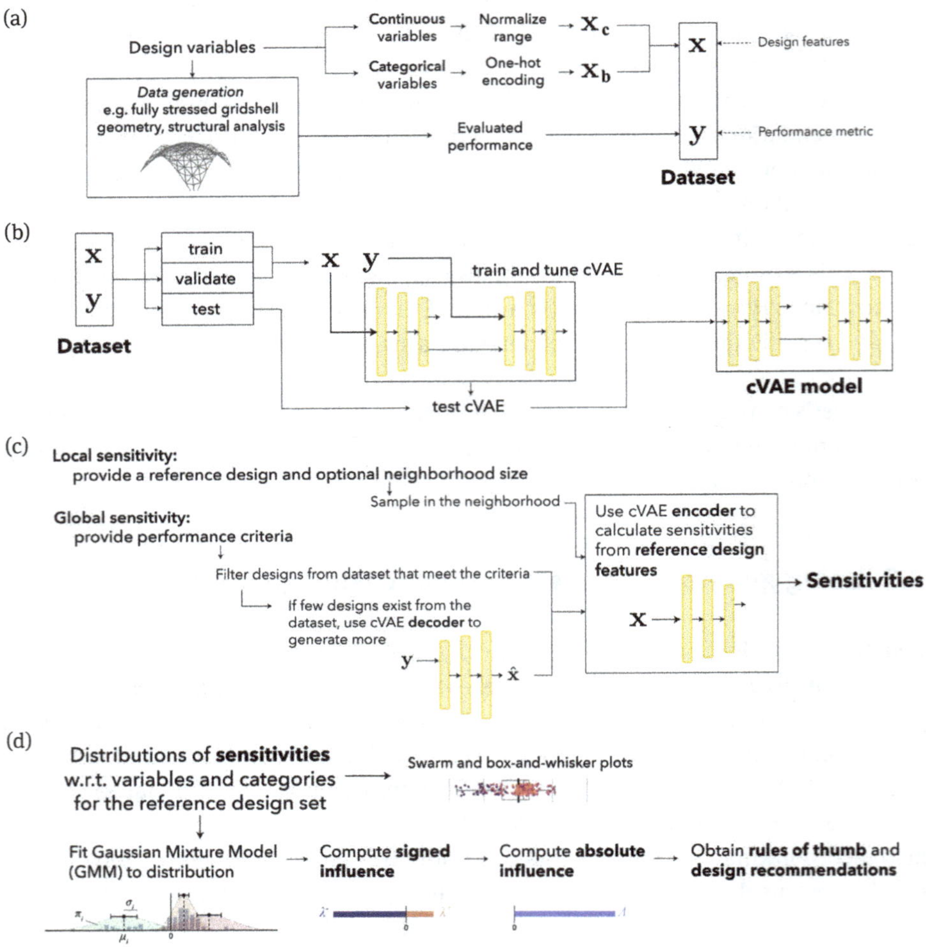

Fig. 1: Proposed method for extracting influence metrics from a mixed variable design space via cVAEs: (a) Data generation; (b) Build cVAE model; (c) Sensitivity analysis; (d) Influence metrics.

3.1 cVAE architecture

The basic deep learning model used in this study is the variation of the cVAE proposed by Balmer, Kuhn et al. 2022. In light of having to solve both a forward as well as an inverse problem, we forgo feeding the conditional **y** to the encoder and instead let it

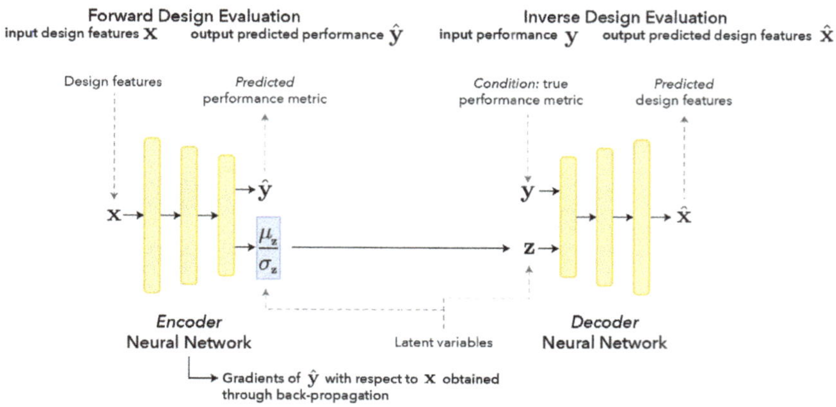

Fig. 2: The cVAE architecture enables solving both forward and inverse design problems. Adapted from Balmer et al. 2022.

predict the performance metrics together with a latent vector in two separate terminal nodes (Fig. 2).

Both the encoder and decoder consist of Multi-Layer Perceptron (MLP) Blocks with fully-connected layers, leaky-ReLU activation, and batch normalization. The total loss function L_{total} for training the cVAE is the sum of the reconstruction loss L_{des}, accuracy of predicted performance metrics L_{perf}, KL divergence L_{KL} , and decorrelation L_{cov} between $\mathbf{y}$ and $\mathbf{z}$, with hyperparameters w_i determining the contribution of each term:

$$L_{\text{total}} = w_1 \cdot L_{\text{des}}(\mathbf{x}, \hat{\mathbf{x}}) + w_2 \cdot L_{\text{perf}}(\mathbf{y}, \hat{\mathbf{y}}) + w_3 \cdot L_{\text{KL}}(\mathbf{z}) + w_4 \cdot L_{\text{cov}}(\mathbf{y}, \mathbf{z}) \quad (1)$$

Further details on background theory as well as the implementation may be found in Balmer, Kuhn et al. 2022.

3.2 Handling mixed variables: Pre-processing and obtaining sensitivities

The mixed variable space requires a new process for handling of continuous and discrete variables, summarized in Fig. 3. Discrete variables are common in AEC design and can be ordinal (e. g. integer) or categorical (unordered); this paper focuses on the latter.

Proper pre-processing is essential before passing design features to the cVAE to ensure comparable scales of sensitivities across variable bounds and units. Continuous variables are normalized to a range of [0, 1], and categorical variables are one-hot encoded. This variable pre-processing can be decoded later to retrieve the design variables in their original ranges and physical units.

For continuous variables, signs of sensitivities are correlated with an increase of the continuous variable value. For example, a positive gradient of embodied carbon

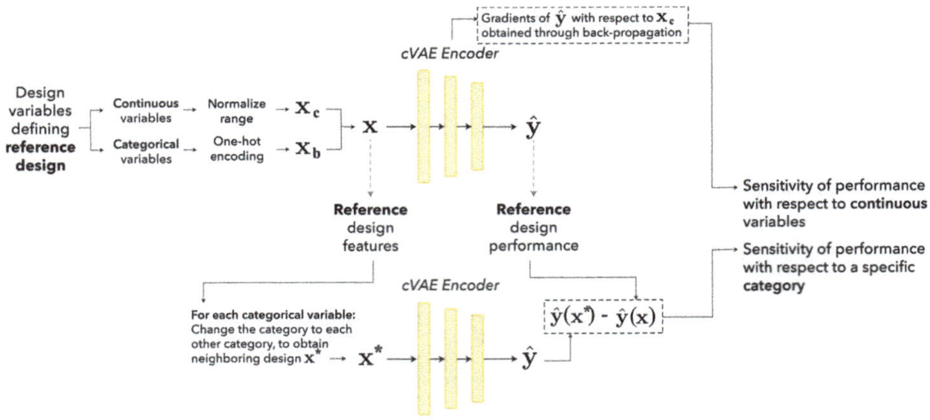

Fig. 3: Obtaining sensitivities of performance relative to continuous and categorical variables from a trained cVAE model at a reference design with design feature inputs **x**.

with respect to density indicates that an increase in density correlates with an increase in embodied carbon. This gradient can be obtained directly from the trained cVAE using AD on the encoder.

Categorical variables receive different sensitivity treatment. We extend the concept of sensitivity analysis for categorical variables, focusing on the impact of switching between specific categories. Rather than examining the overall sensitivity of embodied carbon to material choice, we propose a more relevant metric: the sensitivity of embodied carbon when transitioning from one category (e. g., steel) to another (e. g., timber). To achieve this, we introduce a separate evaluation method using gradients to assess performance with respect to categorical variables (Fig. 3).

3.3 Quantifying design variable influence

Presenting sensitivities in swarm and box-and-whisker plots (Balmer, Kuhn et al. 2022) can give the designer a sense of design variable importance.

Here, we utilize Gaussian mixture models (GMM) to algorithmically extract meaningful and rigorous insights from the data distributions, employing an established technique for approximating multimodal data as a learned weighted sum of normal distributions (Bischof and Kraus 2022).

The GMM returns means, variances, and relative weights of each distribution, which we utilize to compute the proposed influence and uncertainty metrics, summarized in Fig. 4. The absolute magnitude of these metrics is less meaningful than their relative magnitudes.

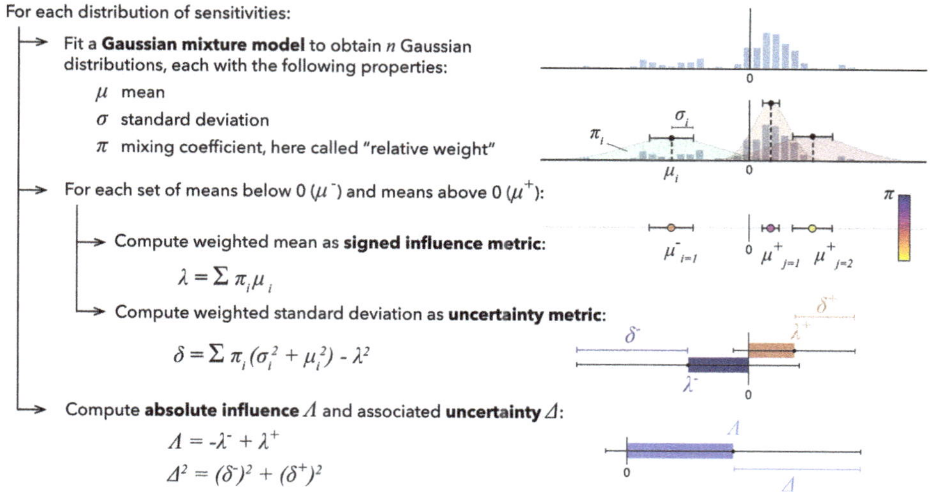

Fig. 4: Procedure for computing influence metric of a given distribution of sensitivities. For minimization performance metrics such as embodied carbon, negative influence metrics indicate decisions that improve performance, and vice-versa.

3.4 Example application

We evaluate the usefulness of our method by applying it to the design of a gridshell (Fang and Mueller 2021) containing a mixed variable design space (Tab. 1, Fig. 5). The effects of each variable on performance (embodied carbon) is somewhat intuitive, making it a good example application for evaluating the method. We employed standard material properties for steel (S235 per EN1993) and timber (C24 per EN1995), while for embodied carbon we use 1.23 kg CO2e/kg for recycled steel section and 0.51 kg CO2e/kg for timber glulam (no carbon storage) (Jones and Hammond 2019).

4 Results and reflections

The results of applying the proposed method on the example application are presented here. Figure 6 presents scatter plots of the objective-variable relationship, providing an overview of the generated design space. While this EDA aids in understanding performance trends, 2D projections can hinder insights, particularly for categorical variables where data points tend to overlap.

The trained cVAE model exhibits good predictive performance (coefficient of determination $R^2 = 0.917$, embodied carbon RMSE = 4786 kg CO2e). For brevity, training and model details are provided in a separate digital appendix (https://demifang.github.io/AAG2023/).

Tab. 1: Parameters of gridshell case study. Output is an embodied carbon estimate under fully stressed design assumptions.

Design variables	*Topology of gridshell:* Discrete {A, B, C, D, E, F, G} *Subdivisions:* Discrete, integer range varies by topology. (This variable is a designer input and a proxy for the "density" of the gridshell topology. A comparable metric, total length of all bars in the structure, is used in the sensitivity analysis to fairly compare "density" performance across topologies.) *Material (all gridshell bars):* Discrete {steel, timber}
Geometry generation	Topology is vertically projected onto a continuous shell of constant geometry.
Energy assumptions/ criteria	*Loading:* 3 kN/m^2 area load, applied as line loads – Full area – Quarter area for asymmetric loading *Strength:* designed to utilization ratio of 70 % *Serviceability:* maximum deflection of $L/200 = 6.7$ cm
Cross-section shapes	*Steel:* circular tube with diameter-to-thickness ratio of 20:1 – Series: outer diameters at 1-cm increments (4 cm to 37 cm) *Timber:* solid square section – Series: widths at 6-cm increments (2 cm to 68 cm)
Cross-section sizing	Sections are auto-sized to the smallest series section meeting the engineering design criteria. Auto-sizing occurs in two groups: edge bars and interior bars.
Performance	Embodied carbon (under fully stressed design assumptions), calculated by multiplying mass and material embodied carbon coefficient

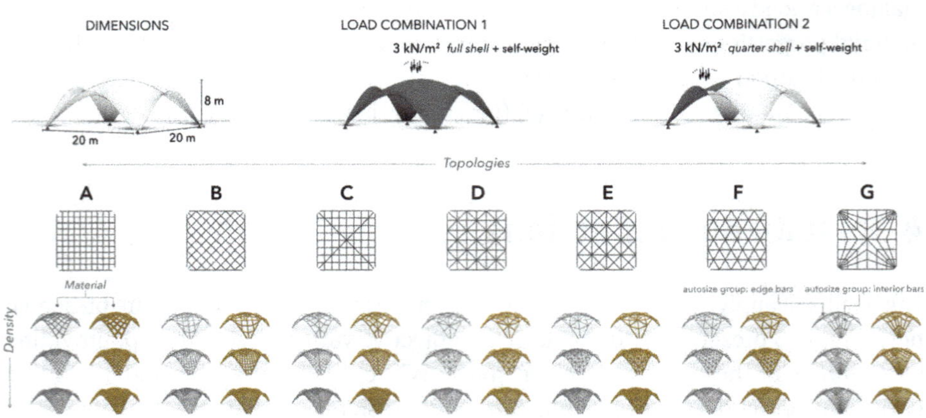

Fig. 5: Design space and engineering assumptions of gridshell example.

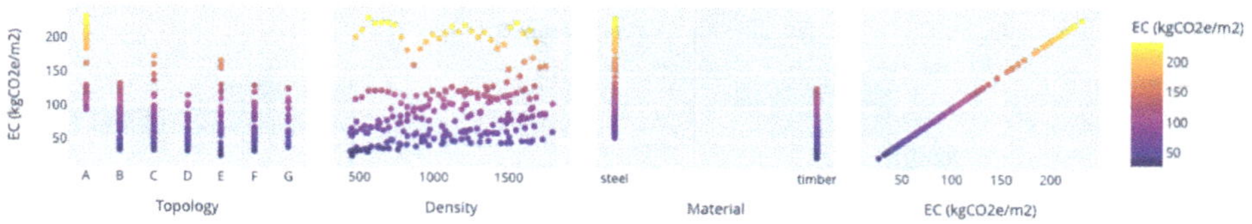

Fig. 6: Design space and engineering assumptions of gridshell example.

4.1 Local sensitivities: Action-oriented insights from a specific design

Local sensitivities assist decision-making when a designer is considering design decisions from a reference design. Rather than display the performance of adjacent designs and direct designers to select a specific design, sensitivities are presented as action-oriented insights, emphasizing the influence of each design decision.

We present local sensitivities and example interfaces for four reference designs in Fig. 7. For a chosen reference design, relative local sensitivities are shown as insights on best and worst design decisions. The results mostly agree with expert intuition (choosing topologies aligning with principal stress directions reduce embodied carbon, and switching to timber is beneficial) while giving more precise quantifications of these relative decision impacts. The density variable leads to results that are potentially counter-intuitive: in all cases, increasing the density (number of subdivisions) of the grid worsens performance. This result is likely due to the discrete nature of available cross sections. Our method can thus expand an experience-based understanding of the design problem based on data. This is especially useful in high-dimensional design spaces.

One can also understand the relative performance of the reference design with respect to its local neighborhood: for Design 4 in Fig. 7d, several design decisions have a large negative sensitivity, indicating that there are a variety of ways to significantly improve performance. We can more rigorously understand decision-making insights in the local neighborhood by evaluating the sensitivity of designs within a neighborhood of the original variable space of the reference design. An example is shown (on one of Fig. 7's reference designs) in Fig. 8. Relative magnitudes of sensitivity within each neighborhood are more meaningful than relative or absolute magnitudes of sensitivity across neighborhoods due to scaling of categorical sensitivities.

Examining neighborhood sensitivities contextualizes the degree to which local insights can be generalized in the larger neighborhood. For example, we can consider Design 1 across Fig. 7 and Fig. 8. The local sensitivities in Fig. 7a show that for Design 1, switching from current topology C to topologies B or F reduce embodied carbon. However, Fig. 8 indicates that in the specified neighborhood, this is almost always true for a switch to F but not for B.

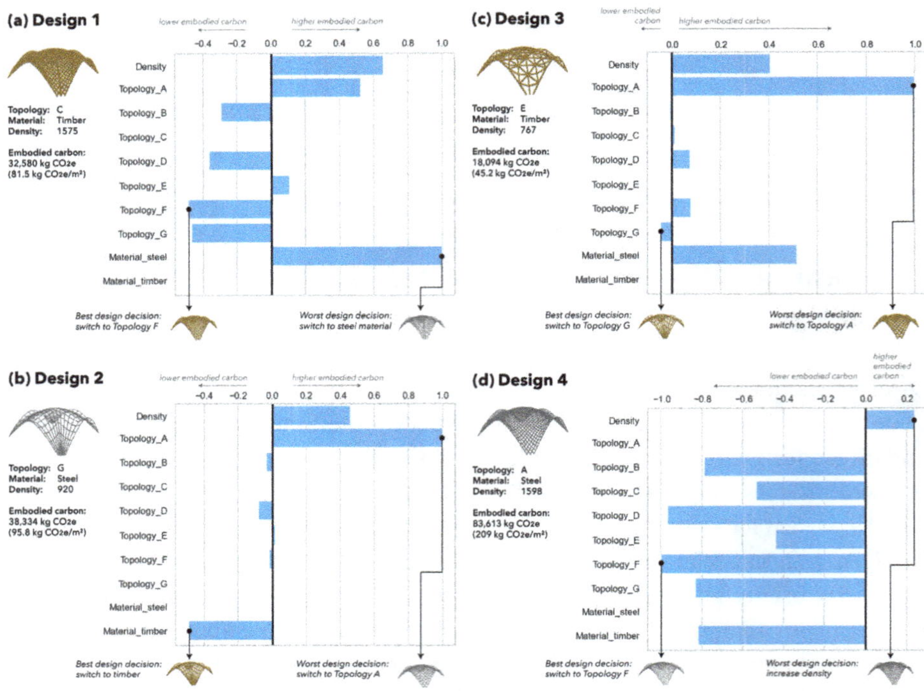

Fig. 7: Local sensitivities of 4 reference designs, annotated with design interpretation. The sensitivities of the categorical variables are scaled to match the order of magnitude of the continuous sensitivity individually for each design, so the magnitudes of sensitivity are not directly comparable across designs.

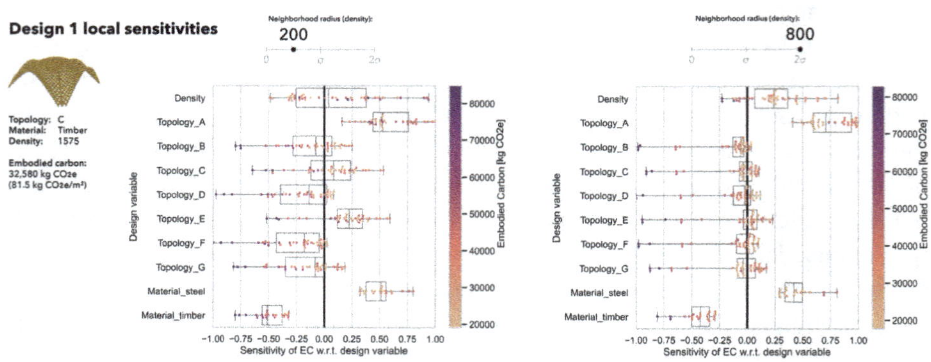

Fig. 8: Local sensitivities of a reference design, in neighborhoods of varying sizes. The radius is applied to the continuous variable density, while a full factorial sampling is done for the categorical variables. The number of designs does not increase with increasing neighborhood radius because the same number of designs are sampled (random uniform) regardless of neighborhood size.

In general, computing local sensitivities – either at a reference design (Fig. 7) or in the neighborhood around it (Fig. 8) – already improve on EDA by giving specific decision-making feedback.

4.2 Global sensitivities: Data-driven rules of thumb

Sensitivities at the global scale provide more generalized insights on decision-making for the design problem. These generalized insights are similar to “rules of thumb” typically employed by designers during more manual design. We extend the box-and-whisker swarm plots used by Balmer, Kuhn et al. 2022, enhancing the decision-making process with performance filtering. In Fig. 9, we illustrate this approach in the context of the gridshell case study. By adjusting the performance filter, we control the number of designs included in the analysis. Using the cVAE decoder, we generate additional designs that align with the filtered performance criteria, along with their corresponding sensitivities. Focusing on the top 5 % performing designs (Fig. 9d) allows for more targeted insights regarding influential design decisions among already high-performing designs, compared to considering the entire design space (Fig. 9a).

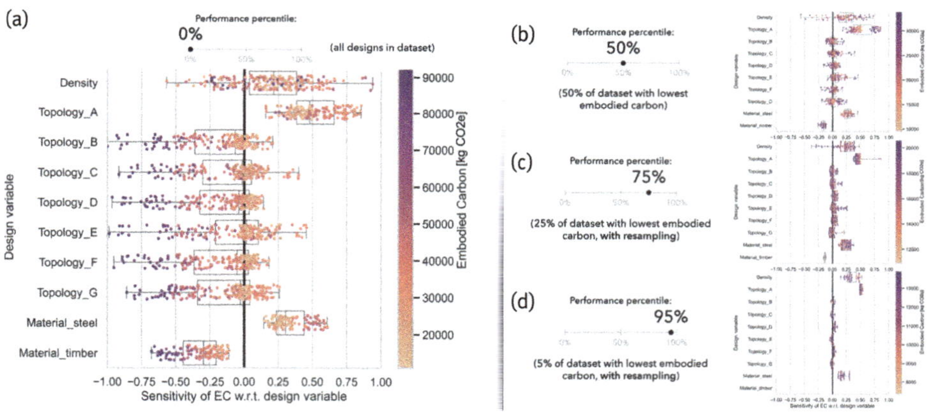

Fig. 9: Global sensitivities of the dataset at varying performance filters.

4.3 Quantifying the influence of design variables

The Gaussian Mixture Model (GMM) fitting approach described in Sec. 3.3 can be applied to any distribution of sensitivities; for brevity, we only show results on global sensitivities. The resulting means, standard deviations, and relative weights determined by the GMM fitting are shown in Figure 10. This representation reduces some of the complexity of the same distribution’s box-and-whisker swarm plot (Fig. 9a).

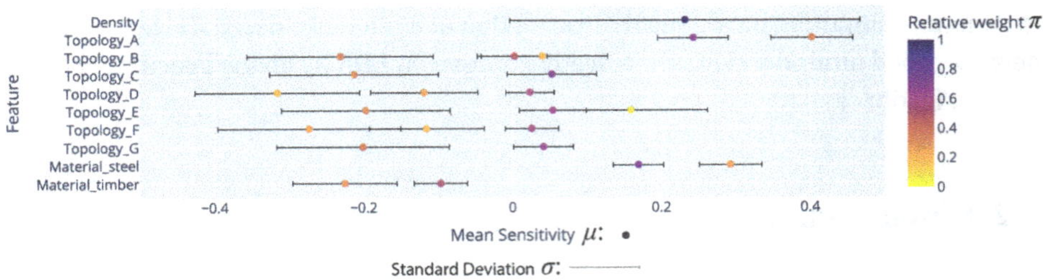

Fig. 10: Results of the GMM fitting on global sensitivity distributions determined by the cVAE. For many decisions, the data are distributed in multiple modes, which is captured in the GMM fit and represented here graphically. This indicates that the design decision can have a variety of impacts on performance, depending on location in the design space.

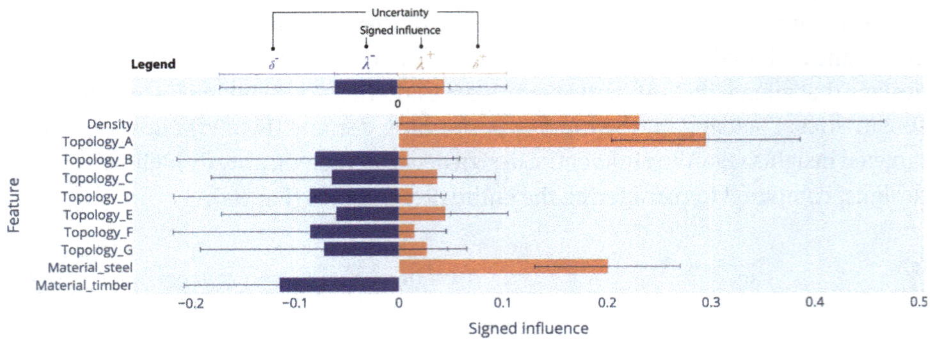

Fig. 11: Signed influence metrics of design variables on embodied carbon.

Signed influence metrics are provided in Fig. 11, computed by combining the results of the GMM model for each decision (Fig. 4). Negative influence metrics λ^- indicate a variable's influence on enhancing performance (lower embodied carbon), while positive influence metrics λ^+ on worsening performance (higher embodied carbon). δ^- and δ^+ represent the associated uncertainty metrics for each influence metric, respectively. By identifying variables with higher λ^+ and lower δ^+ values across the design space, one can understand that switching to Topology A overwhelmingly (in magnitude and certainty) worsens performance, followed closely by switching to steel. Switching to timber is the most influential design decision (in magnitude and likelihood) for improving performance, though not by much more than switching to topologies other than A. Increasing density generally worsens performance but with high uncertainty. We can confirm this by revisiting the EDA (Fig. 6) and noticing that changes in embodied carbon are relatively flat for a given topology of increasing density, an insight that was harder to identify without Fig. 11.

The influence metrics of material categories steel and timber are unexpectedly asymmetric though expected to be symmetric (Fig. 6). The authors identified the cVAE

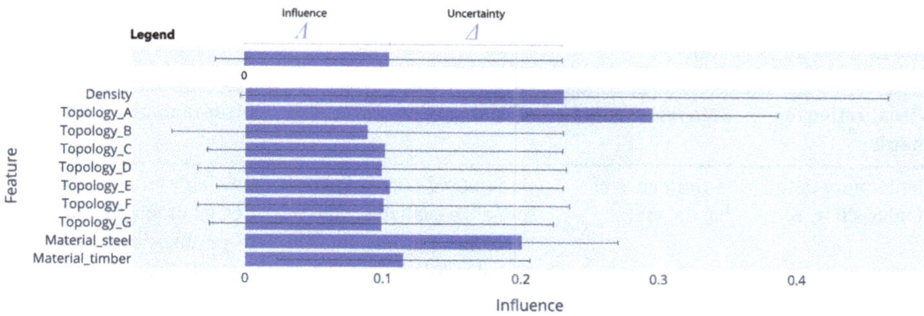

Fig. 12: Absolute influence metrics of design variables on embodied carbon.

Tab. 2: Table 2. Dashboard of absolute influence (categories aggregated into categorical variables).

	Density	Topology	Material
Aggregated absolute influence (Λ)	0.231	0.127	0.158
Aggregated uncertainty (Δ)	0.235	0.126	0.081

model encoder's prediction error to not be a zero-mean Gaussian, propagating into asymmetries in sensitivities and influence during the process of calculating λ. Solutions for overcoming this limitation are discussed in Sec. 5.

Finally, absolute influence metrics Λ and their associated uncertainty Δ are presented in Fig. 12. We can also present this as a "dashboard" (Tab. 2). These metrics further abstract previously available insights, approaching classic "rules of thumb", and might be useful for designers as general insights in the earliest stages of decision making. For example, the metrics indicate that choice of topology and density can be just as or more influential than material choice (Fig. 12), while material choice more certainly impacts performance than density or topology (Tab. 2).

4.4 Reflections

The example application of the proposed method demonstrated alignment with expected insights and rules of thumb. A summary of advantages and disadvantages associated with each visualization or metric is presented in Table 3. We observe tradeoffs between exhaustive representation of sensitivities and interpretability of sensitivities for decision-making insights. Overall, we recommend the GMM summary (Fig. 10) and signed influence metrics (Fig. 11) as representations that balance these properties.

Tab. 3: Parameters of gridshell case study. Output is an embodied carbon estimate under fully stressed design assumptions.

Visualization or metric	Data represented	Advantages	Disadvantages
Exploratory Data Analysis (Fig. 6)	Performance of full dataset	Understand range of performance available; Observe some prominent global trends	Unwieldy for multivariate design problems; Hard to get decision-making insights, especially locally
Bar plot (Fig. 7)	Sensitivities at 1 reference design	Get clear decision-making insights from the reference design	Insights are only applicable at a single reference design
Box-and-whisker swarm plot (Fig. 8, 9)	Distribution of sensitivities at multiple designs	Swarm representation exhaustively depicts the distribution	Can be hard to interpret swarm representation for decision-making; Box-and-whisker plot does not account for multi-modal behavior from local effects
Gaussian Mixture Model summary (Fig. 10)	Distribution of sensitivities at multiple designs	Simplifies complexity of swarm representation to a few key values; Accounts for possible multi-modal behavior from local effects	Reduced data resolution
Signed influence metrics (Fig. 11)	Distribution of sensitivities at multiple designs	Provides a quick snapshot on how each variable improves OR worsens performance across a set of designs	Drastically removes nuance from the GMM summary
Absolute influence metrics (Fig. 12)	Distribution of sensitivities at multiple designs	Most immediately interpretable; closest to a "rule of thumb"	No longer captures whether a decision improves or worsens performance

5 Conclusions

This paper proposes new metrics for quantifying the influence of design decisions, using sensitivities of a trained cVAE for architectural applications. The proposed method successfully serves as an automated, data-driven way to reveal insights for design decision-making. The cVAE is newly proposed for not only generating designs but also primarily extracting sensitivities of mixed variables in a design space, information that was not previously accessible from data generation alone.

The proposed influence metrics combine qualitative and quantitative representations to communicate actionable decision-making insights. We show a range of visualizations and metrics, discussing their respective limitations and advantages.

Ultimately, designers can choose levels of resolution and abstraction that best suit their design-decision-making process.

This approach has some limitations. The cVAE requires an upfront, time-consuming process of generating data and training (although it is instantaneous to query once trained). Model accuracy should be assessed in each application. Even models trained to reasonable accuracy provide approximate sensitivities; in our example, the model produced asymmetries in influence metrics that are expected to be symmetric. This can be addressed in future work by enforcing symmetric sensitivities or reducing prediction errors in the back-end model, and continuing to verify the intuition of the influence metric formulation in other design problems. Furthermore, the dataset in this case study contained a full factorial set of structurally viable designs; the cVAE is assumed to have learned to produce structurally viable designs. Future applications should consider an explicit representation or a condition for ensuring generated designs are structurally sound. Additionally, for the proposed influence metric, the GMM fitting necessarily assumes normal distributions for identifying multimodal behavior.

In future work, more elements of the cVAE can be harnessed for design, such as the latent space for correlation and neighborhood sampling, and the decoder for generating new design concepts. The method can be useful to inform and drive design problems beyond the case study presented here, such as investigating the material- and topology-related design decisions most influential in reducing carbon emissions in multi-story conventional buildings.

References

Balmer, V. M., S. V. Kuhn, R. Bischof, L. Salamanca, W. Kaufmann, F. Perez-Cruz, and M. A. Kraus. 2022. Design Space Exploration and Explanation via Conditional Variational Autoencoders in Meta-Model-Based Conceptual Design of Pedestrian Bridges. arXiv: 10.48550/arXiv.2211.16406

Bischof, R., and M. A. Kraus. 2022. Mixture-of-Experts-Ensemble Meta-Learning for Physics-Informed Neural Networks. In *Proceedings of 33. Forum Bauinformatik.*

Brown, N. C., and C. T. Mueller. 2018. Gradient-Based Guidance for Controlling Performance in Early Design Exploration. In *Proceedings of the International Association for Shell and Spatial Structures (IASS) Symposium.* Boston, MA.

Cho, S., and H.-S. Jung. 2003. Design Sensitivity Analysis and Topology Optimization of Displacement–Loaded Non-Linear Structures. *Computer Methods in Applied Mechanics and Engineering* 192 (22): 2539–53. DOI: 10.1016/S0045-7825(03)00274-3

Danhaive, R., and C. T. Mueller. 2021. Design Subspace Learning: Structural Design Space Exploration Using Performance-Conditioned Generative Modeling. *Autom. Constr.* 127: 103664. DOI: 10.1016/j.autcon.2021.103664

Fang, D., and C. Mueller. 2021. Flow-Informed Topology Design: Evaluating the Conformity of Structural Topologies with Vector Fields. In *Proceedings of the International Association for Shell and Spatial Structures (IASS) Symposium.* Surrey, UK.

Hu, Y., L. Anderson, T.-M. Li, Q. Sun, N. Carr, J. Ragan-Kelley, and F. Durand. 2020. DiffTaichi: Differentiable Programming for Physical Simulation. arXiv: 10.48550/arXiv.1910.00935

Jones, C., and Hammond, G. 2019. Inventory of Carbon & Energy v3.0. https://www.circularecology.com/embodied-energy-and-carbon-footprint-database.html

Regenwetter, L., A. Heyrani Nobari, and F. Ahmed. 2022. Deep Generative Models in Engineering Design: A Review. *J. Mech. Des.* 144 (7): 071704. DOI: 10.1115/1.4053859

Saltelli, A., M. Ratto, T. Andres, F. Campolongo, J. Cariboni, D. Gatelli, M. Saisana, and S. Tarantola. 2008. *Global Sensitivity Analysis: The Primer.* John Wiley & Sons.

Jelena Milošević and Ognjen Graovac

An Approach to Designing Architectural Structures Using 3D Graphic Statics

Abstract: Graphic statics is a well-established method for designing diverse typologies of architectural structures. Although this method has been in use since the 19th century, the advent of digital technologies contributed to its current, more widespread usage. This paper presents the application of 3D graphic statics for finding structural forms. Parametric design tools based on 3D graphic statics for explorations in a generative design process was employed. The explorations were implemented by the 3D Graphic Statics plugin developed for Grasshopper visual programming editor for the commercial computer-aided design (CAD) system Rhinoceros, commonly used by architects. The design outcomes confirm the effectiveness of the approach based on geometrical operation for designing efficient structural forms composed of elements in perfect geometrical equilibrium with external forces. Furthermore, graphical methods are intuitive and facilitate rapid statical calculations in the conceptual stage of the architectural design process. Moreover, the description of the design approach also allows for reflection on its methods, outcomes, and the interactions of designers with design tools throughout the design process. Finally, this study informs architects about the increasingly sophisticated design tools made available by digital technology to support architectural design and how these tools compare to older design mediums.

Keywords: architectural design, computational design tools, spatial structures, form-finding, graphic statics, design research

1 Introduction

Graphic Statics (GS) includes a set of methods for efficiently designing diverse structural typologies in architecture including arches, vaults, and bridges. The method is based on geometrical operation, and the principle that forces in the structural elements are in equilibrium with the external forces. The method is intuitive and facilitates rapid statical calculations (in comparison to, for example, Finite Element Analysis), which makes it relevant for the conceptual design stage.

There are only indications that the Romans and later the Byzantines were familiar with specific graphic techniques. Even though the late Gothic is full of geometrically intricate structures like vaults, arches, rosettes, and buttresses, very few recorded sources provide thorough information on the used graphic techniques. Moreover, until the 20th century, no techniques could reliably describe the equilibrium state of Gothic vaults. Several ancient concepts came back into prominence during the Renaissance and Baroque periods, and structural design approaches based on mathematics and

https://doi.org/10.1515/9783111162683-032

graphics underwent extensive development. The equilibrium state of arches and domes based on a closed polygon of forces is graphically described by Simon Stevin's rule on the parallelogram of forces (1586). More elaborate elliptical forms were also built during the Renaissance and Baroque periods, along with significant structures including the St. Peter's Church (1506) by Bramante and Michelangelo, and Santa Maria del Fiore (1436) by Brunelleschi (1436).

The contributions of various authors allowed GS to become an established scientific discipline. Hooke (1676) described the connection between the geometry of the tensioned chain and compressed arch with the anagram: *As hangs the flexible line, so but inverted will stand the rigid arch.* Gregory independently came to the same conclusion and developed effective equations in 1689. Poleni (1743) calculated the equilibrium state of the dome of St. Peter's during the restoration using an inverted chain model. Rankine introduced the idea of the reciprocity of the diagram of forces and form in 1864. Maxwell established geometric procedures for constructing reciprocal diagrams in 1864, while Culmann (1866) established mathematical proofs. Also, Cremona (1879, 1880) shaped GS as a method for solving specific structural problems in 1890. In addition to these studies, the thrust line theory was developed in the 19th century (Mosely 1843; Heyman 1989; Benvenuto 1991). Milankovitch (1907) contributed to the field by formulating a comprehensive thrust line theory and a mathematical knowledge of the equilibrium state of structures.

Gaudi instrumentalized his knowledge of graphical methods for determining structures to develop distinctive architectural design methodology at the beginning of the 20th century (Burrey et al. 2005). He applied graphic and experimental methods in designing and constructing his unconventional equilibrium forms (Huerta 2003). Gaudi's work and design approach came back in focus in recent years due to the emergence of complex architectural forms and interest in generative form-finding. The term "form-finding" refers to the process of finding the best shape that best represents (or approximates) the state of static equilibrium design (Lewis 2003) as an alternative to random free-form. In the later decades of the 20th century, physical models as instruments for form-finding were replaced by digital tools, facilitating further development and implementation of more complex concepts.

Although GS represents an intuitive technique for the exploration of structural forms in two dimensions (Zalewski and Allen 1998), three-dimensional graphic static (3DGS) methods were also developed (Foppl 1892). The application of 3DGS methods was constrained because of rather complicated procedures that require knowledge of several specific rules of projective geometry. Developments in computational technologies motivated the recent interest of researchers in this topic. The modeling of vaults as a discrete network of forces in equilibrium with gravity loads were proposed based on Heyman's (1982) principle (O'Dweyer 1999), expanded further through the Thrust Network Analysis (TNA) (Block 2009), various related methods (Angelillo et al. 2010; Fraternali 2010), and Thrust Surface Method (Fraddosio 2020). In addition, in recent years, many authors have explored the topic of 3DGS. The results of the literature analysis (Fig. 1) show increased interest in GS in recent years, covering a wide specter of topics from new methods and design tools to digital fabrications.

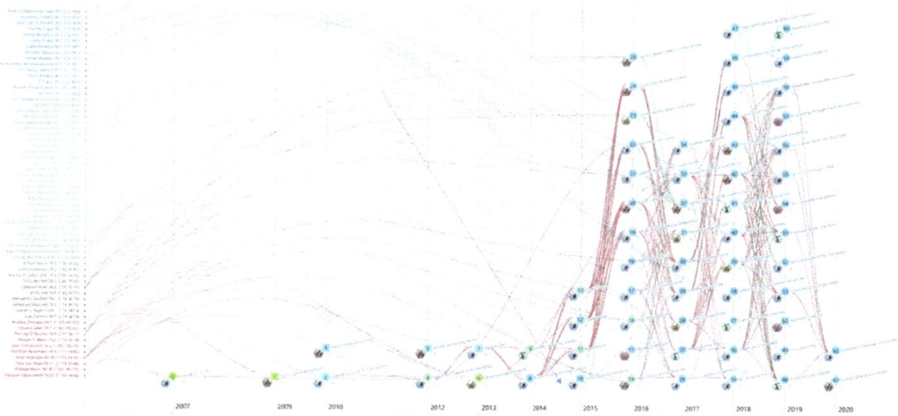

Fig. 1: Literature analysis including information on number of papers (N), citation (C, M), number of authors, their impact, and connections; thematic frame; type of publication.

Various authors developed computational tools. Tools such as software modules for computational-aided design (CAD), which is standardly used in architecture, have increased interest in using GS in the early design stages. An example is *RhinoVault* by Rippmann (2012; 2014) developed as a plug-in for *Rahinoceros* commercial CAD system for 3D modeling. Using the same benefits of Graphic Statics, *RhinoVAULT* uses reciprocal diagrams to provide an intuitive, rapid form-finding tool for structural design in three dimensions. The application of the tool in structural design is best exemplified by Armadillo Vault, constructed for the Architecture Biennale in Venice in 2016 (Rippmann et al. 2016). Also, a form-finding tool developed for *Rhinoceros* using the principles of 3DGS is *PolyFrame* by Akbarzadeh and Nejur (2018). This computational tool allows the construction and manipulation of reciprocal polyhedrons for compression-only structural form-finding. Finally, the *3D Graphic Statics* by Graovac (2019), a parametric tool for the *Rhinoceros/Grasshopper* that facilitates a parametric approach to structural design based on 3DGS, is tested in this paper, in an educational context, for its use in early design stage.

A previous brief overview of the GS history illustrates the constant development of the method supported with new theoretical knowledge and technological advances, as well as recurring interest in its architectural application. This suggests that it is relevant for architects to learn about GS methods and tools.

2 Design Application

The premise of this study is that using the 3DGS computational design tool, which is integrated into a standard CAD modeler, throughout the architectural design process enables designers to explore a variety of effective structural forms and broaden their design vocabulary intuitively and interactively. To test the previous premise, design by research method was used in an educational context. The first part of the research was a study conducted within the course Research Methodologies in the first year of Ph.D. studies at the University of Belgrade – Faculty of Architecture (UB–FA). The study focused extensive review of the development of GS methods and procedures and aimed to test that knowledge practically by developing a computational design tool based on 3DGS. The second part of the research included user study of the *3D Graphic Statics* plugin for *Rhinoceros/Grasshopper* by Graovac (Fig. 2) through design experiments conducted with the group of master architectural students at the UB–AF within the elective course Form-finding of Spatial Structures.

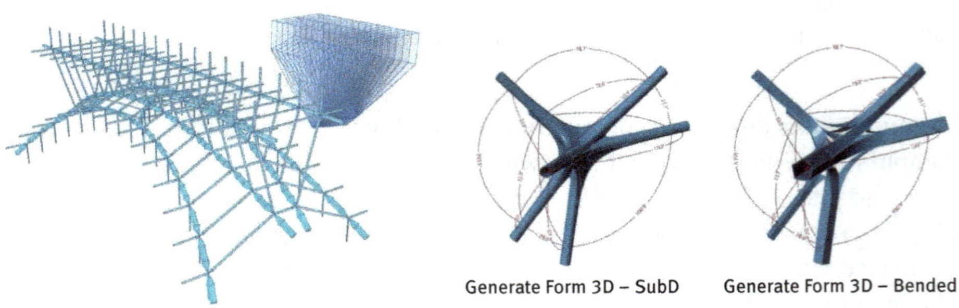

Fig. 2: An example of form-finding of a bridge structure using 3D Graphic Statics plug-in.

2.1 Computational Implementation of 3DGS Concepts

The developed plugin implements common GS concepts (Clarke 2018). In GS, the forces acting on a structure are represented by lines or vectors, and their magnitudes and directions are accurately depicted. The graphical approach allows visualization and analysis of complex interactions of forces in a structure without complex mathematical calculations. The fundamental principle of graphic statics is based on the equilibrium of forces. By constructing reciprocal force and form diagrams, the magnitude and direction of unknown forces can be determined, and the overall equilibrium of the structure can be found. Traditional application of GS implies using 2D reciprocal diagrams; however, the method was extended in 3D using the principle of the equilibrium of polyhedral forms and is known as 3D/polyhedral/spatial GS. While traditional GS primarily focuses on 2D structures, 3DGS allows for analyzing more complex spatial structures (including trusses, frames, and arches). Table 1 summarizes the differences between the two graphical representations and equilibrium representations in 2DGS and 3DGS.

Tab. 1: Differences between key graphical representations in 2DGS and 3DGS.

	2DGS	3DGS
Graphical representations	**Force diagram/polygon** Represents a magnitudes and directions of the forces acting on a structure. It is constructed by drawing lines or vectors to scale, where the length of each line represents the magnitude of the corresponding force, and the direction of the line represents the direction of the force.	**Force diagram/polyhedral** They represent the magnitudes and directions of forces acting on a 3D structure. The edges or faces of the polyhedron correspond to the forces acting on the structure, and their lengths or areas represent the magnitudes of those forces.
	Form diagram/polygon (also called funicular polygon, or an influence line diagram) Represents the distribution of internal forces within a structure. It shows the path of the resultant force for a given set of external lodes applied to the structure.	**Funicular diagram/polyhedral** Represents the flow or forces through a spatial structure and shows the paths of resultant forces for different loading conditions.
Equilibrium	**Closure of polygon** The lines forming force polygons in 2DGS must close to indicate force equilibrium. The closure of the force polygon ensures that the sum of forces and moments acting on the structure is zero.	**Closure of polyhedral** Force polyhedral in 3DGS must satisfy equilibrium conditions. The closure of the force polyhedron ensures that the sum of forces and moments acting on the structure is zero. The polyhedral must be convex.

In GS, force and from diagrams are two key graphical representations of the forces acting on a structure and aid in determining the internal forces and stability. The external and internal forces acting on the structure are considered to construct a force diagram. External forces include applied loads, reactions at supports, and other externally applied forces. Internal forces result from the resistance of the structure to external forces and are usually represented by lines within the structure. The form diagram is constructed considering a particular force or load applied to the structure. The lines or vectors representing the internal forces are drawn to scale and connected to form a polygon, known as a form polygon or form diagram. The form diagram provides insights into the load paths and the magnitude of internal forces at different points along the structure. By analyzing different load cases and constructing corresponding form diagrams, engineers can understand how the structure responds to different loading conditions and optimize its design for strength and stability.

The force and the form diagram are fundamental tools in GS, allowing analysis and understanding of the equilibrium of forces in structures (Fig. 3) without the need for complex mathematical calculations. Furthermore, these graphical representations facilitate the visualization and optimization of structures, leading to efficient and safe designs. Since force and form diagrams are topological duals, design explorations could be done by manipulating either force or form diagrams.

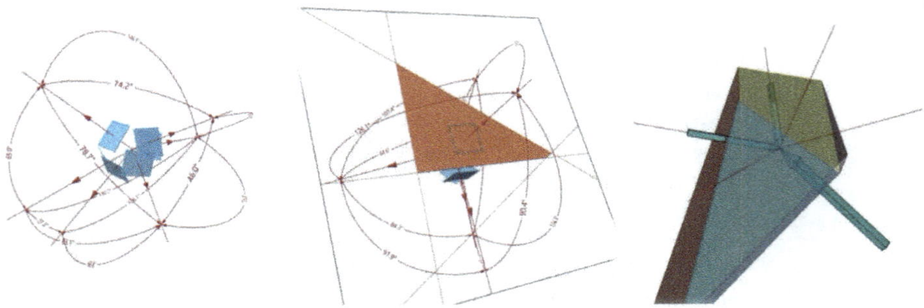

Fig. 3: Representation of statical equilibrium in 3DGS.

Unlike 2D methods, 3DGS is challenging to perform manually, so it was implemented in the *3D Graphic Statics* computational tool for generating funicular structures. The development of the tool had several goals, including automation of the process of force diagrams' construction and acceleration of the design exploration process by facilitating rapid production of alternative formal solutions. The tool is implemented as a set of components within the *Grasshopper* environment for visual programming. It uses the advantages of real-time geometry preview, which facilitates form exploration based on simulation results. The increased popularity of the parametric modeling design approach caused the development of diverse software; however, *Rhinoceros/Grasshopper* interface and strong community support set this system apart, and the development of the tool as a module within this system facilitates exploration connecting of *3D Graphic statics* with other *Grasshopper* tools. Most of the implemented procedures are based on the research conducted by Block Research Group (2009–2023) at ETH Zurich and Polyhedral Structures Laboratory (n.d.) at Weitzman School of Design, University of Pennsylvania. These approaches are repackaged and generalized in the new plugin for design.

Implemented form-finding algorithm of 3D Graphic Statics incorporates two phases: preparation (phase 1) and simulation (phase 2). Phase 1 includes subsequent steps:

Step 1.1 Construct a node (point) inside each polyhedron.
Step 1.2 Determine the interconnections of polyhedral and connect the nodes with lines.

Phase 2 is iterative problem-solving and includes the next steps:

Step 2.1 Rotate the lines around the center so they are parallel to the side normal.
Step 2.2 Connect the separated nodes with the arithmetic mean.
Step 2.3 Repeat until all corner deviations are below tolerance.

Implemented algorithm enables interactive structural form-finding by manipulating both form and force diagrams in parametric settings (Fig. 4). Regarding previous, the plug-in provides a set of components including components for subdividing polyhedral cells, form-finding processes, computing forces, generating cross sections.

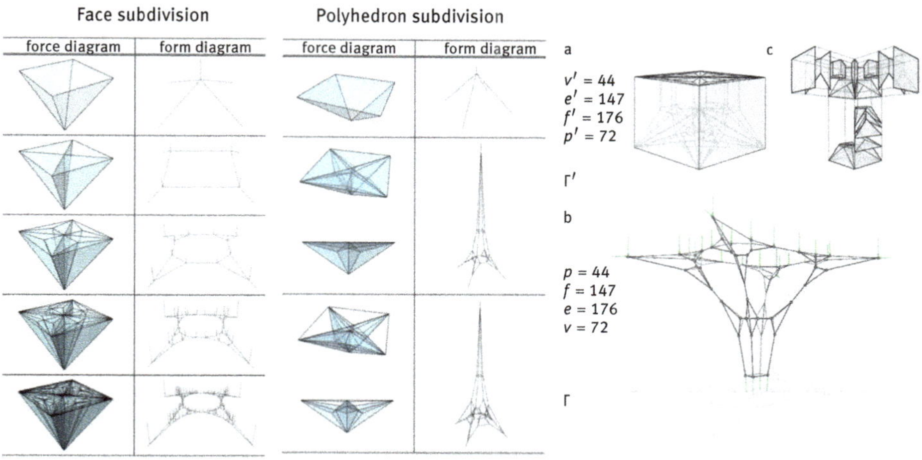

Fig. 4: Form-finding using implemented 3D Graphic Statics algorithm.

In the general structure of the plugin (Fig. 5), components are organized in several pallets (corresponding form-finding algorithm). The first pallet contains pre-processing components which enable initial geometry description and manipulation. The second pallet is the processor and contains components that implement constructing methods and solvers. The third pallet comprises components for post-processing that enable extraction and representation of geometric data. Besides, the plugin contains two additional pallets of utilities and presets containing components that are generally useful for diverse modeling operations. With this digital tool, architects, engineers, students, and researchers can create a parametrically driven system to generate optimal structures of aesthetics.

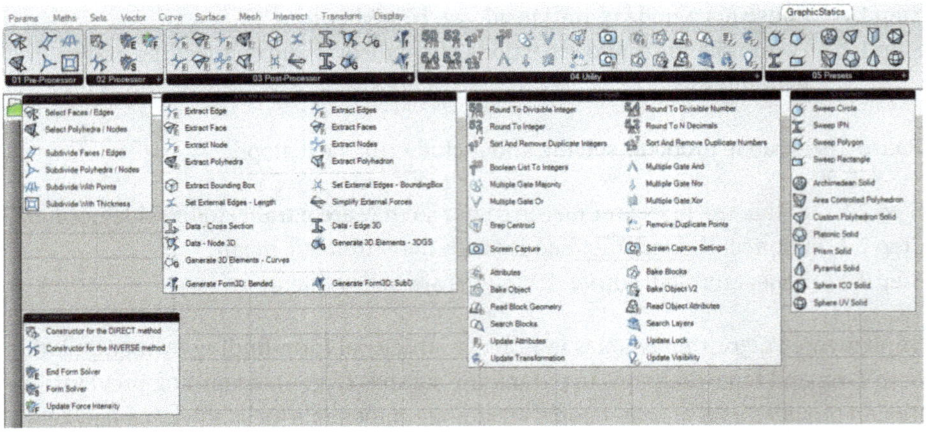

Fig. 5: Structure of 3D Graphic Statics plugin for Grasshopper.

A Grasshopper definition for form-finding of equilibrium structures that uses 3D Graphic Static components (Fig. 6), includes following steps:

Step 1 Set parameter system that generates polyhedral.
Step 2 Convert polyhedral to 3DGS data.
Step 3 Use one of the shape simulation solvers.
Step 4 Based on one known force, match the intensities of the others.

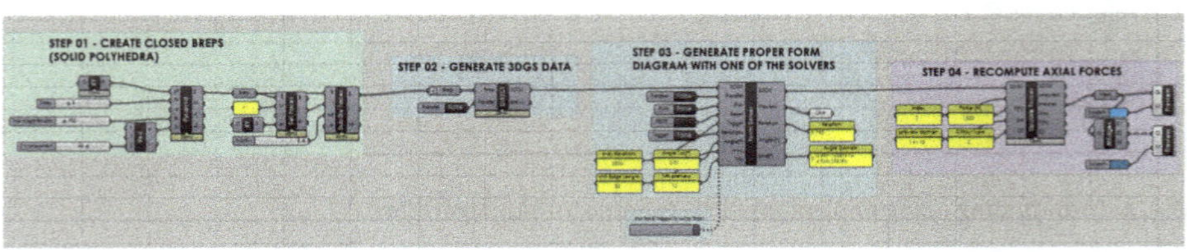

Fig. 6: Grasshopper definition for form-finding equilibrium structures using 3D Graphic statics components.

2.2 Educational/Design Applications

The architectural design application of 3DGS and the developed tool were tested in an educational context. The design project was implemented as a task within the elective course for master architectural (MA) students. The course introduces students to form-finding methods and tools and their application in designing spatial structures (particularly form-active structures) in architecture. Since the course is primarily theoretical, the practical design activity was organized as time constrained workshop during second part of the autumn semester of the academic year 2022/2023. Students working in groups were assigned to design lightweight spatial structures. The workshop's goal was also to connect PhD research and MA education and engage students in a collaborative manner of work. The learning outcomes of this practical course activity were to acquire theoretical and methodological knowledge on 3DGS in architecture, improve digital modeling skills, and exceed themselves in suggesting novel design proposals.

Workshop was implemented as a one-day block of classes that included three parts:

- A lecture on the theoretical basis of graphic statics,
- Instruction related to the application of the 3D Graphic Statics plugin, and
- Designing of spatial structure.

This organization facilitated students to gain procedural experience related to the specific design process and acquisition of competencies through working on a specific design task with the help of workshop tutors. Furthermore, problem-solving design activity motivated the development of knowledge and skills in a creative environment. Also, students took advantage of digital tools to produce interactive, collaborative products. In the continuation of this paper, three design proposals produced by students during the workshop are presented. The results of the workshop, and three design proposals, are presented in Fig. 7.

Previous user study generally validates the practical application of the 3DGS in the early stage of the architectural design process. However, based on the artifacts produced during the workshop and conversation with participants following potentials and limitations were identified concerning the application of the 3D Graphic Statics plugin for the design:

- Using a tool based on 3DGS provides valuable insights into the behavior and stability of spatial structures, allowing designers to optimize their designs.
- The tool enables the production, testing, and elaboration of diverse design concepts.
- The tool enables rapid and easy form explorations and the creation of alternative designs.
- The plugin represents an intuitive design tool that enables structural evaluations adequate for the conceptual design stage.
- Implementing 3DGS as a plugin of a standard CAD environment familiar to architects makes this tool more receptive to designers.
- The tool can support the creative process and lead to the emergence of structurally rational and aesthetically pleasing forms.
- The workshop enabled students to familiarize themselves with 3DGS, improve modeling skills, and creatively apply these competencies to the design activity.

Variation

Class students: Đorđe Milisavljević, Luka Marković, Lazar Žikić and Danilo Radovanović

Scope: Design of a canopy with the rational number of nod connections

Process: (1) Selection of global geometries and experimentations with several types of closed bodies (e. g., cubes and pyramids); (2) Selection of capped pyramid as a geometry for further elaboration; (3) Design elaboration through further form segmentation and generation of new structural formations; (4) Optimization of number of rods

3DGS application: Using plugin for production of design alternatives and fast structural evaluation

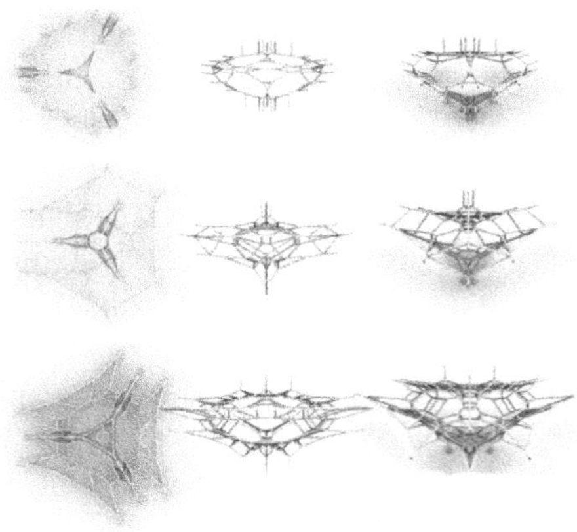

Modularity

Class students: Petar Rakić, Đorđe Stanojlović, Nemanja Papić and Sara Brkić

Scope: Design of a modular canopy and exploration of construction efficiency

Process: (1) Form-finding of a modular element; (2) Design explorations of a architectural composition made of modular elements; (3) optimizing number of rods and modular units

3DGS application: Using plugin for form-finding of a modular element and fast structural evaluation

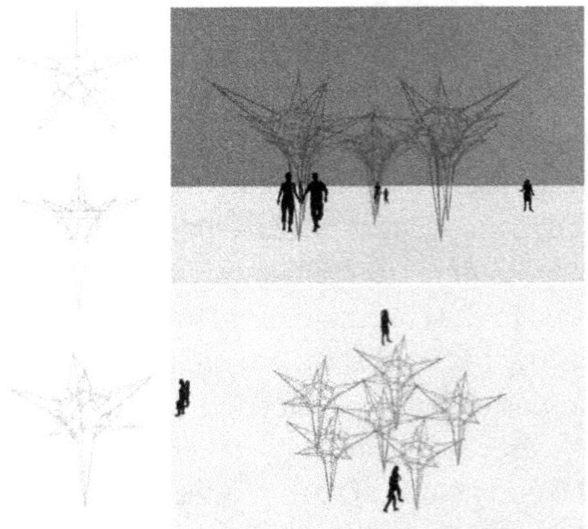

Assembling – disassembling

Class students: Jovana Lazarević, Tijana Pejić, Tamara Bojović, Filip Vasić and Jovana Stakić

Scope: Design traveling pavilion out of a demountable kit of architectural elements

Process: (1) Form-exploration by varying parameters concerning polyhedral (number of sides, number of divisions); (2) Selection of a geometry for further elaboration based on fabrication efficiency demands; (3) Design elaboration and production of the catalogue of elements (rods and nod connections)

3DGS application: Using plugin for design explorations and fast structural evaluation

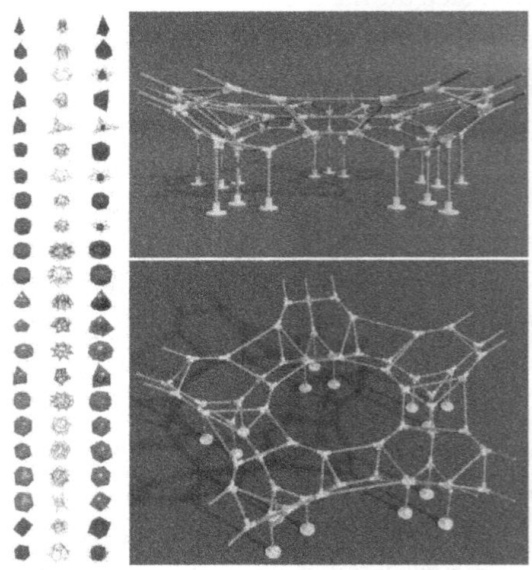

Fig. 7: Topics and projects produced during workshop.

- Different levels of digital modeling skills and time constraints affected the work process and result. A more personalized approach to teaching students computational modeling skills should be considered in the future.
- Organization of workshops at masters' studies based on the results of research conducted within Ph.D. studies proved to be very motivating for students.
- Learning by working on a design problem interested architectural students in structural engineering concepts and familiarizing them with the tool that could be applied in work on different architectural design projects in the future.

3 Conclusion

This paper contributes to the studies of using computational tools in creative design processes. By focusing 3DGS method, its computational implementation and practical application study illustrate an approach to designing architectural structures. The design results support the efficacy of the geometrical operation-based method for creating structural forms made up of parts that are perfectly in geometrical equilibrium with external forces. This approach is a sustainable way of designing complex spatial structures, particularly from-active structural typologies.

Standard CAD software supports architectural design creation through geometric operation without information on the tectonic quality of obtained forms. On the other hand, structural analysis is usually performed based on these geometries in specialized engineering software after the design is established. This analytical step requires a relatively long time, and structural analysis feedback may ask for design modifications. Since the structural behavior of complex forms is difficult to predict, the possibility of early feedback on the structure is essential.

In contrast, generative form-finding, that become lately widespread in architectural design due to the development of various digital tools, enables a structurally informed design process. Interacting with structures characterized by a force-geometry relationship, the designer can directly observe structural behavior while exploring possible forms. This encourages an exploratory approach to design and supports unconventional solutions that integrate and respond to design intent. The structural and dimensional evaluation of the form is not an afterthought but an essential part of the design process.

The field of GS continues to evolve, and some of the future perspectives of the development of this field that could be considered may include integration with computational methods relying on hybrid approaches that leverage the benefits of both graphical and computational methods. Previous could facilitate more efficient and accurate analysis of complex structures. Also, developing (specialized) digital software tools for GS will likely continue. Besides tools that can be integrated within CAD systems, tools that can be integrated with the Building Information Modelling (BIM) systems could be developed to provide real-time feedback during the design process and enable seamless integration of GS with other engineering disciplines. Furthermore, advancements in computer graphics and visualization techniques could enable research of more immersive and interactive force and form diagrams using virtual reality (VR) and augmented reality (AR) technologies.

Furthermore, GS principles could be applied to design, analyze, and optimize advanced structural systems that involve more complex geometries and behaviors, including tensegrity structures, deployable structures, and adaptive structures. GS can play a role in analyzing and optimizing structures for material efficiency and structural performances and developing sustainable solutions, contributing to the overall goal of sustainable, environmentally conscious design. Moreover, GS can be applied in the design and optimization of 3D-printed structures for the production of innovative and efficient designs that leverage the unique capabilities of digital fabrication processes. Finally, although GS has a research history and strong theoretical foundations, future perspectives include further research and educational efforts to explore and expand the application of GS in various fields, including architecture.

References

Akbarzadeh, A. and Nejur, A. (2018). Food4Rhino. PolyFrame. https://www.food4rhino.com/en/app/polyframe

Benvenuto, E. (1991). *An introduction to the history of structural mechanics.* Springer Verlag, New York.

Block Research Group, ETH Zurich. (2009–23). http://www.block.arch.ethz.ch/brg/research

Block, P. (2009). Thrust Network Analysis. PhD thesis, Massachusetts Institute of Technology.

Burry, J., Felicetti, P., Tang, J., Burry, M., and Xie, M. (2005). Dynamical Structural Modeling: A Collaborative Design Exploration. *International Journal of Architectural Computing,* 3(1), 27–42. DOI: 10.1260/1478077053739595

Clarke, G. S. (2018). *The Principles of Graphic Statics* (Classic Reprint). London: Forgotten Books.

Cremona, L. (1879). *Le Figure Reciproche nella Statica Grafica.* Milano: Ulrico Hoepli.

Cremona, L. (1890). *Graphical Statics: Two Treatises on the Graphical Calculus and Reciprocal Figures in Graphical Statics.* Translated by Thomas Hudson Beare. Oxford: Clarendon Press.

Culmann, C. (1864/1866). *Die graphische Statik.* Zurich: Verlag von Meyer & Zeller.

Föppl, A. (1892). *Das Fachwerk im Raume.* Leipzig: Verlag von B. G. Teubner.

Fraddosio, A., Lepore, N., Piccioni, M. D. (2020). Thrust Surface Method: An innovative approach for the three-dimensional lower bound Limit Analysis of masonry vaults. *Engineering Structures,* 202, 109846. DOI: 10.1016/j.engstruct.2019.109846.

Fraternali, F. (2010). A thrust network approach for the equilibrium problem of unreinforced masonry vaults via polyhedral stress functions. *Mechanics Research Communications,* 37 (2), 198–204.

Graovac, O. (2019). Food4Rhino. 3D Graphic Statics. www.food4rhino.com/en/app/3d-graphic-statics

Heyman, J. (1982). *The masonry arch.* Ellis Horwood, Chichester.

Heyman, J. (1989). Hooke's cubico-parabolical conoid. *Notes and Records of the Royal Society,* 52 (1), 39–50.

Hooke, R. *A description of helioscopes, and some other instruments. London,* 1676.

Huerta, Santiago. (2003). El cálculo de estructuras en la obra de Gaudí. *Ingeniería Civil,* 130: 121–33.
Lewis, W. J. (2003). *Tension Structures: Form and Behaviour,* Thomas Telford.
Milankovitch, M. (1907). Theorie der Druckkurven. *Zeitschrift für Mathematik und Physik,* 55, 1–27.
Moseley, H. (1843). *The Mechanical Principles of Engineering and Architecture.* London.
O'Dwyer, D. (1999). Funicular analysis of masonry vaults. *Computers and Structures,* 73, 187–97.
Polyhedral Structures Laboratory. (n.d.). Publications. https://psl.design.upenn.edu/publications/
Rippmann M., Van Mele T., Popescu M., Augustynowicz E., Méndez Echenagucia T., Calvo Barentin C., Frick U., and Block P. (2016). The Armadillo Vault: Computational design and digital fabrication of a freeform stone shell, *Advances in Architectural Geometry 2016,* 344–63.
Rippmann, M. (2014). Food4Rhino. RhinoVAULT. https://www.food4rhino.com/en/app/rhinovault
Rippmann, M., Lachauer, L., and Block, P. (2012). Interactive vault design. *International Journal of Space Structures,* 27 (4), 219–230.
Zalewski, W., and Allen, E. (1998). *Shaping Structures.* New York: John Wiley & Sons.

Hua Chai, Xinjie Zhou, Tianyi Gao, Yanmin Zhou, Philip F. Yuan

Rationalizing Principal Stress Line Networks Using an Agent-Based Modelling Approach

Abstract: The principal stress line (PSL) network, which consists of two sets of orthogonal curves visualizing internal forces, provides an idealized material distribution scheme for structural design under given boundary conditions. While there have been some excellent cases utilizing PSL in structural designs, the application of PSL has been limited due to a lack of effective methods for generating and rationalizing PSL into efficient structural networks. This study proposes an agent-based modelling (ABM) approach that enables designers to generate and tailor PSL networks during the conceptual design phase. This rationalization process is carried out in two consecutive steps, with two corresponding agent systems being developed. Firstly, a "degeneration agent" system is created to identify degenerate points on the designed surface where the orthogonality of the PSL network becomes abnormal. A topological skeleton is then drawn, dividing the surface into a series of regions containing only conjugate PSL curves, using the identified degenerate points. The second step involves drawing and optimizing the layout of PSL in these regions using the "spacing agent" system, according to user-defined design objectives. The proposed method is validated through the design of floor slabs and freeform shells. This research could provide architects with a more intuitive and flexible approach to dealing with PSL during the design exploration stage.

Keywords: principal stress line, agent-based modelling, degenerate points, density control, shell structure

1 Introduction

Architectural form with both aesthetics and performance is the long-standing pursuit of architects in the form exploration stage. Advancements in structural form-finding methods, such as particle-spring systems [1], topology optimization [2], graphic statics [3], have accelerated the exploration and application of structures that integrate form and performance to a high degree. In this context, principal stress line (PSL) networks offer significant potential for efficient structural design due to their excellent geometric features and structural properties [4,5]. PSL networks consist of two sets of orthogonal curves, one in the maximum principal stress direction and the other in the minimum principal stress direction. These curves form an orthogonal line network, visually representing the internal force flow within a continuum structure under given boundary conditions. PSL networks provide an idealized material distribution scheme for a given structural domain [6]. The architectural design and construction based on PSL

https://doi.org/10.1515/9783111162683-033

have been proven to be material-saving and structurally efficient [7], enabling more sustainable ways of building.

1.1 Related works

In terms of architectural applications, the ribbed slab of the Gatti Wool Factory by Pier Luigi Nervi is one of the most famous projects inspired by the visualization of the internal force flow [8] (Fig. 1, left). Another example of a built structure with principal force trajectory is the zoology lecture hall at the University of Freiburg, Germany [9] (Fig. 1, middle). The application of PSLs in architectural design is limited due to architects lacking efficient methods for generating and optimizing them. In recent years, several tools have emerged that allow architects to perform finite element analysis and generate PSLs. Grasshopper Plug-ins such as Karamba3D and Millipede can generate PSLs in a parametric design environment, effectively promoting the application of PSL-based design exploration. In this context, PSLs has become increasingly popular in the digital fabrication research in pursuit of more efficient shells [10] or slabs [11]. Vejrum and Jensen (2022) presented the design and digital fabrication process of a complex ribbed concrete slab that implements the "Nervi system" [12]. The pattern of the CIAB Pavilion, developed by Zaha Hadid Architects, also follows the trajectories of principal stresses [13] (Fig. 1, right). However, the generated PSL networks by these tools are often unfavourable due to problems such as stress line discontinuity and self-intersection. The software's limited ability to control the PSLs makes it difficult to makes it difficult for designers to directly apply the generated results to structural design.

Fig. 1: Design practice inspired by PSL networks includes Gatti Wool Factory Floor System [8] (left), Zoology lecture hall at the University of Freiburg [9] (middle), and CIAB pavilion [13] (right).

The generation of PSL networks heavily relies on the seeding scheme, which determines the placement of seed points to achieve the desired layout of PSLs. Tam and Mueller (2015) optimized the generation of PSLs using an iterative stress line interpolation method and a rule-based correction process [14]. Church (2021) describes two seeding schemes: the neighbor seeding method and the farthest point seeding method, aiming to achieve a relatively evenly spaced network layout [4]. Both studies require an additional post-selection process to control the density of PSLs.

1.2 Research question and objectives

The primary challenge in generating and controlling a PSL network lies in the fact that the network does not always form an orthogonal grid, as desired in the ideal state. Upon observing the PSL network, it becomes evident that certain locations exist where the PSLs in the maximum and minimum stress directions are not orthogonal. These locations are known as degenerate points of the PSL network, which are points where the principal stresses are equal, leading to the absence of principal directions. In other words, all directions can be considered as principal directions [15]. Due to the irregular topology of the PSL network at degenerate points, it becomes challenging to devise a universal solution for both seedings and density control.

Degenerate points in the PSL network can be classified into two basic types: wedge (point A in Fig. 2) and trisector (point B in Fig. 2) [16]. Additionally, compound degenerate points can be formed by merging these basic types, resulting in complex topological PSL networks. For example, a center-type degenerate point (point C in Fig. 2) can be generated in regions where stresses experience discontinuity, such as locations under concentrated loads or supports. This type of degenerate point often results from the merging of two wedges.

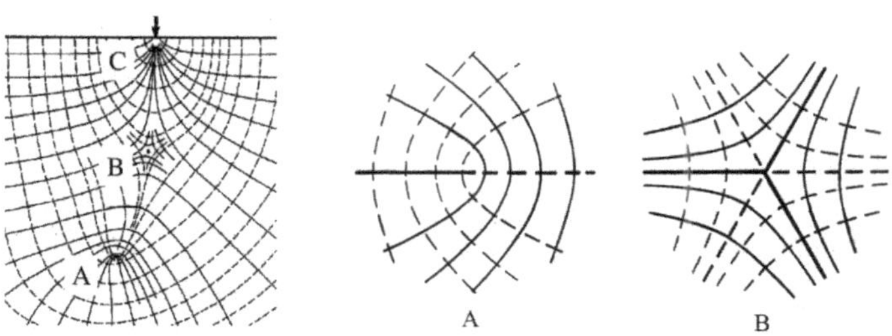

(a) (b)

Fig. 2: The degenerate points of PSL network: (a) point A is wedge point, point B is trisector point, C is a degenerate point under concentrated load [15]; (b) the PSLs networks around isotropic point A and B exhibit different geometric features [17].

To address the limitations in generating and controlling PSLs, this research aims to develop a method for designing and regulating PSL networks by further rationalization. The approach employed is an agent-based modelling (ABM) approach, which facilitates the identification of degenerate points in the PSL network and the regulation of PSL density. Through ABM, the generated network density can be regulated by iteratively adjusting the position of seed points, effectively integrating the seeding and density control process. The customizable behavior of agents in the ABM system allows for direct correlation of results with various design considerations, whether related to geometry, structure, or fabrication. This enables enhanced flexibility in regulating PSL networks. Although the proposed method is developed specifically for PSLs, it is equally applicable to other principal trajectories, such as principal bending moments.

2 Methods

The PSL networks only deviate from orthogonality at specific degenerate points. Consequently, the PSLs passing through these degenerate points, which define the topological skeleton of the PSL network, can divide the designed surface into areas without degenerate points, ensuring that the networks within these areas remain orthogonal. Building upon this principle, this research first develops an ABM system to identify the degenerate points within the PSL network. Subsequently, the PSLs passing through these degenerate points are generated to partition the surface. Another ABM system is then devised to facilitate PSL generation and density control within these divided areas.

The ABM systems are implemented using the ABxM framework, an open-source platform designed for exploration with agent-based systems [18]. Geometric modelling, agent system simulation, and visualization tasks are performed using Rhino3D and Grasshopper. The Grasshopper plugin Karamba3D is utilized for structural analysis, enabling the generation of principal stress directions and values.

3 PSL generation

Accurately generated PSLs are the basis of this study. The drawing of the PSLs starts from a seed point and iteratively seeks for the next point at a given step size by estimating the movement direction from the principal stress field. Since the finite element analysis performed with Karamba3D outputs principal stress directions and values corresponding to the input mesh vertices, the generation of PSLs need to first deal with data interpolation – calculate the data for any point with the known discrete values. This research utilized the Inverse Distance Weighting (IDW) interpolation method to calculate the principal stress directions and values for a seed point. The IDW method assigns weights to the values associated with mesh vertices within a certain radius of the seed point, based on their distance. These weighted values are then interpolated to determine the principal stress directions and values at the seed point (Fig. 3a).

For each starting point that is not located on the edges, the seed points are iteratively traced in two opposite directions simultaneously. This tracing process generates a Principal Stress Line (PSL) by sequentially connecting all seed points. The Fourth Order-

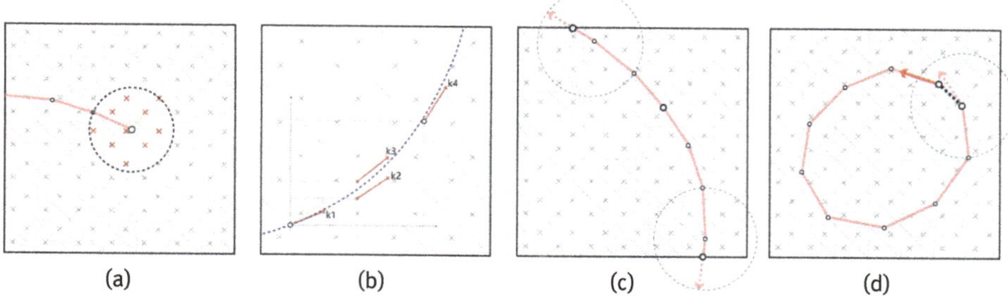

Fig. 3: The generation of a PSL: (a) interpolation method; (b) seed point approximation with RK4 method; (c) edge situation; (d) loop situation.

Runge Kutta method (RK4) is employed to estimate the next seed point at a given step size, based on the principal stress directions at the current seed point (Fig. 3b). While the Euler method is commonly used in PSL generation, the RK4 method is considered to provide better approximation accuracy [19]. The tracing process terminates under two conditions: reaching the design boundary (Fig. 3c), or when the distance from the starting point becomes smaller than the step size, indicating the formation of a loop (Fig. 3d).

4 ABM-based PSL network design

The PSL network is generated through a two-step process, utilizing corresponding agent systems. Initially, the "degeneration agent" system is employed to accurately locate the degenerate points on the designated surface. Subsequently, a PSL skeleton is delineated, dividing the surface into distinct regions characterized by exclusively orthogonal PSL curves. In the second step, the "spacing agent" system is introduced to efficiently draw and optimize the layout of PSLs within the identified regions. This process is meticulously guided by user-defined design objectives, ensuring a tailored outcome. To exemplify the efficacy of the agent-based algorithms, we employ a rectangular slab structure, supported at its four corners and subjected to gravitational forces, as a case study (Fig. 4).

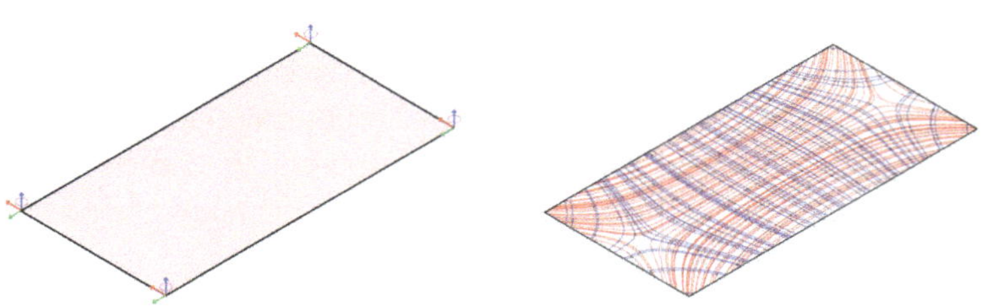

Fig. 4: Left: Selected case slab structure; right: PSL network generated with random seedings.

4.1 Degeneration agent system

The primary challenge of the degeneration agent system lies in identifying a target from the characteristics of degenerate points that can be progressively approached. Degenerate points are characterized by the equality of maximum and minimum principal stresses, resulting in an absolute difference of 0, which is the smallest possible value among all positions. To address this, we frame the problem of identifying isotropic points as a minimization problem. The objective is to minimize the difference, denoted as ΔS, between the maximum and minimum principal stresses at a given point, which is a function of the point coordinates (U, V) within the parameter space of the design surface.

4.1.1 Agent definition

A "degeneration agent" is instantiated as a random point on the design surface. Each agent's position attribute is represented by the coordinates (U, V) in the parameter space of the design surface. This allows us to determine the corresponding ΔS at each agent's current position. In each iteration, every agent calculates a moving vector, V_{m}, based on its defined behavior. The length of V_{m} is determined by a step size, d, weighted by ΔS. Since the objective of this agent system is to minimize ΔS, the length of each agent's movement decreases as it approaches the goal. This behavior is similar to a gradient descent algorithm. Once the length of V_{m} falls below a predetermined threshold, the agent is considered to have reached a degenerate point. Then the agent is terminated and excluded from the subsequent iterations.

4.1.2 Agent behavior

The agent's behavior aims to minimize the ΔS value by continuously seeking the direction where ΔS decreases. To simplify the direction search process, the agent is restricted to move along four predefined options: the maximum and minimum principal stress directions, as well as their opposite directions. Assuming the agent is positioned at a non-degenerate point, it undergoes small step movements in all four directions during each iteration. The direction in which ΔS decreases the most is then selected as the agent's moving direction. This behavior results in the agent's trajectory appearing as a zigzag line with gradually decreasing step sizes. It is important to note that since the data is estimated through interpretation, the final degenerate point found is an approximation of the actual degenerate point.

For the slab case, two degenerate points can be identified. A 3D contour map is employed to visually observe the movement of agents under the aforementioned behavior. In this map, the X and Y axes correspond to the x and y values of the design surface, while the Z-axis represents the ΔS value at corresponding points. By overlapping the design surface with the 3D map, the agent's trajectory can be observed as it gradually descends towards the bottom of the valley, resembling a gradient descent motion (Fig. 5).

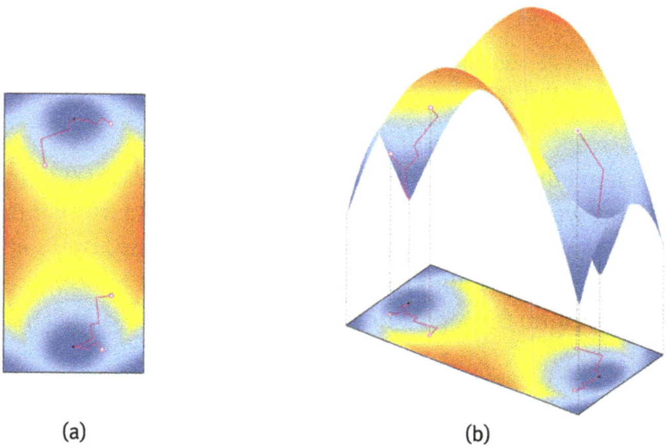

Fig. 5: A 3D contour map is used to better visualize the searching paths of degeneration agents.

4.2 Degenerate points-based surface subdivision

The objective is to partition the surface into regions where only orthogonal networks exist. The topological skeletons formed by PSLs passing through degenerate points are then used to divide the design surface, creating a structured environment for ABM-based PSL network design. This paper focuses on discussing two fundamental types of degenerate points: wedges and trisectors. These types can be easily distinguished based on their geometric features, as observed from the results of the degeneration agents. Wedges exhibit two intersections of PSLs in the maximum and minimum directions, whereas trisectors have only one.

Regarding the wedge points, the PSL network maintains its orthogonal feature, except for the lines that pass through the exact points. Since it is nearly impossible to reach the exact points, obtaining an orthogonal network is relatively straightforward by excluding the area in which these points lie from the design area. However, it's important to note that the topology of the network around wedge points should be further classified into two types: lemon and monster, based on the underlying tensor fields. In the scope of this paper, a detailed discussion of these types is deferred to avoid the need for extensive explanations to tensor fields. Fortunately, the subsequent spacing agent system can effectively operate in areas with wedges.

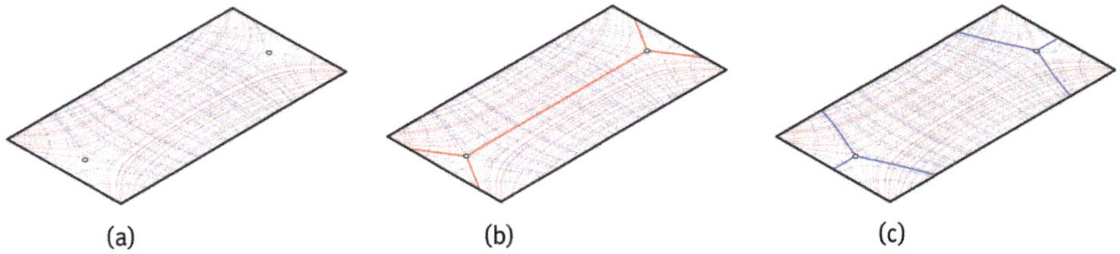

Fig. 6: Surface subdivision with degenerate points: (a) the two trisector points; (b) the maximum PSLs and (c) the minimum PSLs that pass through the trisectors.

For trisector points, both the maximum and minimum PSLs radiate from the point in three directions. After excluding wedge points, these PSLs can divide the surface into regions where only orthogonal networks exist. In the case of the slab, the two isotropic points are both trisectors. Each of these points exhibits three branches in both the principal stress directions, which form the topological skeletons as depicted in Fig. 6.

4.3 Spacing agent system

The spacing agent system is employed to generate PSLs and control their density within the divided area. To illustrate, in the case of the maximum PSL network, the slab is divided by three branches in the maximum principal direction at each isotropic point, while three branches in the minimum direction serve as guide curves for seeding.

4.3.1 Agent definition

The spacing agents are defined as a series of seed points that move along the guide curve. Each agent is associated with a PSL in the maximum direction that passes through the point it occupies. The agents only interact with their neighbors on the same guide curve. Agents located at the starting and ending points of the guide curve maintain their positions unchanged during the simulation. In each iteration, each agent obtains a moving vector, V_S, in the parameter domain of its guide curve through interaction with its neighbors. The system is considered converged once the sum of the V_S values for all agents is smaller than a pre-set threshold.

4.3.2 Agent behavior

The spacing agent's behavior can be customized according to the design requirements. This study takes the goal of evenly spacing the PSLs as an example. For the agents that are not located at the ends of their guide curve, three behaviors are defined to balance the spacing between the PSLs associated with each of the agents.

First, a spacing behavior is designed to directly average the spacing of the PSLs. The calculation of the average spacing D between the PSLs of one spacing agent and its two neighbors is used to determine the difference between the two D values, denoted as ΔD. The spacing behavior aims to reduce ΔD by moving the agent towards the neighbor with a larger D value. The length of the motion vector in each iteration is determined by the step size t, which is weighted by ΔD. As a result, the agent gradually reduces its movement while achieving spacing balance. Second, an adding behavior is developed to introduce a new agent to the system when the D value between any two adjacent agents exceeds a pre-set threshold. The new agent is added in the middle of the two adjacent agents on the guide curve. Third, a removing behavior is used to delete an agent from the system when the D value between one agent and any of its neighbors falls below a pre-set threshold. The removing and adding behaviors work together to maintain the density of PSLs within a desirable range, preventing the generation of layouts that are either too sparse or too tight. The combined action of these three behaviors results in a highly uniform distribution of PSLs in the case slab, as shown in Fig. 7.

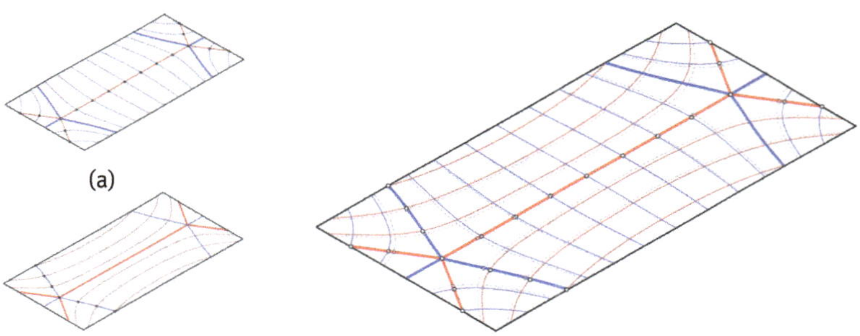

(a) (b) (c)

Fig. 7: (a, b) The spacing agents are initialized as evenly distributed points on the guide curves; (c) the balanced layout before (dashed lines) and after (solid lines) the ABM simulation.

4.4 Stress value weighted PSLs spacing

Based on the customizable features of ABM behaviors, different considerations that influence the layout of the PSL network can be incorporated. In this research, we conducted a simple test to regulate the spacing of PSLs based on local stress values. The density of PSLs is higher in areas with larger local stress values and lower in areas with smaller local stress values. This objective is achieved by incorporating the local stress values as weights in the calculation of the D values for adjacent agents. Therefore, the purpose of the spacing agent is to minimize the difference in weighted D values between each agent and its two neighbors. The strength of the effect of local stress values on the layout can be adjusted by a user-defined weight ratio. We further tested this method on a case slab, and the results demonstrate the difference between stress value-weighted and evenly distributed networks (Fig. 8).

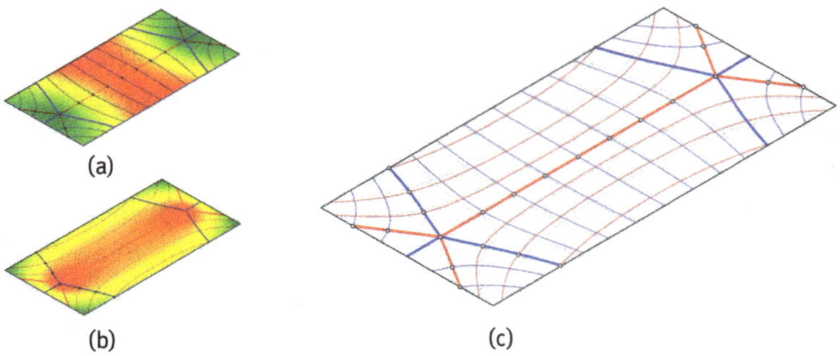

Fig. 8: (1-2) The stress values distribution is used to influence the layout of the PSLs; (3) the difference between evenly spaced (dashed lines) and stress weighted (solid lines) PSL network.

5 Design explorations

5.1 Implementations with different geometries

To test the effectiveness of the proposed PSL network design method, three different scenarios were selected: a floor slab under different boundary conditions, a simple shell, and a freeform shell. The objective was to achieve evenly spaced PSL networks for each scenario.

In the case of the floor slab, it is supported by two columns positioned in the middle. Due to stress concentration at each support, a degenerate point is generated, which is essentially a combination of two wedge points. The notable geometric feature is that the maximum PSL is concentrated at the degenerate point, while the minimum PSL forms a loop around it. By excluding the area where these points are located from the design area (by using a minimum PSL close to the degenerate point), it is easy to obtain an orthogonal network. Additionally, using the degeneration agent system, two trisector points can be identified. By passing through these trisector points, two adjacent loops in the minimum direction and four radial curves in the maximum direction can be obtained as skeletons. The final PSL network consists of closed loops and radial curves, effectively achieving homogenized spacing (Fig. 9).

The simple shell is utilized to assess the effectiveness of this method on hyperbolic surfaces. The shell is supported at its four corners under its self-weight. Considering the symmetry of the form and boundary conditions, it is conceivable that there exists

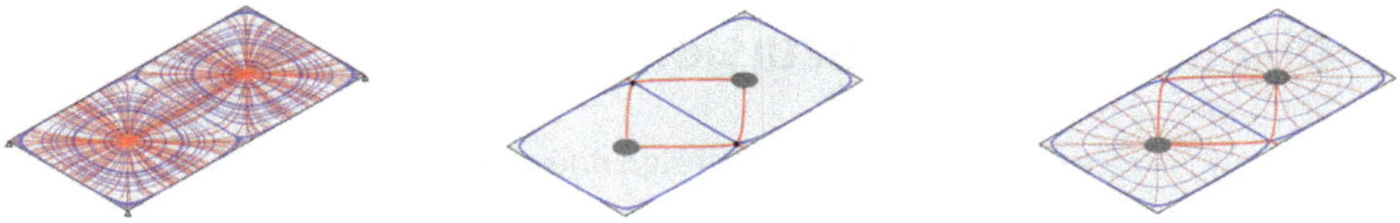

(a) (b) (c)

Fig. 9: Generation of a PSL network for the case slab with two supports in the middle: (a) PSL generated with evenly distributed seed points, (b) topological skeleton derived from degeneration agent system, (c) Evenly distributed PSL generation using spacing agent system.

an isotropic point at the center of the shell. This isotropic point corresponds to a saddle-type point resulting from the merging of two trisectors. Consequently, the PSLs passing through this point in both the maximum and minimum directions exhibit four branches (Fig. 10).

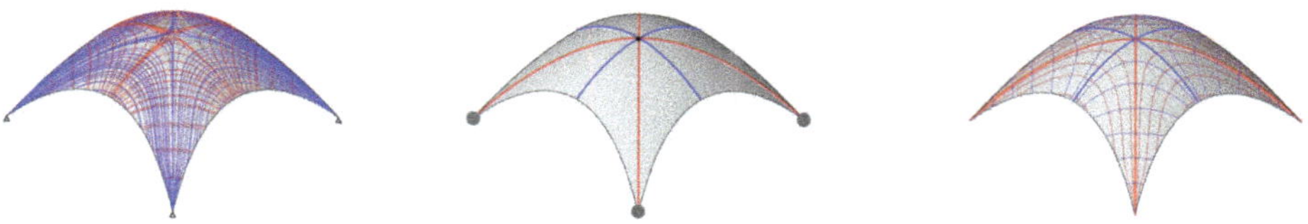

(a) (b) (c)

Fig. 10: Generation of a PSL network for the simple shell: (a) PSL generated with evenly distributed seed points, (b) topological skeleton derived from degeneration agent system, (c) Evenly distributed PSL generation using spacing agent system.

The freeform shell is supported at two corners and one edge. Stress concentrations occur at the point supports and edge support, which are excluded from the simulation. Although it may seem complex, the topology of the PSL network on the remaining surface has also been found to be primarily influenced by two trisector points (Fig. 11).

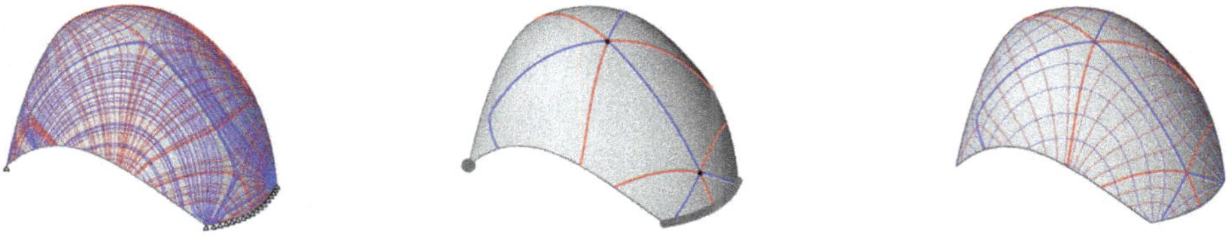

(a) (b) (c)

Fig. 11: Generation of a PSL network for freeform shell: (a) PSL generated with evenly distributed seed points, (b) topological skeleton derived from degeneration agent system, (c) Evenly distributed PSL generation using spacing agent system.

6 Discussion and conclusion

While previous research on PSL network design does not consider the topology of the networks, this research rationalizess the networks by exploring their topology and the underlying formation mechanism. The introduction of the ABM algorithm has made the density control of PSL networks more flexible. The proposed method is intuitive and user-friendly for architects, effectively facilitating the application of PSL in design explorations. Furthermore, the proposed method can be extended to other principal trajectories, such as principal bending moments.

The next step involves exploring the application of this method to a wider range of geometries and boundary conditions. Subsequent studies will also focus on the materialization of PSL-based structures using robotic construction methods, with a specific emphasis on rid-reinforced timber slabs and shell structures (Figure 12).

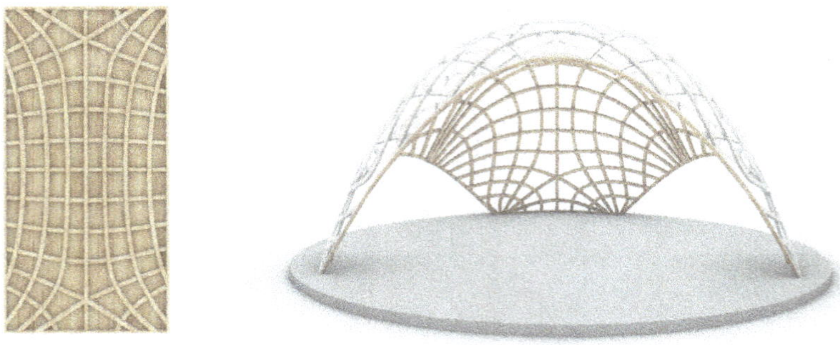

Fig. 12: Application scenarios of PSL in rid-reinforced timber slabs and shell structures.

Acknowledgements

This research is funded by National Key R & D Program of China (2022YFC3802203), National Postdoctoral Program for Innovative Talents of China (BX20220228), China Postdoctoral Science Foundation (2022M722399), Shanghai Post-doctoral Excellence Program (2022752), the Shanghai Municipal Science and Technology Major Project (2021SHZDZX0100) and the Fundamental Research Funds for the Central Universities. We'd like to thank Tobias Schwinn, Fabian Kannenberg, Luis Orozco, Lasath Siriwardena from ICD Stuttgart for their support on the development of agent systems.

References

[1] Kilian, A., and Ochsendorf, J. (2005). Particle-spring systems for structural form finding. *Journal of the international association for shell and spatial structures,* 46(2), 77–84.

[2] Xie, Y. M. (2022). Generalized topology optimization for architectural design. *Architectural Intelligence,* 1(1), 2.

[3] Akbarzadeh, M., Van Mele, T., and Block, P. (2016). Three-dimensional graphic statics: Initial explorations with polyhedral form and force diagrams. *International Journal of Space Structures,* 31(2-4), 217–26.

[4] Church, M. (2021). Investigation of isostatic slabs in timber. Delft University of Technology.

[5] Kwok, T. H., Li, Y., and Chen, Y. (2016). A structural topology design method based on principal stress line. *Computer-Aided Design,* 80, 19–31.

[6] Cascone, F., Faiella, D., Tomei, V., and Mele, E. (2021). Stress lines inspired structural patterns for tall buildings. *Engineering Structures,* 229, 111546.

[7] Liu, Q., Xu, R., Zhou, Y., Ge, J., Yuan, S., Long, Y., and Shi, T. (2021). Metamaterials mapped lightweight structures by principal stress lines and topology optimization: Methodology, additive manufacturing, ductile failure and tests. *Materials & Design,* 212, 110192.

[8] Halpern, A. B., Billington, D. P., and Adriaenssens, S. (2013). The ribbed floor slab systems of Pier Luigi Nervi. In *Proceedings of IASS Annual Symposia,* 23, 1–7. International Association for Shell and Spatial Structures (IASS).

[9] Antony, F., Grießhammer, R., Speck, T., and Speck, O. (2014). Sustainability assessment of a lightweight biomimetic ceiling structure. *Bioinspiration & Biomimetics,* 9(1): 016013.

[10] Tam, K. M. M., and Mueller, C. T. (2017). Additive manufacturing along principal stress lines. *3D Printing and Additive Manufacturing,* 4(2), 63–81.

[11] Bedarf, P., Szabo, A., Zanini, M., Heusi, A., and Dillenburger, B. (2022). Robotic 3D printing of mineral foam for a lightweight composite concrete slab. In *Proceedings of the 27th CAADRIA Conference,* Sydney, NSW, Australia.

[12] Vejrum, P., and Jensen, M. (2022). Design and construction of a ribbed concrete slab based on isostatic lines. In *Structures and Architecture A Viable Urban Perspective?* (1247–54): CRC Press.

[13] (n.d.). CIAB pavilion. Karamba3D. Retrieved March 17, 2023, from https://karamba3d.com/projects/ciab-pavilion/

[14] Tam, K. M. M., and Mueller, C. T. (2015). Stress line generation for structurally performative architectural design. In *Proceedings of ACADIA 2015,* 95–109.

[15] Galybin, A. N., and Mukhamediev, S. A. (2004). Determination of elastic stresses from discrete data on stress orientations. *International Journal of Solids and Structures, 41* (18/19), 5125–42.

[16] Delmarcelle, T., and Hesselink, L. (1994, October). The topology of symmetric, second-order tensor fields. In *Proceedings Visualization,* 94, 140–47. IEEE.

[17] Hutchinson, H. J., Nye, J. F., and Salmon, P. S. (1983). The Classification of Isotropic Points in Stress Fields. *Journal of structural mechanics,* 11(3), 371–81.

[18] Nguyen, L., Schwinn, T., Groenewolt, A., Maierhofer, M., Zorn, M. B., Stieler, D., Siriwardena, L., Kannenberg, F., Menges, A. (2022). *ABxM.Core: The Core Libraries of the ABxM Framework* [Source code]. DOI: 0.18419/darus-2994

[19] Liu, Z., Moorhead, R., and Groner, J. (2006). An advanced evenly-spaced streamline placement algorithm. *IEEE transactions on visualization and computer graphics,* 12(5), 965–72.

David Forster, Fabian Kannenberg, Malte von Scheven, Achim Menges, Manfred Bischoff

Design and Optimization of Beam and Truss Structures Using Alternative Performance Indicators Based on the Redundancy Matrix

Abstract: In structural optimization processes, a common goal is to limit deflections or stresses through topological changes, shape adaption or cross-sectional adjustments. Beyond these well-established performance indicators, alternative measures for the assessment of structures based on the redundancy matrix can be used in the design and optimization process. This contribution shows the extension of the redundancy calculation to three-dimensional beam structures in detail. Using the concept of redundancy for the design of structures, one goal is to homogeneously distribute redundancy within a structure in order to make an overall collapse due to failure of individual elements less likely. Furthermore, the sensitivity towards imperfections is quantified by the redundancy matrix, offering the opportunity to design connections of substructures, such that no constraint forces are introduced during the assembly process. Those two concepts are showcased exemplarily within this contribution. The method is embedded into a computational co-design framework, which allows for quick, interactive feedback on design changes to strengthen the interplay between the design and engineering process.

Keywords: structural assessment, redundancy matrix, structural optimization

1 Introduction

The shortage of raw materials and the increasing necessity of creating a built environment for a growing population implies huge responsibility for architects, engineers and all other disciplines of the building sector. A well-known way to tackle the challenge of using less material in the design of structures comes with structural optimization. A well-established concept is to use the mass of the structure as the objective function to be minimized with side constraints like keeping displacements and stresses within defined limits (Haftka et al. 1990). This implies that the structure is analyzed with defined load cases which yield the calculation of stresses and deformations. In addition to the linear analyses, especially for slender structures, geometrically nonlinear effects like buckling are important to consider in the assessment of structures. Besides, an ultimate load analysis is commonly used to define the load bearing capacity of a structure. In contrast to these well-established approaches to structural assessment,

https://doi.org/10.1515/9783111162683-034

the present contribution uses alternative performance indicators to analyze structures and formulate design requirements. Those indicators are based on the redundancy matrix, first described by Bahndorf (1991). This concept is by no means replacing the above-mentioned methods, but rather complements the options. The redundancy matrix quantifies the distribution of the statical indeterminacy in the structure and by this offers the possibility to assess the structure's robustness and its sensitivity towards imperfections and is furthermore used in the field of adaptive structures (Geiger et al. 2020). We investigate the application of alternative performance indicators in the context of co-design, described by Knippers et al. (2021), meaning that the methods aim to be used integratively by multiple disciplines in the building sector, e. g. manufacturing alongside the classical ones like design and engineering. In that context, we show an example of using coreless filament wound (CFW) structures (Gil Pérez et al. 2022). Coreless filament winding is a manufacturing process that enables the fabrication of highly differentiated, lightweight, resource-effective, and high-performing building components and its application has been demonstrated for long-span and multi-story building systems (Menges et al. 2022). A CFW component is additively manufactured by freely spanning fiber rovings between anchor points, the resulting fiber net itself can form e. g. lattice components (Duque Estrada et al. 2021). CFW structures represent a particularly interesting application for alternative performance indicators since they are sensitive to imperfections and the redundancy distribution can be examined across multiple levels of detail.

Within this contribution, in Sec. 2 we recap the equations to calculate the redundancy matrix and showcase the details with a simple three-dimensional beam example. The optimization details and the integration of the design aspect are also covered in this chapter. Section 3 shows three different case studies. The first one deals with the assembly process, whereas the other two examples point toward a robust structural design in a spatial frame and a discrete gridshell structure. Section 4 summarizes the work and indicates possible future applications of the presented method.

2 Methodology

2.1 Redundancy matrices and statical indeterminacy

The derivation of the redundancy matrix will first be described in matrix notation for plane truss structures and then extended to three-dimensional beam structures. For a complete step-by-step derivation, the reader is referred to Scheven et al. (2021). In plane truss structures, the only deformation mode is the normal elongation.

The relation between external forces $\mathbf{F}_{\text{ext}}$ and internal forces $\mathbf{N}$ can be written as

$$\mathbf{A}^{\mathrm{T}}\mathbf{N} = \mathbf{F}_{\text{ext}}, \tag{1}$$

$\mathbf{A}^T$ being the equilibrium matrix, capturing the topological and geometrical information of the discrete truss structure, and its transpose $\mathbf{A}$ being the compatibility matrix. Using the matrix $\mathbf{C}$, which includes each individual member stiffness on the main diagonal, the relation between internal forces and elastic elongations $\Delta\mathbf{l}_{\text{el}}$ can be stated as follows:

$$\mathbf{N} = \mathbf{C}\Delta\mathbf{l}_{\text{el}}. \tag{2}$$

The elastic elongations are calculated as the difference between the total elongation $\Delta\mathbf{l}$ and an initial, prescribed elongation $\Delta\mathbf{l}_0$. Initial prescribed elongations can be imagined as a member being imperfectly manufactured and therefore squeezed into the system with a certain force. The remaining necessary relation between total elongations and nodal displacements $\mathbf{d}$ reads:

$$\Delta\mathbf{l} = \mathbf{A}\mathbf{d}. \tag{3}$$

Equations (1) to (3) are used to solve for the unknown nodal displacements $\mathbf{d}$, neglecting external loads $\mathbf{F}_{\text{ext}}$, leading to the desired load-independent measure. The redundancy matrix is then defined as the map between initial and elastic elongations:

$$\Delta\mathbf{l}_{\text{el}} = \Delta\mathbf{l} - \Delta\mathbf{l}_0 = -\mathbf{R}\Delta\mathbf{l}_0 \qquad \mathbf{R} = \mathbf{1} - \mathbf{A}(\mathbf{A}^T\mathbf{C}\mathbf{A})^{-1}(\mathbf{A}^T\mathbf{C}). \tag{4}$$

It can be seen in eq. (4), that the redundancy distribution is independent of external loads and therefore a property of the structure only described by topology, shape, stiffness of the members and the support conditions. The main diagonal entries describe the constraint that is introduced on each individual member by the surrounding structure and quantify it with values between zero and one for truss structures.

The redundancy matrix can be used to quantify the distribution of the statical indeterminacy n_s within a structure. Classical formulas, see e. g. Maxwell (1864), describe the degree of statical indeterminacy as an integer number for the whole structure:

$$n_s = n_e - (2n_n - n_c). \tag{5}$$

Within this formula, n_e describes the number of elements, n_n is the number of nodes and n_c captures the number of fixed displacement degrees of freedom. Formally, the degree of statical indeterminacy describes the missing number of equilibrium equations that are necessary to calculate all internal forces of a structural system. In contrast to this consideration as an integer number, the redundancy matrix shows the spatial distribution of the statical indeterminacy within a structure on its main diagonal. The degree of statical indeterminacy can be calculated as the trace of the redundancy matrix:

$$n_s = \text{tr}(\mathbf{R}). \tag{6}$$

If the redundancy of an element equals one, this element does not contribute to the load transfer of the system and can be removed without any impact on the load-bearing behavior of the system. If the redundancy of an element is zero, the element must not be removed because a deletion would produce at least a kinematic substructure and therefore structural failure. Elements with zero redundancy are indispensable for the load transfer.

The extension to two-dimensional beam structures can be reviewed in detail in von Scheven et al. (2021), practical examples are already mentioned in Ströbel (1995). In contrast to trusses, for three-dimensional beams, six deformation modes are present. An eigenvalue decomposition of the element stiffness matrix of a three-dimensional beam element leads to a consistent description of the matrices **A** and **C**. The eigenvalues $\boldsymbol{\lambda}$ and eigenvectors $\boldsymbol{\phi}$ that are used to form those two matrices read as follows:

$$\boldsymbol{\lambda} = \begin{bmatrix} \frac{EA}{L} & 0 & 0 & 0 & 0 & 0 \\ 0 & \frac{3EI_z}{L} & 0 & 0 & 0 & 0 \\ 0 & 0 & \frac{3EI_y}{L} & 0 & 0 & 0 \\ 0 & 0 & 0 & \frac{GI_T}{L} & 0 & 0 \\ 0 & 0 & 0 & 0 & \frac{EI_y}{L} & 0 \\ 0 & 0 & 0 & 0 & 0 & \frac{EI_z}{L} \end{bmatrix} \qquad \boldsymbol{\phi} = \begin{bmatrix} -1 & 0 & 0 & 0 & 0 & 0 \\ 0 & \frac{2}{L} & 0 & 0 & 0 & 0 \\ 0 & 0 & -\frac{2}{L} & 0 & 0 & 0 \\ 0 & 0 & 0 & -1 & 0 & 0 \\ 0 & 0 & 1 & 0 & -1 & 0 \\ 0 & 1 & 0 & 0 & 0 & -1 \\ 1 & 0 & 0 & 0 & 0 & 0 \\ 0 & -\frac{2}{L} & 0 & 0 & 0 & 0 \\ 0 & 0 & \frac{2}{L} & 0 & 0 & 0 \\ 0 & 0 & 0 & 1 & 0 & 0 \\ 0 & 0 & 1 & 0 & 1 & 0 \\ 0 & 1 & 0 & 0 & 0 & 1 \end{bmatrix}.$$

The redundancy matrix for three-dimensional beams has six rows and six columns per element, as opposed to one single row and column for truss systems. The individual contributions of the deformation modes to the redundancy are ordered in the same way as in the matrix above. The first entry refers to the elongation mode, the second and third entries refer to the coupled bending and shear modes in local y- and local z-direction respectively. The fourth eigenvector represents the torsional mode and the last two represent the pure bending modes.

Figure 1 shows a three-dimensional structure, consisting of an angled cantilever with an additional support in global Z-direction at the far end. The degree of statical indeterminacy is one. The respective redundancy matrix

$$\mathbf{R} = \begin{bmatrix} 0.0013 & 0.0 & 0.0 & 0.0 & -0.0013 & -0.0013 & 0.0 & 0.0 & 0.0009 & 0.0 & -0.0009 & 0.0 \\ 0.0 & 0.0 & 0.0 & 0.0 & 0.0 & 0.0 & 0.0 & 0.0 & 0.0 & 0.0 & 0.0 & 0.0 \\ 0.0 & 0.0 & 0.0 & 0.0 & 0.0 & 0.0 & 0.0 & 0.0 & 0.0 & 0.0 & 0.0 & 0.0 \\ 0.0 & 0.0 & 0.0 & 0.0 & 0.0 & 0.0 & 0.0 & 0.0 & 0.0 & 0.0 & 0.0 & 0.0 \\ -0.3394 & 0.0 & 0.0 & 0.0 & 0.3394 & 0.3394 & 0.0 & 0.0 & -0.2399 & 0.0 & 0.2399 & 0.0 \\ -0.3394 & 0.0 & 0.0 & 0.0 & 0.3394 & 0.3394 & 0.0 & 0.0 & -0.2399 & 0.0 & 0.2399 & 0.0 \\ 0.0 & 0.0 & 0.0 & 0.0 & 0.0 & 0.0 & 0.0 & 0.0 & 0.0 & 0.0 & 0.0 & 0.0 \\ 0.0 & 0.0 & 0.0 & 0.0 & 0.0 & 0.0 & 0.0 & 0.0 & 0.0 & 0.0 & 0.0 & 0.0 \\ 0.1131 & 0.0 & 0.0 & 0.0 & -0.1131 & -0.1131 & 0.0 & 0.0 & 0.0799 & 0.0 & -0.0799 & 0.0 \\ 0.0 & 0.0 & 0.0 & 0.0 & 0.0 & 0.0 & 0.0 & 0.0 & 0.0 & 0.0 & 0.0 & 0.0 \\ -0.3394 & 0.0 & 0.0 & 0.0 & 0.3394 & 0.3394 & 0.0 & 0.0 & -0.2399 & 0.0 & 0.2399 & 0.0 \\ 0.0 & 0.0 & 0.0 & 0.0 & 0.0 & 0.0 & 0.0 & 0.0 & 0.0 & 0.0 & 0.0 & 0.0 \end{bmatrix}$$

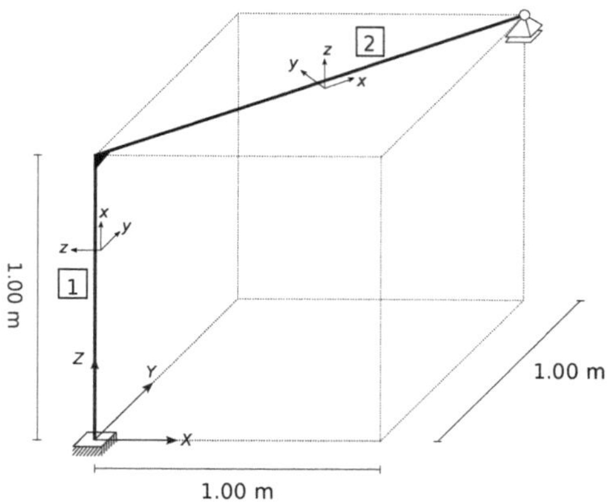

$E = 2.1 \cdot 10^8$ kN/m^2

Square hollow section:

$A = 30.4 \cdot 10^{-3}$ m^2

$I = 1.202 \cdot 10^{-5}$ m^4

Fig. 1: Three-dimensional cantilever with additional support.

shows that the numbers on the main diagonal sum up to one as described before (deviation due to rounding). The contribution of the torsional mode of element 1 is zero. The respective entry in the redundancy matrix can be found in the fourth column in the fourth row. This means that a prescribed torsion of the element is not constrained by the surrounding structure, since the additional support at the far end only fixes displacements in the global Z-direction.

2.2 Optimization using redundancy matrices

Structural optimization can basically be divided into three sub-categories, namely topology optimization, shape optimization and cross-sectional optimization, see Ramm et al. (1998). With regards to truss or beam structures, topology optimization answers the question of which node is connected to which node. Shape optimization defines the spatial coordinates of the nodes and cross-sectional optimization adjusts for example the thickness of a section. Using the distribution of statical indeterminacy as the objective function, changes within those categories offer the possibility to modify $\mathbf{C}$ and $\mathbf{A}$ in eq. (4) and thus redistribution of statical indeterminacy within the structure can be performed. In this work, we present an example using shape optimization; the design variables are denoted as $\mathbf{s}_1$. In a second example, the thickness of the cross-section is varied within a cross-sectional optimization. The design variables are defined as $\mathbf{s}_2$ and a similar optimization can be executed. The results are shown in Sec. 3.2, the optimization problem is defined as:

$$\min_{\mathbf{s}} f(\mathbf{s}), \quad f(\mathbf{s}) = R_{\max} - R_{\min}, \quad \mathbf{s}_1 = \begin{bmatrix} x_1 \\ \vdots \\ x_n \\ y_1 \\ \vdots \\ y_n \\ z_1 \\ \vdots \\ z_n \end{bmatrix}, \quad \mathbf{s}_2 = \begin{bmatrix} t_1 \\ \vdots \\ t_n \end{bmatrix}, \tag{7}$$

$R_{\max}$ being the maximum value of the redundancy of all elements and $R_{\min}$ being the minimum value. The optimum is obtained in MATLAB with the sequential quadratic programming (SQP) implementation (Nocedal and Wright 2006).

There exist already applications, where the redundancy distribution is used in the context of designing structures. In Wagner et al. (2018), the redundancy matrix is used in the design of adaptive structures, Kou et al. (2017) show the application in the context of robustness.

In Ströbel (1995), it is stated that the redundancy should be distributed homogeneously within a structure in order to prevent the collapse of the structure or of substructures in case of failure of specific elements, which is obtained with the optimization as stated in eq. (7). In case that a genuinely homogeneous redundancy distribution is achievable, the redundancy fraction of each element can be calculated by the ratio between the degree of statical indeterminacy and the number of elements. Since this might not always be possible, it is necessary to minimize the range and not just maximize the minimal value $R_{\min}$. For a two-dimensional frame, depending on the ratio between the bending stiffness and the normal stiffness, the different redundancy distributions reflect the importance of the elongation and the bending action in the load transfer, see von Scheven et al. (2021). For a rather homogeneous redundancy distribution, bending and elongation are of similar importance. As seen above, the total degree of statical indeterminacy gives only little insight into the load-bearing behavior. In order to design robust structures, the spatial distribution is of central importance.

Another aspect, where the distribution of statical indeterminacy can be used, is the assembly process. Regarding on-site structural assembly of pre-fabricated modules by connecting them with additional elements, it is desirable if the system can compensate for manufacturing inaccuracies. For truss systems, only imperfections in length are relevant. In a scenario, where substructures are to be connected, the optimal assembly sequence would include the additional elements in such a way that little to no constraint is introduced and imperfections can be compensated through displacements of the already assembled structure. In an optimal case, the additional elements would have zero redundancy.

2.3 Design integration

Interactive design methods require fast feedback to inform the designers and guide them in their decision-making process. Feedback on the redundancy distribution is particularly useful for topology exploration of the global design, independent of specific load cases. The calculation of the redundancy matrix can be made relatively fast (Tkachuk et al. 2023), making it suitable for quick feedback and interactive design methods such as agent-based modeling and simulation (Stieler et al. 2022). This has been investigated by Maierhofer and Menges (2019) in the context of adaptive truss structures. The codebase for calculating the redundancy distribution for truss and beam structures has been implemented as a C# library that can be referenced in CAD environments such as Rhino/Grasshopper or interactive platforms like Unity.

2.4 Application context

In the context of CFW, the method can be applied across different scales. Firstly, on the global design level, allowing for investigation of the overall topology and shape of a structure. Secondly, we can examine the redundancy distribution of CFW building elements on the component level. The fiber net, a result of the interaction of the free-spanning fiber rovings (see Fig. 2), can be conceptualized as a beam structure and the redundancy contribution of each fiber segment can be computed (see Fig. 3). In contrast to the global design, optimization of the fiber net is non-trivial, as the fiber net results from the interaction of the sequentially laid fiber rovings and the final form, the number of nodes and the topology of the structure emerge only at the end (Menges et al. 2022). Therefore, the optimization of the fiber net needs to interface with a simulation of the fiber-fiber interaction. As part of the co-design framework for CFW building systems, the calculation can take cross-section input from other domains via the fiber data object model (Gil Pérez et al. 2022). Figure 3 shows the simulated fiber-fiber interaction and the resulting fiber net in the winding frame in the left. The middle plot shows the redundancy distribution in color scheme with preliminary assumptions that include constant cross-sections. The plot on the right shows the redundancy distribution in color scheme with the measured cross-sections from the physical testing specimen. Significant differences are visible, underlining the sensitivity of the manufacturing process.

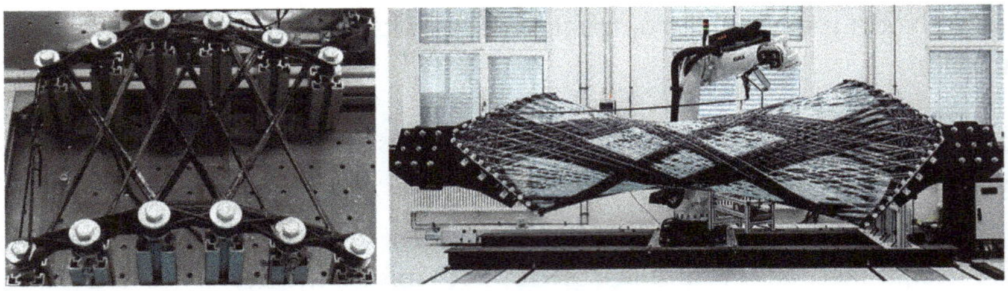

Fig. 2: Examples of coreless filament wound fiber nets: small-scale mechanical testing specimen (left), full-scale building component (right), © ICD/ITKE University of Stuttgart.

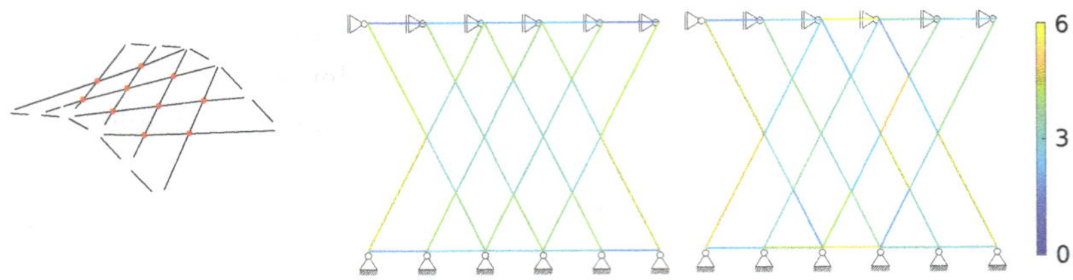

Fig. 3: Result of the winding sequence shown in the winding frame (left). Redundancy distribution with constant cross-sections (middle). Redundancy distribution with cross-sections measured from specimen S3-0 (right); data taken from Gil Pérez et al. (2023). Color spectrum for the redundancy distributions on the right.

3 Numerical examples and discussion

3.1 Specific redundancy for optimized assembly process

As described in Sec. 2.2, during the assembly process it can be of importance that only little constraint forces are introduced. Figure 4 a) shows two pre-fabricated parts of a truss bridge, both of which have a degree of statical indeterminacy of $n_S = 1$. Those pre-fabricated parts will be connected by three elements on-site and the vertical supports at the inner part of the final structure as well as the horizontal support at the right outer node are to be removed during the assembly process. The assembly sequence is shown from a) to d). The first additional member is shown in b). At the same time, the aforementioned horizontal support is removed. The additional member has zero redundancy. The second member, connecting the two parts diagonally makes one vertical support obsolete. In the final step, shown in d), another horizontal member is added and no more supports remain in the inner part of the structure. Throughout the assembly process, all members that were added have zero redundancy. Thus, the assembly process can fully compensate for possible manufacturing imperfections of the connecting elements and no constraint forces are introduced. It should be noted, that every additional member thereafter would raise the overall degree of statical

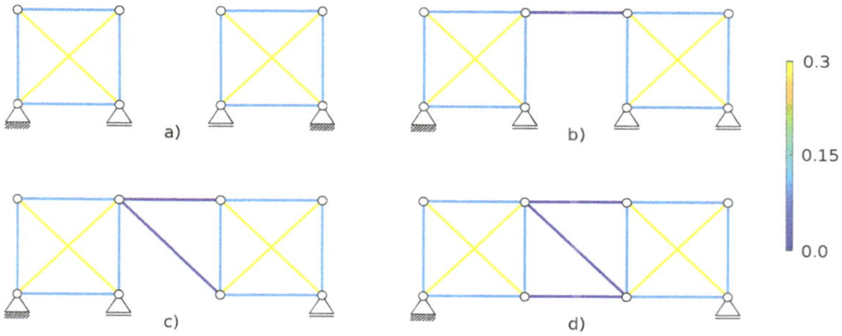

Fig. 4: Assembly sequence from configuration a) to d) with zero constraint forces introduced during the process; color spectrum showing the redundancy distribution.

indeterminacy and therefore, constraint forces would be introduced in case of imperfect manufacturing.

This simple two-dimensional truss example showcases, that it is possible to account for imperfections in the assembly and manufacturing process through knowledge of the spatial distribution of the statical indeterminacy of a system.

3.2 Homogeneous redundancy distribution

The first example is a spatial frame structure, shown in Fig. 5. The structure consists of 59 elements with a constant cross-section and a length of 5 m per element. The connections of the elements are rigid and at the four bottom corner nodes, the displacements are fixed by supports. The initial redundancy distribution can be seen in Fig. 5 on the left, with a range of $R_{\text{max,init}} - R_{\text{min,init}} = 1.05$. An optimization is performed, using the spatial coordinates of the nodes as the design variables, see eq. (7), such that the nodes can change position by 3.0 m in each spatial direction. Furthermore, the structure was kept to be symmetric during the optimization. One can see on the right that the redundancies are distributed homogeneously within the structure after changing the nodal positions, the final values only ranging negligibly from $R_{\text{max,opt}} = 3.154$ to $R_{\text{min,opt}} = 3.149$. In accordance to Sec. 2.2, the small range of the redundancy distribution reflects that the effect of removing any single element is of similar influence rather than one element having a large effect on the load-bearing behavior when being removed.

As a second example, the triangulated base geometry of the segmented timber shell from the Landesgartenschau Exhibition Hall (Krieg et al. 2015) is used in order to test the method on a geometrically more complex structure. The center points of the individual timber plates are connected by lines. By this, we generate a large-scale discrete gridshell to test the presented method without referring to the original timber shell and its planar plate segments. The structure consists of 243 nodes that are connected by 638 elements, see Fig. 6. The structure is supported on 34 nodes and has two openings that are curved. At the supports, only the displacements are set to zero. Using the cross-sectional

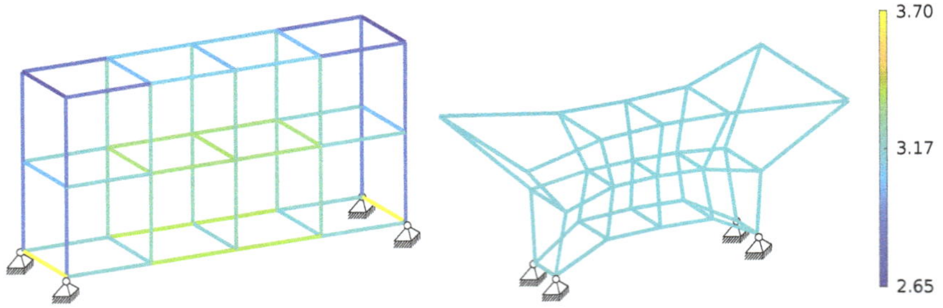

Fig. 5: Redundancy distribution of initial configuration (left) and optimized configuration with new nodal positions (right), geometry of a spatial frame structure shown in color spectrum.

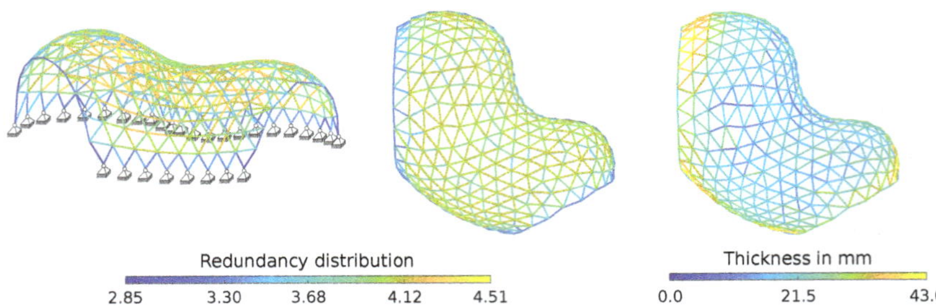

Fig. 6: Redundancy distribution of initial configuration (left) and optimized configuration with new cross-sectional thickness (middle) in color spectrum. Cross-sectional thicknesses in color spectrum (right).

thickness as the design variable, see $\mathbf{s}_2$ in eq. (7), the initial thickness was set to 10 mm using a hollow circular section with 244.5 mm diameter. The goal is to homogenize the redundancy as much as possible within the structure. As it is described in Sec. 2.2, the distribution of the redundancy in the overall structure is a measure of its robustness. This means that in areas with high redundancy fractions, many alternative load paths are available for the load transfer. In Fig. 6 on the left, one can see that the elements that are connected to the supports as well as the elements that build the curved openings are the ones that have the lowest redundancy fraction. The color spectrum ranges from the minimum to the maximum redundancy value of the initial configuration, which is $R_{\text{max,init}} - R_{\text{min,init}} = 1.66$.

In Fig. 6, the optimized redundancy distribution is shown in the middle. The maximum and minimum numbers are closer compared to the initial configuration with equal cross-sections for all elements. The range in the optimized configuration is $R_{\text{max,opt}} - R_{\text{min,opt}} = 0.82$, with the values shown in the color spectrum. On the right, the thickness of the cross-sections is shown in the colorspectrum. One can see that in order to homogenize the redundancy within the structure, the thickness of the cross-sections in the outer parts of the structure is increased, whereas the elements with smaller thickness are present in the inner part of the structure.

4 Conclusion and outlook

The redundancy matrix captures information about the spatial distribution of the degree of statical indeterminacy within a structure. This information can be used to define different goals for the design process of structures. Using the information about statical indeterminacy in the assembly process, the first example showed that constraint-free sequences can be determined. The application in the design of robust structures was showcased in the second and third example. To overcome the shortcomings of the restriction to linear calculations, an extension to non-linear analyses and to ultimate load analyses are subject to present research. Future work includes consideration of additional side constraints such as manufacturing requirements, especially in the context of CFW fiber nets. As the six modes of contribution for three-dimensional beams pose a challenge for visualization, the previously developed visual analytics method for fiber nets (Abdelaal et al. 2022) could be extended to include information about the redundancy distribution.

The presented method is universally applicable to truss and beam structures and has been demonstrated on a variety of geometries. It opens a field of alternative measures for structural assessment, beyond well-established parameters like e. g. stresses and displacements, that are applicable to various building systems across scales.

Acknowledgements

This research was supported by the Deutsche Forschungsgemeinschaft (DFG; German Research Foundation) under Germany's Excellence Strategy – EXC 2120/1 – 390831618. The authors would like to thank Mathias Maierhofer for sharing the codebase of the initial truss implementation and student assistant Henrik Jakob for his support. The fiber net simulation was developed by Valentin Noah Hartmann at IPVS-MLR, University of Stuttgart.

References

Abdelaal, M., F. Amtsberg, M. Becher, R. D. Estrada, F. Kannenberg, A. S. Calepso, H. J. Wagner, G. Reina, M. Sedlmair, A. Menges, and D. Weiskopf (2022). Visualization for architecture, engineering, and construction: Shaping the future of our built world. *IEEE Computer Graphics and Applications 42* (2), 10–20.

Bahndorf, J. (1991). *Zur Systematisierung der Seilnetzberechnung und zur Optimierung von Seilnetzen*. Ph. D. thesis, Universität Stuttgart, Stuttgart.

Duque Estrada, R., F. Kannenberg, H. J. Wagner, M. Yablonina, and A. Menges (2021). Integrative design methods for spatial winding. In O. Baverel, C. Douthe, R. Mesnil, C. Mueller, H. Pottmann, and T. Tachi (Eds.), *Advances in Architectural Geometry 2020*, Paris, pp. 286–305. Presses des Ponts.

Geiger, F., J. Gade, M. von Scheven, and M. Bischoff (2020). Anwendung der Redundanzmatrix bei der Bewertung adaptiver Strukturen. In B. Oesterle, M. von Scheven, and M. Bischoff (Eds.), *Baustatik – Baupraxis 14,* pp. 119–28. Institute for Structural Mechanics, University of Stuttgart.

Gil Pérez, M., C. Zechmeister, F. Kannenberg, P. Mindermann, L. Balangé, Y. Guo, S. Hügle, A. Gienger, D. Forster, M. Bischoff, C. Tarín, P. Middendorf, V. Schwieger, G. Gresser, A. Menges, and J. Knippers (2022, April). Computational co-design framework for coreless wound fibre-polymer composite structures. *Journal of Computational Design and Engineering 9* (2), 310–29.

Gil Pérez, M., C. Zechmeister, F. Kannenberg, P. Mindermann, L. Balangé, Y. Guo, S. Hügle, A. Gienger, D. Forster, M. Bischoff, C. Tarín, P. Middendorf, V. Schwieger, G. T. Gresser, A. Menges, and J. Knippers (2023). Object model data sets of the case study specimens for the computational co-design framework for coreless wound fibre-polymer composite structures.

Haftka, R. T., Z. Gürdal, and M. P. Kamat (1990). *Elements of structural optimization* (Second revised edition ed.). Dordrecht: Springer Netherlands. OCLC: 851381183.

Knippers, J., C. Kropp, A. Menges, O. Sawodny, and D. Weiskopf (2021, September). Integrative computational design and construction: Rethinking architecture digitally. *Civil Engineering Design* 3 (4), 123–35.

Kou, X., L. Li, Y. Zhou, and J. Song (2017). Redundancy Component Matrix and Structural Robustness. *International Journal of Civil and Environmental Engineering 11* (8), 1155–60.

Krieg, O. D., T. Schwinn, A. Menges, J.-M. Li, J. Knippers, A. Schmitt, and V. Schwieger (2015). Biomimetic lightweight timber plate shells: Computational integration of robotic fabrication, architectural geometry and structural design. In P. Block, J. Knippers, N. J. Mitra, and W. Wang (Eds.), *Advances in Architectural Geometry 2014,* Cham, pp. 109–25. Springer International Publishing.

Maierhofer, M. and A. Menges (2019). Towards integrative design processes and computational design tools for the design space exploration of adaptive architectural structures. In *International Conference on Emerging Technologies in Architectural Design (ICETAD2019),* Toronto, Canada.

Maxwell, J. C. (1864, April). On the calculation of the equilibrium and stiffness of frames. *The London, Edinburgh, and Dublin Philosophical Magazine and Journal of Science 27* (182), 294–9.

Menges, A., F. Kannenberg, and C. Zechmeister (2022). Computational co-design of fibrous architecture. *Architectural Intelligence 1* (1), 6.

Nocedal, J. and S. J. Wright (2006). *Numerical optimization* (2nd ed.). Springer series in operations research and financial engineering. New York: Springer.

Ramm, E., K. Maute, and S. Schwarz (1998). Conceptual design by structural optimization. In *Proceedings of EURO-C 1998,* R. de Borst, N. Bicanic, H. Mang, G. Meschke (eds), Badgastein, Austria, March 31 - April 3, 1998, Balkema, Rotterdam, pp. 879–96.

Stieler, D., T. Schwinn, S. Leder, M. Maierhofer, F. Kannenberg, and A. Menges (2022). Agent-based modeling and simulation in architecture. *Automation in Construction* 141, 104426.

Ströbel, D. (1995). *Die Anwendung der Ausgleichungsrechnung auf elastomechanische Systeme.* Ph. D. thesis, Universität Stuttgart, Stuttgart.

Tkachuk, A., T. Krake, J. Gade, and M. von Scheven (2023). Efficient Computation of Redundancy Matrices for Moderately Redundant Truss and Frame Structures. DOI: 10.48550/arXiv.2303.03945

von Scheven, M., E. Ramm, and M. Bischoff (2021). Quantification of the redundancy distribution in truss and beam structures. *International Journal of Solids and Structures 213,* 41–49.

Wagner, J. L., J. Gade, M. Heidingsfeld, F. Geiger, M. von Scheven, M. Böhm, M. Bischoff, and O. Sawodny (2018, August). On steady-state disturbance compensability for actuator placement in adaptive structures. *at - Automatisierungstechnik 66* (8), 591–603.

List of Contributors

Aly Abdelmagid
Laboratoire GSA, ENSA Paris Malaquais, Université PSL, Paris, France

Martin Antemann
Design-to-Production – Erlenbach, Switzerland, **info@designtoproduction.com**

Saqib Aziz
University of the Arts Berlin, Department for Structural Design and Engineering, Germany

David Bailly
Institut für bildsame Formgebung, RWTH Aachen University, Germany

Olivier Baverel
Laboratoire Navier, Ecole des Ponts, Univ Eiffel, CNRS, Marne-la-Vallée, France

Boris Belousov
Intelligent Autonomous Systems Group, TU Darmstadt, Germany, **boris@robot-learning.de**

Manfred Bischoff
Institute for Structural Mechanics, Cluster of Excellence IntCDC, University of Stuttgart, Germany

Philippe Block
Institute of Technology in Architecture, Block Research Group, ETH Zürich, Switzerland

Kilian Bruckner
Dipartimento di Ingegneria Civile e Ingegneria Informatica University of Rome TorVergata, Italy

Hua Chai
College of Architecture and Urban Planning, Tongji University, Shanghai, China

Daniel Chauhan
WSP UK Ltd., London, United Kingdom

Tzu-Ying Chen
Institute of Building Structures and Structural Design, Cluster of Excellence IntCDC, University of Stuttgart, Germany

Pierluigi D'Acunto
School of Engineering and Design, Technical University of Munich, Germany

Renaud Danhaive
Massachusetts Institute of Technology, Cambridge, USA

Benjamin Dillenburger
Digital Building Technologies, ETH Zürich, Switzerland

Daria Dordina
Institut für Geometrie, TU Dresden, Germany

https://doi.org/10.1515/9783111162683-035

John Douglas
Alutec Qatar

Cyril Douthe
Laboratoire Navier, Ecole des Ponts, Univ Eiffel, CNRS, Marne-la-Vallée, France

Rebeca Duque Estrada
Institute for Computational Design and Construction, Cluster of Excellence IntCDC, University of Stuttgart, Germany

Ahmed Elshafei
Centro de Matemática, Universidade do Minho, Campus de Gualtar, Portugal; Laboratoire GSA, ENSA Paris Malaquais, Université PSL, France

Carl Eppinger
CITA, Royal Danish Academy – Architecture, Design and Conservation, Denmark

Mehrzad Esmaeili Charkhab
Digital Design Unit, TU Darmstadt, Germany

Demi Fang
MIT, Cambridge, USA

Iman Fayyad
Syracuse University School of Architecture, USA

David Forster
Institute for Structural Mechanics, Cluster of Excellence IntCDC, University of Stuttgart, Germany

Tianyi Gao
Shanghai Research Institute for Intelligent Autonomous Systems, Tongji University, Shanghai, China

Ilaria Giannetti
Dipartimento di Ingegneria Civile e Ingegneria Informatica University of Rome TorVergata, Italy

Marta Gil Pérez
Institute of Building Structures and Structural Design, Cluster of Excellence IntCDC, University of Stuttgart, Germany

Tomoyuki Gondo
Graduate School of Engineering, University of Tokyo, Japan

Hasti Valipour Goudarzi
CITA, Royal Danish Academy – Architecture, Design and Conservation, Denmark

Ognjen Graovac
Faculty of Architecture, University of Belgrade, Serbia

Serena Gugliotta
WSP UK Ltd., London, United Kingdom

Sei Hayashi
Graduate School of Engineering, University of Tokyo, Japan

Gerhard Hirt
Institut für bildsame Formgebung, RWTH Aachen University, Germany

Ahmed Hussein
Laboratoire GSA, ENSA Paris Malaquais, Université PSL, Paris, France

Florin Isvoranu
Geometric Computing Laboratory, École Polytechnique Fédérale de Lausanne, Switzerland

Fabian Kannenberg
Institute for Computational Design and Construction, Cluster of Excellence IntCDC, University of Stuttgart, Germany

Walter Kaufmann
Institut für Baustatik und Konstruktion, ETH Zürich, Switzerland

Laura Kiesewetter
Institute for Computational Design and Construction, Cluster of Excellence IntCDC, University of Stuttgart, Germany

Jan Knippers
Institute of Building Structures and Structural Design, Cluster of Excellence IntCDC, University of Stuttgart, Germany

Axel Körner
Institute of Building Structures and Structural Design, Cluster of Excellence IntCDC, University of Stuttgart, Germany

Robby Kraft
is.d, Structure and Design, Department of Design, Faculty of Architecture, University of Innsbruck, Austria

Michael A. Kraus
Institute of Structural Engineering, ETH Zürich, Switzerland

Sophia V. Kuhn
Institute of Structural Engineering, ETH Zürich, Switzerland

Uday Kusupati
Geometric Computing Laboratory, École Polytechnique Fédérale de Lausanne, Switzerland

Evan Levelle
Front Inc.; Via Inc.

Ayoub Lharchi
CITA, Royal Danish Academy – Architecture, Design and Conservation, Denmark

Wenjun Liu
MAS DFAB, ETH Zürich, Switzerland; Fab-Union, Shanghai, China

Yuxi Liu
Digital Design Unit, TU Darmstadt, Germany

Leslie Lok
Rural-Urban Building Innovation Lab, College of Architecture, Art, and Planning, Cornell University, USA

Daniel Lordick
Institut für Geometrie, TU Dresden, Germany

Rupert Maleczek
is.d, Structure and Design, Department of Design, Faculty of Architecture, University of Innsbruck, Austria

Mohammad Mansouri
Laboratoire LACTH, ENSAP Lille, Université de Lille, Villeneuve d'Ascq, France

Alison Martin
Weaver and Independent Researcher, Italy

Samim Mehdizadeh
Digital Design Unit, TU Darmstadt, Germany

Achim Menges
Institute for Computational Design and Construction, Cluster of Excellence IntCDC, University of Stuttgart, Germany

Romain Mesnil
Ecole des Ponts ParisTech, Marne-La-Vallée; Laboratoire Navier, Ecole des Ponts, Univ Eiffel, CNRS, Marne-la-Vallée, France

Andrea Micheletti
Dipartimento di Ingegneria Civile e Ingegneria Informatica University of Rome TorVergata, Italy

Cyrill Milkau
Hochschule für Technik und Wirtschaft Dresden, Germany

Jelena Milošević
Faculty of Architecture, University of Belgrade, Serbia

Anahita Mirani
Laboratoire GSA, ENSA Paris Malaquais, Université PSL, Paris, France

Ioanna Mitropoulou
Digital Building Technologies, ETH Zürich, Switzerland

Sarah Mokhtar
Massachusetts Institute of Technology, Cambridge, USA

Caitlin Mueller
Massachusetts Institute of Technology, Cambridge, USA

Klara Mundilova
Computer Science & Artificial Intelligence Laboratory, MIT, Cambridge, USA

Takara Muto
Graduate School of Engineering, University of Tokyo, Japan

Andrei Nejur
School of Architecture, University of Montreal, Canada

Paul Nicholas
CITA, Royal Danish Academy – Architecture, Design and Conservation, Denmark

Moritz Niebler
Design-to-Production – Erlenbach, Switzerland , **info@designtoproduction.com**

Seri Nishimoto
Graduate School of Engineering, University of Tokyo, Japan

Marta Orsz
CIAUD - Research Centre for Architecture, Urbanism and Design, Lisbon School of Architecture, Universidade de Lisboa, Portugal

Diego Padilla Philipps
WSP UK Ltd., London, United Kingdom

Stefana Parascho
Laboratory for Creative Computation, École Polytechnique Fédérale de Lausanne, Switzerland

Mark Pauly
Geometric Computing Laboratory, École Polytechnique Fédérale de Lausanne, Switzerland

Jan Peters
Intelligent Autonomous Systems Group, TU Darmstadt, Germany

Thorsten Pofahl
Lehrstuhl für Tragkonstruktionen, RWTH Aachen University, Germany

Jamie Queisser
Staab Architekten GmbH, Berlin, Germany

Keyan Rahimzadeh
Front Inc., **keyanr@gmail.com**

Mette Ramsgaard Thomsen
CITA, Royal Danish Academy – Architecture, Design and Conservation, Denmark

Francesco Ranaudo
Block Research Group, ETH Zürich, Switzerland

Matthew T. Reiter
Civil and Environmental Engineering, Cornell University

Lisa-Marie Reitmaier
Institut für bildsame Formgebung, RWTH Aachen University, Germany

Yingying Ren
Geometric Computing Laboratory, École Polytechnique Fédérale de Lausanne, Switzerland

Katja Rinderspacher
Institute for Computational Design and Construction, Cluster of Excellence IntCDC, University of Stuttgart, Germany

Gabriella Rossi
CITA, Royal Danish Academy – Architecture, Design and Conservation, Denmark

Lorenzo Santelli
Eckersley O'Callaghan

Lotte Scheder-Bieschin
Institute of Technology in Architecture, Block Research Group, ETH Zürich, Switzerland

Fabian Scheurer
Design-to-Production – Erlenbach, Switzerland, **info@designtoproduction.com**

Malte von Scheven
Institute for Structural Mechanics, University of Stuttgart, Germany

Eike Schling
Department of Architecture, The University of Hong Kong, Hong Kong, China

Christoph Schlopschnat
Institute for Computational Design and Construction, Cluster of Excellence IntCDC, University of Stuttgart, Germany

Danilo Schneider
Hochschule für Technik und Wirtschaft Dresden, Germany

Tobias Schwinn
Institute for Computational Design and Construction, Cluster of Excellence IntCDC, University of Stuttgart, Germany

Alex Seiter
Lehrstuhl für Tragkonstruktionen, RWTH Aachen University, Germany

Shermin Sherkat
Institute for Computational Design and Construction, Institute for Control Engineering of Machine Tools and Manufacturing Units, University of Stuttgart, Germany

Yuta Shimoda
Jun Sato Structural Engineers Co., Ltd., Tokyo, Japan

Lasath Siriwardena
Institute for Computational Design and Construction, Cluster of Excellence IntCDC, University of Stuttgart, Germany

Lior Skoury
Institute for Computational Design and Construction, University of Stuttgart, Germany

Evy Slabbinck
Design-to-Production – Erlenbach, Switzerland, **info@designtoproduction.com**

Konrad Sonne
CITA, Royal Danish Academy – Architecture, Design and Conservation, Denmark

Lawson Spencer
Robotic Construction Lab,
College of Architecture, Art, and Planning,
Cornell University, USA

Kai Suto
Nature Architects Inc., Tokyo, Japan

Seiichi Suzuki
Geometric Computing, Laboratory École Polytechnique Fédérale de Lausanne, Switzerland

Tomohiro Tachi
Graduate School of Arts and Sciences, University of Tokyo, Japan

Kenryo Takahashi
Institute of Building Structures and Structural Design, University of Stuttgart, Germany

Martin Tamke
CITA, Royal Danish Academy – Architecture, Design and Conservation, Denmark

Oliver Tessmann
Digital Design Unit, TU Darmstadt, Germany

Gene Ting-Chun Kao
Block Research Group, ETH Zürich, Switzerland; Ansys, Inc., Sweden

Zlata Tošić
Institut für Geometrie, TU Dresden, Germany; Faculty of Civil Engineering and Architecture in Univercity in Niš, Serbia

Martin Trautz
Lehrstuhl für Tragkonstruktionen, RWTH Aachen University, Germany

Sylvain Usai
Design-to-Production – Erlenbach, Switzerland,
info@designtoproduction.com

Tom Van Mele
Institute of Technology in Architecture, Block Research Group, ETH Zürich, Switzerland

Krittika Walia
Ecole des Ponts ParisTech, Marne-La-Vallée, France

Zongshuai Wan
Department of Architecture, The University of Hong Kong, China

Hui Wang
Visual Computing Center, King Abdullah University of Science and Technology, Jeddah, Saudi Arabia

Jingwen Wang
MAS DFAB, ETH Zürich, Switzerland; Laboratory for Creative Computation, École Polytechnique Fédérale de Lausanne, Switzerland

Dylan Wood
Institute for Computational Design and Construction, Cluster of Excellence IntCDC, University of Stuttgart, Germany; School of Architecture & Environment, University of Oregon, Eugene, USA

Andreas Wortmann
Institute for Control Engineering of Machine Tools and Manufacturing Units, University of Stuttgart, Germany

Thomas Wortmann
Institute for Computational Design and Construction, University of Stuttgart, Germany

Zachariah Wynne
WSP UK Ltd., London, United Kingdom

Philip F. Yuan
College of Architecture and Urban Planning, Tongji University, Shanghai, China

Christoph Zechmeister
Institute for Computational Design and Construction, Cluster of Excellence IntCDC, University of Stuttgart, Germany

Xinjie Zhou
College of Architecture and Urban Planning, Tongji University, Shanghai, China

Yanmin Zhou
Department of Control Science & Engineering, Tongji University, Shanghai, China; Frontiers Science Center for Intelligent Autonomous Systems, Shanghai, China

Sasa Zivkovic
Robotic Construction Lab,
College of Architecture, Art, and Planning,
Cornell University, USA

www.ingramcontent.com/pod-product-compliance
Lightning Source LLC
Chambersburg PA
CBHW061152240926
51846CB00006B/989
* 9 7 8 3 1 1 1 1 6 0 1 1 5 *